THE GEOLOGIC TIME SCALE AND THE GEOLOGIC COLUMN

| | | | Millions of yr | | |
Era[a]	Period	Epoch	Duration	Before Present	Major Physical Events
Cenozoic	Quaternary	Holocene	0.01		Continental glaciations
		Pleistocene	3		
Mesozoic	Tertiary	Pliocene	4		
		Miocene	19		
		Oligocene	12		
Paleozoic		Eocene	16		Yellowstone volcanics begin
		Paleocene	11		
				65	
					Beginning of Rocky Mountains
	Cretaceous		71		
					North America separates from Eurasia
	Jurassic		54		
	Triassic		35		North America begins to separate from Africa Atlantic basin originates
				225	
	Permian		55		Climax of Appalachian mountain building
	Pennsylvanian[d]		45		
	Mississippian[d]		20		
	Devonian		50		
Precambrian	Silurian		35		
	Ordovician		70		Beginning of Appalachian Mountains
	Cambrian		70		
				570	
	Precambrian[e]				Oldest dated rocks (±3.8 billion yr ago) Origin of earth (±4.5 billion year ago)

Source: S. Judson, M. E. Kauffman, and L. D. Leet, *Physical Geology,* 7th ed., © 1987 by Prentice-Hall, Inc., Upper Saddle River, N.J. Reprinted by permission of Prentice-Hall, Inc.

[a]Duration on approximately uniform time scale.

[b]This column does not give the complete time range of the forms listed. For example, fish are known from pre-Silurian rocks and obviously exist today; but when the Silurian and Devonian rocks were being formed, fish represented the most advanced form of animal life.

Geology for Engineers and Environmental Scientists

ALAN E. KEHEW

WESTERN MICHIGAN UNIVERSITY

Geology for Engineers and Environmental Scientists

Third Edition

Upper Saddle River, New Jersey 07458

Library of Congress Cataloging-in-Publication Data

Kehew, Alan E.
    Geology for engineers and environmental scientists / Alan E. Kehew.—3rd ed.
      p. cm.
    Includes bibliographical references and index.
    ISBN 0-13-145730-6
     1. Engineering geology.   I. Title.

TA705.K38 2006
624.1'51—dc22

                                  2005051525

Acquisitions Editor: *Chris Rapp*
Production Editor: *Heather Meledin/Progressive Publishing Alternatives*
Project Manager: *Dorothy Marrero*
Editorial Assistant: *Sean Hale*
Director of Marketing, Science: *Patrick Lynch*
Manufacturing Manager: *Alexis Heydt-Long*
Manufacturing Buyer: *Alan Fischer*
Senior Managing Editor, Art Production and Management: *Patricia Burns*
Managing Editor, Art Management: *Abigail Bass*
Art Production Editors: *Jess Einsig/Eric Day*
Illustrations: *Gex/Laserwords*
Art Director: *Jayne Conte*
Cover Designer: *Bruce Kenselaar*
Cover Image Credit: *AP Wide World Photos*

Pearson Education Ltd., *London*
Pearson Education Australia Pty. Ltd., *Sydney*
Pearson Education Singapore, Pte. Ltd.
Pearson Education North Asia Ltd., *Hong Kong*
Pearson Education Canada, Inc., *Toronto*
Pearson Educación de Mexico, S.A. de C.V.
Pearson Education—Japan, *Tokyo*
Pearson Education Malaysia, Pte. Ltd.
Pearson Education, Inc., *Upper Saddle River, New Jersey*

*This book is dedicated to my wife, Kay
and to my parents, Betty Kehew and the late Dick Kehew,
for their love and support over the years,
and to my children, Melissa, Michelle, and Liz*

CONTENTS

CHAPTER 3

Minerals

CHAPTER 4

Igneous Rocks and Processes

C H A P T E R 5

Sedimentary Rocks and Processes

C H A P T E R 6

Metamorphic Rocks and Processes

CHAPTER 7

Mechanics of Rock Materials

CHAPTER 8

Structural Deformation of the Earth's Crust and Earthquakes

C H A P T E R 9

Weathering and Erosion

C H A P T E R 10

Soils, Soil Hazards, and Land Subsidence

CHAPTER 11

Groundwater

CHAPTER 12

Subsurface Contamination and Remediation

CHAPTER 13

Mass Movement and Slope Stability

CHAPTER 14

Rivers

CHAPTER 15

Oceans and Coasts

C H A P T E R 16

Glaciers, Permafrost, and Deserts

PREFACE

The content and technical level of a geology text for engineers is a matter of broadly differing opinion. The type of geology course required for engineering students at universities around the United States varies from introductory physical geology, earth science, engineering geology, or to no course at all. This situation results from differences in philosophy, engineering curricula, and geology faculty available to teach such a course. *Geology for Engineers and Environmental Scientists* is an attempt to provide a broad introduction to the earth sciences with an emphasis on engineering and environmental applications of the science.

The concept for the first edition evolved from a course I taught at the University of North Dakota. The enrollment included civil and geological engineering students who were required to take the course, and mechanical and other engineering majors who took the course as a science elective. This course was the only geology course in the civil engineering curriculum, but, for the geological engineering majors, it had to serve as a foundation for numerous future geology courses. The academic level of the students varied; the majority of the students completed the course in their sophomore year. As a result, most had taken little, if any, engineering mechanics, soil mechanics, or hydrology.

These conditions made structuring the course very difficult. My basic approach was to cover the essentials of physical geology and also introduce the students to the engineering and environmental applications of geology. To do this, many of the traditional topics of physical geology were omitted or only briefly discussed. On the other hand, it was apparent that students could not gain a full understanding of the applied aspects of geology without a good foundation in rocks, minerals, and structural geology. Several physical geology texts were tried, and in addition, an engineering geology text was used one year. All these proved to be unsatisfactory since they failed to cover part of the desired course content. Students requested a reference other than the lecture presentations for either the basic geology or the engineering aspects, depending on which type of text was used.

I eventually concluded that the only solution for this course was to write a text combining the two areas at an introductory level. The first edition, entitled *General Geology for Engineers*, was the result. Its objective was to demonstrate the importance of geology to engineers by including introductory mechanics, hydraulics, and case studies that illustrate interactions between geology and engineering. The approach was mainly nonquantitative, although appropriate basic formulas were introduced where necessary. Enough material was presented so that individual instructors could use the book to achieve their own personal compromise between physical geology and engineering geology.

Although numerous changes have been made in the second and third editions of this textbook, its basic purpose remains the same. The book is intended to provide an introduction to geology for engineering students, with a focus on applications of geology they are likely to use in their professional careers. The change in title of the text, from *General Geology for Engineers* to *Geology for Engineers and Environmental Scientists,* reflects my opinion that environmental scientists could also benefit from a geology course that emphasizes the applications of the science, particularly those involving environmental problems and solutions. Curricula in environmental science commonly require at least one course in geology. It seems logical to me that the most useful type of geology course would be one in which the environmental applications would be discussed from a technical perspective. Ultimately, we must address environmental problems from a technical standpoint to solve them.

With this objective in mind, a chapter dealing with subsurface contamination and remediation was added in the second edition. In this third edition, one additional change was to move this chapter to a position in the text directly after the chapter on groundwater. There have been numerous other reorganizations and additions in the third edition. An emphasis has been placed on processes, both in the surface and subsurface. The nature and interrelationships of earth systems has been placed near the beginning of the book in Chapter 2, following an introductory chapter that summarizes some of the challenges involved in living on this planet that we humans have to face in the present and future. New material has been added on mineral, energy, and water resources. New examples and case histories are dispersed throughout the text, particularly dealing with geologic hazards and their consequences to the human population.

The editorial staff in geology at Prentice Hall, headed by Patrick Lynch, has been helpful, enthusiastic, and supportive. The explosion of the internet since the last edition and the vast amount of information provided by governmental agencies such as the U.S. Geological Survey, has certainly enhanced the availability of documents and images. Heather Meledin of Progressive Publishing Alternatives, the project manager, guided the book through production with patience and a positive attitude.

Several colleagues, including John Van Brahana, University of Arkansas; John Hole, Virginia Tech; Kenneth Windom, Iowa State University; and David Nash, University of Cincinnati, reviewed the second edition and provided detailed constructive feedback prior to the writing process. These reviews helped to focus my efforts immensely.

Alan E. Kehew

Chapter Opening Photographs

Chapter 1   A drilling rig exploring for oil in the Williston Basin, North Dakota. (A. E. Kehew.)

Chapter 2   Geysir, the Icelandic geyser from which the name is derived. (A. E. Kehew.)

Chapter 3   Asbestiform amphibole fibers from Libby, Montana. (Greg Meeker, U.S. Geological Survey.)

Chapter 4   Columnar jointing at Devil's Postpile, California. (University of Colorado.)

Chapter 5   An abrupt contact between the cross-bedded Coconino Sandstone and the overlying Toroweap Formation, Grand Canyon. (A. E. Kehew.)

Chapter 6   Migmatite, a metamorphic rock containing thin layers of light-colored granitic igneous rock (Park County, Colorado). (C. C. Hawley, U.S. Geological Survey.)

Chapter 7   Highly jointed Precambrian igneous rock mass, south Sinai Peninsula, Egypt. (A. E. Kehew.)

Chapter 8   Normal fault, northern Arizona. (A. E. Kehew.)

Chapter 9   Exfoliation results in the separation of thin slabs of rock from a rock mass along planes parallel to the surface. (Sierra Nevada Mountains, California.) (N. K. Huber, U.S. Geological Survey.)

Chapter 10   Subsidence in the San Joaquin Valley, California, caused by withdrawal of groundwater. (Thomas Holzer.)

Chapter 11   Vasey's Paradise, a series of springs emanating from limestone rock units in the Grand Canyon. (A. E. Kehew.)

Chapter 12   Photo of hazardous fume ventilation pipe at elementary school in Bowling Green, Kentucky. (A. E. Kehew.)

Chapter 13   Landslide of March 3, 1995, which destroyed 9 houses in La Conchita, California. (U.S. Geological Survey.)

Chapter 14   Aerial view of downtown Grand Forks, North Dakota, looking eastward toward East Grand Forks, Minnesota, during the 1997 flood. (North Dakota Geological Survey.)

Chapter 15   Hopewell Rocks, Bay of Fundy at low (a) and high (b) tide. (Tourism New Brunswick, Canada.)

Chapter 16   An active rock glacier moving downslope, Copper River region, Alaska. (U.S. Geological Survey.)

Introduction

Challenges of the 21st Century

With the recent arrival of the new century, it is natural to contemplate the changes that might take place in the next 100 years. In the last century, humankind experienced incredible change in every field of human endeavor. The explosion in technology led us from a society that traveled by train and horseback to one in which space travel was commonplace, communication was facilitated by a worldwide network of computers, and phenomenal advances marked every field of science and technology. A strong case can be made for the importance of geology—the subject of this book—in the advances of the last century. The great developments in manufacturing, agriculture, communication, transportation, and other fields required location and exploitation of the earth's mineral and energy resources. Geology developed as a science to provide these resources for industrial growth. Industrialized countries like the United States became rich and powerful in part because of their ability to find and utilize mineral and energy resources.

As the last century waned, an unexpected consequence of mineral and energy resource development and consumption became apparent—other types of natural resources were impacted. Soil, water, and air became increasingly contaminated and productive agricultural lands were damaged by erosion and salinization. Natural geologic processes—earthquakes, floods, landslides, and storms—continued to take a high toll of human life and property, exacerbated by population growth and expansion into more hazardous coastal and mountainous areas. It became apparent that geologists, along with their counterparts in chemistry, biology, and engineering, were needed to identify and solve some of the environmental problems that they had inadvertently helped to create. Graduates with master's degrees in the earth sciences are now as likely to find employment in environmental consulting firms as they are to join petroleum and mineral extraction firms

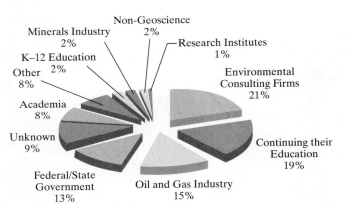

◀ FIGURE 1.1
Employment trends of recent recipients of master's degrees in geology. *Source:* American Geological Institute.

(Figure 1.1). The long-term trends in geological employment in the United States are apparent in Figure 1.2, which compares the years 1986 and 2000. During that 14-year period, jobs in the petroleum industry dropped from 50% to 30%, while environmental jobs increased from less than 10% to approximately 15%. In addition, much of the increase in the government job sector can be attributed to environmental regulatory positions. Underdeveloped countries are far behind the United States in environmental protection and remediation—a situation that must be addressed as their economies and standard of living grow.

As we move farther into the 21st century, many challenges face the human race if we hope to maintain the standard of living in the developed countries and raise the standard of living for citizens of underdeveloped countries, all while preserving and, in many cases, restoring environmental quality.

Four of these challenges are critical to the efforts of engineers and environmental scientists:

- Population growth
- Energy
- Water
- Climate change

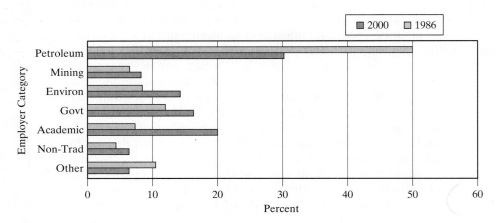

▲ FIGURE 1.2
Changes in employment category for geologists between 1986 and 2000. *Source:* American Geological Institute.

Geology is obviously directly involved in the provision of energy and water of sufficient abundance and quality for the human population. Global climate change is a complex interdisciplinary dilemma that could potentially have huge impacts on humankind. Only by understanding the magnitudes and causes of past climate changes—the realm of geologists—can we begin to understand the impacts we may be currently imposing on climate and predict the direction of future climate change. Population is a challenge for engineers and environmental scientists because it greatly impacts the other three.

To successfully meet these challenges, engineers and environmental scientists must have a wide array of tools in their intellectual toolkit. Because all four challenges involve earth systems, geology is one of the most critical tools. The earth is a dynamic planet with intricately interconnected systems in its solid fraction at and below the surface, along with its oceans and atmosphere. In this text we will consider the solid earth in greatest detail, but also touch upon the oceans and atmosphere as they influence processes on land.

Population Growth

The United Nations estimated that the earth's population reached 6 billion just about 2 months before the beginning of the 21st century. In contrast, the estimated population was 1.65 billion in 1900 and just about 1 billion in 1800. These accelerating numbers of people represent the phenomenon of exponential growth, in which the increase in some quantity is a function of the amount that is there at the beginning of the period. Compound interest is a good illustration of exponential growth. For example, a savings account with an initial balance of $1000 that grows at a rate of 5% for 5 years will be worth $1284, for an increase of $284. However, if the initial balance was $2000 and the account grows at the same interest rate for 5 years, it would be worth $2568, an increase of $568. Exponential growth is described by the following equation:

$$N = N_0 e^{rt} \tag{1.1}$$

where

N = the amount of the quantity after time t (in yr in this example)

N_0 is the amount of the quantity at time t_0

e is the exponential function

r is the rate of growth (expressed as a percentage in this example)

If the growth rate, r, remains constant, the increase in quantity N follows a smoothly increasing curve (Figure 1.3). Human population growth follows the exponential function, but the growth rate does not remain constant. Figure 1.4 shows the increase in population between 1750 and 2050. The steepening of the curve around 1950 resulted from decreasing

▶ FIGURE 1.3
The exponential function.

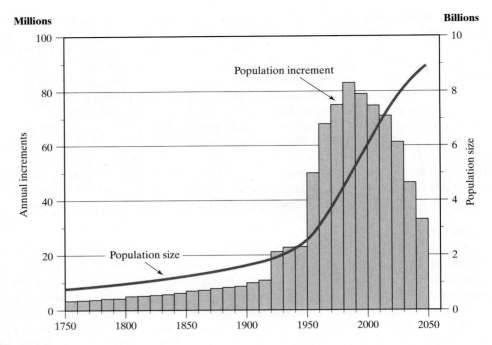

▲ FIGURE 1.4
World population growth, 1750–2050. *Source:* United Nations Population Division.

mortality rates in the less developed countries, causing the growth rate to rise. Birth rates, the other component of growth rate, peaked in the 1960s and began to decline (Figure 1.5). The population increments shown in Figure 1.4 are the difference between births and

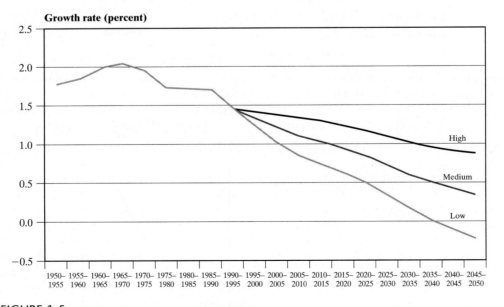

▲ FIGURE 1.5
World population growth rate: past estimates and high-, medium-, and low-fertility assumptions, 1950–2050. *Source:* United Nations Population Division.

Population (in billions)

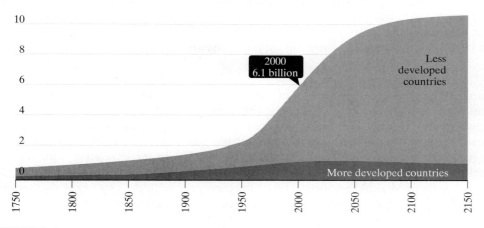

▲ FIGURE 1.6
World population growth, 1750–2150. *Source:* United Nations, *World Population Prospects, the 1998 Revision*; and estimates by the Population Reference Bureau.

deaths. With the medium fertility assumption, total world population is projected to stabilize at just above 10 billion people around the year 2200 (Figure 1.6).

Another useful application of the exponential function is the *doubling time*. Using equation (1.1), set N equal to $2N_0$ at time $t = T_d$, the doubling time,

$$2N_0 = N_0 e^{rT_d}$$

By canceling N_0 from both sides and taking the natural log of both sides,

$$\ln 2 = rT_d$$

and

$$T_d = \frac{\ln 2}{r} = \frac{0.693}{r} \tag{1.2}$$

When dealing with percentages, the doubling time is expressed as

$$T_d \sim \frac{69.3}{r}(\%) \sim \frac{70}{r}(\%) \tag{1.3}$$

At the current growth rate of 1.3%, doubling time is therefore about 53 years. The stabilization of population predicted for 2200 is based on the assumption of continually declining growth rates.

The contribution of the less developed countries to population growth is dramatically apparent in Figure 1.6. Currently, about 80% of the world's population lives in less developed countries. This will rise to 90% in 2050. A final trend worth mentioning about population is the increasing urbanization of the world's populace. One hundred years ago, the population was largely rural. Today, nearly 50% of the population lives in urban areas. Concentration of people in cities, along with accommodating the 3 billion additional people who will share our planet by 2050, will require a huge expansion of urban infrastructure, including water and sewage facilities. When these cities are located, or expand into, areas that are subject to earthquakes, floods, storm surges, tsunamis, or landslides, the human toll of these inevitable events will increase, even in countries that can afford the safest construction standards.

Population growth is an obvious preface to a discussion of energy and water resources. Can we provide equitable levels of these resources to our current population, much less to the billions that will soon be joining us? If that is not a tough enough challenge, can we also simultaneously reduce environmental degradation and mitigate geologic hazards for the world's burgeoning population?

Energy Resources

We are currently living near the end of a period of human history that, in the future, may be known as the Fossil Fuel Age. During this period of time, the great Industrial Revolution was fueled by energy derived from easily extractable fossil fuels, including, in chronological order of importance, coal, oil, and natural gas. Figure 1.7 shows the role of these energy sources in the United States since the mid-19th century. Crude oil is the most vulnerable of these fuels. Despite extensive exploration, the rate of discovery of new reserves (commercially producible resources) is not keeping pace with the rate of production. Most experts do not believe this situation will change. We are currently near a critical point in

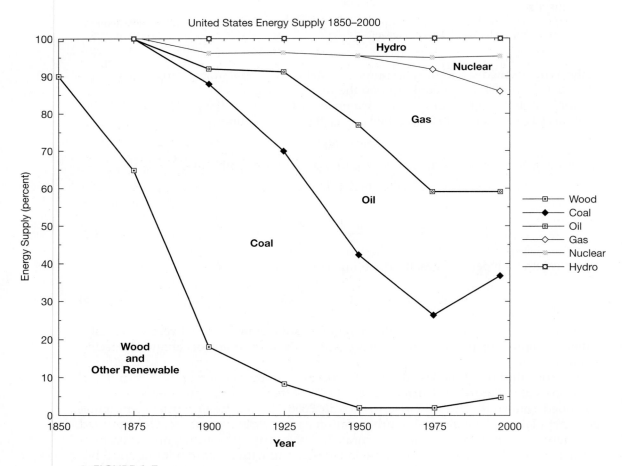

▲ FIGURE 1.7

Trends in U.S. energy usage since 1850. Coal replaced wood as the dominant energy source at the beginning of the Industrial Revolution. Oil and gas supplemented coal during the 20th century. By the end of the 20th century, these three fuels represented 80% of the U.S. energy supply. *Source:* James R. Craig, David J. Vaughan, and Brian J. Skinner, *Resources of the Earth: Origin, Use, and Environmental Impact,* 3rd Edition, © 2001. Reprinted by permission of Pearson Education, Inc., Upper Saddle River, N.J.

▶ FIGURE 1.8
The Big Rollover, the point at which worldwide oil production begins to decline. *Source:* Modified from U.S. Geological Survey Open File 00-320.

worldwide petroleum production, a point known as the Hubbert Peak (after M. K. Hubbert, who proposed this transition point) or the Big Rollover (Figure 1.8). This is the point at which worldwide production is no longer able to meet demand. Prices will rise steeply as production begins to decline on this curve. The worldwide rollover is similar to rollovers that have occurred in smaller producing areas such as the United States and Canada. Depending on which of several predictions is correct, the Big Rollover will occur sometime between 2000 and 2020.

Oil, like all natural resources, is not uniformly distributed on the planet. Countries in the Persian Gulf region contain most of the world's proven oil reserves (Figure 1.9). Most of these countries belong to OPEC (Organization of Petroleum Exporting Countries), an organization that manages oil production to control oil prices. Because of the declining reserves in the non-OPEC countries, OPEC oil will become even more important in the next

▶ FIGURE 1.9
Distribution of the world's proven oil reserves in 2000. *Source:* James R. Craig, David J. Vaughan, and Brian J. Skinner, *Resources of the Earth: Origin, Use, and Environmental Impact*, 3rd Edition, © 2001. Reprinted by permission of Pearson Education, Inc., Upper Saddle River, N.J.

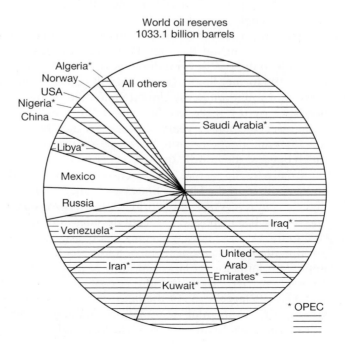

▲ FIGURE 1.10
Decline in oil reserves to the year 2020, assuming current production rates remain constant during the period. Non-OPEC nations are depleting their reserves at a faster rate. *Source:* James R. Craig, David J. Vaughan, and Brian J. Skinner, *Resources of the Earth: Origin, Use, and Environmental Impact*, 3rd Edition, © 2001. Reprinted by permission of Pearson Education, Inc., Upper Saddle River, N.J.

several decades (Figure 1.10). By 2020, the bulk of the remaining reserves will lie in OPEC's hands. This suggests that the importance of OPEC nations in terms of energy resources can only increase in future decades. Although Figure 1.10 does not consider future oil discoveries that will inevitably be made, these discoveries will probably be in proportion to the respective reserves in the OPEC and non-OPEC nations and the overall trend will not change.

The production and use of natural gas lagged behind petroleum because of the difficulty in transporting the fuel. When natural gas pipelines and transport ships became available in the past several decades, demand for natural gas rose dramatically. The countries of the former Soviet Union hold the greatest amount of natural gas reserves. Gas reserves are abundant in the United States, although it produces only about two-thirds of its needs. Current reserves in the United States will last approximately 9 years, and worldwide reserves total 65 years of present usage. New discoveries will undoubtedly extend this interval; like petroleum, however, gas production is unlikely to sustain current levels much beyond the middle of the 21st century.

Coal, a fossil fuel that has been used for hundreds of years, continues to be an important energy source in the United States and other countries. The United States, in fact, has the highest coal reserves in the world. Coal is now used primarily for generation of electricity, and its role has increased in recent decades after the oil crisis of the 1970s. The reserves of coal in the United States will conceivably last several hundred years. However, coal is environmentally the least acceptable fossil fuel for several reasons. First, the burning of coal in power plants injects large amounts of pollutants, including greenhouse gases, into the atmosphere. Second, the mining of coal either by underground or surface mining methods disrupts large areas of the land surface. Reclamation is expensive and does not prevent all environmental degradation. Acid mine drainage, caused by the exposure of sulfide minerals to air and water to form sulfuric acid, is a major problem. Mining also negatively affects aquifers used for water supply in mining regions. Finally, coal mining is a dangerous job. Though they have been reduced in recent years, accidental deaths attributed to roof collapse and explosions and diseases such as black lung disease are still common.

The use of alternative fossil fuels such as tar sands and oil shales will no doubt increase in the future. Significant quantities of these materials exist in various places. However, extraction of petroleum from the heavy hydrocarbon fractions in these deposits is expensive and will create environmental problems. As energy prices rise, however, greater utilization of these sources will likely take place.

Fossil fuels are not the only available energy resources, although they account for about 85% of energy usage. Nuclear power, once touted as a cheap, clean alternative to fossil fuels, is the second-largest source of electrical power generation, after coal, in the United States. The use of nuclear power has declined significantly, however, due to fears about safety and the disposal of high-level nuclear waste. High-level reactor waste is currently stored on a temporary basis at reactor sites. The current U.S. plan is to develop a disposal site at Yucca Mountain, Nevada, for all commercial high-level nuclear waste. Legal and technical challenges remain before this site becomes a reality. Public concern is focused on the great length of time that high-level waste must be isolated (10,000 years) and on safety issues involved in transportation of the waste by truck or train to the site.

There are a wide variety of potential energy sources to replace fossil fuels, which must come on line at least partially during the current century. These include renewable energy sources such as the sun, wind, tides, and geothermal heat. Hydroelectric power, generated by water flowing from reservoirs through dams, will probably not grow rapidly because of the environmental effects of damming large rivers. Eventually, new technologies must be developed. Nuclear fusion could theoretically produce nearly infinite amounts of energy if the daunting technical problems to utilizing this source can be solved.

Even this abbreviated discussion of energy resources leaves no doubt that energy will be one of the great challenges of the 21st century. The application of geological knowledge, both for production and for environmental protection, will be critical for most current and medium-term solutions to the energy supply. Engineers and environmental scientists will have to be at the forefront of the quest to extend fossil fuel energy sources and develop their replacements.

Water Resources

In addition to energy resources, the increasing future world population will need food and water. Unlike energy, in which alternatives to present energy sources are theoretically possible, there is no alternative to water. The human population, at whatever level that it reaches, must use the currently available amount of freshwater for domestic, industrial, and agricultural purposes. There is no other viable source, excluding the unlikely possibility, however, that energy becomes so cheap that desalinization of seawater becomes economically feasible.

On a global basis, the amount of freshwater accounts for only 2.8% of the total amount of water (Table 1.1). Of this small percentage, water in the largest contributor, ice caps and glaciers, is mostly unavailable, leaving lakes, streams, and groundwater as potentially exploitable sources of freshwater. There are two further problems with these sources of water. First, they are not uniformly distributed on the earth, creating water-rich and water-poor areas. Second, groundwater may be difficult or expensive to extract or poor in quality, so as to make it not acceptable for its intended purpose.

Table 1.1	Distribution of Water on the Earth	
	Water volume, km^3	Percentage of total water
Freshwater lakes	125×10^3	0.009
Saline lakes and inland seas	104×10^3	0.008
Average in stream channels	1×10^3	0.0001
Vodose water (includes soil moisture)	67×10^3	0.005
Groundwater within depth of half a mile	4.2×10^6	0.31
Groundwater—deep lying	4.2×10^6	0.31
Ice caps and glaciers	29×10^6	2.15
Atmosphere	1.3×10^3	0.001
World ocean	1320×10^6	97.2

Source: U.S. Geological Survey

Prediction of trends in water usage and availability is inherently difficult because of the many factors that could affect these estimates. The International Food Policy Research Institute developed a model for global food and water resources for the year 2025, a time when the population should reach about 8 billion people, about a third more than inhabit the earth now. Because food production and water resources depend upon many technological and governmental policy factors, three scenarios were developed in the model— one in which current trends are continued (Business-As-Usual), one in which a water crisis develops, and one in which changes to water usage are implemented that create sustainable water resources and food production for the human population. Huge changes in water usage have already taken place; five times more land is allocated to irrigated agriculture now compared to the beginning of the 20th century. One of the costs of this impressive increase was the loss of about half the world's wetlands, along with the wildlife habitat that wetlands contain.

The Business-As-Usual scenario is one of benign neglect. Current policies and trends regarding water use and development persist into the future without significant changes. Increases in efficiency in irrigation and water use occur, but not to the extent possible or across the board in developed and developing countries. Water consumption will increase significantly in this scenario (Figure 1.11). Consumption of water for nonirrigation needs will increase to a much greater degree than consumption of water for irrigation because of insufficient investment and prioritization of irrigation systems. The increase in irrigation consumption is less than the increase in potential irrigation consumption in the developing countries. In other words, there is a bigger gap between actual irrigation and the amount of irrigation that could potentially take place, creating an agricultural water scarcity. In turn, the water scarcity for agriculture will threaten the adequacy of food supplies. Food imports will increase in developing countries, straining their ability to finance the imported food. Hunger among the world's poorest citizens will increase.

If economic conditions deteriorate in the coming decades, the Water Crisis scenario could materialize. Governmental investments in water resources, agriculture, and irrigation would decline, as would agricultural efficiency. Water withdrawals for irrigation would be greater than in the Business-As-Usual scenario, but because of inefficiency, gains in food production would not keep pace with population growth. Water demand in the growing urban areas would be very high and governments would be unable to increase access to treated water for domestic and industrial use. Rising food prices would be the result of the

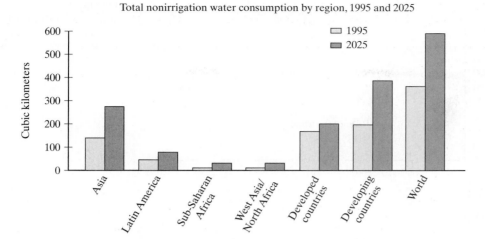

Total nonirrigation water consumption by region, 1995 and 2025

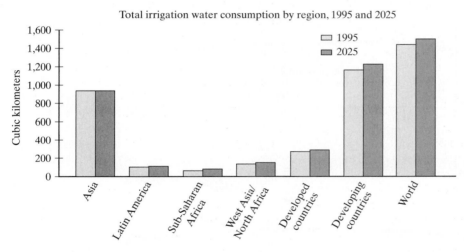

Total irrigation water consumption by region, 1995 and 2025

▲ FIGURE 1.11

Water consumption for nonirrigation uses and irrigation, by area between 1995 and 2025 assuming a Business-As-Usual scenario. *Source:* International Food Policy Research Institute.

Water Crisis scenario (Figure 1.12), reaching 40% for rice, 80% for wheat, and 120% for maize. Developing countries would be hard hit by these price increases, and per capita consumption of cereal grains would be lower in 2025 than 1995 levels. The human and economic consequences of this scenario would be severe, and the growing gulf between developed and developing countries would lead to political unrest and global economic instability.

The most optimistic scenario for world water resources is the Sustainable Water scenario. In this model, adequate water would be available for agricultural purposes and all urban households would be connected to treated water. These favorable outcomes would be realized through greater agricultural efficiencies driven by a strategy of higher water prices and governmental investments in improved irrigation technology

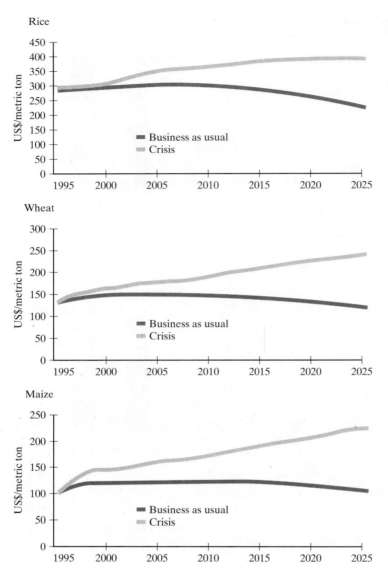

◀ FIGURE 1.12
World prices for rice, wheat, and maize in the Business-As-Usual and Water Crisis scenarios. *Source:* International Food Policy Research Institute.

and crop research to improve yields. Groundwater usage will be regulated to sustainable yields. With improved efficiency, water consumption will be lower than in the Business-As-Usual scenario (Figure 1.13). The water saved is diverted to wetlands and other environmental purposes, thus enhancing environmental quality. Food production is slightly higher and food prices are lower in the Sustainable Water scenario relative to the Business-As-Usual model.

The International Food Policy Research Institute models indicate that sustainable water resources can be achieved in 2025 for a world population of 8 billion people, but only if appropriate policy changes are implemented. Obviously, a high degree of international cooperation will be necessary to achieve this goal. Even if the Sustainable Water Resources model can be achieved, it must be extended to accommodate another 2 billion people at the stable population level of 10 billion. The continuing challenge to use our finite water resources wisely as the population grows will be a daunting task for humankind.

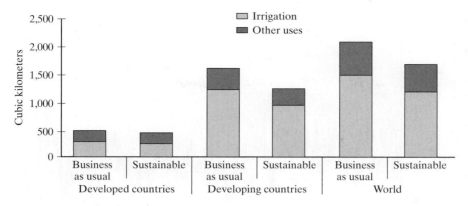

▲ FIGURE 1.13

Total and irrigation water consumption, by region, in the Business-As-Usual and Sustainable scenarios in 2025. *Source:* International Food Policy Research Institute.

Climate Change

The threat of worldwide climate change driven by the burning of fossil fuels is a scenario that has received increased attention in the past few years. The potential effects are ominous—melting of the polar ice caps to raise sea level and inundate low-lying coastal areas; intensification of damaging storms; and alteration of regional weather patterns with disastrous implications for agriculture. Concern and research into these potential threats are certainly warranted. Geologists play a major role in any discussion of climate change, based on their perspective on past climate. For climate is constantly changing, with or without human interference. Geologists have found increasingly sophisticated ways to interpret the causes and effects of past climatic changes. Our purpose here is merely to introduce some of the major issues of climate change—both natural and human induced.

In recent decades, tremendous amounts of data on paleoclimates have been collected. One of the best sources of information has been the thick ice sheets on Greenland and Antarctica. Cores of ice extracted from these ice sheets have yielded data on climates back through several hundred thousand years of geologic time. The results have shown surprising fluctuations of climate—drastic changes in temperature over very short intervals of time, decades in some cases. Attempts to explain these fluctuations, aside from major climatic changes that appear to be driven by changes in the earth's orbital movements, focus on interactions between the atmosphere and the oceans. The complexities of these interconnected systems cannot yet be fully understood. However, there is strong evidence that oceanic circulation plays a major role in climate change. Deep oceanic currents act as a huge conveyor belt to transport heat and salt between the poles and the equator. It also appears that the conveyor belt operates in two dominant states—on and off. When the conveyor belt is on, warm moderate climates, such as those of the last 10,000 years of geologic time, prevail. When the conveyor belt shuts down, sometimes very suddenly, the planet is plunged into a distinctly colder period. The causes for these switches have not yet been conclusively determined.

The current concern for climate change is based on the functioning of the earth's atmosphere as a "greenhouse." As radiation from the sun is absorbed by the earth, it reradiates energy back toward space. Gases within the atmosphere, including water vapor, carbon dioxide, methane, and others absorb this radiation, warming the earth's atmosphere in the process. Ice-core records, supplemented by direct measurements during the past 100 years, indicate that the earth's temperature has increased (Figure 1.14). Although there

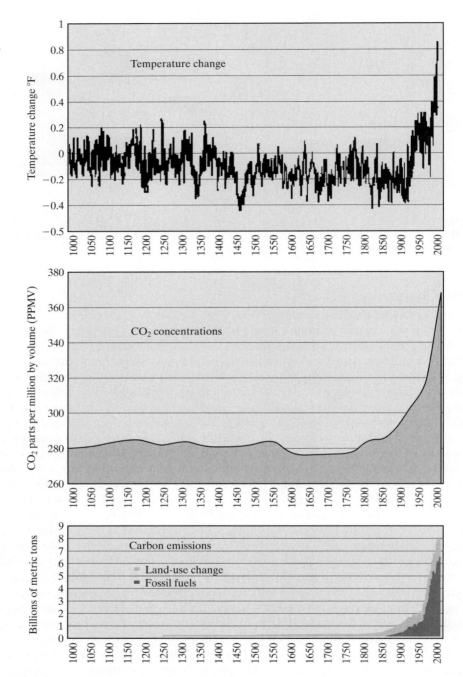

▲ **FIGURE 1.14**
Reconstructed changes in temperature, CO_2 concentration in the atmosphere, and carbon emissions from land clearing and fossil-fuel combustion for the last 1000 years. The data represent actual measurements for the past century and estimates obtained from paleoclimate reconstructions using ice-core analyses and other sources for earlier data. *Source:* U.S. Global Change Research Program, 2000.

▶ FIGURE 1.15
Projected temperature rises in the United States into the 21st century based upon the two leading climate models. *Source: From U.S. Global Change Research Program, 2000.*

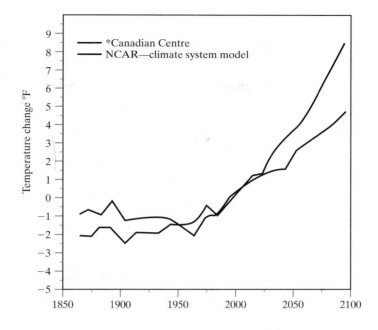

is some controversy about the data, the Intergovernmental Panel on Climate Change (IPCC), which produced the temperature curve shown in Figure 1.14, concludes that the decade of the 1990s was one of the warmest, if not the warmest, during the past 1000 years. Possible causes for this temperature rise include the increases in CO_2 content in the atmosphere, which is the likely result of the burning of wood during clearing of lands in heavily forested areas such as the Brazilian rain forest and burning of fossil fuels for heating and energy purposes (Figure 1.14).

Climate changes and effects are projected into the future using climatic models, with the assumption of increasing concentrations of greenhouse gases in the atmosphere. For example, the average temperature in the United States is projected to rise between 3° and 9°F in the 21st century, using various models (Figure 1.15). Along with the rise in temperature, consequences of rapid global warming could include melting glaciers and ice sheets, rising sea levels, changes in precipitation patterns and ecosystems, and more frequent and intense storms. These phenomena would have significant impacts on the human population. However, the sophistication of climate models has not progressed to the degree that they can make accurate predictions of all the interactions and feedbacks in the complex atmospheric and oceanic systems that control climate. One scenario, in fact, suggests that global warming could actually flip the switch of oceanic deep circulation and plunge the earth into a colder period. This model is conceptually illustrated in Figure 1.16. An increase in freshwater flux to the North Atlantic could dilute surface water to the extent that it would no longer sink to the bottom to drive the oceanic conveyor belt. Such an increase in freshwater flux could be driven by melting glaciers in Greenland and by the melting of sea ice in the Arctic Ocean. In Figure 1.16, we would currently be on the upper line moving to the right, with greater freshwater flux. At some critical threshold, the conveyor belt would be shut down or severely weakened and the temperature would drop suddenly into the cold mode corresponding to decreased oceanic deep circulation. This mode would persist until freshwater flux decreases (lower line) and another threshold would be crossed, allowing deep-water circulation to recommence. At that point, the temperature would rise to the upper line again.

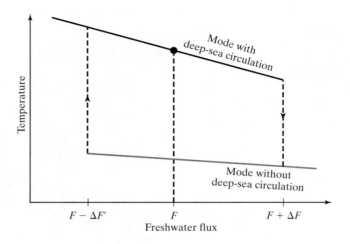

▲ FIGURE 1.16
Conceptual model of global temperature as a function of deep ocean circulation. When circulation is vigorous (upper line) climatic conditions are similar to the present. When an increasing freshwater flux to the North Atlantic, perhaps caused by global warming, reaches a threshold, circulation ceases and temperature drops to a lower stability state. As the lower curve is followed right to left, freshwater flux decreases and oceanic circulation resumes at another threshold value. Temperature then rapidly returns to its upper stability state. *Source:* Reprinted with permission from Climate Shock: Abrupt Changes over Millennial Time Scales, *Physics Today*, v. 55, no. 12, pg. 32 by Eduoard Bard, 2002, American Institute of Physics.

Although this scenario is interesting, we have no idea how close we are to either of the critical thresholds. Only through advances in climate modeling will we be able to make an informed guess about the future of the earth's climate. This challenge will be in the forefront of scientific research in the next few decades, along with the challenge of dealing with the impacts of climate change, whatever they might be.

Application of Geology to Engineering and Environmental Science: Meeting the Challenges

Solutions to the looming problems that could be caused by population increases, energy and water shortages, and climate change will be solved by current and future generations of scientists and engineers through research supported by governmental and private organizations. Engineering is the utilization of science in the solution of practical or technological problems. Practical problems that are most likely to require the application of geology include problems found in the traditional areas of construction, mining, petroleum development, and water resources development and management. Innovative solutions to environmental problems such as waste disposal are likely to occupy engineers and environmental scientists more and more in the future. An excellent example is the critical problem mentioned earlier concerning the disposal of high-level nuclear waste. The solution to this problem is being addressed by an intensive collaboration between engineers, geologists, and other environmental scientists. The increasing population of the world is going to compel the construction of more roads and buildings; it will necessitate the location and extraction of more metallic, nonmetallic, and energy resources; it will increase the risk of loss of life and property in natural disasters; and it will intensify the pressure on our natural environment. For all these reasons, geology will continue to play a critical role in engineering and environmental science in the immediate and distant future.

▶ FIGURE 1.17
Collection of a soil sample by using a truck-mounted auger. *Source:* Photo courtesy of USDA Soil Conservation Service.

Geology and the Construction Site

In the realm of civil engineering, there are two basic ways in which geology must be considered. First, the geologic setting of the construction site must be described and characterized. This is accomplished by mapping the distribution of geological materials and by collecting and testing samples of rock and soil from the surface as well as from below the surface of the site. The problem with these procedures is that natural materials are rarely homogeneous or continuous. Because of the high costs involved in drilling, sampling, and testing during site investigation (Figure 1.17), the prediction of certain engineering properties of the geologic materials and their response to engineering activity often relies on a thorough understanding of the origin and geologic history of the rocks and soils present. The amount of geologic investigation that is necessary depends upon the cost, design, purpose, and desired level of safety of the project. A dam across a canyon above a city requires a much more detailed investigation than a typical suburban housing development. Failures of large engineering projects, although very rare, often can be attributed to an inadequate understanding of the geology of the site.

The second major interaction between geology and engineering concerns the threat of hazardous geologic processes. This problem is often more relevant to the location of the structure rather than to its design or construction. The list of hazards includes spectacular geologic events such as volcanic eruptions, tsunamis, earthquakes, landslides, and floods (Figure 1.18). Other less sensational processes, such as land subsidence, soil swelling, frost heave, and shoreline erosion, result in billions of dollars of property damage in the United States every year.

The critical task in dealing with geologic hazards is how to predict where and when these events will take place. Often, the problem of when the hazardous process will occur is more difficult than where. Although geologists have made great strides in understanding the causes and effects of natural hazards, the ability to predict the place and, particularly, the time of a hazardous event remains an elusive goal.

An important characteristic of natural hazards is that there is usually a statistical relationship between the magnitude and frequency of the event. *Magnitude* is an indication of

▲ FIGURE 1.18
Damage from the Big Thompson flood of 1976 in Colorado, an event of high magnitude and low frequency. *Source:* W. R. Hansen; photo courtesy of U.S. Geological Survey.

the size or intensity of an event; the amount of energy released by an earthquake, the size of a landslide, or the amounts of flow in a flood are examples of magnitude. The *frequency*, on the other hand, is a measure of how often, on the average, an event of a particular size will occur. Frequency can be expressed as the probability of occurrence within a specific period, or the *recurrence interval* of the event. A flood with a 1% chance of taking place within a period of any one year has a recurrence interval of 100 years.

Careful recording of floods and other types of hydrologic data indicates that there is an inverse relationship between magnitude and frequency (Figure 1.19). These data can be fitted to the equations of theoretical statistical distributions with the result that the probability of events of high magnitude and low frequency can be predicted. Other geologic processes, including earthquakes and landslides, seem to follow a similar pattern. Engineers

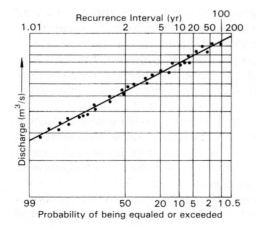

◀ FIGURE 1.19
The relationship between magnitude and recurrence interval for floods from a hypothetical river. Each point represents the magnitude and calculated recurrence interval of a particular flood.

▶ FIGURE 1.20
Economics of the design of hydraulic structures. The optimal cost of the structure is associated with an event of moderate recurrence interval. *Source:* From J. E. Costa and V. R. Baker, *Surficial Geology: Building with the Earth,* © 1981. Reprinted by permission of John Wiley & Sons, Inc., New York.

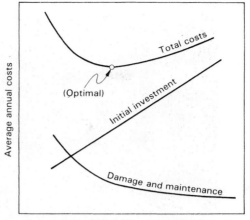

Average annual costs

Total costs

(Optimal)

Initial investment

Damage and maintenance

Average replacement period
(recurrence interval)

should understand this relationship and remember not to overlook the possibility of a destructive event of large magnitude. Critical structures such as nuclear power plants, schools, and hospitals must be designed to withstand large, infrequent events.

When certain or extensive loss of life is not a prime consideration in an engineering project, the economics of the project must also be considered in terms of the magnitude-frequency relationship. For example, a bridge that is built to withstand a flood of very high magnitude and low frequency would have high initial costs but low maintenance costs because the smaller, more frequent floods would cause little damage. Alternatively, the bridge could be built to withstand only small floods. In this case, the initial construction costs would be lower, but the maintenance costs would be higher because smaller floods that might damage the bridge are more frequent than large floods. In order to minimize the long-term costs of the structure, both initial costs and maintenance must be considered (Figure 1.20). The optimal total cost would be realized by designing the bridge for a flood of moderate recurrence interval.

Development of Natural Resources

The search for mineral deposits and petroleum is carried out by geologic exploration. When a resource is located, it must be delineated and assessed in order to determine if it is economically recoverable. If it is determined to be so, the extraction of the resource is conducted by mining and petroleum engineers. These engineers must have a sound knowledge of geology because of the geologic conditions of the location and the characteristics of the deposit.

Mining engineers supervise mine design and operation, including reclamation of the site (Figure 1.21). Environmental scientists are heavily involved in the reclamation process, both from industry and regulatory positions. The geologic and hydrologic setting of the deposit and the site influence many decisions that must be made in the process.

The production of petroleum often involves drilling through great thicknesses of rock (Figure 1.22). The drilling process requires the petroleum engineer to be familiar with all relevant physical and chemical conditions that may be encountered at various depths. This geologic knowledge must encompass a broad range of solutions to problems associated with rocks, liquids, and gases that are present in the subsurface. If a petroleum reservoir is located, petroleum engineers must develop it in the most efficient and economic manner possible. Extensive information must be gathered concerning the size, shape, and

▲ FIGURE 1.21
Extraction of coal from an open-pit mine in central North Dakota. About 20 m of overburden has been removed above the coal bed.

▲ FIGURE 1.22
A drilling rig exploring for oil in the Williston Basin, North Dakota.

rock characteristics of the reservoir. New technologies are constantly being tested to produce the maximum amount of oil possible from each reservoir.

Water Resources and the Environment

The development and protection of water resources is becoming one of society's most important concerns. Future increases in water supply are going to rely more heavily upon

▲ FIGURE 1.23
Crop irrigation places a heavy demand upon water resources. *Source:* Photo courtesy of USDA Soil Conservation Service.

groundwater than surface water. The occurrence, movement, and quality of groundwater is therefore a matter of great significance. Engineers and geologists together are developing techniques to manage the utilization of groundwater supplies to ensure their availability for the future. Overuse of groundwater for irrigation or other needs can lead to depletion and exhaustion of the supply (Figure 1.23).

Along with the need to wisely develop groundwater resources is the increasing necessity of protecting the chemical quality of groundwater. Past, and perhaps present, waste disposal and land-use activities have contaminated groundwater supplies in all parts of the United States. Development of new methods to rehabilitate groundwater reservoirs will require the talents of a wide variety of environmental scientists and engineers. Geology will be critical to the effort because an understanding of the subsurface geologic environment is necessary for dealing with any aspect of groundwater occurrence or movement.

References and Suggestions for Further Reading

BARD, E. 2002. Climate shock: abrupt changes over millenial time scales. *Physics Today*, 55 (no. 12):32–38.

COSTA, J. E. and V. R. BAKER. 1981. *Surficial Geology—Building with the Earth*. New York: John Wiley.

ROSEGRANT, M. W., X. CAI, and S. A. CLINE. 2002. Global water outlook to 2025. International Food Policy Research Institute. www.ifpri.org/media/water2025.htm.

U.S. GLOBAL CHANGE RESEARCH PROGRAM. 2000. *Our changing climate*. www.gcrio.org/NationalAssessment/overpdf/.

CHAPTER 2

The Earth and Its Systems

People study geology—the science of the earth—for different reasons. Even among those who become professional geologists, there is a great diversity in purpose and objective. For engineers, an understanding of geology is a tool that is utilized in the design and construction of any structure that is in contact with the earth or that interacts with its natural materials in any way. For environmental scientists, the geology of a site controls both the distribution and movement of contaminants below the ground surface and the selection of methods used for environmental cleanup.

Unfortunately, the application of geology to engineering and environmental science is not a simple task. One cannot refer to a manual for the solution to a geologic problem. Instead, the solution must come from experience and judgment, qualities that are based on a sound knowledge of the fundamental principles of geology. For this reason, engineers and environmental scientists must approach geology as they approach chemistry or physics: as a science that must be assimilated by thorough and systematic study.

In recent years, the way in which geologists and other scientists view and study the earth has dramatically changed. Geology, like other sciences, used to be a separate and distinct field of study—with minimal interactions with chemistry, physics, and biology, except where necessary to address a particular problem. The crux of the revolution in geologic thinking is considering the earth as a dynamic system, one that is constantly changing and one in which components or subsystems interact with each other to produce the changes that we can observe or changes that occurred in the past. The study of these past changes, using evidence of many types that has been left behind, is a big part of what geologists do. Environmental scientists have played a big role in this revolution because you can't explain what is happening in one part of the environment—the atmosphere, for example—without knowing what is going on in the oceans or other parts of the environment.

Although it is enticing to think of the earth system as a whole, an approach that is encouraged by viewing the planet from outer space (Figure 2.1),

▶ FIGURE 2.1
The earth as it looks from space. The lithosphere, hydrosphere, and atmosphere are all clearly visible. At their interface is the biosphere. Hurricane Andrew, about to strike the Louisiana coast, is visible in this image taken on August 25, 1992. *Source:* NASA.

where realizing the relationships between the land, the sea, and the atmosphere is unavoidable, in practice, however, it is necessary to examine individual components of the systems in detail, without losing sight of the fact that all components interact with all other components. One way to subdivide the system is to think of the earth system as being composed of four components, or spheres—the lithosphere, atmosphere, hydrosphere, and biosphere (Figure 2.2). These systems overlap with each other in more ways than we can probably even conceive. As biological organisms, we live in the biosphere, but our life is dependent upon all of the other components. Because this is a geology text, we will spend more time on the lithosphere than the other components. We cannot forget, however, that we live in the biosphere, which lies at the interface of the three other subsystems of the earth. We spend most of our time on the earth's solid surface or in structures that rest on the uppermost surface of the lithosphere, but our species is critically dependent upon processes occurring in the atmosphere and hydrosphere.

The Third Rock from the Sun

The earth is a good example of an *open system*, one that receives matter and energy from its surroundings. The role of the Sun's energy in earth systems will be summarized later in this chapter. As we all know, the earth does not exist alone in space, but is itself part of a larger system, the solar system. It is difficult to understand the earth as it is today without knowing how it evolved throughout its history.

The Solar System

The planet on which we live is only one of nine planetary bodies that revolve around the Sun in the solar system (Figure 2.3). To put Earth in its proper perspective, we should briefly examine the characteristics of our planetary neighbors. One of the most basic observations is that there are two groups of planets: four smaller planets orbiting close to the Sun and four larger planets occupying the outer reaches of the solar system. The ninth planet, Pluto, is somewhat of an exception to this grouping. The inner, or *terrestrial*, planets are dense bodies mainly composed of iron and silicate (containing the element silicon) rocks. The four *giant* planets are much lighter in density because of their gaseous composition.

▲ FIGURE 2.2
The biosphere, atmosphere, hydrosphere, and lithosphere, four overlapping components of the earth system. *Source:* Modified from Fred T. Mackenzie, *Our Changing Planet: An Introduction to Earth System Science and Global Environmental Change*, 2nd ed., © 1998, p. 10. Reprinted by permission of Pearson Education, Inc., Upper Saddle River, N.J.

Although many ideas have been proposed to explain the origin of the solar system, this question is still a matter of debate. In the most commonly accepted current hypothesis, the history of the solar system begins about 4.6 billion years ago with a large diffuse mass of gas and dust slowly rotating in space, a *solar nebula*. The matter in the solar nebula was just a tiny part of the primordial material produced in a huge explosion, the *Big Bang*, which is thought to have occurred about 15 billion years ago. About 10 billion years after the Big Bang, our solar nebula began to contract, due to gravitational forces, and to increase its rotational velocity. At some point, a concentration of matter formed at the center of the nebula. Compression of matter raised its temperature to the point at which nuclear fusion was initiated. Thus the Sun, a body composed of 99% hydrogen and helium, was born. The matter rotating around the newly formed Sun gradually cooled and condensed into the nine planets. The heavier elements in the solar system were inherited

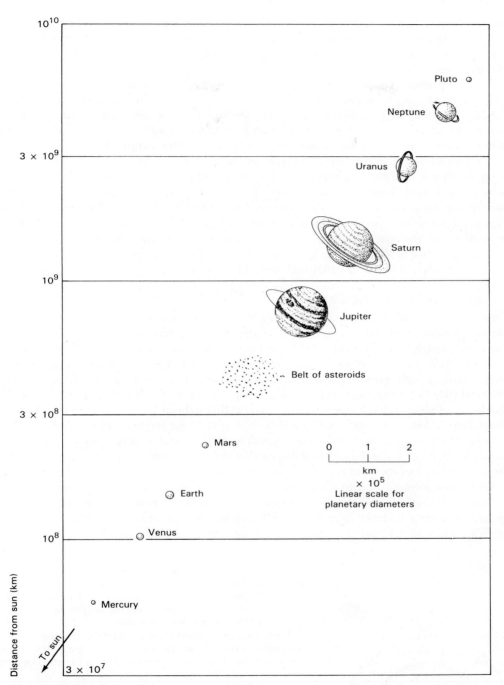

▲ FIGURE 2.3
Relative sizes and positions of the planets.

from earlier cycles of star formation and destruction during the time between the Big Bang and the formation of the solar system. Destruction of stars is accompanied by a giant explosion known as a *supernova*, which provides enough energy for the formation of the heavy elements.

As the planets began to develop through the repeated accretion of smaller bodies that crashed into them and added to their mass, their final composition was controlled by their initial position in the nebula with respect to the Sun. Because the temperature was highest in the vicinity of the Sun and decreased with distance away from the Sun, the terrestrial planets were formed of material with relatively high boiling points. Thus Mercury, Venus, Earth, and Mars are dense bodies of iron and silicate rocks. In fact, four elements—iron, oxygen, silicon, and magnesium—make up about 90% of these planets. Volatile elements were carried away from the inner part of the nebula by matter streaming from the Sun and perhaps by the inadequate gravitational pull of the terrestrial planets before they attained their total mass. Volatile gases like water, methane, and ammonia were driven to the cooler regions of the giant planets. Jupiter, Saturn, Uranus, and Neptune retained their volatiles because of their greater gravitational attraction, and they are therefore more similar in composition to the original solar nebula.

Differentiation of the Earth

After the initial condensation of the earth, its internal structure was very much different than at present. It probably consisted of a homogenous accumulation of rock material. At this point, heating of the earth must have taken place. The impacts from smaller bodies of accreting matter in the solar nebula, *planetesimals*, were major contributors to the rise in temperature—their kinetic energy was transformed into heat upon collision with the growing planet. Compression of the newly condensed matter and the energy released by the decay of radioactive elements disseminated throughout the earth also helped raise the temperature. Many scientists believe that during the first 500 million years of the earth's accretion from smaller bodies, a giant impact might have taken place. The impacting body may have been so large that it tilted the earth's axis to its current angle of 23° and brought the temperature of the young planet to or near the melting point. The debris blasted from the collision accreted within the earth's orbit to form the moon. The consequence of the temperature rise was that the melting point of iron and nickel was exceeded. These dense molten elements formed immiscible globules in the magma ocean that existed at the earth's surface and began to migrate toward the center of the earth (Figure 2.4). Gradually,

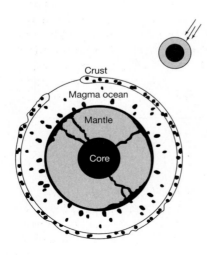

◄ FIGURE 2.4

Sinking of molten iron and nickel toward the center of the earth to form its core. During early melting or partial melting of the earth due to impacts of planetesimals from space, iron and nickel segregated into dense, immiscible globules, which sank to form the core at the center of the earth. Planetesimals may have already been differentiated by radioactive heating prior to impact. *Source:* John A. Wood, *Solar System, The,* 2nd ed., © 2000, p. 77. Reprinted by permission of Pearson Education, Inc., Upper Saddle River, N.J.

▶ FIGURE 2.5
Internal structure of the earth.
Source: From S. Judson, M. E. Kauffman, and L. D. Leet, *Physical Geology*, 7th ed., © 1987 by Prentice Hall, Inc., Upper Saddle River, N.J.

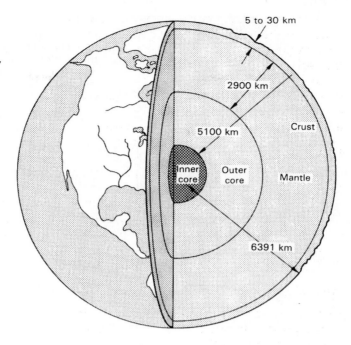

a molten iron core began to develop at the earth's center. This process released gravitational energy in the form of heating, leading to more melting and partial melting. The end result was a density-stratified planet (Figure 2.5). The dense core was overlain by a mantle composed of iron and magnesium silicate rocks. The lightest materials accumulated in a very thin layer near the surface to form the crust.

The age of a rock sample can be determined by the abundance of certain radioactive elements. This method will be explained later in this chapter. Although the oldest rocks on the earth are about 4 billion years old, the planet itself is considered to be about 4.6 billion years old. Our estimate for this age comes from dating meteorites that fall to Earth. These rocky fragments consistently yield ages of 4.5 to 4.6 billion years. They are considered to be composed of material formed at the time of origin of the solar system. The oldest rocks dated from the moon have similar ages, suggesting that the giant impact that formed the moon occurred very early in the earth's history. Why are there no rocks on Earth from the period of time between 4.6 and 4.0 billion years ago? The answer is that the period of time between the accretion of the earth at about 4.6 billion years and the formation of the oldest existing rocks at 4.0 billion years may be accounted for by the stratification or *differentiation* of the earth. The original rocks may have been melted and their components redistributed into the core, mantle, and crust during this interval.

The Earth's Heat Engine

After differentiation of the earth produced a planet stratified by density and temperature with a partially molten core of iron-nickel alloy, a cooler mantle composed of silicate rocks, and a rigid crust made up of the least dense rock materials, the planet began to function as a giant heat engine, a mode of operation that continues to the present day. The engine is powered by the flow of heat from the partially molten core to the cold outer surface. The inferred distribution of temperature, pressure, and density in the earth is shown

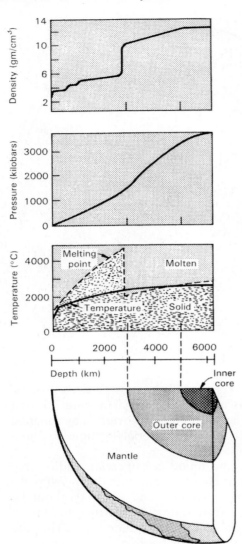

◀ FIGURE 2.6
Physical conditions in the interior of the earth. The inferred temperature profile of the earth passes above the melting point curve in the region of the earth's outer core. *Source:* From E. A. Hay and A. L. McAlester, *Physical Geology: Principles and Perspectives*, 2nd ed., © 1984 by Prentice Hall, Inc., Englewood Cliffs, N.J.

in Figure 2.6. To fully understand the operation of this engine, it is necessary to take a closer look at the internal layers of the earth. The boundary shown between the crust and mantle in Figure 2.5 was detected by a rapid increase in the velocity of seismic waves traveling through the earth. The boundary itself, the *Mohorovicic discontinuity*, or the Moho for short, has been recognized since the early 20th century. The Moho ranges in thickness between 5 and 10 km beneath the ocean floors to about 40 km beneath the continents. The sharply increasing velocity of seismic waves at this interface probably represents a change in rock type to denser silicate rocks in the mantle below.

In the 1960s and 1970s, a new view of the earth's internal structure began to emerge (Figure 2.7). This view allowed geologists to integrate data and observations from many diverse fields into a unifying theory of the earth's behavior, called *plate tectonics*. This theory now forms the basis of our understanding of the earth's large-scale internal behavior

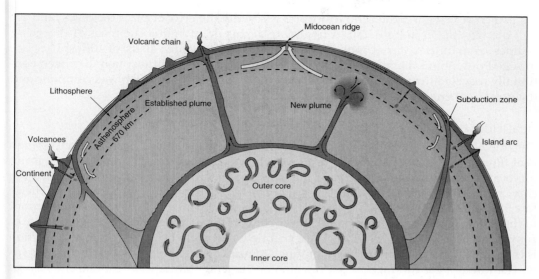

▲ FIGURE 2.7
Internal structure of the earth, showing location of the lithosphere and asthenosphere, formation and movement of rising mantle plumes, descending cold slabs, and several types of boundaries of lithospheric plates. *Source:* Jon P. Davidson, Walter E. Reed, and Paul M. Davis, *Exploring Earth: An Introduction to Physical Geology*, 2nd ed., © 2002. Reprinted by permission of Pearson Education, Inc., Upper Saddle River, N.J.

and also has many implications about geological processes that we can observe at the surface as well.

The new model considers the earth's molten outer core to be a very active zone of circulation. Movement is generated in the liquid outer core by *convection* (Figure 2.8). The liquid heated at the boundary of the solid inner core will become less dense than the surrounding liquid and a blob of it will slowly rise, as in a lava lamp. These rising currents are then modified by the rotation of the earth on its axis. Most currents in the outer core tend to terminate at the mantle boundary, where heat is then transferred by *conduction*. However, isolated vertical currents, or *plumes*, are able to continue rising through the mantle to the base of the crust. Circulation in the upper mantle is more vigorous because a zone lying between 400 and 670 km beneath the surface is very close to the melting point (Figure 2.6) and therefore plastic and easily deformable by hot, low-density plumes rising through it. This zone of plastic deformation is called the *asthenosphere*. The more rigid zone above it, including the entire crust and uppermost mantle, is called the *lithosphere*. The interaction between mantle plumes rising through

▶ FIGURE 2.8
Convective heating of water in a pan may be similar to convection cells of mantle material that rise to the base of the lithosphere, spread laterally, and then descend.

the asthenosphere and the rigid lithosphere is really the crux of plate tectonics, and we will discuss this in more detail. A final note at this point is that the mass of the rising plumes of mantle material must be balanced, so that colder slabs of lithosphere are dragged down into the mantle (Figure 2.7). Thus, the earth's lithosphere has been continually recycled through the mantle since differentiation of the earth was complete and the great heat engine began to operate.

Heat Flow

The unequal distribution of volcanoes, hot springs, geysers, and other indications of high heat flow suggests that the flow of heat from the interior of the earth to the surface is not uniform. We have already mentioned the two major sources of internal heat—the molten core and crustal radioactivity. Heat flows whenever a temperature gradient exists; therefore, heat flows from the earth's core to its cooler surface. Three physical processes account for heat transport; these include conduction, convection, and radiation. The importance of convection in heat transport through the mantle has already been described. Conduction is the atomic transfer of energy by vibrating atoms to adjacent atoms vibrating less rapidly. The rate of heat conduction is a function of the temperature gradient and the *thermal conductivity* of the rocks through which the heat is flowing. Rocks are poor heat conductors; fluctuations of temperature due to solar effects at the surface, for example, are felt only in the upper few meters of soil. By itself, conduction is therefore inadequate to explain the distribution of heat flow in the earth. In fact, heat that began to move outward from the center of the earth at the time of its origin would not yet have reached the surface. Radiation of heat is unlikely through rocks; therefore, convection must play a critical role in heat transport.

The second major source of internal heat in the earth is radioactive decay. Since differentiation, radioactive elements have been concentrated in the rocks near the earth's surface—in particular, in granitic rocks that make up the majority of the earth's continental crust. Thus, radioactive heat production is most significant in the continental crust. Heat flow measured at the surface is the sum of radioactive production from the crust and deep heat flow from the core.

Variations in Heat Flow

Heat-flow measurements have been made both on the continents and in the ocean basins. Continental heat-flow measurements are made by inserting temperature probes into holes drilled into the walls of mines and tunnels. In the oceans, temperature probes are driven into bottom sediments to measure the temperature. In both areas, measurements of temperature changes with increasing depth are used to determine the temperature gradient. When the temperature gradient is multiplied by the measured thermal conductivity of the rock or sediment, the heat flow is obtained.

Heat-flow values throughout the world can be related to the geologic setting of the point of measurement. In the ocean basins, high heat flow is found beneath the midoceanic ridges (Figure 2.7). Midoceanic ridges are thought to be the surface manifestations of rising mantle convection currents. Volcanism in these areas forms new lithosphere, which moves laterally away from the ridges. Heat flow declines with distance from midoceanic ridge crests and reaches minimum values over trenches. Trenches develop over subduction zones, where cold lithospheric slabs are dragged or pushed downward into the mantle. Subduction zones (Figure 2.7) constitute the descending plumes of convection current cells; thus heat flow is low in these areas. Iceland is a place where volcanism on a midoceanic ridge has been so intense that the ridge rises above sea level to form an

► FIGURE 2.9
Geysir, the Icelandic geyser from which the name is derived. *Source:* Photo courtesy of the author.

island. High heat flow in Iceland is immediately visible in the form of active volcanoes, geysers (a term that comes from a well-known example in Iceland called Geysir), and hot springs (Figure 2.9).

On the continents, heat flow also varies, even though the radioactive heat component of total heat is relatively constant. Areas of high heat flow have had recent tectonic and volcanic activity. The Basin-and-Range Province of western North America is one such area (Figure 2.10). Volcanism and recent faulting have been common in the Basin-and-Range Province, so it is possible that a rising convection current or plume is present beneath this province. The high heat flow in this region is under study for possible production of *geothermal energy*. The Geysers geothermal plant in California (Figure 2.11) already produces 600 MW of electrical power from geothermal energy. Low continental heat flow occurs in geologically old regions, like the Canadian Shield, that have been tectonically inactive for millions of years.

The Earth's Magnetic Field

A consequence of circulation of molten medal in the outer core is the earth's magnetic field, because metal conducts electricity. A dynamo is a device that generates electricity by moving a conductor through a magnetic field. The current generated in the dynamo will also create a magnetic field. Just how the earth's dynamo originally started is still somewhat unclear; however, if a small magnetic field was initially present, the principle of a *self-exciting dynamo*, one in which the movement of a conductor through a small magnetic field could create an electrical current that would in turn produce a stronger magnetic field, is often invoked. The magnetic field of the earth acts like a dipolar bar magnet, with its negative pole in the northern hemisphere (Figure 2.12). The axis of the bar magnet is roughly aligned with the earth's axis of rotation. Magnetic objects orient themselves along the magnetic *lines of force* radiating outward from the poles. The discovery of this phenomenon led to the development of the compass, whose positive end points to the magnetic north pole.

The vertical angle between a line of force and the earth's surface is known as magnetic *inclination*. This angle is zero, or horizontal, at the magnetic equator and 90°, or vertical,

(a)

Lithosphere

Mantle plume

(b)

◀ FIGURE 2.10
(a) Map of the western United States showing location of the Great Basin area of the Basin-and-Range Province. *Source: From E. A. Hay and A. L. McAlester, Physical Geology: Principles and Perspectives,* 2nd ed., © 1984 by Prentice Hall, Inc., Upper Saddle River, N.J. (b) Diagram of possible basin-and-range structure caused by rising mantle plume beneath.

▲ FIGURE 2.11
Steam wells at The Geysers geothermal area, California. *Source:* Photo courtesy of U.S. Geological Survey.

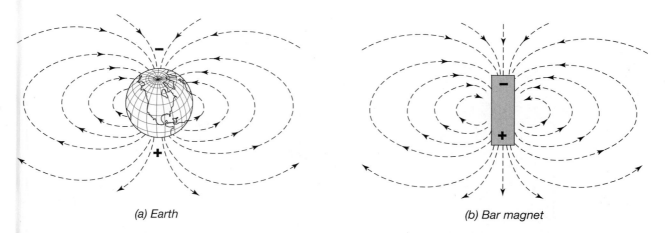

(a) Earth　　　　　*(b) Bar magnet*

▲ FIGURE 2.12
Comparison of the earth's magnetic field to a bar magnet, with the negative pole in the northern hemisphere. The axis of the magnetic field is inclined 10.9° from the rotational axis. The positive needle of a compass points to magnetic north, which is close to geographic north. *Source:* Robert J. Lillie, *Whole Earth Geophysics: An Introductory Textbook for Geologists and Geophysicists*, 1st ed., © 1999. Reprinted by permission of Pearson Education, Inc., Upper Saddle River, N.J.

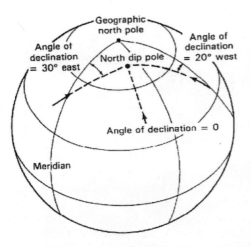

◀ FIGURE 2.13
Declination is the angle measured at any point on the
earth's surface between the magnetic north pole and
the geographic north pole. *Source:* From S. Judson,
M. E. Kauffman, and L. D. Leet, *Physical Geology*, 7th ed.,
© 1987 by Prentice Hall, Inc., Upper Saddle River, N.J.

at the magnetic pole. Because the magnetic and geographic poles do not exactly coincide,
there is a small angle between the direction of a compass needle pointing to the magnetic
north pole and the direction of a line pointing to the geographic north pole. This angle is
known as *magnetic declination* (Figure 2.13).

The earth's magnetic field has not remained constant since the inception of the magnetic field. Periodically, the magnetic field weakens and reverses itself so that the magnetic north becomes magnetic south and vice versa (Figure 2.14). During a magnetic
reversal of this type, compass needles would actually point south! Many magnetic reversals have been identified in rocks of different ages near the earth's surface. The study of
magnetic reversals has actually provided a way to determine the age of the rocks when a

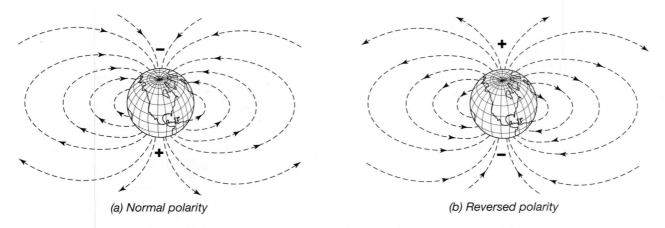

(a) Normal polarity   *(b) Reversed polarity*

▲ FIGURE 2.14
Configuration of the earth's magnetic field under normal and reversed polarities. *Source:* Robert J. Lillie,
Whole Earth Geophysics: An Introductory Textbook for Geologists and Geophysicists, 1st ed., © 1999. Reprinted by
permission of Pearson Education, Inc., Upper Saddle River, N.J.

known sequence of reversals can be recognized. Rocks that are magnetized in the current direction of the earth's field are known as "normal," and rocks that are magnetized with an opposite polarity are known as "reversed."

Paleomagnetism

As molten rock erupting from volcanoes cools, the atoms in magnetic minerals, one of which is known as magnetite, become aligned in the direction of the earth's magnetic field. This process occurs at a temperature called the *Curie point*, a temperature well below the melting point of the mineral. Once magnetism is acquired by the rock, it is retained unless the rock becomes heated above the Curie point.

Measurement of a rock's magnetic declination and inclination can be made; such measurements are said to indicate the *paleomagnetism* of the rock. Paleomagnetic studies have yielded dramatic results concerning the history of the earth. For example, from declination and inclination values it is possible to determine the position of the magnetic (and therefore geographic) pole at the time of formation of the rock. One of the most fascinating results of paleomagnetic studies is that the magnetic poles have apparently moved great distances through geologic time. For several reasons, movement of the magnetic poles over great distances is not thought to be possible. Therefore, to explain the paleomagnetic results, it was proposed that the continents containing the rocks, rather than the magnetic poles, have moved over the earth's surface during geologic time. This was one of the main lines of evidence supporting an early predecessor of the theory of plate tectonics called *continental drift*.

The second form of evidence for continental drift was gathered when *magnetometers* were towed behind research ships crossing the oceans. These instruments were able to measure the magnetism of the rocks beneath the sea floor. When these ships crossed midoceanic ridges, a remarkable magnetic pattern was discovered (Figure 2.15). At the crest of the midoceanic ridge, rocks of normal (that is, in the orientation of the earth's present magnetic field) magnetism were encountered. However, as the ship moved away from the midoceanic ridge crest, linear bands of rock were crossed that were magnetized in alternately reversed and normal polarities. Even more exciting was the realization that these magnetic bands or stripes were symmetrical on opposite sides of the ridge crests. This was strong evidence that new oceanic crust was formed by volcanism at the ridge crest in the prevailing polarity of the earth's magnetic field. The new oceanic crust was then split apart, moving in opposite directions away from the ridge. When a magnetic reversal occurred, the rocks forming at the ridge crests would be of that polarity. Thus the magnetic bands were symmetrical about the ridge. This discovery became nearly irrefutable evidence for the formation of new lithosphere at midoceanic plate margins. This theory became known as *sea-floor spreading*. It is considered to be one of the foundations of plate tectonics.

Major Features of the Earth's Surface

The earth's surface can be divided into two dominant topographic features: the continents and the ocean basins (Figure 2.16). The rocks that lie beneath the continents differ greatly from the rocks of the ocean basins. Continental rocks are low in density and include the oldest rocks on the earth. These are the lowest density materials concentrated by differentiation and subsequent recycling of the lithosphere. The ocean basins are composed of denser, younger rocks. Volcanic eruptions continually produce new oceanic lithosphere at places where mantle plumes reach the upper part of the mantle and cause volcanism in the crust. In later sections of this book, the rocks of the continents and ocean basins will be described in more detail.

▲ **FIGURE 2.15**
Symmetrical magnetic patterns across the Mid-Atlantic Ridge, indicating that bands of newly formed rocks were magnetized at the ridge axis and that the bands were split apart during lateral movement in opposite directions. The black and stippled magnetic bands represent normal polarity, and the white bands beyond represent reversed polarity. *Source:* From E. A. Hay and A. L. McAlester, *Physical Geology: Principles and Perspectives*, 2nd ed., © 1984 by Prentice Hall, Inc., Englewood Cliffs, N.J.

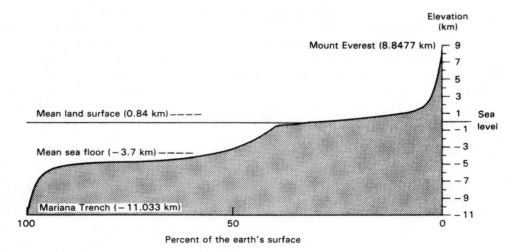

▲ **FIGURE 2.16**
The two predominant levels of the earth's surface: the continents and ocean basins. *Source:* From W. K. Hamblin, *The Earth's Dynamic Systems*, 4th ed., © 1985 by Macmillan Publishing Co., Inc., New York.

The Earth's Gravity

The force of gravity can be expressed as

$$F = \frac{mM}{r^2} G \tag{2.1}$$

where F is the gravitational attraction between two bodies, m and M are the masses, r is the distance between the bodies, and G is the universal gravitational constant. For a body at the surface of the earth, r and M become the radius of the earth and its mass, respectively. If the earth were perfectly spherical, nonrotating, and uniform in density, the gravitation acceleration, or the acceleration of a freely falling body, would be constant anywhere upon the earth's surface. Gravitational acceleration varies from place to place, however, and it is these minor variations that make gravity measurements a very useful tool in geology.

Measurement

The unit of measurement used in gravity studies is the Gal, which equals an acceleration of 1 cm/sec². The average value of g (gravitational acceleration) at the earth's surface is about 980 Gal, although the measured values vary from place to place. These variations are attributed to (1) rotation of the earth, (2) topography, and (3) variations in density of near-surface rocks. Rotation affects gravity by causing a bulge at the earth's equator that is due to centrifugal force. The radius of the earth is thus greater at the equator than at the poles. The radius also varies from highlands to lowlands or to ocean basins. These departures from the ideal ellipsoidal shape of the earth cause minor changes in gravitational attraction across the surface (Figure 2.17). Finally, lateral changes in density of near-surface rocks can affect the value of g. The value of g would be greater over a region of dense near-surface rocks such as basalt than over lighter granitic rocks at the same elevation. The magnitude of these variations is on the order of milliGals (0.001 Gal) or fractions of a milliGal. These minute variations are measured by instruments called *gravity meters*. A gravity meter consists of a weight suspended on a spring that expands or contracts according to

▶ FIGURE 2.17
Variations in gravitational attraction are caused by the earth's ellipsoidal shape, elevation differences, and differences in density of crustal materials. *Source:* From E. A. Hay and A. L. McAlester, *Physical Geology: Principles and Perspectives*, 2nd ed., © 1984 by Prentice Hall, Inc., Englewood Cliffs, N.J.

▲ FIGURE 2.18
Horizontal deflection of a plumb bob by the mass of a prominent mountain range.

the local gravitational field. Despite this simple principle, gravity meters are extremely sophisticated, expensive instruments because of the precision they must attain.

After a field measurement is taken, it must be corrected before it will yield geologically
useful information. One major correction resolves the effects of local topographic variations.
For example, the presence of a nearby mountain range will deflect a pendulum from its vertical position because of the gravitational attraction caused by the mass of the mountains
(Figure 2.18). The *Bouguer correction* adjusts the reading to represent a slab of uniform thickness
above a hypothetical plane at sea level called the *geoid* (Figure 2.19). Also, a correction is applied to compensate for the material between the geoid and the point of measurement. This is
known as the *free-air* correction. The final values are compared to an expected value, which is

Topographic correction

Bouguer correction

Free-air correction

◄ FIGURE 2.19
Corrections to measured gravity readings are necessary to discover subsurface geological features. The
topographic correction removes the effect of local
variations in elevation. The Bouguer correction com-
pensates for a uniform slab of material between the
geoid and the point of measurement. The free-air correction results in a reading that would be obtained if
the meter were at sea level. *Source:* From S. Judson,
M. E. Kauffman, and L. D. Leet, *Physical Geology*, 7th ed.,
© 1987 by Prentice Hall, Inc., Upper Saddle River, N.J.

calculated for an ideal earth, the spheroid, with a homogeneous crust. Any differences between the corrected measurement and the expected value are due to local variations in density caused by the composition of near-surface rocks. These differences are called *gravity anomalies*.

The study of gravity anomalies has led to important theories about the earth's crust. For example, young mountain ranges typically yield large negative anomalies. This implies that rocks of low density extend to some depth below the mountain range. Ocean basins have positive anomalies, which indicates that the rocks beneath the ocean basins are denser than average. Thus, the density of near-surface rocks is inversely proportional to their topographic position; regions of dense rocks usually lie below sea level, and mountain ranges are composed of lighter rocks. For example, a Bouguer anomaly profile across the United States (Figure 2.20) is drawn over a profile showing

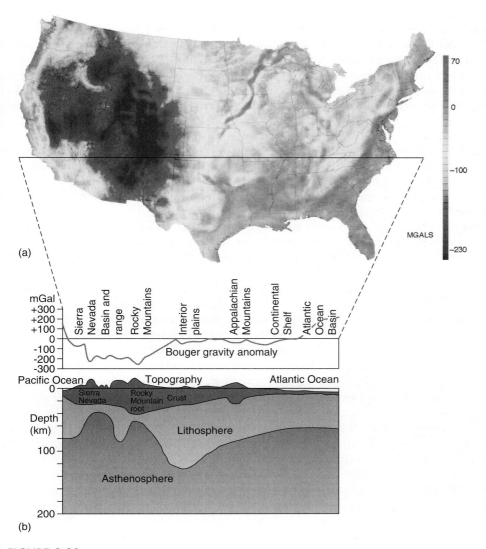

▲ FIGURE 2.20
Bouguer anomaly profile across the United States. The anomalies are negative over the Rocky Mountains because the crust is thick and the asthenosphere below it is hot and therefore lower in density than normal. The thin, cold, oceanic crust of the Atlantic Ocean leads to a positive anomaly. *Source:* Jon P. Davidson, Walter E. Reed, and Paul M. Davis, *Exploring Earth: An Introduction to Physical Geology*, 2nd ed., © 2002. Reprinted with permission of Pearson Education, Inc., Upper Saddle River, N.J.

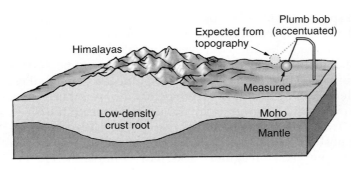

◀ FIGURE 2.21

Comparison of the measured versus calculated deflections of a plumb bob near the Himalayas. The measured value is less because of the low-density root of the mountain range. *Source:* Jon P. Davidson, Walter E. Reed, and Paul M. Davis, *Exploring Earth: An Introduction to Physical Geology*, 2nd ed., © 2002. Reprinted by permission of Pearson Education, Inc., Upper Saddle River, N.J.

the internal structure. Negative anomalies are present over the Rocky Mountains because of the low-density "root" of continental crustal material in this area. Anomalies become positive over the Atlantic Ocean, which is underlain by thin, dense oceanic crust. Gravity measurements also have many practical applications. Masses of igneous rock that contain ore bodies often differ in density from the rocks that surround them. In addition, sedimentary rock structures that trap petroleum may cause characteristic gravity anomalies.

Isostasy

The discovery that mountains had roots of low density came about by accident during the survey of India by Sir George Airy in the nineteenth century (Figure 2.21). Surveyors correctly assumed that the plumb bob set up on a plain near the Himalayas would be deflected laterally because of the mass of the high mountains projecting above the plains. The surprise came when the amount of deflection turned out to be much less than calculated, when the mountains were assumed to be simply resting on crustal material of higher density. The only way that this could be explained was that the mountains were underlain by low-density crust that extended significantly below the surface of the plain. It has since been verified that all mountain ranges have low-density crustal roots extending below them. In effect, mountain masses are "floating" in the denser mantle rock like icebergs, which have a lower density than water, float in the ocean. Ship captains are well aware that the thickness of the iceberg below the water is much greater than its projection above the water. This concept is known as the theory of *isostasy*. Although mantle rocks are rigid solids with respect to short-term deformations such as the propagation of seismic waves, they may behave as a viscous fluid over long periods of geologic time. The mountain mass is supported by buoyant forces in the mantle similar to the buoyancy that supports any floating object.

Implied in the theory of isostasy is the idea that crustal masses establish a state of dynamic equilibrium with the mantle. Thus, the dense basalts of the ocean basins lie at a low elevation in comparison with lighter granitic mountain ranges, which float higher in the mantle. Any change in the mass of a crustal body will be reflected in changes in this *isostatic equilibrium*. For example, as mountains are slowly eroded, mass is removed and the low-density root is uplifted and is decreased in size because less support is required (Figure 2.22). Thus, the Rocky Mountains have deep roots in comparison with the old, eroded Appalachian Mountains, whose roots are much thinner than they were during the period following their formation (Figure 2.20). Isostatic equilibrium is also illustrated by the advances and retreats of glaciers. The heavy load of a glacier imposed upon the crust causes the crust to sink deeper into the mantle. When the ice melts and retreats, however, the load is removed much faster than the crust and mantle can reestablish isostatic equilibrium. The slow uplift of the crust after glacial retreat is called *isostatic rebound*. Slow uplift of the crust is still going on, 10,000 years after the last glaciation (Figure 2.23).

▶ FIGURE 2.22

Isostatic equilibrium. (a) A mountain range with a low-density crustal root. (b) Isostatic equilibrium is maintained by reduction in the size of the root as the mountain mass is slowly eroded. *Source:* From S. Judson, M. E. Kauffman, and L. D. Leet, *Physical Geology,* 7th ed., © 1987 by Prentice Hall, Inc., Upper Saddle River, N.J.

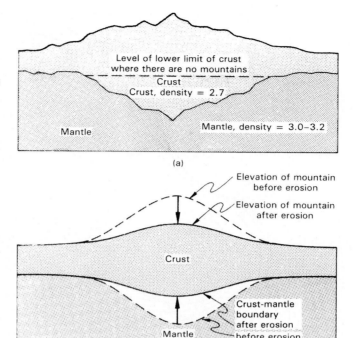

Plate Tectonics

One of the greatest accomplishments in geology during the past few decades has been the formulation of a new theory called *plate tectonics*. This theory revolutionized the science of geology because it provided a comprehensive model that linked the internal behavior of the earth to its surface features. With plate tectonics, the distribution of earthquakes and volcanoes makes perfect sense. Plate tectonics integrated and replaced the earlier theories of continental drift and sea-floor spreading, incorporating heat flow, gravity, and magnetism into the new paradigm. The synergistic integration of fields of study that had previously been studied within a narrow focus led to many new discoveries and advances.

According to the plate tectonics model, the uppermost part of the earth is divided into the upper, rigid lithosphere, which includes the older divisions of crust and upper mantle, and the underlying, warm, plastic asthenosphere. The lithosphere is broken into about 12 major *plates*, which slowly move laterally over the earth's surface (Figure 2.24). Plates are rigid slabs of lithosphere that move as a unit. Active geologic processes are concentrated, for the most part, at plate boundaries. Movement of lithospheric plates is driven by the convective flow of material in the asthenosphere (Figure 2.24). Rock in the asthenosphere is soft and plastic because it is near its melting point. Upwelling plumes of hot mantle material impact the base of the lithosphere and spread laterally outward, cracking apart, or *rifting*, the lithosphere in the process (Figure 2.25). Volcanoes erupt to form new lithosphere in the cracks formed as the plates split apart. The plates move until they reach a zone of downward flow in the asthenosphere. Here, the lithospheric plates are carried down, or *subducted*, into the asthenosphere and may become partially melted.

▲ FIGURE 2.23
Uplift in Scandinavia during the last 5000 years caused by glacial retreat. The contours are the elevations in meters of a shoreline that was at sea level 5000 years ago. *Source:* After R. K. McConnell, Jr., 1968, Viscosity of the mantle from relaxation time spectra of isostatic adjustment, *Journal of Geophysical Research*, 73: 7090.

Evidence for Plate Tectonics

Continental Drift

One of the earliest precursors to the plate tectonics theory was put forth in 1912 by Alfred Wegener, a German meteorologist who proposed that the continents had moved great distances laterally across the earth's surface through geologic time. The shapes of the coastlines of Africa and South America, which would fit together like the pieces in a jigsaw puzzle if the continents were closer, was one of the main lines of evidence for this idea. Wegener's theory, which became known as *continental drift*, proposed that all the continents were once assembled into a huge continental mass called *Pangaea* (Figure 2.26) and that they then drifted apart to their present positions.

In addition to noting the fit of continents, Wegener summarized geologic evidence suggesting that the continents were once joined together (Figure 2.27). This evidence included several types. First, mountain belts, structural trends, and rock types found on different continents would be continuous if the continents were assembled as in Pangaea. Second, the distribution of numerous fossil species found on several continents could be explained if the continents had been originally joined. These species included land plants and animals that could not have crossed an ocean to colonize widely separated continents.

▲ FIGURE 2.24
Major plates of the lithosphere.

▶ FIGURE 2.25
Rifting caused by rising plumes in the astheno-sphere. Subduction of lith-osphere at the other end of the plate.

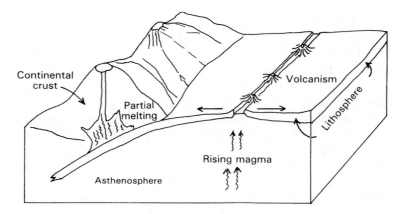

Third, rock types that form under specific climatic conditions were encountered in regions whose present climate is very different. For example, rocks containing coal and tropical plants were discovered in polar regions, and rocks formed by glaciation were mapped near the present equator. The locations of these rocks would be difficult to explain unless the land masses had moved into different climatic zones.

◄ FIGURE 2.26
Assembly of the continents at the 900-m depth of the ocean bottom to form Pangaea. *Source:* From E. Bullard et al., 1965, *Phil. Trans. Royal Soc.*, 1088: 41–51.

Line of 900-m depth

Zones of gap

Zones of overlap

Wegener's hypothesis was largely discounted by scientists of the day because of the implausible mechanism to which he attributed the drift of continents. Wegener suggested that the granitic rock of the continents had moved laterally through the stronger, denser mantle rock. Other scientists quickly pointed out the mechanical impossibility of this process, and Wegener's ideas were ignored by most geologists for several decades.

The Theory of Sea-floor Spreading

In the first decades after World War II, tremendous advances in exploration of the oceans were realized. Topographic surveys of the ocean floor using sounding devices discovered the immense midoceanic ridges positioned near the center of the ocean basins (Figure 2.28). Rock samples from the ridge crests were composed of fresh basaltic lava covered by very little sediment. The thickness of bottom sediment above bedrock increases laterally away from the ridge crests, implying an increase in the age of the oceanic crust with distance from the midoceanic ridges.

The most dramatic evidence pertaining to the origin of the ocean basins was obtained from paleomagnetic studies of the oceanic crust. The symmetrical bands of alternately magnetized volcanic crustal rock (Figure 2.15) provided important evidence for the *sea-floor spreading* theory, which proposed that new crustal material was formed by volcanic eruptions at the crests of the midoceanic ridges and that slow lateral movement of the crust away from the ridges was occurring. Other geophysical evidence suggested that the

▶ FIGURE 2.27
Evidence for continental drift as proposed by Alfred Wegener. *Source: From E. A. Hay and A. L. McAlester, Physical Geology: Principles and Perspectives,* 2nd ed., © 1984 by Prentice Hall, Inc., Englewood Cliffs, N.J.

Geologic: ▢▢▢▢▢ Trends of Pre-Cenozoic fold belts

Paleontologic: Limits of the distribution of Mesosaurus; a shallow water Mesozoic reptile

Paleoclimatic: Regions of late Paleozoic glaciation

● S   Spitzbergen: fossil ferns from late Paleozoic rocks

layer in motion consisted of the crust and upper mantle, a layer that was subsequently named the *lithosphere*. This rigid slab could move laterally above a hot plastic layer called the *asthenosphere*. Thus, Wegener's basic premise was vindicated: Continents do indeed move laterally, but as part of a thicker, rigid slab that slides along above the weak asthenosphere.

Plates and Plate Margins

Plates can be thought of as slabs of lithospheric material that generally have interiors with boundaries marked by highly active geologic processes, including earthquakes and volcanism. New lithosphere is being formed at *divergent plate boundaries*, including the midoceanic ridges, and lithosphere is being destroyed or deformed at *convergent plate boundaries*, where plates collide. A third major type of boundary, *transform faults*, marks locations where plates are sliding laterally in opposite directions. Different geologic processes characterize each type of plate margin.

Divergent Plate Boundaries

Divergent plate margins occur where rising convection currents in the asthenosphere nearly reach the earth's surface. The rising mantle material causes extremely high heat

◀ FIGURE 2.28
Discovery of midoceanic ridges
was an important step in the development of the sea-floor
spreading hypothesis. Dots show
positions of earthquakes along
the crest of the Mid-Atlantic
Ridge. *Source:* From B. C. Heezen,
1960, The rift in the ocean floor,
Scientific American (Oct.)
pp. 98–110.

flow, which causes the lithosphere to bulge upward. This expansion causes tension in the lithosphere, which cracks apart parallel to the boundary. Large linear blocks subside downward along faults to form *rift valleys* or *grabens* along the crest of the rift zone. Molten magma rises from the asthenosphere to fill the cracks, often reaching the surface as volcanic eruptions. The net result is that the plates on either side of the boundary are pushed apart and move in opposite directions. As they move in opposite directions, new lithosphere is created by volcanic activity, and the process continues. Divergent boundaries can occur on the continents and in the ocean basins. Some of the characteristics of continental rifts are shown in Figure 2.29. The hot, low-density rocks produce a pronounced negative gravity anomaly over the rift.

Divergent plate boundaries in the ocean basins are marked by the midoceanic ridges. These are similar to continental rift zones, except that only oceanic crust is broken apart (Figure 2.30). The ridges usually have a rift valley at the crest and abundant volcanic activity. Crests of the ridges are often displaced by *transform faults*, which will be described later. These faults pass laterally into fracture zones. Iceland is unique in that the midoceanic ridge rises above sea level. The central rift zone, containing a series of grabens, is clearly visible (Figure 2.31). Because new oceanic lithosphere is being continually created at midoceanic ridges, the age of the ocean floor increases with distance from the midoceanic ridge in both directions. This was clearly demonstrated by the discovery of symmetrical bands of magnetic anomalies flanking the midocean ridges. Based on the ages of the magnetic reversals represented by these bands, the age of the ocean floor can be mapped (Figure 2.32).

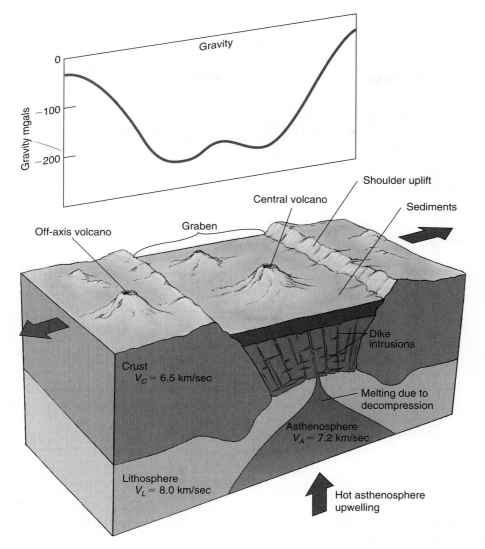

▲ FIGURE 2.29

Characteristics of continental rifts, including uplift due to heating of the lithosphere and upwelling of the asthenosphere. The decrease in density of the rocks below the rift causes a negative gravity anomaly. A graben develops as the center of the rift zone drops downward along faults. *Source:* Jon P. Davidson, Walter E. Reed, and Paul M. Davis, *Exploring Earth: An Introduction to Physical Geology*, 2nd ed., © 2002. Reprinted by permission of Pearson Education, Inc., Upper Saddle River, N.J.

Rifting in the Red Sea has produced a divergent plate boundary separating the African and Arabian plates (Figure 2.33). Originally, the Red Sea rift was a continental rift, but it became an oceanic rift as the ocean flooded into the rift as the plates spread apart. This plate boundary joins two other divergent boundaries at a *triple point*. The eastern arm of the triple point is an oceanic rift zone passing through the Gulf of Aden, and the southern arm is a series of continental rifts known as the East African rift zones. These rifts contain active volcanoes such as Mount Kilimanjaro and huge lakes.

Convergent Plate Boundaries

Convergent plate boundaries mark the zone of contact or collision where plates move toward each other. One of the plates, consisting of oceanic lithosphere, will be forced

▲ FIGURE 2.30
Divergent plate boundary in an ocean basin. The actual boundary is marked by a midoceanic ridge.

▲ FIGURE 2.31
Flank of a graben associated with one branch of the Mid-Atlantic rift in Iceland. The tilted block to the right of the river is part of a downdropped central block of graben. *Source:* Photo courtesy of the author.

downward below the overriding plate, a process known as *subduction*. The two general configurations of convergent margins are shown in Figure 2.34. If both plates are composed of oceanic lithosphere, the margin is known as an *oceanic convergent margin*. Alternatively, one plate may be composed of continental lithosphere; the margin is then known as a *continental convergent margin*. Both types have a descending lithospheric slab with a *trench* near the point of contact between the two plates. Trenches are linear troughs in the ocean floor that run along the plate boundary; they are sites of the greatest water depths in the oceans. Another similarity between the two types is that melting of the overlying rocks of the asthenosphere is associated with the descending slab. This melting may be the result of release of water from the descending slab into the surrounding asthenosphere. The effect of water would be to lower the melting point of the asthenosphere so that the rock would melt at the existing temperatures at that depth. Magma produced by melting rises toward the surface because it is less dense than the overlying rock and finally reaches the surface through volcanoes that build up above the melting point.

There are also many differences between continental and oceanic convergent boundaries. Volcanism in oceanic convergent margins produces a line of volcanoes known as an *island arc*. These volcanoes must build up from the great depths of the sea floor and therefore constitute a line of islands located over volcanic centers rather than a continuous land

m.y. = millions of years	
0–2 m.y.	58–66 m.y.
2–5 m.y.	66–84 m.y.
5–24 m.y.	84–117 m.y.
24–37 m.y.	117–144 m.y.
37–58 m.y.	144–208 m.y.

▲ FIGURE 2.32
The age of the ocean floor defined by magnetic anomalies in the volcanic rocks that were formed at the midoceanic ridge and carried in opposite directions by plate movement. *Source:* Lee R. Kump, James F. Kasting, and Robert G. Crane, *Earth System, The,* 2nd ed., © 2004. Reprinted by permission of Pearson Education, Inc., Upper Saddle River, N.J.

mass. Under some circumstances, lithospheric tension occurs over the subducting slab and a *back arc basin* develops with a spreading ridge similar to a midocean ridge. When compressional forces dominate, the back arc basin does not form.

The continental convergent margin involves the subduction of oceanic lithosphere beneath continental lithosphere, in which continental crust makes up the upper part of the lithosphere. Melting over the subducting slab also occurs in this situation, but the line of volcanoes occurs on land instead of in the ocean. The Cascade volcanoes and the Andes in South America (Figure 2.35) are good examples of volcanism associated with continental convergent margins.

Sediments eroded from continental or oceanic volcanic chains can play an important role in convergent plate margins. These sediments tend to accumulate in and near the trench, where subduction is taking place (Figure 2.36). As the lower plate is descending into the asthenosphere, the sediment is scraped off the plate, forming an *accretionary wedge*. Because of the compressional forces at the point of plate convergence, sediment in the accretionary wedge is highly deformed, with structures such as thrust faults and folds that we will describe in detail in Chapter 8. Sometimes the rocks in the accretionary ridge are uplifted above sea level and can be directly observed in the island arc or continental plate.

Transform Boundaries

The third major type of plate boundary is the *shear*, or *transform*, boundary. Lithospheric plates are sliding past each other along vertical fractures, known as *transform faults,* at these boundaries. Transform faults can connect segments of spreading ridges, spreading

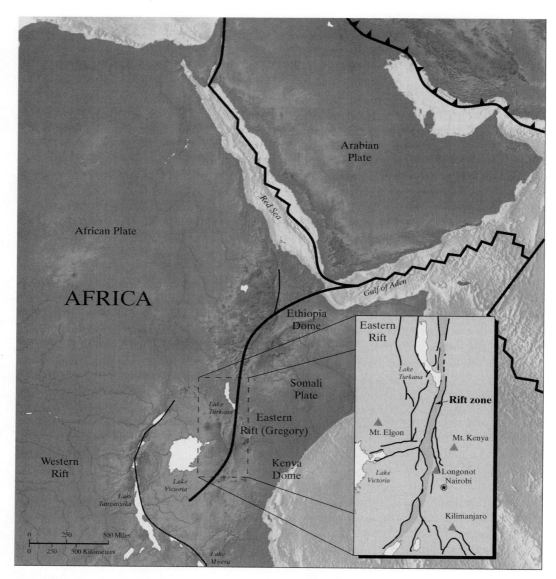

▲ **FIGURE 2.33**
Three divergent plate boundaries meeting at a triple point. The Red Sea rift separates the African and Arabian Plates and the Gulf of Aden rift separates the Arabian Plate from the Somali Plate. The southern-most boundary is composed of discontinuous continental rift valleys in East Africa.

ridges with subduction zones, or segments of subduction zones. Figure 2.37 shows a transform fault between two spreading ridge segments. The transform fault, with the relative movement of the two plates shown, ends at the spreading ridge, even though *fracture zones* continue along the trend of the transform fault beyond the ridge segments on both sides. Transform faults are present in both oceanic and continental lithosphere. The San Andreas fault, notorious for its damaging earthquakes in California and the even bigger earthquakes that could occur in the future, is one of the best examples of a continental transform boundary, separating the Pacific Plate from the North American Plate. The relative motion of the plates, with the Pacific Plate moving northwestward relative to the North

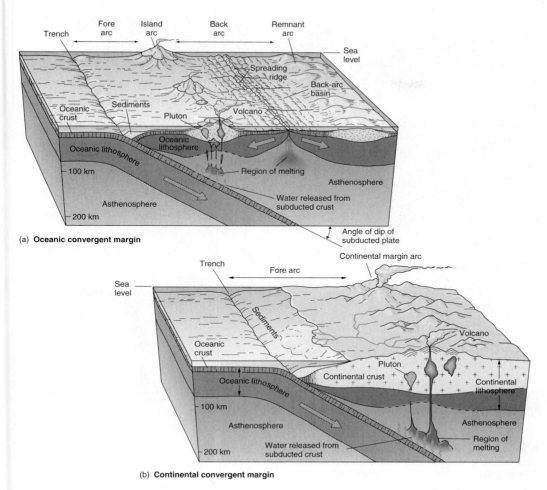

▲ FIGURE 2.34

Oceanic and continental convergent margins. (a) Island arcs and back arc basins are typical features of oceanic convergent margins, as two oceanic plates converge. (b) Continental convergent margins develop when oceanic lithosphere is subducted beneath a continental plate margin. *Source:* Jon P. Davidson, Walter E. Reed, and Paul M. Davis, *Exploring Earth: An Introduction to Physical Geology*, 2nd ed., © 2002. Reprinted by permission of Pearson Education, Inc., Upper Saddle River, N.J.

American Plate, is shown on Figure 2.38. The San Andreas fault connects the subduction zone of the Juan de Fuca Plate with a spreading ridge in the Gulf of California.

Interesting and important aspects of plate tectonics have not been mentioned in this brief introduction. We will discuss some of these in later chapters of the text, especially the association of earthquakes and volcanoes with plates and plate boundaries.

Earth Surface Systems

Just as the earth's interior can be thought of as a giant heat engine causing circulation of molten or nearly molten materials in the core and mantle—as well as continual construction and destruction of lithosphere in plate tectonic processes—the earth's surface is a dynamic place as well. Here, another engine drives the circulation of the oceans and the atmosphere and wears down crustal material uplifted by plate interactions. The energy source for this powerful engine comes through space to the earth from the sun. Most of the

▲ FIGURE 2.35

Convergence of the Nazca and South American plates. Melting in the asthenosphere above the subducting plate explains the volcanoes of the Andes. *Source:* Jon P. Davidson, Walter E. Reed, and Paul M. Davis, *Exploring Earth: An Introduction to Physical Geology*, 2nd ed., © 2002. Reprinted by permission of Pearson Education, Inc., Upper Saddle River, N.J.

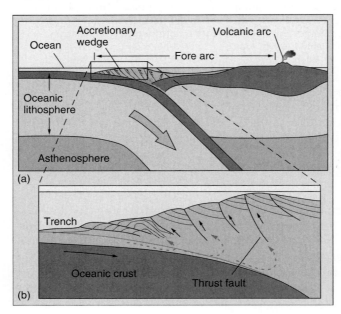

◀ FIGURE 2.36

(a) Development of an accretionary wedge at an oceanic convergent plate margin. (b) Enlargement of the wedge to show the deformation caused by compression and uplift of the wedge. *Source:* Jon P. Davidson, Walter E. Reed, and Paul M. Davis, *Exploring Earth: An Introduction to Physical Geology*, 2nd ed., © 2002. Reprinted by permission of Pearson Education, Inc., Upper Saddle River, N.J.

▶ FIGURE 2.37
Location of a transform boundary between two ocean spreading ridges. The continuations of the transform fault beyond the ridge crests are called *fracture zones*.

FIGURE 2.38
The San Andreas fault, a prominent transform fault connecting the subducting Juan de Fuca Plate with the East Pacific Rise, a spreading ridge that has caused rifting in the Gulf of California. The portion of California west of the San Andreas fault is moving northwestward with the Pacific Plate, and the area east of the fault is part of the North American Plate.

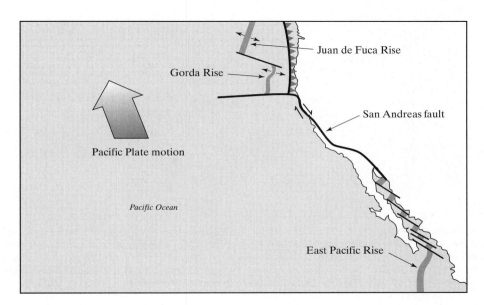

surficial geologic phenomena that we will be discussing in this text—weathering, mass wasting, erosion, and deposition—all function because of interactions among the oceans, atmosphere, and energy from the sun.

Atmospheric Circulation

Solar energy arriving at the earth is not equally distributed. Rays that arrive at and near the equator strike the planet directly, which concentrates the heat contained within a small area (Figure 2.39). In contrast, rays that arrive at the surface near the poles strike the earth in an oblique fashion, which distributes the same amount of energy over a larger area. This creates a heat imbalance on the earth's surface, with more heat and higher temperatures at the equator than at the poles. Whenever there is a temperature difference, heat will flow from

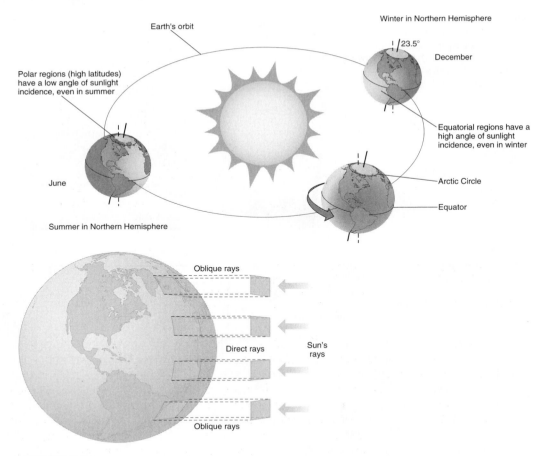

▲ FIGURE 2.39

Unequal distribution of solar energy at the surface creates a temperature difference between the equator and the poles. Tilting of the earth's axis of rotation creates the seasons, as solar radiation is greater when either the Northern or Southern Hemisphere is tilted toward the sun and less when a particular hemisphere is tilted away from the sun. *Source: Jon P. Davidson, Walter E. Reed, and Paul M. Davis, Exploring Earth: An Introduction to Physical Geology, 2nd ed., © 2002. Reprinted by permission of Pearson Education, Inc., Upper Saddle River, N.J.*

areas of higher to lower temperature, and one might expect that air masses would flow from the equator to the poles. There are a few complications, however. First the earth is tilted on its axis of rotation (Figure 2.39). This means that during part of the year, each hemisphere is tilted toward the sun part of the time—when it receives more radiation than the other hemisphere—and away from the sun part of time—when it receives less radiation than the other hemisphere. We know these times as summer and winter.

The equator to poleward circulation is broken into a series of cells spread over the globe. At the equator, where heating is most intense, air masses rise because they are less dense than surrounding air. They also become charged with water vapor evaporated from the oceans. As the air masses rise from the surface, air moves along the surface from the north and south to replace the rising air masses. This creates a *convergence zone* (Figure 2.40), which in this case is called the *intertropical convergence zone (ITCZ)*. Air masses rise in the ITCZ to the top of the troposphere, where they are diverted poleward. Because water vapor condenses and falls from clouds that form as the air is rising, the air masses are becoming dryer as they move toward the poles. At about 30° north and south, the air masses begin to sink back toward the surface, producing a zone of high

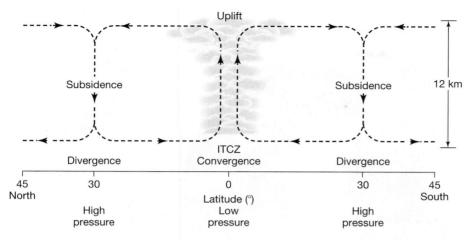

▲ FIGURE 2.40

Vertical cells of atmospheric circulation between the equator and 30° north and south generated by intense heating at the equator. Deserts occur in the divergence zones that are set up at about 30° north and south of the equator. *Source:* Lee R. Kump, James F. Kasting, and Robert G. Crane, *Earth System, The*, 2nd ed., © 2004. Reprinted by permission of Pearson Education, Inc., Upper Saddle River, N.J.

pressure. When they descend to near the surface, the flow splits apart, creating a *divergence zone*. The dry descending air at about 30° north and south produces two bands of deserts at those latitudes. Air moving toward the poles from this divergence zone collides with air that sinks at the poles and flows toward the equator to create two more cells on each side of the equator, one between 30° and 60° north and south, and one between 60° and 90° (Figure 2.41). The convergence zone at 60° north and south latitude creates an area of instability because warm air from the equator is forced over the cold air moving from the poles. This interface is known as the *polar front*, which is important because it generates storms that can produce precipitation.

One final wrinkle in the atmospheric circulation is caused by the rotation of the earth. The west to east rotation imparts an apparent deflection to air masses moving in a north or south direction. This phenomenon, the *Coriolis effect*, will cause an air mass moving north in the northern hemisphere to be deflected to the east. It is really an apparent deflection because the air mass is just moving northward, but the earth is rotating to the east at the same time. The Coriolis effect modifies the movement of air masses in both hemispheres, as shown in Figure 2.41.

The general circulation of the atmosphere just described explains a lot about global precipitation patterns (Figure 2.42). Rising air masses in the ITCZ produce a band of high precipitation along the equator, and divergence around 30° north and south creates some of the world's greatest deserts. When water vapor in the air masses condenses, it falls as either rain or snow. Other factors, however, influence precipitation. Whenever an air mass moves over a continent, it must rise in elevation. When it crosses a mountain range, the rise is generally enough to cause precipitation on the windward side of the mountain range (Figure 2.43). This type of precipitation is called *orographic precipitation*. It explains why the westerly flow of air from the Pacific Ocean over the mountains along the western coasts of the United States and Canada dumps its moisture on the windward side. On the eastern side of the mountains, the air masses are dry and arid to semi-arid conditions prevail. The orographic effect is also obvious in India (Figure 2.42), where the greatest rainfall amounts on earth, the monsoon rains, result from the flow of air masses over the Himalayas. The global distribution of deserts (Figure 2.44) is a function of both divergence zones and orographic effects.

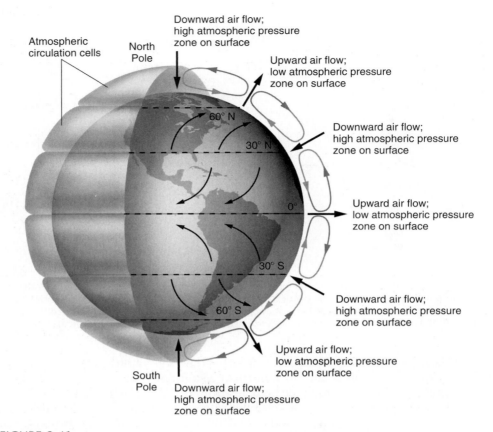

▲ FIGURE 2.41
Circulation cells of the atmosphere associated with convergence and divergence. *Source:* Jon P. Davidson, Walter E. Reed, and Paul M. Davis, *Exploring Earth: An Introduction to Physical Geology*, 2nd ed., © 2002. Reprinted by permission of Pearson Education, Inc., Upper Saddle River, N.J.

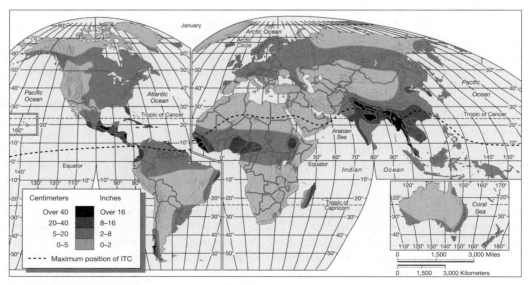

▲ FIGURE 2.42
Precipitation amounts over land in July. *Source:* From T. McNight, *Physical Geography: A Landscape Appreciation*, 6th ed., © 1999, by Prentice Hall, Inc., Upper Saddle River, N.J.

▶ FIGURE 2.43
Generation of orographic precipita-
tion as air masses move over moun-
tain ranges. *Source:* Jon P. Davidson,
Walter E. Reed, and Paul M. Davis,
*Exploring Earth: An Introduction to
Physical Geology*, 2nd ed., © 2002.
Reprinted by permission of Pearson
Education, Inc., Upper Saddle River, N.J.

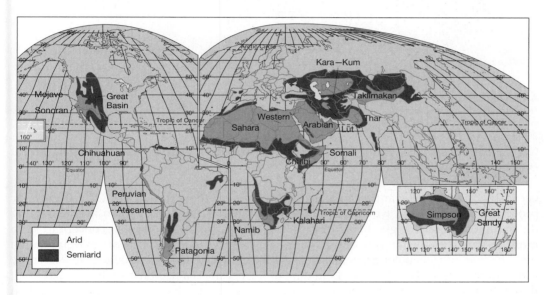

▲ FIGURE 2.44
Distribution of arid and semi-arid regions on the earth. *Source:* From R. W. Christopherson, *Geosystems: An
Introduction to Physical Geography*, 3rd ed., © 1997 by Prentice Hall, Inc., Upper Saddle River, N.J.

The Hydrosphere and Oceanic Circulation

The occurrence and circulation of water on our planet is intricately linked to the biosphere, atmosphere, and lithosphere. Water is critical to life on earth, the only planet in
the solar system to contain liquid water. The circulation of the atmosphere drives the cy-
cling of water from the oceans to the continents and back to the oceans. This important
cycle is known as the *hydrologic cycle* (Figure 2.45). Because of its atmospheric link, it is
also a product of the sun's energy contribution to the earth. The small amount of freshwater in this cycle, about 3%, sustains most life-forms on Earth outside the oceans. The
lithosphere becomes part of the hydrologic cycle because of the component of water that
infiltrates below the surface and becomes part of the reservoir of water we call groundwater. As most surface-water sources are approaching the limits of their exploitation by
humans, groundwater becomes very important for expansion of freshwater sources. The
role of water and the hydrologic cycle is important to many of the topics in this book.
Within the interior of the earth, water plays a role in volcanism, earthquakes, and tectonics.

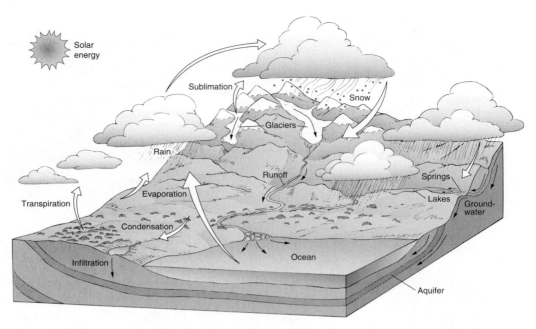

▲ FIGURE 2.45
The hydrologic cycle. *Source:* Jon P. Davidson, Walter E. Reed, and Paul M. Davis, *Exploring Earth: An Introduction to Physical Geology*, 2nd ed., © 2002. Reprinted by permission of Pearson Education, Inc., Upper Saddle River, N.J.

On the earth's surface, water contributes to nearly every surface geologic process and every geologic hazard.

In keeping with our emphasis on global systems and cycles, the circulation of the oceans is critically relevant to the earth's climate and, therefore, to its surficial geologic processes. In Chapter 1, we described the function of oceanic circulation in maintaining the earth's climate as a giant conveyor belt. Here, we expand upon that concept to get a more complete understanding of the oceans. An initial step is to divide oceanic circulation into shallow and deep components, because these two types of circulation are driven by different processes. The shallow circulation of the oceans is a response to the circulation of the atmosphere, modified by the configuration of the ocean basins and continents. When wind blows over the surface of the ocean, friction between moving air and water sets up ocean currents. The pattern of shallow ocean circulation is therefore similar to wind patterns. In the ITCZ, strong easterly (blowing from the east) winds set up strong equatorial currents moving in the same direction. When these currents reach blocking land masses—North and South America, for example (Figure 2.46)—the currents are diverted to the north and south. The current that moves north along the eastern coast of North America is known as the Gulf Stream. As it flows north like a huge river in the ocean, the midlatitude winds diverted toward the east by the Coriolis effect steer the ocean currents in the same direction. Thus, the Gulf Stream crosses the northern Atlantic to the coast of Europe. The southward diversion of the current along the European coast sets up a huge circulation cell, called a *gyre*, in the North Atlantic. Similar gyres develop in the other ocean basins.

The shallow circulation of the oceans, even though it only affects the upper several hundred meters of seawater, performs an important function, transporting heat from the equatorial regions to the polar regions. Northern Europe would be a cold, forbidding place if it were not for the moderating effect of the Gulf Stream. Oceanic currents that may be even more important to the earth's long-term climate involve deeper flow. These currents are known as *thermohaline currents* because their driving forces are a combination of

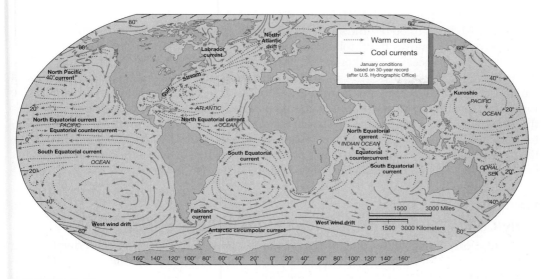

▲ FIGURE 2.46
Shallow ocean currents—caused by wind circulation patterns, the Coriolis effect, and the distribution of the continents. *Source:* Lee R. Kump, James F. Kasting, and Robert G. Crane, *Earth System, The*, 2nd ed., © 2004. Reprinted by permission of Pearson Education, Inc., Upper Saddle River, N.J.

temperature and density, which is partially a function of salt content. In high latitudes, surface waters become cooled by low air temperatures. The surface water can freeze in these extreme conditions, even though seawater has a lower freezing point because of its high salt content. When it does freeze, the salinity increases because of the removal of freshwater to form ice. The resulting water, now denser because of its higher salinity, sinks to form masses of *bottom water*, which only moves along the bottom of the oceans. This cold, dense water is generated in two places—near Antarctica and in the North Atlantic Ocean. The deep circulation patterns are shown in Figure 2.47. The circulation

▲ FIGURE 2.47
General circulation of the deep oceans. Dark shading shows the deepest currents and light shading indicates intermediate return flows. Numbers refer to current fluxes in 10^6 m^3/sec. *Source:* From A. L. Bloom, *Geomorphology*, 3rd ed., © 1998 by Waveland Press, Long Grove, IL.

pattern includes the deep water (shown in gray) and higher-level return flows (shown in white), which are still below the surface currents discussed previously. Research over the past few decades shows that global climate is closely related to this conveyor-like deep circulation of the oceans. Weakening or shutdown of the deep circulation could plunge the planet into a glacial period, in which ice sheets similar to those in Greenland and Antarctica would once again form over Canada and northern Europe, as they have formed many times in recent geologic history. When these midlatitude ice sheets reach their full extents, well south of Chicago in the United States, sea level drops about 100 meters because of the water transferred from the oceans to the continental ice sheets. Although perhaps not as imminent as global warming due to burning of fossil fuels, the next glacial age may not be too far in the future.

Geologic Time

Early Concepts of the Formation of Rocks and the Earth's History

The evolution of the science of geology was intimately involved with our understanding of geologic time. Modern Western Civilization evolved with a strong reliance on literal interpretations of the opening chapters of the Bible. Early scientists applied these concepts to their observations of the natural environment. Accordingly, the earth was thought to be quite young, and its formation was attributed to a series of events that were rapid and cataclysmic in intensity. This school of thought, which became known as *catastrophism*, held that the earth's surface was generally a stable, nonchanging place that had remained quiescent except for violent upheavals during which mountains were thrust upward from the plains and canyons were formed as the earth's surface was ripped apart. The related theory of *neptunism* involved the presence of a primordial ocean from which rocks at the earth's surface precipitated. A specific sequence of precipitated rocks, assumed to have formed as the ocean level dropped, was developed to account for observations of rocks that were made in western Europe.

An important early contribution to geology was made by Nicolaus Steno (1638–1687), a Danish physician working in Italy. Steno observed the rocks exposed in cliffs and in stream banks, and he also studied *sediment*, particles of rock and soil being deposited by streams and in the Mediterranean Sea. He realized that the exposed rocks had many similarities to the modern sediments he had seen, and he concluded that they had, in fact, originated as sediments at some time in the distant past. To formulate these ideas, Steno used true scientific thought because he recognized order in nature.

Steno also stated three principles that are fundamental to modern geology. First, the *Principle of Original Horizontality* states that layers of sediment are always deposited in horizontal sheets, as can be observed when they are seen to be accumulating on the bed of a stream or ocean (Figure 2.48). Thus, when we see layers, or *beds*, of rock that appear to be tilted or folded (Figure 2.49), we must realize that subsequent events have deformed the originally horizontal beds. Second, Steno's *Principle of Original Continuity* states that accumulations of sediment are deposited in continuous sheets up to the point where they terminate against a solid surface (Figure 2.48). There is no better illustration of these two principles than in the Grand Canyon (Figure 2.50), where individual horizontal rock layers can be traced for miles along the side of the canyon. Steno's third principle, known as the *Principle of Superposition*, states that in a vertical sequence of sedimentary rock layers, the oldest layer lies at the base and each successive layer above is younger than the layer it overlies. Although this principle may seem intuitively obvious to us, in Steno's era it required acceptance of the radical idea that many rocks are formed by the gradual deposition of sediment, the same processes that can be observed to be occurring in a stream, lake, or ocean.

Blanketlike without
gaps in continuity of
deposition

Youngest layer

Layers are horizontal

Oldest layer

▲ FIGURE 2.48
Principles proposed by Nicolaus Steno. These principles are essential to an understanding of sedimentary rocks. *Source:* From E. A. Hay and A. L. McAlester, *Physical Geology: Principles and Perspectives*, 2nd ed., © 1984 by Prentice Hall, Inc., Englewood Cliffs, N.J.

A more modern approach to the age and geologic history of the earth can be attributed to the ideas of James Hutton, an 18th-century Scottish farmer. Hutton challenged the ideas of catastrophism and neptunism by proposing that the earth's surface is not a stable, unchanging landscape. Instead, changes are occurring continuously, although mostly at very slow rates. Therefore, valleys are formed by the slow erosion of streams that flow within them, and mountains are uplifted gradually to their elevated positions. Hutton also suggested that the rocks exposed on the landscape had been formed by the same gradual processes operating continuously, now and in the past. In place of neptunism, Hutton advocated *plutonism*, which explains the formation of *igneous* rocks by cooling from a hot, melted state rather than by precipitation from the ocean.

These ideas are incorporated into his *Principle of Uniformitarianism*, which stated that geological processes causing change on the earth are gradual rather than catastrophic and are the same processes that can be observed today. Uniformitarianism was adopted by one

▶ FIGURE 2.49
Beds of sedimentary rocks that have been bent into a fold from their original horizontal orientation. *Source:* A. Keith; photo courtesy of U.S. Geological Survey.

▲ FIGURE 2.50
In the Grand Canyon, horizontal beds of sedimentary rock can be traced for great distances.
Source: J. R. Balsley; photo courtesy of U.S. Geological Survey.

of the most influential geologists of the 19th century, Charles Lyell, whose textbook on geology was the standard of the day. Lyell interpreted uniformitarianism literally, believing that past geologic processes necessarily operated at their modern rates. This actually impeded the progress of the science, because Lyell blocked the acceptance of important new ideas that contradicted his view of uniformitarianism. One good example was the revolutionary new theory of glaciation and ice ages, which was presented to the scientific public by Louis Agassiz in the 1830s. This theory advocated a period of time in the past when great glaciers covered large parts of Europe and North America. Since this required a drastic change from the present climate, Lyell rejected the theory and helped to suppress it for about 20 years, until the evidence for past glaciation became irrefutable. We now interpret the principle of uniformitarianism to mean that the physical and chemical laws governing natural processes have been constant throughout the history of the earth. It does not mean that the rates of natural processes are always constant or that drastic and rapid changes in geologic processes did not take place. The Principle of Uniformitarianism is often summed up by the phrase "the present is the key to the past," because it gives geologists the ability to interpret the origin of ancient rocks in terms of present-day geologic processes. The most important implication of Hutton's idea of uniformitarianism was that the earth and the rocks exposed at its surface are very old. A single exposure along the coast of Scotland emphatically drove this point home. The exposure consisted of horizontal beds of rocks resting upon tilted rocks of a different type (Figure 2.51). Using Steno's principles, Hutton realized that this exposure must represent a complex sequence of events that occurred over a great length of time. First, the lower sequence of rock beds was deposited in horizontal layers of sediment (Figure 2.52). Next, these sediments hardened into rocks and became deformed and tilted into their present orientation. Deformation of this type is associated with major movements of the earth's crust, the type of movement that may result in the lateral compression and vertical uplift of rocks to form a mountain range. It is interesting to imagine the awe that Hutton must have felt as he began to realize that the

▲ FIGURE 2.51
The exposure of rock at Siccar Point, Scotland, studied by James Hutton. The unconformity (plane marked by lens cover and heavy white line) represents a long period of geologic time between the formation of the vertical rock beds in the lower part of the outcrop and the nearly horizontal rock beds overlying them. The geologic events involved in the history of these rock units are shown in Figure 2.52. *Source:* Photo courtesy of John Bluemle.

▶ FIGURE 2.52
Generalized sequence of events making up the geologic history of the rock exposure shown in Figure 2.51. (a) Deposition of the rocks of sequence A. (b) Deformation and uplift. (c) Erosion of the rocks of sequence A. (d) Deposition of the rocks of sequence B forming an unconformity between sequences A and B. (e) Uplift and erosion of both rock sequences as a unit.

▲ FIGURE 2.53
The geologic cycle. Rocks in the earth's crust follow a cycle that includes deposition of sediments, formation of sedimentary rocks, conversion to igneous or metamorphic rocks, and erosion to form sediments.

next event in the history of that outcrop must have been a period of erosion representing an enormous amount of geologic time. A mountain range must have been gradually worn down by erosion to a low-lying plain on which the sediments of rock unit B were deposited, perhaps by a sea gradually spreading over the plain. The surface separating the two rock sequences is called an *unconformity*. The tremendous amount of time that these changes represent must have convinced Hutton that he was right about the great age of the earth.

By studying rock exposures in many areas, Hutton reached the conclusion that rocks in the crust follow a pattern that has been repeated over and over again throughout geologic time. Hutton's chain of events is known as the *geologic cycle* (Figure 2.53). Although the cycle has no definite beginning or end, it is convenient to think first of sediments being deposited in continuous horizontal beds according to the principles of Steno. As successive layers are deposited, the beds near the base of the sequence become compacted and begin the transformation from sediments into *sedimentary rocks*, a process known as *lithification*.

With continued deposition, the sedimentary rocks are deeply buried and subjected to higher temperatures and pressures. Under these conditions, the rocks are deformed; layers may be bent, broken, or heated to the melting point. The melting or partial melting of a rock produces *magma*, a molten liquid that can later cool to form an igneous rock. Magmas can also be produced from sources other than sedimentary rocks. If any type of rock is highly altered by heat and pressure but not to the point of melting, the original rock is changed to such a great degree that it is now called a *metamorphic* rock. The close association of deformation, metamorphism, igneous activity, and mountain belts suggests that these activities are all related. It was obvious to Hutton and most other early geologists that thick sequences of sedimentary rocks are intensely deformed and altered in the process of mountain building (Figure 2.54).

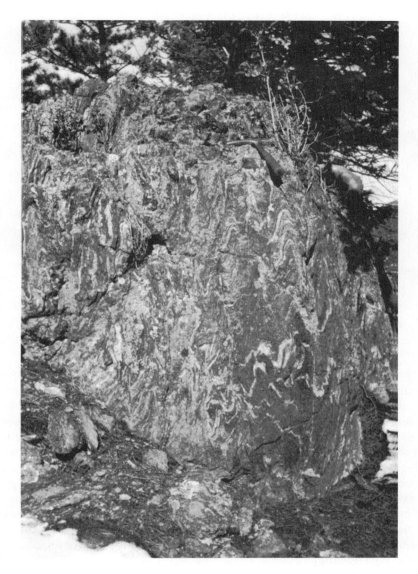

▲ FIGURE 2.54
An exposure of a metamorphic rock produced by alteration under high temperature and pressure.
Source: C. C. Hawley; photo courtesy of U.S. Geological Survey.

The final step in the cycle occurs after the uplift of a mountain belt. Exposed to the destructive agents of gravity, wind, rain, and ice, mountains are gradually lowered by erosion. The sediments produced by erosion are transported by rivers to the ocean, where the cycle can continue in a new location. Given enough time, mountain ranges can be worn down to gentle, low-lying plains. The interior of Canada contains regions of igneous and metamorphic rocks that once must have been at the core of towering mountains but now are reduced to an elevation barely above sea level. In the United States, both the Appalachian and Rocky Mountains are following the path of the geologic cycle (Figure 2.55). The Appalachians are older and thus erosion has reduced them to a lower elevation than the Rockies. We now place the geologic cycle, with its partial explanation of lithospheric processes, in the larger context of plate tectonics.

▲ FIGURE 2.55
The Grand Tetons, a mountain range consisting of igneous and metamorphic rocks. Uplift and erosion have produced the striking, rugged topography. *Source:* Photo courtesy of the author.

The controversy set off by Hutton's ideas raged in scientific circles for many years. It was not until the mid-19th century that uniformitarianism was accepted by the majority of scientists. Unfortunately, uniformitarianism was carried too far by early geologists and delayed acceptance of the ice ages and other ideas that required different conditions than currently exist. The work of Steno, Hutton, and many others paved the way for the recognition of geology as a modern science. The practice of geology requires the observation of rock or sediments and the interpretation of the origin and sequence of events that has led to the present condition. The only change in this approach since the time of Hutton lies in the ever-increasing sophistication and complexity of the tools and instruments that scientists have available to study the evidence found in the natural landscape. In this respect, geology differs from other sciences because the interpretations of past geologic events cannot be verified by experimentation. Geologic study is not unlike the procedure followed by a detective who collects clues and evidence in order to solve a crime.

Relative Time

Geologists are, in part, historians. Our goal is to decipher the physical, chemical, and biological evolution of our planet from its origin to the present time. Like historians who study the history of the human race, geologists must have a method of determining the time at which various events in the earth's history took place. Imagine how difficult it would be to study the history of the United States if there were no concept of time; how could we comprehend the differences between the exploration of the West and the space program of the 20th century if we did not understand the amount of time and technological progress that separates the two periods?

Geologists in some ways face a much more formidable task in the investigation of the earth's history. There are no written records available. The only direct evidence that exists from the geologic past is the sequence of rocks and sediments that we can observe

▲ FIGURE 2.56
Beds of sedimentary rock can be traced across areas in which they are not present by using correlation based on the principles of Superposition and Original Continuity.

at and below the earth's surface. In the 20th century, we have learned to use the natural radioactivity in certain types of rocks and organic materials to directly determine their age; with this method we can determine the *absolute* time of an event. Geologic pioneers like Hutton, however, did not have radioactive age dating available to them. Hutton could only speculate about the amount of time represented by an unconformity in a rock exposure. Confronted with this problem, geologists began to work out the history of the rock record using *relative* time. This method is an attempt to construct a time scale based upon the relative ages of rock units by establishing their relationship to other rock units.

Steno's Principle of Superposition is a useful tool in determining the relative age of sedimentary rocks. In an undisturbed layered sequence, the beds are successively younger from bottom to top. Another method that can be used with sedimentary rocks is *correlation*, the tracing of rock units from one area to another. In addition, the Principle of Original Continuity can be used to extend the correlation of rock units with similar characteristics across areas in which the rocks are not present or are not exposed (Figure 2.56).

More complicated rock exposures require additional techniques to establish relative time. The *Principle of Cross-Cutting Relationships*, for example, is particularly useful in some places. It states that any rock that cuts across or penetrates a second rock body is younger than the rock it penetrates. This principle is often called upon for interpretation of sequences containing igneous rocks that were forced (intruded) into existing rocks in a liquid state and then cooled to a solid state. Because the igneous rocks often cut across the layering of sedimentary or metamorphic rocks, they can be identified as younger than the rocks that enclose them (Figure 2.57). In some places a layer of igneous rocks lies between two beds of sedimentary rocks. In these instances, it is sometimes possible to determine the relative age by examining the boundaries, or *contacts*, of adjacent rock units. If the heat of the molten rock altered the rocks both above and below the igneous rock at the time it was intruded, then the igneous rock penetrated the sequence and is younger than both the overlying and underlying units. If only the upper contact of the sedimentary rock below the igneous rock was altered, then the igneous rock was probably formed as a lava flow on the ground surface and thus is

▶ FIGURE 2.57
A body of igneous rock penetrates and cuts across the bedding of a series of sedimentary rocks. It is therefore younger than the sedimentary rocks.

Altered
sedimentary
rock

Igneous
rock

(a)

Altered
sedimentary
rock

Igneous
rock

(b)

◀ FIGURE 2.58

(a) Sedimentary rocks are altered both above and below their contacts with the igneous rock. Therefore, the igneous rock body is younger than both sedimentary rock units. (b) Only the lower sedimentary rock unit is altered near its contact with the igneous rock. In this case, the igneous rock is a lava flow that is younger than the lower sedimentary rock unit but older than the upper sedimentary rock unit.

◀ FIGURE 2.59

The principles of Original Horizontality, Original Continuity, Cross-cutting relationships, and Superposition can be used to unravel the sequence of events in this rock exposure. They include (1) deposition of rock sequence A; (2) metamorphism of A; (3) uplift and erosion of A to produce an unconformity; (4) deposition of sequence B, (5) deformation, uplift, and erosion of sequences A and B to produce an unconformity; (6) deposition of sequence C; and (7) intrusion of igneous rock body D through sequences A, B, and C.

younger than the rock beneath it but older than the rock above it (Figure 2.58). It is often possible to use several of the principles described here to determine the sequence of events of a rock exposure (Figure 2.59). We therefore would be using relative time to compare the ages of rock units but still have no idea of the absolute age.

However useful the methods described may be, the foundation of relative age dating is based upon *fossils*, the visible remains of organisms preserved in rocks. William Smith (1769–1839), an English civil engineer, was one of the first scientists to recognize the usefulness of fossils as tools for determining relative age and correlation of sedimentary rocks. While supervising the construction of canals and other engineering works over a period of many years, Smith discovered that certain rock units always contained the same group, or *assemblage*, of fossils. The fossilized shells of marine organisms were most useful in this work. By using superposition and other principles, Smith was able to group these rock units in order of relative age on the basis of the correlation of fossil assemblages. The method was later applied throughout Europe, and a *geologic column* was constructed for all sedimentary rocks known in that region (Figure 2.60). The rocks of each *system* contained a distinct and unique assemblage of fossils. Although certain fossils were similar to those in other systems, consistent differences between systems were evident. Rocks beneath the Cambrian system were generally thought to be nonfossiliferous and therefore could not easily be divided. They were given the name *Precambrian*.

The reasons for the changes in fossil characteristics were not known in the time of William Smith. Charles Darwin later provided an explanation for this phenomenon with his theory of evolution. The idea of the systematic change of organisms over long periods of time was consistent with the fossil record.

▶ FIGURE 2.60
The geologic column, showing the names of the rock systems and the regions for which they were named.

	Systems	Where first described
	Quaternary	Europe
	Tertiary	Europe
	Cretaceous	England
	Jurassic	Jura Mts. Europe
Youngest	Triassic	Germany
	Permian	Russia
	Carboniferous	Great Britain
	Devonian	Great Britain (Devonshire)
	Silurian	Great Britain (Wales)
	Ordovician	Great Britain (Wales)
	Cambrian	Great Britain (Wales)
Oldest	Precambrian Era	Nonfossiliferous, therefore not well subdivided

Absolute Time

The construction of the geologic column was the culmination of relative age dating of rocks. Throughout the 19th century, geologists and physicists began to speculate about the age of the earth and the ages of the systems of the geologic column. The estimates became older and older. In the early 20th century, one of the triumphs of modern geology was achieved with the application of radioactivity to the dating of rocks. Finally, a method for determining the absolute ages of certain rocks was available.

Radioactive particles (atoms) are disseminated throughout many types of rocks in trace amounts. These particles decay spontaneously by emitting atomic particles from the nucleus of the atom. Most of the radioactive particles are *isotopes* of elements; that is, they are forms of a certain element that differ slightly in atomic mass from other isotopes of the same element. (Atomic particles are more fully discussed in Chapter 3). Upon radioactive decay, isotopes are altered to different elements by the loss or gain of atomic particles by the nucleus.

A basic assumption of radioactive dating is that each radioactive isotope (uranium 235, for example) will decay at a constant rate. In the example of uranium 235, the number 235 represents the mass of the isotope, that is, the number of protons and neutrons in the nucleus. The radioactive decay rate is often described using the *half-life* of the isotope. The half-life is the amount of time required for the decay of half of any amount of a particular isotope. For example, the half-life of uranium 235 is 0.7 billion years. After this amount of time, only one-half of the original amount of uranium 235 will remain. After two half-lives, only one-fourth of the original amount will be left.

Radioactive decay is an exponential process and as such can be explained by the same mathematical expression as equation (1.1), with one modification: In equation (2.2), N becomes the number of molecules of the radioactive isotope of an element at time t, instead of the number of people in a population, and N_0 is the original number of isotopes prior to decay, rather than the number of people at the beginning of the time interval.

$$N = N_0 e^{-kt} \tag{2.2}$$

The one difference in the two equations is that the term k represents the decay rate instead of the growth rate, and it is negative in this case because molecules of the radioactive isotope are declining instead of increasing like people in a growing population. Half-life is analogous to doubling time and can be determined as follows. Since after one half-life, N/N_0 is equal to $\frac{1}{2}$,

$$\frac{1}{2} = e^{-kt_{1/2}}$$

where $t_{1/2}$ is the half-life. We then have

$$2 = e^{kt_{1/2}} \quad \text{and} \quad \ln 2 = kt_{1/2}$$

Therefore,

$$t_{1/2} = \ln\frac{2}{k} = \frac{0.693}{k}$$

A radioactive isotope that decays to another isotope is called the *parent*; the isotope produced by the decay is known as the *daughter*. To determine the age of a rock, the amount of both parent and daughter isotopes must be measured. This analysis is performed with a *mass spectrometer*, an instrument that can measure minute amounts of matter with great accuracy. The most useful parent-daughter isotope pairs are shown in Figure 2.61. With the exception of carbon 14, all these isotopes can be used for dating very old rocks. The date obtained, however, must be carefully interpreted. It can be considered to be accurate only if the rock has not been substantially altered since its formation. The age of an igneous rock, therefore, would represent the time that it cooled from a magma. Sedimentary and metamorphic rocks present some difficult dating problems. Sedimentary rocks consist of particles eroded from older rocks. The date obtained from an igneous grain incorporated into a sedimentary rock would thus represent the cooling of the igneous rock rather than the deposition of the sediment to form the sedimentary rock. In metamorphism, parent and daughter particles can be separated by heat and partial melting. Therefore, dates obtained from metamorphic rocks may or may not yield the correct age of the metamorphism.

Carbon 14 serves a different function in radioactive dating due to its short half-life of 5730 years. The value of carbon 14 lies in its utilization for dating very recent sediments, particularly those from the last part of the most recent Ice Age. Carbon 14 is a radioactive isotope of carbon produced by the activity of cosmic rays in the earth's atmosphere. It mixes with the normal carbon (carbon 12) in the atmosphere and is taken up by plants and animals, so organic material contains carbon 14 at the same concentration as the atmosphere at the time of the death of the organism. Wood fragments and bone buried by the sediment deposited by glaciers can be used for dating the time of glacial advance. Because of the short half-life, about 75,000 years is the practical limit for dating by the carbon 14 method.

Application of radioactive dating to rocks began soon after the discovery of radioactivity. By dating rocks from all over the world and integrating these dates with the geologic column, scientists were able to construct an absolute chronology of geologic

Radioactive parent nuclide		Stable daughter nuclide		Half-life (years)
Potassium 40	K	Argon 40	Ar	1.3 billion
Rubidium 87	Rb	Strontium 87	Sr	47 billion
Uranium 235	U	Lead 207	Pb	0.7 billion
Uranium 238	U	Lead 206	Pb	4.5 billion
Carbon 14	C	Nitrogen 14	N	5730

◄ FIGURE 2.61
Radioactive isotopes used for geologic radioactive age dating.

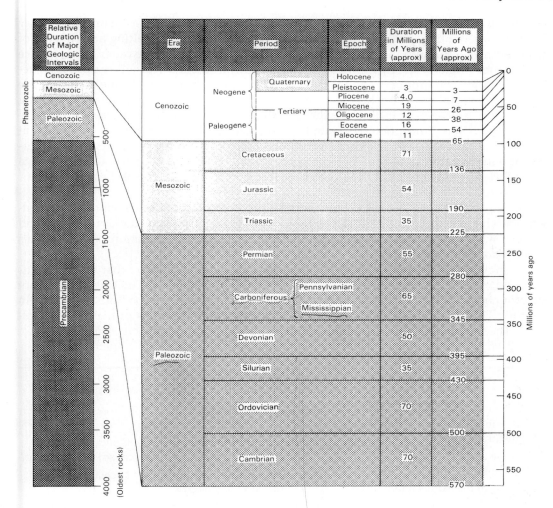

Relative Duration of Major Geologic Intervals	Era	Period		Epoch	Duration in Millions of Years (approx)	Millions of Years Ago (approx)
Cenozoic	Cenozoic	Neogene	Quaternary	Holocene		0
Mesozoic				Pleistocene	3	3
				Pliocene	4.0	7
Paleozoic		Paleogene	Tertiary	Miocene	19	26
				Oligocene	12	38
				Eocene	16	54
				Paleocene	11	65
	Mesozoic	Cretaceous			71	136
		Jurassic			54	190
		Triassic			35	225
	Paleozoic	Permian			55	280
		Carboniferous	Pennsylvanian		65	
			Mississippian			345
		Devonian			50	395
		Silurian			35	430
		Ordovician			70	500
		Cambrian			70	570

▲ FIGURE 2.62
The geologic time scale.

time. The geologic time scale (Figure 2.62) is the result. Geologic time is divided into two eons, Precambrian and Phanerozoic. The Phanerozoic eon is divided into eras, periods, and epochs. The periods are named after the rock systems in the geologic column; thus, finally, geologists had a means for relating evolutionary changes in organisms to absolute time.

The ages of rocks determined by radioactive dating made it necessary to greatly increase our conception of the age of the earth. Phanerozoic time alone, which includes most of the fossil record, is only a small percentage of geologic time (Figure 2.62). The oldest rocks dated extend back to about 4 billion years. If the age of the earth were reduced to a 24-hour scale, the human species would not appear until the last few seconds. Thus, radioactive dating has had a major effect upon our thinking about the age of Earth and the other planets.

Summary and Conclusions

The earth is best studied as a system, one in which subsystems and cycles operate in complex interrelationships. The internal circulation of the earth has been creating and consuming lithosphere since differentiation soon after becoming a distinct planetary body. Differentiation included the formation of the core through melting and descent of iron and nickel through the rocky mantle. Circulation currents in the outer core produce the magnetic field of the earth, which fingerprints volcanic rocks as either normal or reversed as they cool from eruption. The flow of heat from the core heats the mantle to the point that circulation occurs here as well, especially where the rocks are near the melting point. Plumes of high heat flow rise through the mantle and force their way through the lithosphere. New lithosphere is created at divergent plate boundaries by volcanism at continental or oceanic rifts where rising plumes force lithospheric plates apart. The overall formation of lithosphere, along with its destruction and recycling at convergent plate boundaries, is explained by plate tectonics, the unifying model of solid earth behavior that has revolutionized the science of geology.

Complementary systems exist on and near the earth's surface in response to solar energy reaching the planet. These systems involve the circulation of the atmosphere and oceans, which, to a great extent, control our climate. Human alteration of these cycles could lead to dramatic changes in climate.

Although Hutton and other early geologists had begun to suspect that the earth was very old, refuting older nonscientific concepts of the earth's history, its true age was not determined until the method of radioactive age dating was developed in the early 20th century. This dating provided a method of quantifying the divisions of the geologic column. These divisions were incorporated into the geologic time scale.

Problems

1. What is the distribution of elements in the earth? How and why is this distribution different from that of planets such as Jupiter and Saturn?

2. Describe the interval of the earth's history between its accretion and the beginning of plate tectonics.

3. How has the interpretation of the earth's internal structure changed in the past several decades?

4. What is the origin of the earth's magnetic field? How did magnetism of the lithosphere support the model of plate tectonics?

5. How does gravity vary across major surface features of the earth? What is meant by the term *isostatic equilibrium?*

6. Summarize the various lines of evidence that were incorporated into the theory of plate tectonics.

7. Describe each major type of plate boundary. How can these boundaries be recognized at the earth's surface?

8. How is energy distributed by atmospheric circulation? Explain why atmospheric circulation is broken into cells limited to specific latitudes.

9. Discuss the differences between atmospheric and oceanic circulation.

10. What drives deep circulation of the oceans and what is the importance of these deep currents?

11. In what way do the early concepts of catastrophism and neptunism differ from modern scientific thought?

12. What were the contributions of Nicolaus Steno to the science of geology?

13. Why is Hutton's Principle of Uniformitarianism so important to modern geology? How have modern interpretations of uniformitarianism changed since the development of the theory?

14. What forms of evidence are used in relative age dating?

15. Discuss the limitations of radioactive age dating.

16. What is the decay rate of carbon 14? Express your answer in yr^{-1}.

17. In dating using radiocarbon, the amount of ^{14}C remaining at the time of measurement is expressed as a percentage of modern carbon. For example, for a sample that has 30% modern carbon, N/N_0 would be 0.3. Using equation (2.2) and the k value determined in problem 16, what is the age of the sample in years?

References and Suggestions for Further Reading

BLOOM, A. L. 1998. *Geomorphology*, 3rd ed. Waveland Press, Long Grove, IL.

BULLARD, E. J., E. EVERETT, and A. G. SMITH. 1965. The fit of the continents around the Atlantic. *Philosophical Transactions of the Royal Society of London*, 1088:41–51.

CHRISTOPHERSON, R. W. 1997. *Geosystems: An Introduction to Physical Geography*, 3rd ed. Upper Saddle River, N.J.: Prentice Hall.

COSTA, J. E., and V. R. BAKER. 1981. *Surficial Geology—Building with the Earth*. New York: John Wiley.

DAVIDSON, J. P. W. E. REED, and P. M. DAVIS. 2002. *Exploring Earth*, 2nd ed. Upper Saddle River, N.J.: Prentice Hall.

HAMBLIN, W. K. 1985. *The Earth's Dynamic Systems*, 4th ed. New York: Macmillan Publishing Co., Inc.

HAY, E. A., and A. L. McALESTER. 1984. *Physical Geology: Principles and Perspectives*, 2nd ed. Upper Saddle River, N.J.: Prentice Hall.

JUDSON, S., M. E. KAUFFMAN, and L. D. LEET. 1987. *Physical Geology*, 7th ed. Upper Saddle River, N.J.: Prentice Hall.

KUMP, L. R., J. F. KASTING, and R. G. CRANE. 2004. *The Earth System*, 2nd ed. Upper Saddle River, N.J.: Prentice Hall.

LILLY, R. J. 1999. *Whole Earth Geophysics: An Introductory Textbook for Geologists and Geophysicists*. Upper Saddle River, N.J.: Prentice Hall.

MACKENZIE, F. T. 1998. *Our Changing Planet*, 2nd ed. Upper Saddle River, N.J.: Prentice Hall.

McNIGHT, T. 1999. *Physical Geography: A Landscape Appreciation*, 6th ed. Upper Saddle River, N.J.: Prentice Hall.

WOOD, J. A. 2000. *The Solar System*, 2nd ed. Upper Saddle River, N.J.: Prentice Hall.

CHAPTER

Minerals

The minerals, rocks, and soils that occur at and beneath the earth's surface are the materials with which the engineer and environmental scientist must work. Unlike the materials used in buildings and other structures, which have uniform, homogeneous properties, these natural materials are notoriously variable and nonhomogeneous. The task for engineers designing any structure is to evaluate the distribution and properties of the natural materials present at the site and then base the design upon this assessment. It is impossible to evaluate natural materials at specific sites without a general understanding of the physical and chemical characteristics of earth materials. In addition, the engineering properties must be ascertained. In this part of the text, we will concentrate upon the origin and characteristics of the minerals and rocks that make up the earth's crust. In the following sections, we will turn our attention to geologic processes operating both deep within the earth and at its surface.

An in-depth study of geology usually begins with an introduction to minerals. This particular point of departure should not come as a surprise, considering that the earth's solid surface is composed of rocks and soils that are primarily mineral aggregates. An understanding of mountains, volcanoes, earthquakes, and all other geologic phenomena could never be complete without some knowledge of the types and states of matter that make up the solid part of the earth.

Knowledge of minerals is essential for the engineer who deals with earth materials. Minerals are partially responsible for the physical and mechanical properties of rock and soil encountered in mines, tunnels, excavations, and environmental cleanups. Likewise, dams, embankments, and other structures built from earth materials function because engineers have successfully utilized the properties of rock and soil in the project design. In industry, minerals are directly incorporated into chemicals, abrasives, and fertilizers and are processed into thousands of other useful products.

The Nature of Minerals

Minerals are inorganic, naturally occurring solids. In order to define the term *mineral* completely, however, we must further state that minerals are crystalline substances that have characteristic internal structures and chemical compositions that are fixed or that vary within fixed limits. The distinct internal structure of each mineral results in diagnostic external properties that can be used in mineral identification. In subsequent sections, we will examine each component of the definitions of a mineral.

Internal Structure

At the heart of the definition of a mineral is the concept of a regular internal structure. It is this characteristic that distinguishes minerals from other types of solid, inorganic matter. A mineral, therefore, is a chemical compound in which elements react and combine to form a regular arrangement of particles within the solid.

The fundamental unit of each chemical element in the periodic table is the *atom*, the smallest amount of the element that retains its characteristic properties. An atom, however, is composed of even smaller particles. The most common model of atomic structure consists of a central concentration of mass, the *nucleus*, surrounded by concentric shells inhabited by minute charged subatomic particles called *electrons* (Figure 3.1). Electrons are held in the vicinity of a nucleus because of the attraction of opposite electrical charges. By convention, electrons are considered to be negatively charged and the nucleus is considered to be positively charged.

The nucleus of the atom can also be subdivided. Positive charges are concentrated in particles called *protons*. The number of protons in an atomic nucleus defines the *atomic number* of the element. Thus, an atom of sodium has 11 protons in the nucleus balanced electrically by 11 electrons orbiting around the nucleus in three shells (Figure 3.2). Particles that are neither positively nor negatively charged also reside in the nucleus. These *neutrons* have about the same mass as a proton, so the *atomic mass* of an atom is basically the mass of the protons plus the neutrons. The mass of an electron is about three orders of magnitude less than the mass of a proton or neutron and can be neglected in determining the atomic mass. Unlike the atomic number, the atomic mass of an element is not fixed.

Nuclei of an element may vary slightly in the number of neutrons that they contain. *Isotope* is the name given to atoms of the same element that differ in the number of neutrons in the nucleus. For example, hydrogen, with one proton in the nucleus, forms the isotope protium with an atomic mass of approximately 1. With the addition of one neutron to the nucleus, the hydrogen isotope deuterium is formed. Another hydrogen isotope, tritium, contains two neutrons in the nucleus. Even though the properties of these three forms of

▶ FIGURE 3.1
Electron cells around the nucleus of an atom.
Source: From L. D. Leet, S. Judson, and M. E. Kauffman, *Physical Geology*, 5th ed., © 1978 by Prentice Hall, Inc., Englewood Cliffs, N.J.

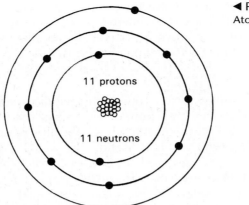

◀ FIGURE 3.2
Atomic structure of sodium.

hydrogen vary slightly, they are all unmistakably hydrogen. The proportions of the three isotopes of hydrogen vary in substances such as water as a function of temperature, age, and other variables. Thus, the measurement of the isotopic content of a particular body of water is an important research tool in hydrology and other branches of the earth sciences.

A final type of atomic particle with which we must be familiar is the *ion*. Ions form when an atom gains or loses electrons, so the number of electrons is then no longer equal to the number of protons. When an ion is formed, the atomic particle then becomes a positively charged *cation* or a negatively charged *anion*. The formation of ions is caused by the tendency of atoms to attain a state in which the outer electron shell is completely filled with electrons. Thus, atoms like sodium with a small number of electrons in their outer shells will readily lose them and become cations. Elements that lack only one or two electrons to complete their outer shells will gain electrons and become anions. The charge of a cation or anion, called its *oxidation number*, is determined by the number of electrons that it gains or loses. The oxidation number of the sodium ion is +1 because sodium loses the single electron from its outer shell.

In a mineral, atoms or ions are arranged in an orderly fashion to form the regular internal structure. Among the reasons that particles assume a certain structure are *chemical bonds* between adjacent atoms or ions in the internal framework. The presence of interparticle bonds indicates that minerals are no different from any other type of *chemical compound*. As compounds, the elemental composition of minerals can be described by chemical formulas that specify the exact proportions of elements in the mineral. *Quartz*, for example, has the formula SiO_2, meaning that there are two oxygen ions for every silicon ion in the structure.

Two types of bonds are of particular importance for geologic applications (Figure 3.3). An *ionic bond* is formed when one atom donates one or more electrons to an atom of another element. The first atom becomes a cation and the second, an anion. As illustrated in Figure 3.3, when an electron is transferred from an atom of sodium to an atom of chlorine, both ions then have achieved a stable state because their outer shells are filled with electrons. The resulting compound is called sodium chloride (common table salt), has the formula NaCl, and is given the mineral name *halite*. Elements that do not have as strong a tendency to lose or gain electrons may form *covalent bonds*, in which electrons are shared between atoms to complete the outer shell. Examples of covalent bonding are methane (Figure 3.3) and water. The type and number of bonds that particles in a mineral form with adjacent particles have a strong influence on the properties of the mineral. In fact, two entirely different minerals can have the same chemical composition. *Diamond* and *graphite*, for example, are both composed only of carbon. In diamond, the carbon atoms form a three-dimensional framework with

Table salt: NaCl

Na^{+1}ion
(1 electron transferred
away, leaving outer
shell with 8)

10 electrons total

Electrical
attraction

18 electrons total

Cl^{-1}ion
(1 electron gained,
thus filling outer
shell with 8)

+1 CHARGE

−1 CHARGE

(a)

Methane: CH$_4$

Hydrogen atom
(1 electron only)

Carbon atom
(4 electrons in outer
shell + 4 shared electrons
from hydrogen)

(b)

▲ FIGURE 3.3
Common types of atomic bonding in minerals. (a) Ionic bonding (electron transfer). (b) Covalent
bonding (electron sharing). *Source:* From E. A. Hay and A. L. McAlester, *Physical Geology: Principles and Perspectives*, 2nd ed., © 1984 by Prentice Hall, Inc., Englewood Cliffs, N.J.

▶ FIGURE 3.4
The internal structure of (a) diamond and
(b) graphite, both composed entirely of carbon.
Source: After Linus Pauling, *General Chemistry*, © 1953
by W. H. Freeman and Co., San Francisco.

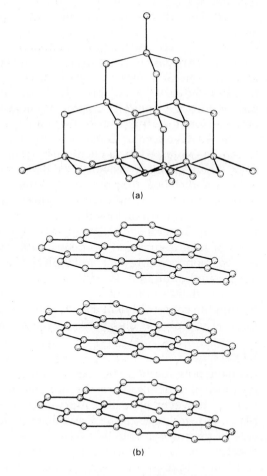

(a)

(b)

each atom bonded strongly to four adjacent atoms (Figure 3.4). The graphite structure consists of strong bonds between each carbon atom and three neighbors, all within a single plane. This results in sheets or layers of graphite with much weaker bonds *between* layers than *within* layers. Because of these differences in internal structure, diamond is the hardest mineral known and graphite is one of the softest.

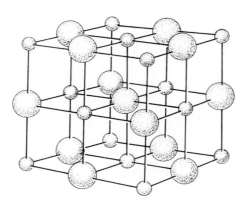

◀ FIGURE 3.5
Structure of halite, showing arrangement of chloride ions (large spheres) and sodium ions (small spheres). *Source:* From S. Judson, M. E. Kauffman, and L. D. Leet, *Physical Geology*, 7th ed., © 1987 by Prentice Hall, Inc., Englewood Cliffs, N.J.

The regular internal structure of minerals that we have alluded to justifies their description as *crystalline*. A crystalline solid is one in which a regular arrangement of atoms is repeated throughout the entire substance. The smallest unit of this structure is called the *unit cell*. The relative size of the ions plays an important role in the manner in which ions are packed in a particular structure. The unit cell of halite is composed of six chloride ions surrounding each sodium ion. The repetition of the unit cell in three directions produces a cubic structure, as shown in Figure 3.5. The presence of a regular, unique internal structure for each mineral was suspected by mineralogists for several centuries before proof was obtained in the early part of the 20th century. In the process called *X-ray diffraction*, X-rays are passed through mineral samples. Photographic images produced by the X-rays as they emerge from the sample show the regular pattern of atomic particles in the structure.

As we have seen in the cases of quartz and halite, the chemical composition of a mineral can be expressed by a chemical formula. For some minerals, the chemical formula is fixed. There are minerals, however, that have a variable composition. If two ions have similar size and charge, they may occupy the same types of sites in a mineral structure without altering the structure and without greatly affecting the properties of the mineral. Magnesium and iron form cations that have the same charge and the same approximate size. In the silicate mineral *olivine*, magnesium and iron can substitute freely. The varying proportions of magnesium and iron constitute a *solid-solution series*. Thus, olivine can range in composition from Mg_2SiO_4 to Fe_2SiO_4. These two end members represent olivine with all magnesium or all iron. In addition, any intermediate composition is possible. A sample of olivine with 50% iron and 50% magnesium would have the formula $(Mg_{0.5}, Fe_{0.5})_2SiO_4$. Although the composition of a solid-solution-series mineral is variable, it varies within fixed limits. The limits for olivine are the magnesium and iron end members.

Crystals

We have now reached the point in our discussion where we can consider the exterior appearance of minerals. Normally, minerals will be present as irregular grains in a rock. Occasionally, however, we will be lucky enough to find a mineral specimen bounded by smooth, plane surfaces arranged in a symmetrical fashion around the specimen (Figure 3.6). This type of sample is said to be a *crystal*. The plane, exterior surfaces, or *crystal faces*, are manifestations of the internal structure of the mineral.

The unique internal arrangement of atoms in each mineral determines the possible crystal faces that can be present. Crystals form slowly by adding unit cells of the mineral to the outer faces. Only when crystals are allowed to grow in this manner in an uncrowded environment, that is, unhindered by the growth of neighboring crystals, are well-shaped crystal faces developed. These conditions are somewhat rare, explaining

▶ FIGURE 3.6
A crystal of garnet. The exterior surface of the specimen is composed of smooth, planar crystal faces arranged in a regular geometrical pattern. *Source:* Photo courtesy of the author.

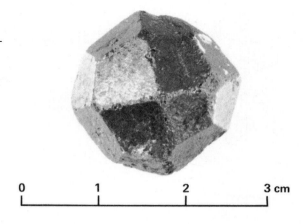

▶ FIGURE 3.7
Crystals of the same mineral, no matter what size, have the same angle between corresponding crystal faces. *Source:* From E. A. Hay and A. L. McAlester, *Physical Geology: Principles and Perspectives*, 2nd ed., © 1984 by Prentice Hall, Inc., Englewood Cliffs, N.J.

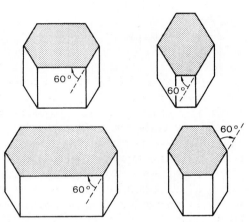

the occurrence of most minerals as irregular grains. Despite the lack of crystal development, the characteristic internal structure is always present in any mineral.

Crystals have been treasured as gems for thousands of years. The scientific study of crystals took great strides when Nicolaus Steno recognized in 1669 that the angles between corresponding crystal faces of the same mineral are always the same. This holds true no matter what size crystals are measured. Steno's observation is now known as the *Law of Constancy of Interfacial Angles* (Figure 3.7).

The location of crystal faces on a crystal is quantified by their orientation with respect to the *crystallographic axes*, which are imaginary lines drawn through the crystal connecting the centers of corresponding crystal faces on opposite sides of the crystal. Despite the large number of possible crystal-face orientations that have been observed in natural crystals, all crystals have been grouped into six *crystal systems* based upon their geometrical properties. The definition of each system is based upon the number of crystallographic axes and the angles between the axes. Illustrations and examples of the crystal systems are presented in Figure 3.8.

Physical Properties

Because minerals are rarely present as crystals, we must use criteria other than the shape and arrangement of crystal faces for identification. Although there are about 2000 minerals known, only a few are abundant in the most common rocks of the earth's crust. The

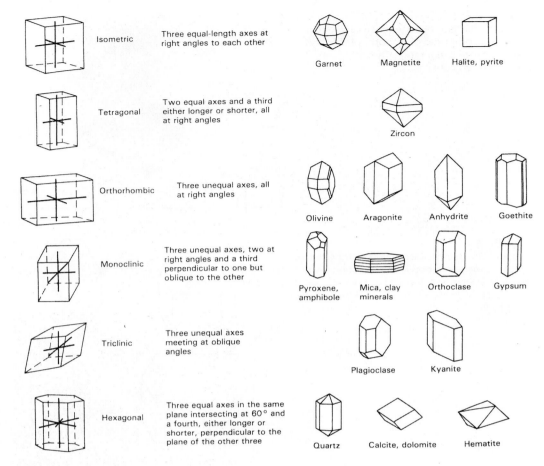

▲ FIGURE 3.8
The six crystal systems with representative minerals and crystal forms. *Source:* From E. A. Hay and
A. L. McAlester, *Physical Geology: Principles and Perspectives*, 2nd ed., © 1984 by Prentice Hall, Inc.,
Englewood Cliffs, N.J.

common rock-forming minerals can be identified by their *physical properties*, which are
characteristics that can be observed or determined by simple tests.

Cleavage and Fracture

The internal structure of a mineral is responsible for an extremely useful group of properties involving its strength. For example, there are two ways in which a mineral can break.
First, the breakage can take place along regularly spaced planes of weakness caused by
weak bonds in those directions within the internal structure. This type of breakage is
called *cleavage* and the planes are called *cleavage planes*. The second type of breakage occurs
in minerals that lack preferred directions of weakness. The surfaces of rupture are more irregular in these minerals and the type of breakage is known as *fracture*.

Cleavage planes can sometimes be confused with crystal faces because both develop
in response to the internal arrangement of atoms in a mineral. Crystal faces, however, do
not necessarily represent planes of weak bonds. For example, quartz has no cleavage but
often exhibits well-developed crystal faces. Cleavage is classified as *perfect* if the cleavage
planes are level and smooth. *Biotite* and *muscovite*, the micas, are examples of minerals
with excellent cleavage in one direction (Figure 3.9). Cleavage can be developed in as

▲ FIGURE 3.9
Muscovite, showing excellent cleavage in one direction. Samples split readily parallel to the cleavage planes into very thin plates. *Source:* Photo courtesy of the author.

▶ FIGURE 3.10
Halite, a mineral with well-developed cubic cleavage. *Source:* Photo courtesy of the author.

many as six directions, creating a variety of cleavage types. Several common minerals develop cleavage in three directions. When the three planes are mutually perpendicular, the cleavage form is called *cubic* (Figure 3.10). If the three directions of cleavage do not intersect at right angles, *rhombohedral cleavage* can be formed (Figure 3.11).

Most fractures are relatively rough and irregular. Fracture is the form of breakage displayed by materials composed of atoms that are evenly spaced and equally attracted to adjacent particles in all directions. Occasionally, when materials of this type break, the surface is marked with smooth, concentric depressions (Figure 3.12). This type of fracture is called *conchoidal fracture*. Quartz sometimes displays this surface feature.

Hardness

The strength of atomic bonds, along with the size and the density of the packing of atoms or ions in a mineral, determines its *hardness*. The physical test used to determine the hardness of a mineral involves scratching it with various materials. A mineral with a high hardness value cannot be easily scratched by most substances. A soft mineral, on the

◄ FIGURE 3.11
Rhombohedral cleavage in the mineral calcite. The three directions of cleavage intersect at angles of 75° and 105°. *Source:* Photo courtesy of the author.

◄ FIGURE 3.12
Obsidian, a type of volcanic glass that produces conchoidal fracture upon breakage. *Source:* Photo courtesy of the author.

other hand, can be scratched by any harder material. Tests of this type have shown that minerals vary quite widely in hardness. A relative scale used to compare minerals is the *Mohs Hardness Scale* (Table 3.1). The scale is made up of 10 levels arranged in order of increasing hardness, each represented by a common mineral. It is important to remember, however, that the scale is relative and the hardness increments between the levels are not necessarily equal.

Color and Streak

It would simplify mineral identification greatly if each mineral had a unique color that never varied from sample to sample. Unfortunately, this is not the case; many minerals

Table 3.1 The Mohs Hardness Scale	
Mineral	**Common Object**
1. Talc	
2. Gypsum	Fingernail $\left(2\frac{1}{2}\right)$
3. Calcite	Penny (3)
4. Fluorite	
5. Apatite	Knife blade $\left(5\frac{1}{2}\right)$
6. Orthoclase	File (6), window glass (6)
7. Quartz	Unglazed porcelain streak plate (7)
8. Topaz	
9. Corundum	
10. Diamond	

occur in a wide variety of colors. The property of color is caused by the absorption of selective wavelengths of white light. The color of a mineral is associated with the wavelengths that are not absorbed. These portions of the spectrum are reflected from the surface of the mineral. The absorption of light energy is related to the configuration of electrons around the nucleus in certain elements. In the transition elements, the outer electron shell contains one or two electrons before the adjacent inner shell is filled. Upon stimulation by light energy, the loosely held electrons in the outer shell vibrate easily and in doing so readily absorb energy. Even a minute percentage of such elements present in a mineral as an impurity can supply color to the mineral. For example, pure quartz is colorless, but only a small amount of manganese or titanium can impart a pink hue to the mineral. Quartz of this color is known as rose quartz.

The color of finely powdered mineral particles produced by scraping the specimen across a porcelain (*streak*) plate is often more diagnostic than the bulk color of the mineral. The mineral *hematite*, for example, produces a reddish brown streak, even though the sample may have a metallic gray appearance. The limitation of a streak plate is that it can only be used on minerals with a hardness less than 7.

Luster

Luster is a property that results from the manner in which light is reflected from a mineral. All minerals can be classified with respect to luster as either *metallic* or *nonmetallic*. Metallic luster is caused by high surface reflectivity of light by the opaque minerals, minerals that strongly absorb light. The native metals such as gold and silver, as well as numerous other minerals (Figure 3.13), have the appearance that we normally associate with metals because of their metallic luster.

The remaining minerals have various types of nonmetallic luster. These include the brilliant, reflective luster of diamond and other gems known as *adamantine luster*. One of the most common varieties of nonmetallic luster is *vitreous luster*, which is best illustrated by common glass. Quartz is a mineral with vitreous luster. Other minerals have luster that can be described as *greasy, waxy, pearly,* or *dull* (earthy).

Specific Gravity

Density provides a simple way of identifying some minerals that look similar simply by lifting the specimen. When the density is expressed as the ratio of the weight of a mineral to the weight of an equal volume of water, it is called the *specific gravity*. It is most useful for minerals composed of heavy elements, such as iron, nickel, and lead. The mineral galena (Figure 3.13), which contains lead and sulfur, has a specific gravity of 7.57. A sample of

◀ FIGURE 3.13
The mineral galena has metallic luster. Cubic cleavage is also evident. *Source:* Photo courtesy of the author.

galena is much heavier than a specimen of quartz of about the same size because quartz has a specific gravity of 2.65.

Other Properties

A few other properties may be useful for identifying specific minerals. Examples are magnetism, radioactivity, and taste, smell, or touch. *Magnetite*, an iron-bearing mineral, is the best example of a magnetic mineral. Highly magnetic varieties of magnetite called *lodestone* were used as the first compasses in the early days of navigation.

Mineral Groups

Minerals are classified according to chemical composition and structure. The composition of the most common rock-forming minerals is limited by the abundance of elements in the earth's crust. In fact, only eight elements constitute about 98% of the weight of the earth's crust (Table 3.2). Most of the minerals are members of a group characterized by combinations of the two most abundant elements, oxygen and silicon. This group is called the *silicate group* because all its members contain a specific structural combination of silicon and oxygen, even though most silicate minerals also contain other elements. Similarly, the

Table 3.2 Elemental Abundance in the Earth's Crust		
Element	**Symbol**	**Weight Percent**
Oxygen	O	46.60
Silicon	Si	27.72
Aluminum	Al	8.13
Iron	Fe	5.00
Calcium	Ca	3.63
Sodium	Na	2.83
Potassium	K	2.59
Magnesium	Mg	2.09

Source: From Brian Mason, *Principles of Geochemistry,* © 1958 by John Wiley & Sons, New York.

▶ FIGURE 3.14
The silica tetrahedron shown from (a) top and (b) side views. A central silicon ion is surrounded by four oxygen ions. *Source:* From S. Judson, M. E. Kauffman, and L. D. Leet, *Physical Geology*, 7th ed., © 1987 by Prentice Hall, Inc., Englewood Cliffs, N.J.

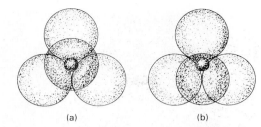

(a)   (b)

other major mineral groups are composed of specific compositional units, usually anions or ion groups. We will begin our description of these groups with the silicates because of their abundance in rock-forming minerals.

Silicates

Every silicate mineral contains a basic structural unit called the *silica tetrahedron*. As shown in Figure 3.14, its structure involves a central silicon cation with an oxidation number of +4 surrounded by four oxygen anions, each with an oxidation number of −2. The four oxygen anions form the corners of a four-sided geometric form called a *tetrahedron*. In its isolated state, the silica tetrahedron has a charge of −4. To form a mineral with an overall neutral charge, the negative charge must be balanced by adding cations or by forming linked tetrahedra in which the oxygen ions are shared by adjacent tetrahedra. The vast number of silicate minerals are formed by utilizing both of these mechanisms.

The silicate group is subdivided by the way in which silica tetrahedra interact within the structure. The most basic structural form consists of isolated tetrahedra with no sharing of oxygens. Olivine (Figure 3.15) is an example of a mineral with isolated tetrahedra. Magnesium and iron are added in various proportions to balance the charge, forming the solid-solution series that we previously considered. The physical properties of olivine are listed along with other minerals in Table 3.3.

In other subdivisions of the silicate group, the linking pattern of silica tetrahedra becomes successively more complex. One common type of linkage produces chains of tetrahedra. Two types of chains are formed: The *pyroxene* group consists of silicate minerals with single chains (Figure 3.16), and the *amphibole* group contains double chains of silica tetrahedra (Figure 3.17). *Augite* and *hornblende* are common representatives of the pyroxene and amphibole groups, respectively. Hornblende has a particularly recognizable form of cleavage with two directions intersecting at angles of 56° and 124°.

The *sheet silicates* are composed of thin layers, or sheets, of silica tetrahedra in which three oxygens are shared with adjacent tetrahedra (Figure 3.18). These sheets are described as tetrahedral sheets because the silicon atoms are surrounded by four oxygens. In addition, the sheet silicates contain other structural types of sheets. For example, octahedral sheets (Figure 3.18) contain aluminum, magnesium, or other cations coordinated by the six oxygens or hydroxyl atoms. The geometric figure defined by the arrangement of oxygen atoms is an octahedron. Important minerals within the sheet silicate group include the micas and the clay minerals. Different minerals are formed by different ways of combining sheets and by variations in charge caused by ionic substitutions within the structures. The clay minerals are especially significant to engineering because of the tendency of some clays to absorb water and swell. The effects of swelling soils on foundations can be very destructive.

One of the most basic structural configurations is the combination of one tetrahedral sheet and one octahedral sheet to form the clay mineral *kaolinite* (Figure 3.19). Aluminum ions fill the octahedral sites. Hydroxyls substitute for some of the oxygens to balance the charge of the structural unit. Adjacent double-sheet layers of kaolinite are

▲ FIGURE 3.15
(a) Photo of the mineral olivine. *Source:* Photo courtesy of the author. (b) Structure of olivine is characterized by isolated silica tetrahedra and regularly spaced magnesium and/or iron ions. *Source:* From
E. A. Hay and A. L. McAlester, *Physical Geology: Principles and Perspectives*, 2nd ed., © 1984 by Prentice Hall,
Inc., Englewood Cliffs, N.J.

attracted to each other by very weak forces called *van der Waals* bonds. When two tetrahedral sheets crystallize and are separated by an octahedral sheet, several minerals can
form. The *smectite* group of clay minerals, which includes the mineral *montmorillonite*,
crystallizes with this structure (Figure 3.20). Substitutions of cations in the smectite
structure commonly include aluminum (+3) for silicon (+4) in the tetrahedral sheets;
and magnesium (+2), iron (+2), or other cations for aluminum (+3) in the octahedral
sheets. The result of these substitutions is that there is a net negative charge on the
structure that must be balanced to maintain electroneutrality. Charge balance is maintained by the incorporation of cations into positions between the layers (Figure 3.20).
These cations, which usually include calcium and sodium in montmorillonite, serve to
balance the positive charge deficiency and, in addition, bond adjacent montmorillonite
layers together. Because the charge imbalance is relatively small for montmorillonite,
the interlayer cations are weakly held and can be exchanged for other cations that may
be brought into contact with the clay. *Cation exchange* is a very important clay property.
In addition to cations, interlayer sites contain variable amounts of water. If water comes
into contact with the clay, water molecules are drawn into the interlayer spaces and the
structure expands with a force sufficient to lift or crack the foundation of fully loaded
buildings.

Table 3.3 Physical Properties of Common Rock-Forming Minerals

Mineral	Chemical Formula	Color	Cleavage Directions	Hardness	Specific Gravity	Other Properties
Silicates						
Augite (pyroxene)	$Ca(Mg, Fe, Al)(Al, Si_2O_6)$	Dark green to black	2 at 90°	5–6	3.2–3.6	
Biotite (mica)	$K(Mg, Fe)_3AlSi_3O_{10}(OH)_2$	Black	1	$2\frac{1}{2}$–3	2.8–3.2	
Garnet	$(Ca, Mg, Fe, Mn)_3(Al, Fe, Cr)_2(SiO_4)_3$	Dark red, brown, green	0	$6\frac{1}{2}$–$7\frac{1}{2}$	3.5–4.3	
Hornblende (amphibole)	$(Na, Ca)_2(Mg, Fe, Al)_5Si_6(Si, Al)_2O_{22}(OH)_2$	Dark green to black	2 at 56° and 124°	5–6	2.9–3.2	
Muscovite (mica)	$KAl_3(AlSi_3O_{10})(OH)_2$	Colorless to pale green	1	2–$2\frac{1}{2}$	2.8–2.9	
Olivine	$(Mg, Fe)_2SiO_4$	Pale green to black	None	$6\frac{1}{2}$–7	3.3–4.4	
Orthoclase (feldspar)	$KAlSi_3O_8$	White, gray, or pink	2 at 90°	6	2.6	
Plagioclase (feldspar)	$(Ca, Na)(Al_2Si)AlSi_2O_8$	White to gray	2 at 90°	6	2.6–2.7	Striations
Quartz	SiO_2	Colorless to white but often tinted	None	7	2.6	
Oxides						
Hematite	Fe_2O_3	Reddish brown to black	None	$5\frac{1}{2}$–$6\frac{1}{2}$	5.26	
Goethite (limonite)	$FeO \cdot OH$	Yellowish brown to dark brown	None	5–$5\frac{1}{2}$	4.37	Limonite is noncrystalline
Magnetite	Fe_3O_4	Black	None	6	5.18	Strongly magnetic
Halides and sulfides						
Halite	$NaCl$	Colorless or white	3 at 90°	$2\frac{1}{2}$	2.16	
Pyrite	FeS_2	Pale brassy yellow	3 at 90°	6–$6\frac{1}{2}$	5.02	Sometimes called "fool's gold"
Chalcopyrite	$CuFeS_2$	Brassy yellow	2 at 90°	$3\frac{1}{2}$–4	4.1–4.3	Ore of copper
Sphalerite	ZnS	Brown to yellow	6 at 120°	$3\frac{1}{2}$–4	3.9	Ore of zinc
Galena	PbS	Lead gray	3 at 90°	$2\frac{1}{2}$	7.54	Ore of lead
Sulfates and carbonates						
Gypsum	$CaSO_4 \cdot 2H_2O$	Colorless to white	1	2	2.32	
Calcite	$CaCO_3$	White to colorless	3 at 75°	3	2.72	
Dolomite	$CaMg(CO_3)_2$	Pink, white, or gray	3 at 74°	3.5–4	2.85	Forms limestome

▲ FIGURE 3.16

(a) Photo of pyroxene. *Source:* Photo courtesy of the author. (b) In pyroxene, single chains of silica tetrahedra are bound together by calcium and magnesium ions. *Source:* From E. A. Hay and A. L. McAlester, *Physical Geology: Principles and Perspectives,* 2nd ed., © 1984 by Prentice Hall, Inc., Englewood Cliffs, N.J.

◄ FIGURE 3.17

Hornblende, a member of the amphibole group of silicate minerals. The distinctive cleavage of hornblende consists of two planes intersecting at angles of 56° and 124°. *Source:* Photo courtesy of the author.

The structure of the micas is very similar to the smectites. The excellent cleavage of muscovite (light mica) and biotite (dark mica) is developed along the planes of the individual layers (Figures 3.9 and 3.21). The micas differ from the smectites in that aluminum ions always substitute for silicon in the tetrahedral layers in the micas, resulting in a larger charge imbalance. Potassium ions fill the interlayer positions in micas and, unlike the smectites, are tightly bound and not exchangeable. *Illite* is a clay mineral similar in structure to muscovite, but it has a structure in which the exchange of interlayer potassium does occur because of variations in ion substitutions.

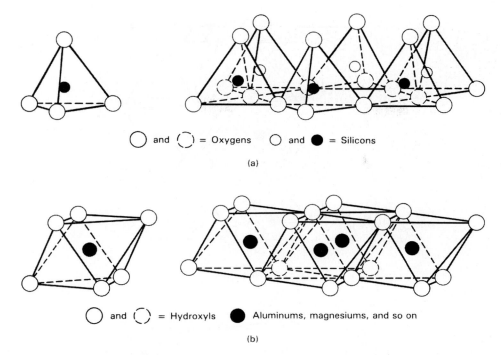

▲ FIGURE 3.18
The sheet silicates are composed of two types of sheets: (a) A tetrahedral sheet of linked silica tetrahedra (single tetrahedron shown at left) and (b) an octahedral sheet (single octahedron shown at left) in which cations are surrounded by oxygens in octahedral coordination. *Source:* From R. E. Grim, *Clay Mineralogy*, 2nd ed., © 1968 by McGraw-Hill, Inc., New York.

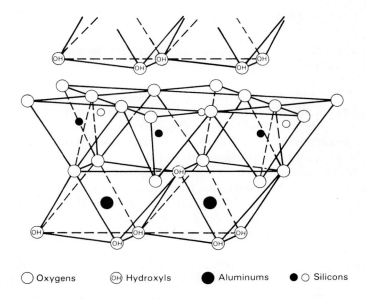

▲ FIGURE 3.19
The kaolinite structure forms layers composed of one octahedral sheet and one tetrahedral sheet. Adjacent layers are held together by van der Waals bonds. *Source:* From R. E. Grim, *Clay Mineralogy*, 2nd ed., © 1968 by McGraw-Hill, Inc., New York.

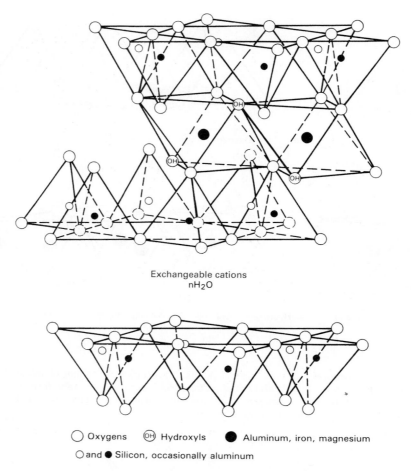

Exchangeable cations
nH₂O

○ Oxygens   ⊕ Hydroxyls   ● Aluminum, iron, magnesium
○ and ● Silicon, occasionally aluminum

▲ FIGURE 3.20
Two tetrahedral sheets and one octahedral sheet form the smectite structure. Layers are separated by exchangeable cations and water molecules. *Source:* From R. E. Grim, *Clay Mineralogy,* 2nd ed., © 1968 by McGraw-Hill, Inc., New York.

The final group of silicate minerals, the *framework silicates,* contain the *feldspars* and quartz, which are the most abundant minerals in the earth's crust. The word *framework* refers to the three-dimensional linking of the silica tetrahedra, forming strong bonds in all directions. There are two important types of feldspar. *Plagioclase feldspar* includes a number of minerals in a solid-solution series with sodium and calcium as end members. Plagioclase can sometimes be recognized by fine, closely spaced lines, or *striations,* on the cleavage faces of the specimen. Striations are caused by defects in the crystal structure that formed as the mineral was crystallizing. *Orthoclase* is the other main type of feldspar. It differs from plagioclase in its lack of striations and in its high potassium content (Figure 3.22).

Quartz (Figure 3.23) is a framework silicate composed entirely of silica tetrahedra in which all four oxygen ions are shared. The development of strong bonds in all directions and the lack of additional elements in the structure result in the absence of cleavage. The color of quartz is extremely variable due to minute amounts of impurities within the structure.

The silicate minerals are used for many purposes in our society. Most gem stones, including emerald, topaz, jade, and turquoise, are silicate minerals. In addition, silicate minerals are mined as ores for various metals and used in a variety of industrial processes. The

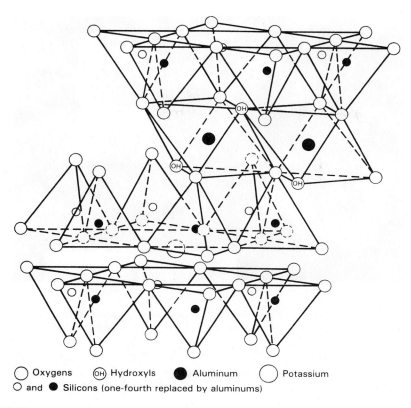

Oxygens   (OH) Hydroxyls   ● Aluminum   ◯ Potassium
◦ and ● Silicons (one-fourth replaced by aluminums)

▲ FIGURE 3.21
The muscovite structure is similar to the smectites except that layers are more strongly bonded by nonexchangeable potassium cations. *Source:* From R. E. Grim, *Clay Mineralogy*, 2nd ed., © 1968 by McGraw-Hill, Inc., New York.

▲ FIGURE 3.22
Orthoclase, one of the two major types of feldspar, is a framework silicate mineral. Note two directions of cleavage at right angles. *Source:* Photo courtesy of the author.

▲ FIGURE 3.23
Quartz, one of the most common silicate minerals, occurs as well-developed crystals (lower left), but more commonly in massive form (upper right). *Source:* Photo courtesy of the author.

clay minerals are particularly valuable in industry and manufacturing, with uses in the production of brick, tile, plastics, rubber, paint, ceramics, and paper.

Oxides

Minerals that form by combination of various cations with oxygen are called *oxides*. Among the oxide minerals are important ore minerals of iron, aluminum, chromium, and other metals. The iron oxide minerals are particularly common. Iron and oxygen can form several different minerals, depending on the oxidation number of the iron ions. In hematite (Figure 3.24) and *geothite*, iron is present in the oxidized (+3) state, while in magnetite, iron is present in both the oxidized and reduced (+2) states.

Halides and Sulfides

Minerals of the halide and sulfide group contain anions of fluorine, chlorine, bromine, iodine, and sulfur as the framework anion. Minerals composed of the first four elements of this list constitute the halide minerals. An example of a halide mineral is halite, or common table salt (Figure 3.10). The sulfide mineral group constitutes a very important source of the metallic ores of iron, copper, zinc, and lead. *Pyrite* (Figure 3.25) is a common mineral often mistaken for gold because of its metallic luster. For this reason, it is sometimes called "fool's gold."

Sulfates and Carbonates

Sulfates and carbonates consist of framework radicals similar to the silica tetrahedra in that an anion group is the basis of the structure. The sulfate anion group consists of sulfur and four oxygen ions (SO_4^{2-}). An important mineral in this group is *anhydrite*, $CaSO_4$. Anhydrite is similar in composition to an even more common calcium sulfate mineral, *gypsum*. Gypsum is a hydrated mineral with the formula $CaSO_4 \cdot 2H_2O$. Gypsum and anhydrite precipitate from solution as seawater evaporates in restricted basins. They are associated with deposits of limestone, dolomite, and halite. Because of their origin, these minerals are often called *evaporites*. They occur in beds of variable thickness that can be

▶ FIGURE 3.24
Hematite, a common iron oxide mineral. *Source:* Photo courtesy of the author.

▲ FIGURE 3.25
Pyrite, an important iron sulfide mineral, can be recognized by its brassy yellow color, cubic cleavage, and metallic luster. The specimen on the left contains thin parallel lines, or striations, on its exterior surfaces. *Source:* Photo courtesy of the author.

very extensive in sedimentary rock sequences. Evaporites accumulated in marine basins that were located in warm climates and had a restricted flow of seawater between the basin and the open ocean. As seawater evaporated from these basins, the water became more saline and evaporate minerals began to precipitate. The distribution of evaporate basins in the United States is shown in Figure 3.26.

▲ FIGURE 3.26
Location of major evaporite deposits in the United States and southern Canada. Sylvite, carnallite, and langbeinite are potassium salts. *Source:* James R. Craig, David J. Vaughan, and Brian J. Skinner, *Resources of the Earth: Origin, Use, and Environmental Impact*, 3rd ed., © 2001. Reprinted by permission of Pearson Education, Inc., Upper Saddle River, N.J.

In carbonate minerals, the basic building block is the carbonate ion (CO_3^{2-}). The most important carbonate minerals are *calcite* (Figure 3.11), which combines calcium with the carbonate ion, and *dolomite*, which contains calcium and magnesium in its structure. Calcite is often found in hot-spring deposits and caves. Large amounts of *travertine* (a variety of calcite) have been deposited at Mammoth Hot Springs in Yellowstone National Park (Figure 5.13) in a beautiful series of terraces and pools. Among the many uses of calcite and dolomite are as building stones and in the production of lime and Portland cement, which is discussed in the next section. Lime is valuable for its acid-neutralizing properties with applications in agriculture and industry, and Portland cement is an essential ingredient of concrete.

Native Elements

Certain minerals consist of a single element. Examples are gold, silver, copper sulfur, and carbon. Elemental gold, silver, and copper are relatively rare in nature; they more commonly occur as components of other minerals. Sulfur is utilized in the production of sulfuric acid and other chemicals. Graphite and diamond (Figure 3.4) are two minerals composed of carbon. The difference in structure between graphite and diamond is largely due to pressure at the time of formation. Diamond forms in high-pressure geologic environments deep within the earth and therefore develops a dense structure. Its great hardness is attributed to a three-dimensional network of strong covalent bonds. Graphite is a soft, greasy mineral with perfect cleavage in one direction.

Use and Misuse of Minerals

Minerals are the raw materials of our industrial society. There is no question that minerals are essential for the products that we use in our daily life. A partial list of minerals used in industry is given in Table 3.4. Some of the minerals listed are rare and were not discussed

Table 3.4 Uses of Minerals in Industry	
	Minerals
Abrasives	Bauxite, garnet, industrial diamond
Aggregates	Diatomite, dolomite, limestone, perlite, pumice, vermiculite
Cements	Asbestos, dolomite, gypsum, limestone
Ceramics	Barite, bauxite, bentonite, beryllium minerals, feldspar, kaolinite, lithium minerals, manganese minerals, pyrophyllite, rare earth minerals, silica, talc, wollastonite, zircon
Drilling fluids	Barite, bentonite, sepiolite and attapulgite
Electronics	Beryllium minerals, graphite, manganese minerals, rare earth minerals, silica
Fertilizers and soil conditioners	Dolomite, limestone, nitrates, phosphates, potassium salts
Fillers, filters, and absorbants	Diatomite, kaolinite, mica, sepiolite and attapulgite, silica, talc, wollastonite, zeolites
Fluxes	Fluorite, limestone
Foundry sands	Bentonite, chromite, graphite, olivine, silica, zircon
Glass	Borax and borates, celestite, feldspar, fluorite, limestone, lithium minerals, silica, soda ash, zircon
Insulation	Asbestos, mica, perlite, pumice
Lubricants	Graphite, molybdenite
Pigments	Barite, iron oxide pigments, titanium minerals
Plasters	Gypsum
Refractories	Bauxite, chromite, graphite, magnesite, pyrophyllite, silica, sillimanite, andalusite, kyanite, zircon

Source: L. L. Y. Chang, *Industrial Mineralogy: Materials, Processes and Uses,* © 2002 by Prentice Hall, Inc., Upper Saddle River, N.J.

in the general introduction to minerals in this chapter. In addition, some of the minerals are not entirely consistent with the definition of minerals in that they may be groups of related minerals or rocks that are dominated by a particular mineral. Minerals, or mineral aggregates, are also essential ingredients of construction materials. Because of the large quantities that are needed, sources must be found close to the point of use to keep construction costs reasonable. Inclusions of minerals with unfavorable properties can cause major problems during construction or in the performance of the project.

Location and extraction of economically recoverable mineral deposits constitutes one of the major career fields for geologists and engineers. The search is worldwide because most minerals occur only in areas where specific types of rocks occur and the geological conditions that control the size, concentration, and depth of a commercial deposit may be very rare. The political stability and infrastructure of the country that contains the deposit are also important factors in its exploitation. Mineral deposits are a major source of wealth for countries that are lucky enough to have rich mineral deposits.

Metals

Out of the 35 metallic elements that are mined for industrial use, 6 elements are present in quantities above 0.1% by weight in the earth's crust. These metals—silicon, aluminum, iron, magnesium, titanium, and manganese—are in sufficient abundance that their supply should not be a problem for the next several centuries. The extraction of metals is economically feasible only when they are concentrated in *ore deposits*. Although the 6 abundant metals occur in many different minerals, a relatively small number constitute ore minerals because these have high concentrations of the metal and can easily be

Table 3.5   Important Ore Minerals of the Abundant Minerals

Metal	Important Ore Minerals	Amount of Metal in the Ore Mineral
Silicon	Quartz (SiO_2)	46.7
Aluminum	Boehmite ($AlO \cdot OH$)	45.0
	Diaspore ($AlO \cdot OH$)	45.0
	Gibbsite ($Al(OH)_3$)	34.6
	Kaolinite ($Al_2Si_2O_5(OH)_4$)	20.9
	Anorthite ($CaAl_2Si_2O_8$)	19.4
Iron	Magnetite (Fe_3O_4)	72.4
	Hematite (Fe_2O_3)	70.0
	Goethite ($FeO \cdot OH$)	62.9
	Siderite ($FeCO_3$)	62.1
	Chamosite ($Fe_3(Si, Al)_2O_5(OH)_4$)	45.7
Magnesium	Magnesite ($MgCO_3$)	28.7
	Dolomite ($CaMg(CO_3)_2$)	13.1
Titanium	Rutile (TiO_2)	60.0
	Ilmenite ($FeTiO_3$)	31.6
Manganese	Pyrolusite (MnO_2)	63.2
	Psilomelane ($BaMn_8O_{18} \cdot 2H_2O$)	46.0
	Rhodochrosite ($MnCO_3$)	39.0

Source: James R. Craig, David J. Vaughan, and Brian J. Skinner, *Resources of the Earth: Origin, Use, and Environmental Impact*, 3rd ed., © 2001. Reprinted by permission of Pearson Education, Inc., Upper Saddle River, N.J.

processed to extract the metal (Table 3.5). Aluminum, after iron the most widely used metal, is a good example—although it is a constituent in common minerals such as the feldspars—it is too difficult to extract and too poorly concentrated for these to be considered ore minerals. Instead, aluminum is obtained from minerals that form by weathering processes in tropical climates. These ores, known as *bauxites*, will be further described in the chapter that covers weathering.

Iron ores occur in a wide variety of geological settings. By far the most important in terms of importance to worldwide production are the rocks known as *banded iron formations* (Figure 3.27). These deposits date from early Proterozoic time—2.8 to 1.6 billion years ago. The distribution of these deposits is shown in Figure 3.28. They are limited to early Proterozoic rocks because of the lack of oxygen in the atmosphere at this time. In the presence of oxygen, iron precipitates in its *ferric* (+3) form as oxide minerals and is immobile in aqueous environments. However, in the low-oxygen, early Proterozoic atmosphere, iron was released by weathering processes and transported in dissolved form in rivers to shallow lakes or seas where it precipitated in concentrated layers. When photosynthesis began to release oxygen to the atmosphere, iron was no longer mobile and the window of opportunity for formation of banded iron formations was closed. The sediments containing the banded iron layers were subsequently metamorphosed and the iron is present in the minerals hematite and magnetite, along with various other silicate and carbonate minerals. The banded iron formations in the Lake Superior region provided the iron ore for the production of steel during the Industrial Revolution in the United States. Banded iron formations throughout the world will supply iron ore well into the future.

The other abundant metals—manganese, magnesium, titanium, and silicon—are widely used in industry and occur in numerous minerals. Even quartz, one of the most common minerals, serves as an ore mineral for silicon, the raw material for computer chips.

▲ FIGURE 3.27
Banded iron formation, Lake Superior region. *Source:* Photo courtesy of the author.

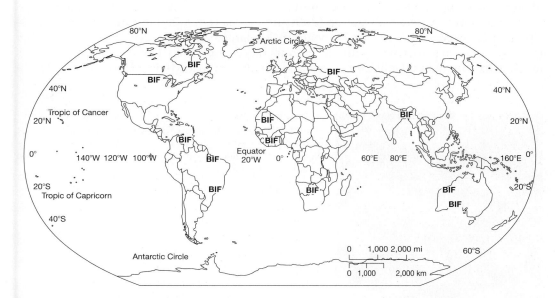

▲ FIGURE 3.28
Location of the banded iron formations. *Source:* James R. Craig, David J. Vaughan, and Brian J. Skinner, *Resources of the Earth: Origin, Use, and Environmental Impact*, 3rd ed., © 2001. Reprinted by permission of Pearson Education, Inc., Upper Saddle River, N.J.

A large group of minerals contains metals that are much less common than the six abundant metals just discussed. Despite their scarcity, these metals are essential for many industrial processes and products. These *geochemically scarce metals* can be divided into four groups: the *ferro-alloy metals*, the *base metals*, the *precious metals*, and the *special metals*. The concentrations of these metals in the crust are so low that extraction is economically feasible only where they are significantly concentrated. The ferro-alloy metals, including chromium, cobalt, molybdenum, nickel, tungsten, and vanadium, are added to iron to create many of the varieties of steel that are currently utilized in manufacturing. The base metals have been essential to human societies since prehistoric times. They include cadmium, copper, lead, mercury, tin, and zinc. These metals and their alloys are critical to countless products—from coins to electrical wires. Brass (copper and zinc), bronze (copper and tin), and pewter (lead and tin) are three of the most ancient base-metal alloys. Lead was the first metal used for pipes in plumbing systems. Gold, silver, and the platinum group metals comprise the *precious metals*. Gold, in particular, has been the standard of wealth for human societies since civilization began (Figure 3.29). Silver has also been highly prized since ancient times. The *special metals* constitute a diverse group of rare metallic elements that have specialized uses in many industries.

Examples of ore minerals for the geochemically scarce metals are given in Table 3.6. Occasionally, gold, silver, and other metals occur as native elements. Although this is more common for the precious metals, copper (Figure 3.30) and other metals also occur in elemental form. Ore minerals containing other elements are much more common hosts for the scarce metals. The geological origins and settings of ore minerals are extremely diverse, leading to a wide variety of mineral assemblages. The ore deposits of the Mississippi Valley lead-zinc deposits are an excellent example. These deposits are associated with sedimentary rocks in former marine basins in the eastern and midwestern United States (Figure 3.31). The geological history of these ore bodies includes dissolution and transport of ore fluids in groundwater solutions. Eventually, the minerals precipitated from solution in bodies of carbonate rocks such as limestone and dolomite. The primary ore minerals include the sulfide minerals sphalerite (ZnS) and galena (PbS). Because these minerals precipitated in large voids in the carbonate rocks, large, well-developed crystals of other minerals occur in association with the ore minerals (Figure 3.32). Large crystals of fluorite, much prized by mineral collectors, are characteristic of the lead-zinc deposits.

◀ FIGURE 3.29
The death mask of Pharaoh Tutankhamen of ancient Egypt is made of 11 kg of gold with inlays of turquoise, lapis lazuli, and carnelian (a variety of chalcedony). *Source:* Photo courtesy of the author.

Table 3.6   Examples of Ore Minerals Containing the Geochemically Scarce Metals

	Examples of Ore Minerals
Sulfide Minerals:	
Copper	Chalcocite (Cu_2S), chalcopyrite ($CuFeS_2$)
Lead	Galena (PbS)
Zinc	Sphalerite (ZnS)
Mercury	Cinnabar (HgS)
Silver	Argentite (Ag_2S)
Cobalt	Linnaeite (Co_3S_4), Co-pyrite ($(Fe,Co)S_2$)
Molybdenum	Molybdenite (MoS_2)
Nickel	Pentlandite ($(Ni, Fe)_9S_8$)
Oxide Minerals:	
Beryllium	Beryl ($Be_3Al_2Si_6O_{18}$)
Chromium	Chromite ($FeCr_2O_6$)
Niobium	Columbite ($FeNb_2O_6$)
Tantalum	Tantalite ($FeTa_2O_6$)
Tin	Cassiterite (SnO_2)
Tungsten	Wolframite ($FeWO_4$), scheelite ($CaWO_4$)
Vanadium	V in solid solution in magnetite (Fe_3O_4)
Native Metal:	
Gold	Native gold
Silver	Native silver
Platinum	Platinum-palladium alloy
Palladium	Platinum-palladium alloy
Iridium	Osmium-iridium alloy
Rhodium	Solid solution in osmium-iridium alloy
Ruthenium	Solid solution in osmium-iridium alloy
Osmium	Osmium-iridium alloy

Source: James R. Craig, David J. Vaughan, and Brian J. Skinner, *Resources of the Earth: Origin, Use, and Environmental Impact*, 3rd ed., © 2001. Reprinted by permission of Pearson Education, Inc., Upper Saddle River, N.J.

Construction and Agricultural Minerals

A number of common minerals are utilized in construction and in agriculture for fertilizer. The actual raw materials that are mined are more correctly considered to be rocks, or mineral aggregates. For example, stone used in construction of buildings, both interior and exterior, consists of rocks quarried for that purpose and used without significant alteration or treatment. In this chapter, we will discuss applications that utilize rocks consisting of one or more dominant minerals, which are extracted from the rock during processing for use in construction. Cement is by far the most important material of this type. *Portland cement*, which is the standard type of cement, uses the mineral calcite, extracted from the rock limestone, as its basic raw material. Limestones that are composed of nearly pure $CaCO_3$ are necessary for the production of Portland cement. To produce the final product, limestone from the quarry is crushed and heated to 1500°C to drive off carbon dioxide. The remaining lime (CaO) reacts during heating with alumina (Al_2O_3) and silica (SiO_2) to produce a glassy material, which is then ground to a fine powder. A small amount of the mineral gypsum is also added prior to sale of the product. When water is added to the powder, chemical reactions produce calcium and aluminum silicate compounds that harden to a material of high strength. Sand is added to Portland cement to produce mortar used in masonry construction, and sand and gravel, or *aggregate*, can be added to produce concrete, the material used for roads and buildings. The characteristics of aggregate, which are relevant to performance of the concrete, will be discussed later.

▲ FIGURE 3.30
Large clast of native copper from the Upper Peninsula of Michigan. In elemental form, copper is weak and malleable. This clast, which is approximately 0.75 m in height, was deformed during erosion and transport by a glacier. *Source:* Photo courtesy of the author.

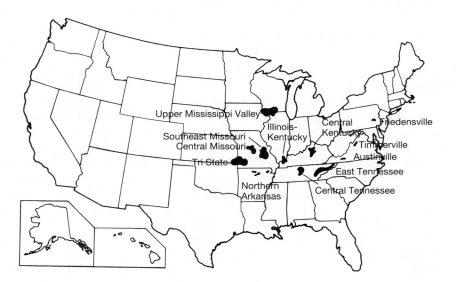

▲ FIGURE 3.31
Location of Mississippi Valley lead-zinc ore deposits in the central United States. *Source:* James R. Craig, David J. Vaughan, and Brian J. Skinner, *Resources of the Earth: Origin, Use, and Environmental Impact*, 3rd ed., © 2001. Reprinted by permission of Pearson Education, Inc., Upper Saddle River, N.J.

▲ FIGURE 3.32
Crystals of sphalerite (dark, smaller crystals), fluorite (cubes), and barite (light-colored sphere), Elmwood Mining District, Smith County, Tennessee. *Source:* Photo courtesy of the author.

Gypsum is another mineral that is essential for modern construction. Gypsum occurs in deposits of sedimentary rocks that formed by evaporation from seawater. In addition to Portland cement, gypsum is the raw material for plaster, which prior to the last century was the most common finished surface for interior walls. Plaster is produced by heating gypsum and driving off some of the water contained in the mineral structure. Today, plaster is manufactured as *plasterboard*, which is composed of plaster inserted between sheets of heavy paper. During manufacture, drying of the plaster creates solid, crystalline gypsum. Production of gypsum in the United States is widespread (Figure 3.33).

The advent of modern agriculture, with dramatic increases in crop yields, has been made possible by the widespread production of fertilizers. Of the three elements contained in fertilizer—nitrogen, phosphorous, and potassium—two are derived from mineral sources. Nitrogen fertilizers now are produced almost exclusively from atmospheric nitrogen. Phosphorous and potassium are both derived from sedimentary rocks of shallow marine origin. Of the two, phosphorous-bearing rocks are more restricted in extent. The principal mineral used for phosphorous production is *apatite,* $Ca_5(PO_4)_3(F, Cl, OH)_2$. Economically significant deposits of phosphorous formed where upwelling ocean currents saturated with phosphate moved across shallow nearshore environments where the apatite was able to precipitate. Fish bones and teeth also contain phosphate and these make a significant contribution to the amount of recoverable phosphorous in some deposits. Large deposits of potassium occur in evaporite sequences in association with calcite, halite, and gypsum (Figure 3.26). The mineral *sylvite,* KCl, is one of several potassium-bearing evaporite minerals. Fortunately, deposits of potassium and phosphorous are widely distributed in the world and the production of fertilizers should not be a problem in this century.

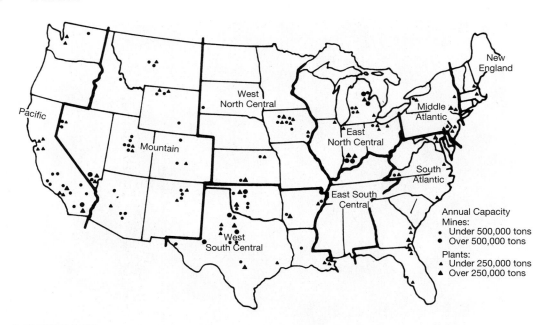

▲ FIGURE 3.33
Gypsum mines and plants in the United States. *Source:* James R. Craig, David J. Vaughan, and Brian J. Skinner, *Resources of the Earth: Origin, Use, and Environmental Impact*, 3rd ed., © 2001. Reprinted by permission of Pearson Education, Inc., Upper Saddle River, N.J.

Health Effects and Environmental Problems

Although minerals are put to an almost endless list of beneficial industrial uses, the extraction and utilization of minerals is not without problems. A major concern lies in the recognition that some minerals are harmful to human health. Chief among these is asbestos, which actually includes several specific minerals. To be called asbestos, the mineral varieties must be composed of a fibrous structure. The most abundantly used asbestos mineral is *chrysotile*, which is a 1:1 sheet silicate mineral that resembles the structure of kaolinite (Figure 3.19). The major difference between the two is that in chrysotile the central ions in the octahedral layer are magnesium. The fibrous structure of chrysotile develops when long sheets roll up, much like a roll of wrapping paper. The other main type of asbestos is represented by the mineral *crocidolite*, which is a member of the amphibole group of double-chain silicate minerals. In crocidolite, the fibers are sharper and needlelike in structure. The properties of asbestos that make it so useful as a construction material include its high tensile strength and its heat resistance. It is an important component in products such as roof shingles, floor and ceiling tile, insulation, brake linings and many, many others. The health effects of asbestos arise when the fibers are inhaled as dust, usually during the manufacturing of asbestos products. The fibers of crocidolite and similar minerals become trapped in the lungs and cannot be broken down by the body. The resulting disease, asbestosis, is deadly in itself and is also associated with lung cancer.

In the 1980s, the recognition of the dangers of asbestos led to its banning from many products and triggered its removal from certain types of buildings, for example, schools. The remodeling or demolition of old buildings became extremely expensive because removal of asbestos requires specialized workers wearing protective clothing and disposal is costly because it must be treated as a hazardous waste. Ironically, this huge effort, costing as much as $100 billion annually in the United States, may have been mostly an overreaction. There has been no evidence that humans exposed to chrysotile asbestos, which makes up more than 90% of the total amount used in building materials, have incurred

▲ FIGURE 3.34
Asbestiform amphibole fibers from Libby Montana. *Source:* Photo courtesy of Greg Meeker, U.S. Geological Survey.

any significant health effects. Asbestos workers appear to be by far the people most significantly threatened by these minerals. Heath problems from dust inhalation are not limited to asbestos. Prolonged breathing of silica, either in crystalline or amorphous form, is responsible for silicosis, a disease of the respiratory tract that is also linked to lung cancer. Silica, which is derived from deposits of sand or sandstone, is widely used in industry for foundry sand, abrasives, filters, and many other purposes.

Mining or extraction of minerals is associated with a wide range of environmental problems. Many of these problems are caused by either the processing of ore minerals or the disposal of waste rock produced during mining. Sulfide ore bodies mined for metallic ores are a particular concern. When exposed to air and water, the reduced sulfur in minerals such as pyrite becomes oxidized, and the oxidized sulfur reacts to form sulfuric acid. Acid mine drainage, which occurs in old underground mines or surface mines, can be a very serious problem, affecting streams, lakes, and groundwater. Waste rock materials derived from mining are called *tailings*, and they are emplaced in huge piles at mine sites or slurried to ponds or lagoons. Because of the huge volume of rock involved, problems can persist for a very long time. The Berkeley Pit, a huge surface mine pit in Butte, Montana, was once mined for rich sulfide ore deposits. After it was abandoned, it slowly began to fill with groundwater and is now a grossly contaminated lake with a pH of 2.7.

High concentrations of metals occurring in soils or air are most likely located near ore *smelters*, where the ore is recovered from the source rock. A range of health problems have been attributed to smelter emissions. In elemental form, after extraction from their respective ore minerals, metals are hazardous in some circumstances. Good examples include lead in paints and gasoline, which can cause health problems if ingested in large enough quantities. Leaded gasoline is now banned in the United States for this reason. Metals disposed of in landfills or waste lagoons can be leached into surface water or groundwater where they can be intercepted by wells or water supply intakes. Strict environmental laws, especially in the developed countries, have made tremendous strides in preventing and mitigating these problems, but some degree of environmental degradation from mineral production is inevitable.

Summary and Conclusions

The rocks of the earth's crust are aggregates of minerals, which are inorganic crystalline solids with a chemical composition that is fixed or varies within fixed limits. Within each mineral, atoms and ions are arranged in a specific structural pattern that determines the properties we can observe on the exterior surfaces of each sample.

When a mineral slowly crystallizes in unobstructed surroundings, crystals form. The smooth, planar crystal faces reflect the internal structure of the mineral and, for a specific mineral, maintain uniform angles between corresponding faces on crystals of any size. The conditions necessary for the formation of crystals are rare, so, instead, minerals usually occur as irregular masses in rocks because of competition for space with other minerals during crystallization.

Identification of most common rock-forming minerals is possible by using observation and simple tests to determine the physical properties of the specimen. The most useful physical properties include color, streak, luster, cleavage, hardness, and specific gravity.

Composition is the main criterion for dividing minerals into classes of chemically similar minerals. The basic structural unit used in classification is the major anion or anion group in the mineral. Silicates, constituting the most abundant group, are defined by the silica tetrahedron, a structural unit of four oxygen anions and one silicon cation. Subgroups of silicates are formed by different arrangements and linking of silica tetrahedra in the structure. Oxides, halides and sulfides, sulfates and carbonates, and native elements are the classes of nonsilicate minerals.

Minerals are indispensable to human society. Metallic ore minerals, in which the metallic element occurs in economically extractable concentrations, are found in an extremely wide variety of geologic settings. Sources of the abundant metals are sufficiently widely distributed that supply will not be a major problem in this century. Other metals critical to industry, however, are much more limited in their occurrence. Mining and extraction of minerals pose an environmental threat and a health hazard in the case of asbestos and other minerals. Mineral deposits of phosphorous and potassium are adequate for agricultural needs for the foreseeable future.

Problems

1. List the characteristics of each major atomic particle.
2. How do ionic bonds and covalent bonds differ?
3. Why are only crystalline solids included in the definition of a mineral?
4. How does the internal structure of a mineral relate to its external appearance?
5. What is mineral cleavage? How could cleavage affect the engineering properties of a rock composed of minerals with strong cleavage?
6. For which common minerals is hardness a very useful diagnostic physical property?
7. Why are the silicate minerals so important?

8. Briefly describe the structure of clay minerals. What is it about these minerals that makes their presence in soil so important to engineering?

9. What is cation exchange?

10. If you allow a bucket of seawater to totally evaporate, various minerals will precipitate. Into which mineral groups do you think most of these minerals will fit?

11. Find five industrial uses for each of the abundant metals.

12. Why are the base metals so important in human history?

13. What elements are needed for agricultural fertilizers? Discuss the occurrence of the two that are extracted from minerals.

14. Why is asbestos considered to be hazardous? Is the massive effort to remove asbestos from buildings really necessary?

References and Suggestions for Further Reading

CHANG, L. L. Y. 2002. *Industrial Mineralogy: Materials, Processes and Uses*. Upper Saddle River, N.J.: Prentice Hall, Inc.

CRAIG, J. R., D. J. VAUGHAN, and B. J. SKINNER. 2001. *Resources of the Earth*, 3rd ed. Upper Saddle River, N.J.: Prentice Hall, Inc.

GRIM, R. E. 1968. *Clay Mineralogy*, 2nd ed. New York: McGraw-Hill.

HAY, A. E., and A. L. MCALESTER. 1984. *Physical Geology: Principles and Perspectives*, 2nd ed. Englewood Cliffs, N.J.: Prentice-Hall, Inc.

JUDSON, S., M. E. KAUFFMAN, and L. D. LEET. 1987. *Physical Geology*, 7th ed. Englewood Cliffs, N.J.: Prentice Hall, Inc.

MASON, B. 1958. *Principles of Geochemistry*. New York: John Wiley.

TENISSEN, A. C. 1983. *Nature of Earth Minerals*. Englewood Cliffs, N.J.: Prentice Hall, Inc.

CHAPTER 4

Igneous Rocks and Processes

Rocks that crystallized below the earth's surface from magma or at the earth's surface from lava fall into the most fundamental category of rocks, the *igneous* rocks. In order to understand the formation of these rocks, we will first discuss igneous processes. When magma reaches the surface, a wide array of possible phenomena occur that can be grouped under the umbrella of volcanism. These processes include not only the formation of the rocks themselves by the cooling of lava but the often hazardous related processes that are characteristic of volcanoes. Throughout history, volcanic eruptions have posed one of the great natural risks facing the human population. These risks persist today, even with our extensive knowledge of volcanic processes. In fact, increasing development encroaching upon potentially active volcanoes has increased the risk to uninformed citizens of countries that cannot afford to do the basic science necessary to understand and predict potential hazards.

From a practical standpoint, a basic understanding of how rocks are formed and how their properties may influence various engineering projects is extremely important. The consequences that result from failure to understand rocks and their properties will be a recurrent theme in this book. Thus, although it may seem academic to study the processes of rock formation, an understanding of these processes must be applied in every major rock-engineering project.

Volcanism

Volcanism is the most obvious igneous process. Major eruptions generate worldwide interest because of their spectacular power and beauty and, as Mount Pinatubo in the Philippines and Mount St. Helens have recently proven so clearly, their danger and destructiveness. In addition to the areas of the earth that have been affected by volcanic activity in historic times, vast regions, including all the ocean basins, are underlain by volcanic rocks that were erupted earlier in geologic time.

Volcanoes and Plate Tectonics

One of the most important applications of the theory of plate tectonics is its use in explaining the distribution of active volcanoes on the earth's surface. Most of these are concentrated in narrow bands along the edges of lithospheric plates. Divergent plate margins are characterized by voluminous outpourings of basaltic lavas at midoceanic ridges (Figure 4.1). These lavas form the foundation of new oceanic crust that fills the void left as the plates move away from each other. Convergent plate margins (Figure 4.1) also generate volcanic activity. Here, the descending plate and/or the rocks above the descending plate in a subduction zone are partially melted. The magmas formed at depth migrate to the surface to form chains of volcanic mountain ranges, such as the Cascades of the northwestern United States, or to form island arcs composed of volcanic accumulations on the sea floor that rise above sea level. The Aleutian Islands of Alaska form an island arc adjacent to a subduction zone. The volcanic rocks of convergent plate margins are usually intermediate in composition, as defined by their silica content. Andesite, named for the Andes Mountains of South America, is a common volcanic rock at convergent plate margins. The location of the major subduction zones, as well as the midoceanic ridges, are shown on Figure 4.2.

Volcanism is not restricted to plate margins. A number of volcanic areas occur at some distance from plate margins. The volcanic activity is attributed to localized zones of high heat flow extending downward, deep into the mantle. These *mantle plumes* are assumed to remain in fixed positions as the plates above move along their current paths. As long as the plate movement is in a straight line, a linear chain of volcanoes is produced. The Hawaiian Ridge–Emperor Seamount chain records the passage of the Pacific plate over the Hawaiian hot spot (Figures 4.3 and 4.4). The bend in the ridge indicates a change in direction of plate movement from northward, the orientation of the Emperor Seamount chain, to northwestward, the bearing of the Hawaiian Ridge. Seamounts in the ridge and seamount chain are extinct volcanoes that have subsided below sea level. The Hawaiian Islands themselves are located near and over the hot spot. Individual islands increase in age from Hawaii, which lies over the plume and is the location of the active volcanoes

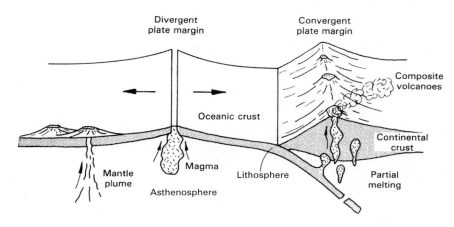

▲ FIGURE 4.1
Volcanism is common at both divergent and convergent plate margins. Fissure eruptions produce new oceanic crust at divergent margins; composite volcanoes of andesitic composition are common at convergent margins, where partial melting of the crust occurs.

▲ FIGURE 4.2
Midoceanic ridges are built by volcanic action at divergent plate margins. Subduction zones are linked to volcanic chains in island arcs or on continental plates.

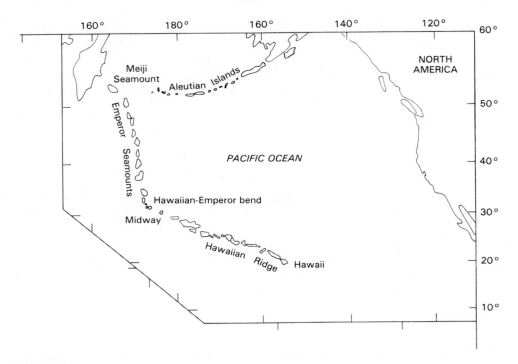

▲ FIGURE 4.3
Map of Hawaiian Ridge–Emperor Seamount chain.

▲ **FIGURE 4.4**

Diagram of Pacific Plate moving over hot spot to form Hawaiian islands. *Source:* From R. I. Tilling, C. Heliker, and T. L. Wright, 1987, *Eruptions of Hawaiian Volcanoes: Past, Present and Future*, U.S. Geological Survey; based on a drawing by the late Maurice Krafft.

Mauna Loa and Kilauea, to Kauai at the northwest end of the chain. The size of these volcanoes is truly impressive. Mauna Kea on the island of Hawaii, for example, rises 4200 m above sea level but extends another 6000 m below sea level to the depth of the seafloor adjacent to the ridge. Therefore, the total height of the volcano, from the seafloor to the summit, makes it the largest mountain on the earth. A comparison of the total size of Mauna Loa to some land-based volcanoes is shown in Figure 4.5. Hot spots are not limited to ocean basins. The Snake River Plain in southern Idaho and the Yellowstone volcanics are related to the migration of the North American Plate over a hot spot.

Eruptive Products

Volcanoes erupt quite a wide variety of materials. Aside from the gases mentioned earlier, eruptive products generally can be divided into *lava* and *pyroclastic* material. Lava, the name for magma after it reaches the surface, flows downslope away from the eruptive vent at a rate dependent on its temperature and silica content. Basaltic lavas, which are

▲ **FIGURE 4.5**

Hot-spot volcanoes in the Hawaiian Islands, when measured from their base on the sea floor to their tops, dwarf volcanoes on land *Source:* John D. Winter, *Introduction to Igneous and Metamorphic Petrology, An*, 1st ed., © 2001, p.48. Reprinted by permission of Pearson Education, Inc., Upper Saddle River, N.J.

Table 4.2   Human Fatalities from Volcanic Hazards in Two Periods of Time. The average fatalities per year increased in the 20th century, although the relative contribution of different types changed from the previous three centuries.

Primary Cause	1600–1899		1900–1986	
Pyroclastic flows and debris avalanches	18,200	(9.8%)	36,800	(48.4%)
Mudflows (lahars) and floods	8,300	(4.5%)	28,400	(37.4%)
Tephra falls and ballistic projectiles	8,000	(4.3%)	3,000	(4.0%)
Tsunami	43,600	(23.4%)	400	(0.5%)
Post-eruption starvation, disease, etc.	92,100	(49.4%)	3,200	(4.2%)
Lava flows	900	(0.5%)	100	(0.1%)
Gases and acid rains	—		1,900	(2.5%)
Other or unknown	15,100	(8.1%)	2,200	(2.5%)
TOTALS	186,200	(100%)	76,000	(100%)
Fatalities per year (average)	620		880	

Source: W. E. Scott, 1989, Volcanic and related hazards, in R. I. Tilling, ed., *Volcanic Hazards*, American Geophysical Union.

Lava Flows and Domes

In rare occurrences, thin, very fluid lava flows have taken human lives. More commonly, however, residents of threatened areas have time to evacuate prior to release of a lava flow into an occupied area. Lava flows are a definite threat to human property, including roads, houses, and other structures. The island of Hawaii is an excellent example of lava flow hazards, although other types of volcanic hazards are low in Hawaii owing to the low-volatile content of basaltic magmas and lavas. The island is constructed of five volcanoes, two of which—Mauna Loa and Kilauea—are currently active. Kilauea is the most active because it is more directly above the hot spot as the Pacific Plate moves northwesterly. An even younger volcano is building up below sea level off the coast of the island south of Kilauea; it may add to the area of the island at such time when it emerges.

The lava flow hazards on the island of Hawaii are shown in Figure 4.30. Kilauea, which has been erupting intermittently from vents along its East Rift since 1982, is studied continuously by scientists at the Hawaiian Volcano Observatory run by the U.S. Geological Survey. Lava flows from these events have occasionally flowed into developed areas and caused significant loss of property (Figures 4.31, 4.32).

Pyroclastic Density Currents

Pyroclastic density currents (Table 4.2) include a range of related phenomena involving the rapid downslope movement of pyroclastic debris and gases. They occur at nearly every composite volcano because of the explosive eruptions and high pyroclastic content. As mentioned earlier, two basic types of density currents can be identified: *pyroclastic flows* and *pyroclastic surges*. Flows have a high proportion of pyroclastic debris relative to gas and are therefore denser. They range in temperature between 300°C and 800°C and attain velocities up to several hundred meters per second. Because of their higher density, they tend to hug the ground and follow existing valleys. Pyroclastic surges are density currents with a lower density and higher gas content. They can be "cold" (less than the boiling point of water) if they have a high content of water vapor derived from groundwater, surface water, or snowmelt. Hot surges constitute the most deadly hazard associated with pyroclastic eruptions. Thousands of people have been killed by incineration and asphyxiation from hot pyroclastic surges. Because of their lower density, surges

▲ FIGURE 4.30
Hazards map of the island of Hawaii. Relative hazard from lava flows ranges from 1 (high) to 9 (low). The thin, solid line marks the boundary between the Mauna Loa and Kilauea volcanoes. The stipple pattern indicates areas covered by pre-1975 historical lava flows. Nearly 30% of the land surface in hazard zone 2, south of the East Rift Zone of Kilauea, has been covered with lava flows since 1955. *Source:* From D. R. Mullineaux, D. W. Peterson, and D. R. Crandall, 1987, U.S. Geological Survey Professional Paper 1350.

can move faster than flows and are not confined by topography. The rapid velocity of these events makes it impossible to escape once they have started. Pyroclastic surges are genetically related to pyroclastic flows and can separate from the denser flows during the event (Figure 4.33).

Directed blasts are very similar to pyroclastic flows and surges, except that a surge moves along the path of the directed blast rather than downslope from the top of the volcano. The most famous example is the lateral blast from the 1980 eruption of Mount. St. Helens. This event is described in Case In Point 4.1. Directed blasts represent a hazard that can occur anywhere around the volcano, without regard for topography.

▲ FIGURE 4.31
Distribution of lava flows from the Pu'u O'o eruption from the East Rift Zone of Kilauea between 1983 and 2003. Flows covered parts of the Royal Gardens subdivision and a major coastal road. *Source:* Hawaiian Volcano Observatory, U.S. Geological Survey.

Lahars

When pyroclastic debris comes in contact with water, snow, or ice on the slopes of a volcano, a high-density slurry can be generated. The resulting lahars move downslope along the paths of existing river valleys. Velocities vary greatly but most do not exceed

▲ FIGURE 4.32
Lava flow covering road along south coast of Hawaii. *Source:* Photo courtesy of the author.

10–20 m/sec. These velocities, much lower than pyroclastic flows and surges, along with easily predicted paths, allow time for evacuation if the threat is recognized, evacuation procedures are in place, and warnings are issued. The lahar hazard is increased, however, because lahars can travel great distances, sometimes as much as several hundred kilometers.

When warnings are not given or heeded, the consequences can be deadly. By most standards, the November 1985 eruption of Nevado del Ruiz in Columbia was not major. The volume of magma erupted was only about 3% of the Mount St. Helens eruption. Lahars generated on the snow-covered peak, however, buried the town of Armero (Figure 4.34), killing 23,000 people in the process. The most tragic aspect of this eruption is that the destruction of Armero could perhaps have been avoided. A hazards map published just one month before the eruption clearly indicated the potential danger for the town. Even more lamentable is that there was a historical record of lahar damage in Armero. Mudflows similar to those of 1985 had inundated the town twice before, in 1595 and 1845, with hundreds dead in both cases. Unfortunately, natural hazard prediction and mitigation is not the highest priority in underdeveloped countries.

A much more active response undoubtedly saved thousands of lives during the 1991 eruption of Mount Pinatubo in the Philippines. This eruption is now thought to be one of the largest eruptions of the century. Small, precursor eruptions and seismic activity began in April and May 1991. A geologic study of the volcano and the surrounding area was begun immediately. This investigation showed that pyroclastic flows were widely distributed around the base of the volcanic cone. As the activity intensified, about 50,000 people were evacuated in stages, including 14,000 personnel and dependents from Clark Air Base, a U.S. Air Force facility. On June 12, a massive explosion occurred, blasting a column of ash to a height of 19,000 m above sea level. Three days later, on June 15, a lateral blast sent huge pyroclastic flows in all

▲ FIGURE 4.33
Formation of a pyroclastic flow and surge from the same original event. (a) Orginal pyroclastic densi-ty current. (b) Gradual separation of flow (dark stipple) and surge (light stipple) by density. (c) The surge outruns the flow and begins to rise as sediment is being deposited from it and the density de-creases. (d) The pyroclastic flow overtakes the surge. This sequence can be repeated if the flow accel-erates. *Source:* W. E. Scott, Volcanic and related hazards, *Volcanic Hazards,* © 1989, American Geophysical Union. Reproduced by permission of American Geophysical Union.

directions down the volcano (Figure 4.35). After this event, the total number of evacuees was pushed to 200,000 people, the largest number ever evacuated for a volcanic eruption. By an unfortunate coincidence, a typhoon was moving through the area at the same time as pyro-clastic flows were moving across the landscape. Torrential rains soaked the ash, making it much heavier. The distribution of pyroclastic flow and lahar deposits is shown in Figure 4.35. The loss of life from the eruption—about 300 people were killed—was immeasurably less be-cause of the ability and willingness to respond to the hazard effectively.

Structural Collapse

Large volcanic cones are inherently unstable masses. Heterogeneous eruptive products may accumulate very rapidly on the steep slopes of the cone. These rocks and sediments may also be subjected to very rapid weakening by weathering processes that are unique to volcanoes. Because most composite volcanoes are very high, they receive substantial

▲ FIGURE 4.34

Hazards map of Nevado del Ruiz, Columbia, published one month before the 1985 eruption. Distribution of mudflows from the eruption are shown. *Source:* From T. L. Wright and T. C. Pierson, *Living with Volcanoes, The U.S. Geological Survey's Volcano Hazards Program*, U.S. Geological Survey Circular 1073.

orographic precipitation and may contain extensive snowfields and glaciers, even near the equator. As rain and snowmelt infiltrate into the rocks of the volcano, they are warmed because of the high heat flow emanating from the magma chamber. This warm water reacts rapidly with minerals in a process called *hydrothermal alteration*. Feldspars and other primary minerals can be converted to clays, thus weakening the rocks that host them. Other conditions on a volcano are ideal for massive collapses—the slopes are steep and rising magma prior to eruptions can produce earthquakes similar to the sequence of events at Mount St. Helens leading up to the lateral blast. The collapse and downslope movement of huge sections of the mountain under these conditions is not unexpected.

A wide variety of slope movements results from structural collapse. One of the most common is the debris flow or avalanche. These are downslope movements of rock debris at highly variable velocities and with highly variable amounts of water. They are generally limited to the volcano itself, but if water is present, the debris can mobilize into fluid lahars that move much farther.

Tephra Falls and Ballistic Projectiles

Hazards of various types accompany tephra falls from volcanic eruptions. Pyroclastic particles are classified according to size as ash (<2 mm), lapilli (2–64 mm), and blocks or bombs (>64 mm). Bombs are hot enough to be viscous at the time of impact. Blocks and bombs pose an obvious threat of injury or death; clasts this size can be carried for as much as 5 km from the vent. A tragic event that epitomizes the notion of bad timing occurred at the active Columbian volcano Galeras in 1993. Many of the world's top volcanologists were attending a workshop on volcanic hazards in the nearby city of Pasto. On one of the field trips at the conference, a group of 12 scientists, journalists, and tourists descended into the crater of the volcano to sample gas emission. By all indications of activity, the volcano appeared to be in a

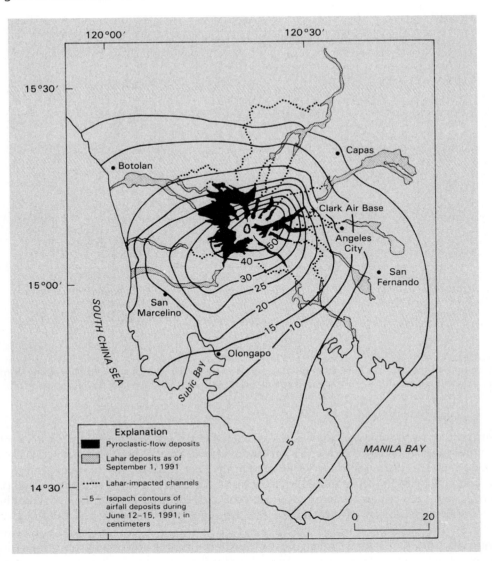

▲ FIGURE 4.35
Map showing pyroclastic and lahar deposits from the 1991 eruption of Mount Pinatubo. Contours show thickness of airfall deposits. *Source:* From E. W. Wolfe, The 1991 eruptions of Mount Pinatubo, Philippines, *Volcanoes and Earthquakes*, 23:5–37.

period of dormancy and the short trip into the crater was not expected to pose more than the normal risks of this dangerous profession. Unfortunately, the volcano began a minor eruption at the very time that the party was beginning its ascent out of the crater. Bombs up to a meter or more in size began to rain on to the floor and sides of the crater. Six members of the party were killed and five more were injured.

Beyond the range of the larger pyroclastic particles, the finer tephra falls have many deleterious effects. Tephra accumulations on roofs can cause collapse, particularly when it becomes wet and waterlogged. This occurred with the Mount Pinatubo eruption, and many of the deaths from the eruption occurred by collapse of roofs buried beneath thick ash accumulations (Figure 4.36). The total accumulations of tephra from the eruption are shown in Figure 4.35. Ash in the air above and downwind from eruptions is damaging to jet engines,

▲ FIGURE 4.36
Thick accumulation of pyroclastics near Mount Pinatubo. *Source:* R. P. Hoblitt, U.S. Geological Survey.

and several flights almost crashed while flying through the tephra plume from Pinatubo. Ash falling on land can damage auto engines, start fires, and kill livestock and vegetation. The ash particulates also cause respiratory problems for humans outside during the airfall.

Volcanic Gases

In addition to water vapor, the most abundant volcanic gas, more hazardous gases are emitted, including carbon dioxide, carbon monoxide, sulfur oxides, hydrogen sulfide, chlorine, and fluorine. Carbon dioxide is denser than air and undetectable by humans. Several well-known releases of carbon dioxide have caused many deaths to humans and animals by asphyxiation. The most famous example of gas release occurred at two lakes in the west African country of Cameroon in the mid-1980s. The lakes involved are located in volcanic maars that are less than 1000 years old. Gases that rise through the volcanic vent conduits accumulate in the bottom of these deep lakes, one of which is called Lake Nyos. The gas, much of which is carbon dioxide, tends to build up at the bottom of the lakes, perhaps because lakes in the tropical climate typical of Cameroon are not subject to yearly overturn as are lakes in temperate climates. On August 21, 1986, Lake Nyos released a large quantity of carbon dioxide as a giant bubble, which caused a splash as high as 100 m on the steep walls of the crater. The mechanism causing the sudden rise of gases from the bottom of the 208-m-deep lake is still not definitely known. At any rate, the gas release formed a lethal cloud of CO_2 that flowed as a density current near the ground surface down several river valleys (Figure 4.37). The momentum of the gas cloud was sufficient to flatten stalks of corn due to the fact that CO_2 is 1.5 times denser than air. When the gas encountered several villages along its path, people were overwhelmed and killed by asphyxiation. The survival rate was less than 1% in the village of Nyos. In all, 1746 people and 8300 livestock were lost. Deaths were caused as far as 23 km from Lake Nyos. A similar event of this type at a nearby lake occurred in 1984, although with a much lower loss of life. This indicates that gas releases must be considered a continuing threat in Cameroon and in other tropical volcanic regions.

A solution to the threat of future gas eruptions of Lake Nyos has been tested and is now in operation. The solution is based on the principle of a self-siphon (Figure 4.38), in which a pipe is installed from the lake bottom to its surface. The self-siphon is initiated by pumping water from the pipe, which lowers the pressure on the gas at depth to the point that it becomes saturated. Bubbles will then form and rise to the surface and, at this point,

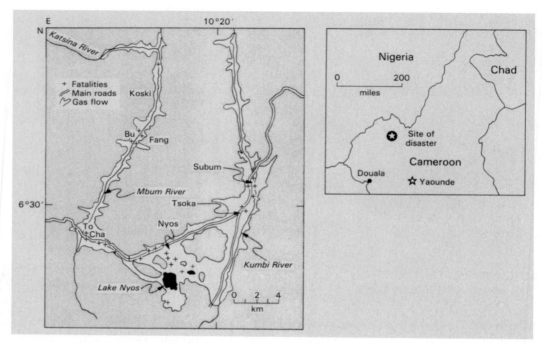

▲ FIGURE 4.37
Map of the Lake Nyos area, showing areas covered by the gas cloud. *Source:* From *Eos, Transactions, American Geophysical Union*, June 9, 1987.

◄ FIGURE 4.38
The principle of a self-siphon to put into operation a controlled release of gas from Lake Nyos. 1. Pumping water to initiate self-siphon. 2. Self-sustained release of gas bubbles. *Source:* From M. Halbwachs, 2001, http://perso.wanadoo.fr/mhalb/nyos/index.htm.

the pump is no longer needed. This technology is probably a controlled version of the spontaneous gas release. Implementation of this technology would keep gas contents in the lake below the level at which spontaneous releases would occur.

Volcanic Earthquakes, Atmospheric Shock Waves, and Tsunamis

Earthquakes are commonly measured at volcanoes as magma moves into the volcano. These events are carefully monitored by scientists to help understand the internal conditions of the volcano and to attempt to predict eruptions. The earthquakes generally are small in magnitude and do not cause extensive damage. The release of huge amounts of pyroclastic material can generate an atmospheric disturbance that propagates outward from the volcano. Damage to structures can occur within several hundred kilometers of the mountain. Tsunamis are huge waves generated in the oceans by sudden movement of faults or various volcanic processes such as structural collapse. These waves travel through the ocean basins at high velocities and cause extensive damage to low-lying coastal areas. The Indian Ocean tsunami of December 2004 provides a chilling example of these phenomena (Chapter 15).

Case in Point 4.1
Effects of a Major Eruption and Predictions for Future Eruptions: Mount St. Helens and Mount Rainer

The Cascade Range in the northwestern United States is a continental volcanic arc related to subduction of the Juan de Fuca Plate (Figures 4.19, 4.23). The active volcanoes in this chain pose a very serious geologic hazard to the region, especially as the development and population growth in the major cities of the region expand outward toward the mountains. The main volcanoes in Washington and Oregon, along with their hazard zones, are shown in Figure 4.39. There is no better illustration of these hazards than the 1980 eruption of Mount St. Helens. We will review the effects of this event and then apply the lessons learned to an equally dangerous, but currently quiescent, volcano, Mt. Rainier.

The eruption of Mount St. Helens on May 18, 1980 terminated a 123-year period of inactivity. Preliminary signals, including earthquake activity, minor eruptions, and the formation of a huge bulge on one side of the mountain, preceded the eruption for months and left no doubt that a major event was possible at any time. A huge effort was made to evacuate the nearby population to safety, but the eruption began in an unpredicted fashion—the north side of the mountain gave way in a massive debris avalanche that buried the upper part of the North Fork of the Toutle River valley beneath 3 billion m^3 of rock, ash, pumice, snow, and ice (Figure 4.40). The sudden release of pressure caused by the debris avalanche unleashed a lateral blast of hot gases and ash. This unexpected directed blast of pyroclastic material devastated the forests to the north of the mountain for a distance of 25 km and affected a total area of 650 km^2. Fifty-seven people were killed, mainly because the lateral blast had not been anticipated. Trees within the blast zone were blown over like toothpicks (Figure 4.41). Subsequent to the lateral blast, ash was erupted vertically into the atmosphere, where it began to drift eastward across the United States as a dense cloud.

Among the most damaging long-term effects of the Mount St. Helens eruption were lahars mobilized from the debris avalanches; within hours after these huge landslides came to rest, waves of fluidized debris moved down the major rivers draining the volcano. The largest lahar swept down the North Fork of the Toutle River into the Cowlitz River, and eventually into the Columbia River (Figure 4.42). The channels of the Toutle and Cowlitz Rivers were filled by the surging mudflows, which occasionally overflowed

▲ FIGURE 4.39
Delineation of hazard zones for the Cascade volcanoes in Washington and Oregon. Circles indicate proximal hazard zones with a radius of 50 km. Hazards that can occur within this zone include structural collapse, debris avalanches, pyroclastic flows and surges, and tephra falls. The shading shows lahar and flooding hazard areas in valleys downstream from the volcano. *Source:* From P. T. Pringle, 1994, Volcanic hazards in Washington—A growth management perspective, *Washington Geology*, 22 (no. 2):25–32.

onto the floodplain, where roads and buildings were buried by the debris (Figure 4.43). In all, 27 bridges and nearly 200 homes were damaged by the lahars. The blast and lahars together destroyed more than 185 miles of highways and roads and 15 miles of railways.

Deposition within the channel itself was also a serious problem. In some places, the capacities of the channels were reduced to only one-tenth of their preeruption value. Thus, the threat of flooding was increased to a much greater degree than before the eruption. Soon after the eruption, flood-abatement programs were initiated. These projects included dredging and levee improvement in flood-prone areas.

▲ FIGURE 4.40
Debris-avalanche deposits on the north flank of Mount St. Helens and extending into the upper por-
tions of the North Fork Toutle River valley. Deposits are about 200 m thick in the area shown.
Source: R. M. Krimmel; photo courtesy of U.S. Geological Survey.

▲ FIGURE 4.41
Blown-down timber in the Green River valley from the Mount St. Helens eruption. *Source:* Washington
Dept. of Natural Resources.

▲ FIGURE 4.42
Map of the Mount St. Helens region showing the extent of debris avalanches and lahars.
Source: Modified from U.S. Geological Survey Circular 850-B.

▲ FIGURE 4.43
Building partially buried by lahar deposits, North Fork Toutle River valley. *Source:* R. L. Shuster; photo courtesy of U.S. Geological Survey.

(a)

(b)

▲ FIGURE 4.44
Mount St. Helens before (a) and after (b) the 1980 eruption. *Source:* Courtesy Cascade Volcano Observatory). (a) Photo by Richard P. Hoblitt. (b) Photo by Lyn Topinka.

The net effect of the Mount St. Helens eruption on the mountain itself was a reduction in elevation from 2950 m before to 2550 m after (Figure 4.44). This eruption created a new urgency to understand the hazards of the Cascade volcanoes and to protect the rapidly growing population in the lowlands below these volcanoes. Recent activity in Mount St. Helens in 2004 and 2005 has again focused interest on this dangerous mountain.

Mount Rainier is the highest and potentially most dangerous volcano in the Cascades, mainly because of the large and rapidly growing population that lives within reach of lahars and other hazards (Figure 4.45). In recent decades, especially after the eruption of Mount St. Helens, research into the geological history of Mount Rainier has intensified. Because of its great height, Mount Rainier is covered with glaciers and snowfields. Snow and ice provide the water necessary for the formation of lahars that can travel long distances. Mount Rainier also has extensive hydrothermal alteration, which weakens the rock and leads to large-scale structural collapses. Lahars that have a high clay content, which can be inherited from hydrothermally altered rocks, have been shown to travel farthest. These two factors make lahars the most feared hazard for this volcano.

Mount Rainier has not erupted for about 150 years. Despite this period of dormancy, recent research has shown that the volcano has experienced between five and seven major eruptions in the past 5000 years of geologic time. Each of these eruptions produced a structural collapse of part of the cone and generated a clay-rich lahar (Table 4.3). The distribution of the deposits of three of these lahars is shown in Figure 4.46. Today, more than 100,000 people live on the deposits of these flows; because lahars are topographically confined, these same areas will again be targeted by lahars, depending on which side the mountain fails in the eruption. The largest of recent lahars was the Osceola mudflow. This lahar was

▲ FIGURE 4.45
Mount Rainier with Tacoma, Washington in the foreground. *Source:* Lyn Topinka; photo courtesy of U.S. Geological Survey.

Table 4.3 Characteristics of the Most Recent Lahars Originating from Mount Rainer. The clay content of these flows indicates that structural collapse was involved in the respective eruptions.

Flow	Clay (percent)	Age (yr, before present)	Drainage	Volume km³	Extent
Central part of Tahoma lahar	3–4	<500	Tahoma Creek; Nisqually River	<0.15	Probably to Elbe
Electron mudflow	6–11	530–550	Puyallup River	0.26	Puget Sound lowland
1000-yr-old lahar	5–12	1050–1000	Puyallup River	Possibly >0.30	At least to Mowich River; possibly to Puget Sound lowland
Unnamed lahar (possibly same as Round Pass mudflow).	4–5	Same as below?	Puyallup River	—	Puget Sound lowland
Round Pass mudflow (main part)	4–5	2170–2710	Puyallup River	—	Puget Sound lowland
Osceola mudflow (probably includes Greenwater lahar)	2–15	4500–5000	White River (main fork and West Fork)	3	Puget Sound lowland
Greenwater lahar (probably part of Osceola mudflow)	<3	—	Main fork White River	—	Puget Sound lowland

Source: K. M. Scott, J. W. Vallance, and P. T. Pringle, 1995, *Sedimentology, behavior, and hazards of debris flows at Mount Rainier, Washington,* U.S. Geological Survey Professional Paper 1547.

▲ **FIGURE 4.46**
Extent of three large lahars from Mount Rainier (dark stippling); the Osceola mudflow (White River), the Paradise lahar (Nisqually River), and the Electron mudflow (Puyallup River). *Source:* National Research Council, 1994

mobilized by a structural collapse of the volcano between 4500 and 5000 years ago during a major eruption. The Osceola mudflow traveled nearly 100 km, into the area that is now part of metropolitan Tacoma, Washington. Recent estimates put the volume of the mudflow at more than 3 km^3. By comparison, the lahar that killed 23,000 people in Armero, Columbia had only about 3% of this volume. An eruption equal to the one that produced the Osceola mudflow would dwarf the damages of the 1980 Mount St. Helens eruption. The broad valleys leading up to Mount Rainier are much more attractive sites for development than the steeper topography between the valleys (Figure 4.47). While adequate warnings and evacuation procedures could perhaps save most lives, the property damage would be unprecedented. As more time passes before this eruption, the hazard will only increase as population growth continues.

▶ FIGURE 4.47
New house under construction in subdivision in Orting, Washington, with Mount Rainier in the background. In the foreground, a spruce stump projects upward from the storm drainage basin. Spruce trees in this area were buried by the Electron mudflow about 500 years ago. *Source:* Photo courtesy of the author.

Intrusive Processes

Intrusion refers to the movement of magma from a magma chamber to a different subsurface location. Bodies of rock formed by the intrusion of magma are called *plutons*. Rocks that make up plutons usually have phaneritic texture because the cooling time was sufficient to allow the formation of large crystals. It is only after erosion has removed the overlying rocks that we are able to observe intrusive rocks at the earth's surface.

Types of Plutons

Plutons differ in terms of size, shape, and composition from the rocks that were intruded by the magma, which are older rocks known as *country rocks*. A major group of plutons is classified as *tabular* because they are thin in one dimension as compared with the other two dimensions (Figure 4.48). If a tabular pluton is roughly parallel to the layering of the country rocks, it is said to be *concordant* and is called a *sill. Discordant* plutons cut across the layering of the country rock. *Dikes* (Figure 4.49) are vertical or steeply dipping tabular igneous bodies that fill planar cracks through which magma was moved to the surface in fissure eruptions. At the close of the eruption, the magma remaining in the crack connecting the magma chamber with the fissure solidifies to become a dike.

Nontabular plutons include *laccoliths, stocks*, and *batholiths*. The shapes of these plutons are thick as well as broad. A laccolith is a relatively shallow concordant pluton that causes doming of the overlying country rocks (Figure 4.50). The rocks above are bowed upward during the intrusion of the magma. Nontabular, discordant plutons are called stocks if their surface exposure covers an area of less than 100 km^2 and batholiths if they are exposed over a larger area. Batholiths are truly immense bodies of rock that form the cores of entire mountain ranges. The Idaho and Sawtooth batholiths in central Idaho are excellent examples (Figure 4.51). The distribution of batholiths in western North America is shown in Figure 4.52.

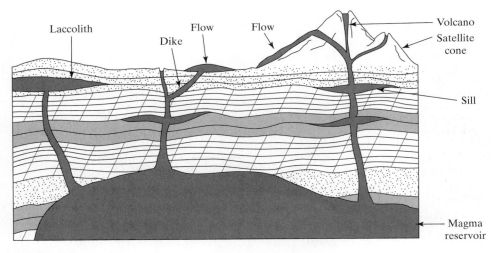

▲ FIGURE 4.48
Types of plutons. *Source:* From S. Judson, M. E. Kauffman, and L. D. Leet, *Physical Geology*, 7th ed., © 1987 by Prentice-Hall, Inc., Upper Saddle River, N.J.

(a)   (b)

▲ FIGURE 4.49
(a) A basaltic dike (dark rock at center) cutting metamorphic rocks near Bar Harbor, Maine.
(b) Granitic dikes cutting Precambrian metamorphic rocks, South Sinai, Egypt. *Source:* Photos courtesy of the author.

Crystallization of Magmas

The great variety in the mineral and chemical composition of igneous rocks suggests that the crystallization of magma is not a simple process. Pioneering experiments by N. L. Bowen in the early 1900s (Bowen, 1922) demonstrated that minerals crystallize sequentially as the temperature drops in a silicate magma and that solid crystals can react with the liquid phase of the magma to form new minerals during the crystallization process. This process is also known as *magmatic differentiation*.

▲ FIGURE 4.50
Laccoliths in Texas exposed by erosion of the overlying sediments. *Source:* C. C. Albritton, Jr.; photo courtesy of U.S. Geological Survey.

▲ FIGURE 4.51
Granitic rocks of the Sawtooth batholith, central Idaho. *Source:* T. H. Kilsgard; photo courtesy of U.S. Geological Survey.

To explain the differentiation process, let us assume that we have a silicate melt of basaltic composition at about 1500°C. As the temperature is slowly lowered, crystals begin to separate from the liquid. There are two crystallization sequences that are observed as the melt cools. The first sequence can be illustrated by the crystallization of plagioclase, which, as indicated in Chapter 3, forms a solid-solution series between calcium-rich and

◄ **FIGURE 4.52**
Map of batholiths in western North America. *Source:* From J. P. Davidson, W. E. Reed, and P. M. Davis, *Exploring Earth*, 2nd ed., © 2002 by Prentice Hall, Inc., Upper Saddle River, N.J.

sodium-rich compositions. The first plagioclase crystals to form are higher in calcium content than the calcium content of the liquid phase. As the mixture continues to cool, the crystals that form have progressively less calcium and more sodium than the original plagioclase crystals. In addition, the earlier-formed crystals react with the liquid by exchanging sodium from the liquid for calcium in the solid crystals. The internal structure of the mineral remains the same during this process. The crystallization of plagioclase follows what is called a *continuous reaction series*, in which the liquid and the crystals continuously change in composition until no liquid remains.

The ferromagnesian minerals follow a second type of crystallization sequence. In this series, olivine is the first ferromagnesian mineral to crystallize. As the temperature decreases, no change in the olivine crystals occurs until a critical temperature is reached. At this point, pyroxene rather than olivine begins to crystallize and the earlier-formed olivine crystals react with the liquid to form pyroxene. These reactions are different from the continuous reaction of plagioclase because entirely new minerals with different internal structures form at specific temperatures. For this reason, the ferromagnesian crystallization

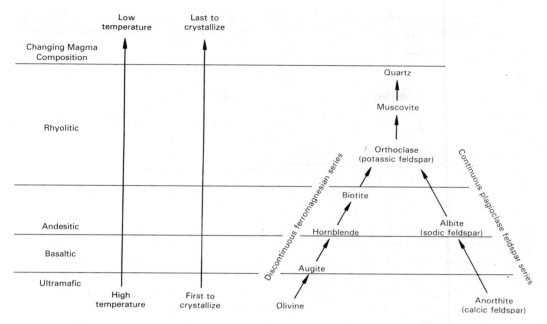

▲ FIGURE 4.53
Bowen's Reaction Series describes the sequence of minerals that crystallize from a basaltic melt.

sequence is called a *discontinuous reaction series*. The same type of reaction occurs between pyroxene and the liquid to form amphibole at a lower temperature.

The entire sequence of mineral crystallization is known as *Bowen's Reaction Series* (Figure 4.53). As the composition of minerals crystallizing from the melt changes, the magma composition changes, or differentiates (Figure 4.53), especially if early crystals can be separated from the melt. This can happen by gravitational settling, which is evident in some intrusions. In other cases, the magma can migrate to a different location after separation from early-formed crystals. In this process of differentiation, a mafic magma could gradually evolve to a felsic magma, from which rhyolitic or granitic rocks could crystallize. Through differentiation, a wide range of rock types could form from a magma originally of mafic or ultramafic composition. Differentiation alone, however, cannot explain the huge volumes of granitic rocks occurring in batholiths. For these plutons to form by differentiation of a mafic magma, there would have to be a gabbroic pluton many times the volume of the granitic batholith. These do not appear to exist. Alternative explanations for the formation batholiths include melting of continental igneous or metamorphic rocks or the melting of mixtures of basaltic mantle rocks and sedimentary rocks associated with subduction.

Properties of Igneous Rocks

When a rock sample is examined, much can be inferred about its origin by observing its color, texture, and mineral content. The term *texture* refers to the size, shape, and arrangement of the mineral grains in a rock. In this chapter we are concerned with igneous rocks, rocks that were formed by cooling and solidification from a molten liquid called *magma*. These rocks are known as *intrusive* igneous rocks if the cooling takes place below the land surface. If the molten liquid flows to the surface, it is called lava; here cooling can also take place to form *extrusive* igneous rocks. In subsequent chapters we will examine the other two fundamental rock types, sedimentary and metamorphic.

Table 4.4	The Texture of the Igneous Rocks
Rock type	**Description and interpretation**
Extrusive rocks	
Glassy	Noncrystalline, very fine-grained; very rapid cooling
Aphanitic	Uniformly fine-grained; rapid cooling
Porphyritic	Large phenocrysts within fine-grained groundmass; two-stage cooling process
Vesicular	Numerous small holes on surface; gas escape during cooling
Intrusive rocks	
Phaneritic	Uniformly coarse-grained; slow, gradual cooling in subsurface

Texture

The texture of igneous rocks (Table 4.4) is variable but usually consists of an interlocking network of mineral crystals that grew from the magma or lava during cooling. If the crystals are too small for observation by the naked eye, the texture is called *aphanitic* (Figure 4.54). These extremely fine-grained rocks are associated with volcanic processes because of the relatively rapid cooling that occurs when a magma reaches the earth's surface during a volcanic eruption and becomes lava. Crystal size is one of the major differences between extrusive, rocks and intrusive rocks. Growth of large crystals in a rock takes long periods of time and conditions of gradual, undisturbed cooling. The intrusive

▲ FIGURE 4.54
Rhyolite, an igneous rock with aphanitic texture. Individual crystals are not visible with the unaided eye. *Source:* Photo courtesy of the author.

▲ FIGURE 4.55
Phaneritic texture is characteristic of coarse-grained rocks. This specimen of granite contains large grains of orthoclase, quartz, and biotite. *Source:* Photo courtesy of the author.

rocks that formed under these conditions have a texture called *phaneritic* (Figure 4.55). Phaneritic rocks are composed of an interlocking network of large crystals that are easily visible.

When a volcano erupts, the outpouring lava is quickly cooled by contact with the atmosphere, which is much cooler than the subsurface *magma chamber* that contained the magma prior to eruption. At times, this cooling can be so rapid that crystallization is prevented. The silica tetrahedra are frozen in place before they can attain an orderly arrangement necessary for mineral growth. Cooling at these rates leads to the formation of *glassy* texture, the glass in this case being a natural, noncrystalline material. *Obsidian* (Figure 3.12) is a good example of a rock composed primarily of volcanic glass.

Several other types of textures are associated with volcanic rocks. Rocks that are composed of crystals of two sizes have a *porphyritic* texture (Figure 4.56). The interpretation of this texture is that crystals began to separate from the magma in the magma chamber or within the conduit leading from the magma chamber to the surface. These earlier-formed crystals grew to a substantial size by the time they reached the surface, enclosed in the still-liquid magma. Once exposed to the air, the remainder of the magma then crystallized rapidly as much smaller crystals. The resulting rock therefore consists of large *phenocrysts*, the early crystals, surrounded by a fine-grained crystalline *groundmass*.

In addition to liquid, magmas contain significant amounts of dissolved gases that are kept in solution by the great pressure exerted on the magma chamber by the overlying rocks. Water vapor is the most abundant gas, although carbon dioxide (CO_2), sulfur dioxide (SO_2), and other gases are present. When a magma reaches the earth's surface, the pressure decreases and the dissolved gases subsequently migrate to the surface of the lava and escape to the atmosphere. The upper surfaces of solidified lava flows are given an appearance much like Swiss cheese by the small holes, or *vesicles*, through which the gases escape (Figure 4.57). These rocks are said to have a *vesicular* texture. The types of igneous rock texture are summarized in Table 4.4.

▲ FIGURE 4.56
Porphyritic texture is composed of large phenocrysts embedded in a finer-grained groundmass.
Source: Photo courtesy of the author.

◄ FIGURE 4.57
A sample of vesicular basalt. As this lava cooled, gases escaped to the air through the holes (vesicles) in the rock. *Source:* Photo courtesy of the author.

Color and Composition

The chemical and mineralogical composition of igneous rocks is a reflection of the composition of the magma from which the rocks crystallized. Magmas are variable in composition, most importantly in the amount of silica (SiO_2) that they contain. The silica content ranges from less than 24% to more than 66%. Rocks that are rich in silica are called *silicic*, or *felsic*, rocks; and those that are low in silica are called *mafic* rocks. Fortunately, color provides a valuable clue for identification of igneous rocks because the silicic rocks are mainly composed of light-colored minerals like quartz and feldspar, whereas the mafic rocks are dark colored because of the abundance of *ferromagnesian* minerals. The dark-colored

► FIGURE 4.58

Classification of the igneous rocks. *Source:* Modified from L. Pirsson and A. Knopf, 1926, *Rocks and Rock Minerals,* John Wiley & Sons, Inc., New York.

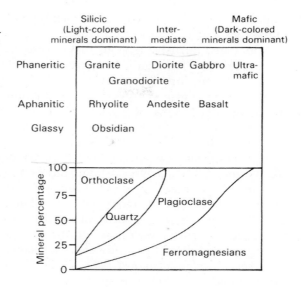

ferromagnesian minerals, which, as their name implies, are rich in iron and magnesium, include olivine, pyroxene, and hornblende. The major igneous rock types fall into categories of high, intermediate, and low silica content. Figure 4.58 shows the principal minerals contained in each of these rock types.

Texture and silica content provide a basis for classifying the igneous rocks. Aphanitic and phaneritic rocks of the same silica and mineral content have different rock names. For example, *granite* and *rhyolite* are both light-colored felsic rocks containing mostly quartz and orthoclase. They receive different names, however, because granite is a phaneritic intrusive rock and rhyolite is an aphanitic extrusive rock. In the intermediate range of silica content (52%–66%), *diorite* and *andesite* are the coarse-grained and fine-grained equivalents, respectively. The common intrusive rock *granodiorite* falls between granite and diorite in silica content. As Figure 4.58 indicates, granodiorite contains more plagioclase feldspar than orthoclase.

On the mafic side of the chart, *gabbro* and *basalt* (Figure 4.58) are important rock types. Basalt is the most common type of volcanic rock present on the earth's surface. The equivalent intrusive rock is gabbro. Rocks that are even lower in silica than gabbro have been discovered in the crust and are inferred to be much more common in the mantle. These rocks, which are largely composed of ferromagnesian minerals, are known as *ultramafic* rocks.

The most common rock with a glassy texture is obsidian (Figure 3.12). Despite its high silica content, obsidian is usually black in color. This dark-color exception to the usual light color of silica-rich rocks is due to the presence of impurities.

Figure 4.58 is a handy guide to the igneous rocks. By observing the color, it is possible to estimate the silica content of an igneous rock and thereby to predict its major minerals. The texture of the rock indicates whether it is intrusive or extrusive. Together, the color and texture can be used to assign the proper rock name.

Engineering in Igneous Rocks

Igneous rocks vary greatly in suitability for various types of engineering projects. An engineering site investigation must answer two questions: First, what rock types are present and how are they distributed; and second, how have the rocks been changed or altered since formation?

The first question deals directly with the topic of this chapter. The geologists and engineers working on the project must determine the nature of the igneous rock, the contacts it has with the adjoining rock types and their conditions, and the mineralogy of the rocks.

Unaltered intrusive igneous rocks generally are very suitable for most types of engineering projects. The interlocking network of mineral crystals gives the rock great strength. These rocks thus provide adequate support for building or dam foundations, can remain stable at high angles in excavations, and require minimal support in tunnels. Because of the dense interlocking of crystals within the rock, very little water can move through. Therefore, unaltered intrusive rocks are well suited for the construction of reservoirs because of the low potential for leakage.

The engineering properties of extrusive rocks are much less uniform. Sequences of extrusive rocks contain pyroclastic materials and lahar deposits, which are much weaker than the interlayered lava flows. These rocks may be susceptible to slope failures in excavations and also provide more variable and generally weaker foundation support. In general, the water-bearing capacity of extrusive rocks is much greater than intrusive rocks. Sequences of basaltic lava flows on the Columbia Plateau and elsewhere are known for their great supplies of groundwater. This same property can render the rocks unsuitable for reservoir or tunnel construction. The presence of lava tubes in a series of basalt lava flows would prove to be an especially serious problem.

The second question concerns the geologic history of the igneous rocks since the time of their formation. Several types of changes have significant engineering implications. Any process that tends to fracture the rocks will cause weakening of a large rock mass with respect to engineering suitability. Under the forces imposed by the interaction of lithospheric plates, fracturing of crustal rocks is quite common. These processes will be discussed in Chapter 8. It is sufficient to note at this point that a network of fractures within a rock mass can greatly increase the potential for failures of natural or excavated slopes and also increase the construction problems of dams, tunnels, and other structures. Fractured welded tuffs were implicated in the failure of the Teton Dam in 1976 in Idaho. According to a probable scenario for the failure of the dam, groundwater flowing through the fractured volcanics came in contact with erodible silt in the core of the dam and washed the silt through fractures in the rock. Larger and larger channels in the silt were eroded, leading to collapse of the embankment (Chapter 11).

Although tectonic fracturing is important, there are several other mechanisms that cause fracturing. Fracturing of the extrusive rocks, as we have seen, occurs during cooling, in addition to any later processes that may take place. An additional class of fracturing mechanisms becomes possible when the rock mass is at or near the earth's surface. These processes are collectively called *weathering*. A detailed discussion of weathering is found in Chapter 9. Rocks can be fractured in the near-surface environment by freezing and thawing as well as by other means. Thus, a mass of igneous rocks generally is more fractured near the surface than at depth.

Weathering produces other changes in the rock besides fracturing. Chemical reactions between the minerals within the rock and air and water gradually form new minerals. Clay minerals are a common product of these alteration processes. The result is a significant loss of strength as the feldspars and ferromagnesian minerals are converted into clay. In warm, humid climates, igneous rock bodies may be mantled with tens of meters of weathered material. The engineering properties of this material are totally different from the properties of unaltered rock.

Summary and Conclusions

Volcanoes erupt many types of materials and produce a variety of landforms. Most active volcanoes are associated with tectonic-plate margins, although intraplate volcanism is known to occur; mantle plumes are thought to be the cause of some of this activity.

Volcanic products include gas, lava, and pyroclastic material. All of these products can be hazardous during eruptions. Lava flows range from high-temperature, low-viscosity flows of basaltic composition to highly viscous, silicic flows. When gas is explosively released during an eruption, pyroclastic materials are abundant. Large pyroclasts fall in the immediate vicinity of the vent, whereas volcanic ash can be transported thousands of kilometers by winds. Occasionally, destructive pyroclastic flows and pyroclastic surges are generated during eruptions. These mixtures of pyroclastic debris and gases flow down the mountainsides at high velocities and with great force. Lava and pyroclastic materials sometimes mix with snow, ice, or water to form lahars, flows that are similar to mudflows in terms of velocity and the mechanics of movement.

Volcanic products can be extruded from fissures or central vents. Fissure eruptions produce highly fluid lavas of basaltic composition that flow great distances. By successive eruptions, vast plateaus composed of thick vertical sequences of flood basalts are built up. Eruptions from central vents, on the other hand, tend to produce conical mountains composed of lava, pyroclastic materials, or both. Shield volcanoes are composed of lava flows only, whereas composite volcanoes contain alternating layers of lava and pyroclastic materials. Cinder cones form when the eruption is limited to pyroclastic material.

Intrusions are formed by migration of magma from a magma chamber to a subsurface location higher in the crust. The resulting plutons range from moderate-size laccoliths and stocks to giant batholiths that form the cores of major mountain ranges. Some batholiths probably formed by melting or alteration of existing rocks rather than by intrusion of magma from a separate magma chamber.

Magma composition and the rocks that crystallize during cooling evolve in a process known as differentiation. The order in which minerals crystallize from a silicate melt is known as Bowen's Reaction Series. This sequence includes both a discontinuous and a continuous series, the latter represented by plagioclase, in which the earlier-formed crystals and the liquid change composition gradually while retaining the internal structure of plagioclase. The ferromagnesian minerals follow the discontinuous series, in which reactions between crystals and liquid produce entirely new minerals at specific temperatures. By separation of crystals from the cooling magma, successively more felsic magmas are produced.

Igneous rocks can be divided into extrusive rocks that crystallized at the earth's surface and intrusive rocks that crystallized below the surface. Texture is the major criterion used to assign rocks to one of these categories. Texture and mineralogical composition form the basis for classification of igneous rocks. Based on mineralogy, composition also refers to the amount of silica in the rock, ranging from the low-silica mafic rocks to the silicic rocks. Each major compositional group (silicic, intermediate, and mafic) has both intrusive and extrusive rock types. With the exception of obsidian, color is a good indication of composition because the silicic minerals like quartz and feldspar are mostly light in color.

Engineering conditions in igneous-rock terrains depend on the composition and type of rock as well as on the amount and type of alteration that has occurred since intrusion or extrusion. Weathering and fracturing are two types of alteration that can weaken the rock and render it less suitable for engineering projects.

Problems

1. If an igneous rock has phaneritic texture, what origin can be presumed for the rock?

2. List all the criteria you might use to identify a volcanic rock.

3. What criteria are used to classify the igneous rocks?

4. How can the distribution of volcanoes on the earth be explained by the theory of plate tectonics?

5. If someone told you you were walking across rocks that cooled from pahoehoe lavas, what would you

know about the characteristics of the lava at the time of eruption?

6. Pyroclastic surges and lahars both involve flow of material from a volcano. What are the differences between the two processes?

7. What factors control the type of volcanic landforms that develop from an eruption?

8. List and discuss the volcanic hazards that are associated with composite volcanoes in the Cascade range.

How and why do these hazards differ from those of the Hawaiian volcanoes?

9. What factors determine whether a dike or a sill will form?

10. Summarize Bowen's Reaction Series.

11. Contrast the general conditions relative to engineering projects that would be encountered in intrusive and extrusive rock terrains.

References and Suggestions for Further Reading

BOWEN, N. L. 1922. The reaction principle in petrogenesis. *Journal of Geology*, 30:177–198.

EICHELBERGER, J., ed. 1987. Lethal gas bursts from Cameroon crater lakes. *Eos, Transactions, American Geophysical Union*, June 9, 1987:568.

DAVIDSON, J. P., W. E. REED, and P. M. DAVIS, 2002. *Exploring Earth*, 2nd ed. Upper Saddle River, N.J.: Prentice Hall, Inc.

HAY, E. A., and A. L. McALESTER. 1984. *Physical Geology: Principles and Perspectives*, 2nd ed. Upper Saddle River, N.J.: Prentice Hall, Inc.

JUDSON, S., M. E. KAUFFMAN, and L. D. LEET. 1987. *Physical Geology*, 7th ed. Upper Saddle River, N.J.: Prentice Hall, Inc.

MULLINEAUX, D. R., D. W. PETERSON, and D. R. CRANDALL. 1987. Volcanic hazards in the Hawaiian Islands, in *Volcanism in Hawaii*, R. W. Decker, T. L. Wright, and P. H. Stauffer, eds. U.S. Geological Survey Professional Paper 1350.

NATIONAL RESEARCH COUNCIL. 1994. *Mount Rainier: Active Cascade Volcano*. Washington, D.C.: National Academy Press.

PRINGLE, P. T. 1994. Volcanic hazards in Washington—A growth management perspective. *Washington Geology*, 22 (no. 2):25–32.

SCOTT, K. M., J. W. VALLANCE, and P. T. PRINGLE. 1995. *Sedimentology, Behavior, and Hazards of Debris Flows at Mount Rainier, Washington*. U.S. Geological Survey Professional Paper 1547.

SCOTT, W. E. 1989. Volcanic and related hazards, in R. I. Tilling, ed. *Volcanic Hazards*. American Geophysical Union, Short Course in Geology, v. 1.

TILLING, R. I., C. HELIKER, and T. L. WRIGHT. 1987. *Eruptions of Hawaiian Volcanoes: Past, Present, and Future*. U.S. Geological Survey.

U.S. GEOLOGICAL SURVEY. 1980. *Hydrologic Effects of the Eruptions of Mount St. Helens, Washington, 1980*. U.S. Geological Survey Circular 850.

WINTER, J. D. 2001. *An Introduction to Igneous and Metamorphic Petrology*. Upper Saddle River, N.J.: Prentice Hall, Inc.

WOLFE, E. W. 1992. The 1991 eruptions of Mount Pinatubo, Philippines. *Earthquakes and Volcanoes*, 23:5–37, U.S. Geological Survey.

WOLFE, E. W., M. O. GARCIA, D. B. JACKSON, R. Y. KOYANAGI, C. A. NEAL, and A. T. OKAMURA. 1987. The Pu'u O'o eruption of Kilauea Volcano, Episodes 1–20, January 3, 1983 to June 8, 1984, in *Volcanism in Hawaii*, R. W. Decker, T. L. Wright, and P. H. Stauffer, eds. U.S. Geological Survey Professional Paper 1350.

WRIGHT, T. L., and T. C. PIERSON. 1992. *Living with Volcanoes. The U.S. Geological Survey's Volcano Hazards Program*. U.S. Geological Survey Circular 1073.

5 CHAPTER

Sedimentary Rocks and Processes

Early in the development of geology as a science, James Hutton and others recognized that rocks exposed on the continents, particularly in mountainous areas, are continuously broken down into small particles under the relentless attack of the atmosphere. These finer constituents of the rock are gradually transported to lower elevations, where they are deposited as sediments. This phase of the geologic cycle leads to the formation of sedimentary rocks, the second major rock group. The paths leading from existing rocks to sediments and finally to sedimentary rocks may be long and complicated. Our understanding of these rocks is aided, however, by using the Principle of Uniformitarianism. By studying the deposition and characteristics of sediments on the earth today, sedimentologists can reconstruct the geologic history of ancient sedimentary rocks.

Although sedimentary rocks constitute less than 10% of the rocks in the crust, their importance is emphasized by the fact that they cover about 70% of the surface of the continents. As a result, sedimentary rocks and modern sediments are the most common materials encountered in construction and waste disposal projects. In addition, sedimentary rocks contain the deposits of petroleum, coal, and other energy resources upon which our society depends.

The benefits of utilizing geology in engineering practice have never been better illustrated than by the work of the English civil engineer William Smith (1769–1839). Smith compiled the first real geologic map, which showed the surficial sedimentary rock units in southern England, as a result of his observations of rocks and fossils in connection with the construction of canals and roads. Since the time of Smith, sedimentary rocks have been intensely studied for clues into the history of the earth as well as to locate the resources they contain.

Origin of Sedimentary Rocks

The processes involved in the formation of sediment from existing rocks, in addition to those processes relating to the transportation and deposition of sediment, will be described in considerable detail in later chapters of this text. For this reason, we briefly introduce these concepts here and concentrate on the characteristics and properties of the sedimentary rocks that result from these processes.

The decomposition and disintegration of igneous, sedimentary, and metamorphic rocks near the earth's surface is the result of physical and chemical *weathering* processes. The products of weathering are detached rock fragments, or *clasts*, that can be transported by gravity, water, wind, or ice from higher elevations to lower elevations (Figures 5.1 and 5.2). The ultimate repository of sediments is the ocean basin, but by the time sediments reach the oceans, they have been reduced to extremely small particles. There are many intermediate points of deposition along the path from mountain peak to ocean basin. These *depositional environments* can include almost any part of the continent or ocean floor. On the continents, depositional environments include alluvial (stream) (Figure 5.3), lacustrine (lake), paludal (swamp), desert (Figure 5.4), and glacial types. Along the continental margins a variety of shallow marine depositional environments can be recognized (Figure 5.5). Finally, sediment can be deposited in deep marine settings away from the influence of continents.

The nature of the depositional environment determines the characteristics of the resulting sedimentary rock. Some of the factors that influence rock type are listed in Table 5.1. These factors leave their geologic signature upon the sediments and resulting rocks that form in a particular environment. One of the most frequent objectives of studies of sedimentary rocks is to reconstruct the conditions of an ancient depositional environment from the rocks that remain.

◀ FIGURE 5.1
The weathering of rocks in mountainous regions commonly begins by the formation of large rock fragments, or talus, that accumulate on slopes (Shenandoah Valley, Virginia). *Source:* J. T. Hack; photo courtesy of U.S. Geological Survey.

▶ FIGURE 5.2
The river plain in the foreground is composed of sediments transported by the stream from uplands in the background (Southern Iceland). *Source:* Photo courtesy of the author.

▲ FIGURE 5.3
Meandering stream at base of a valley, Mono County, California. *Source:* W. T. Lee; photo courtesy of U.S. Geological Survey.

Table 5.1   Important Factors of Depositional Environments
Type of transporting agent (water, wind, ice)
Flow characteristics of depositing fluid (velocity, variation in velocity)
Size, shape, depth of body of water, and circulation of water (in lacustrine and marine basins)
Geochemical parameters (temperature, pressure, oxygen content, pH)
Types and abundances of organisms present
Type and composition of sediments entering environment

▲ FIGURE 5.4
Some deserts are depositional environments for sand transported by strong winds (Sonora, Mexico). *Source:* E. D. McKee; photo courtesy of U.S. Geological Survey.

▲ FIGURE 5.5
Coral reefs constitute important shallow marine depositional environments in tropical areas (Aga Point, Guam). *Source:* J. I. Tracey, Jr.; photo courtesy of U.S. Geological Survey.

A final step is required to convert an aggregate of particles into a sedimentary rock. This step is called *lithification,* and it can be accomplished in several ways. The most basic lithification process is *compaction.* Sediment deposited in a loose condition is gradually compacted to a denser state as the weight of additional material is added from above. Glaciers and tectonic forces can impose additional stress upon the sediment. The result of

▶ FIGURE 5.6
Decrease in porosity with depth in South Louisiana Tertiary sands; data include 17,367 samples averaged for each 1000-ft interval. *Source:* Modified from H. Blatt, G. Middleton, and R. Murray, *Origin of Sedimentary Rocks*, 2nd ed., © 1980 by Prentice Hall, Inc., Englewood Cliffs, N.J.; data from G. I. Atwater and E. E. Miller, 1965, unpublished manuscript.

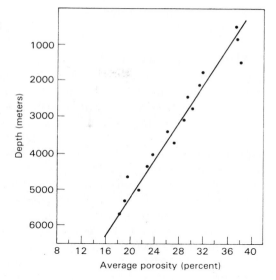

compaction is a reduction in void space between particles. One measure of the relative amount of void space is *porosity* (Chapter 7). In a deep sedimentary basin, decreases in porosity with depth are significant (Figure 5.6). A process that accompanies compaction is the expulsion of water from the void spaces between particles as they are forced closer together. Clay-rich sediments undergo a much greater degree of compaction than sands. Interparticle attractive forces between clay particles also aid in the lithification process.

Along with simple compaction of sediments to a denser state, a wide variety of physical and chemical processes contribute to lithification. Over long periods of time, the pore spaces remaining after compaction may be gradually filled by precipitation of solid material from groundwater circulating through the sedimentary sequence. The filling of void spaces by chemical precipitation is appropriately called *cementation*. Cementation is one of the most effective lithification processes because the chemical cement bonds the particles in the rock together. Most commonly, the cements are composed of either silica (SiO_2) or calcium carbonate ($CaCO_3$). Iron-bearing compounds are also fairly common cements. Rocks cemented by silica are among the hardest and strongest sedimentary rocks known. As such, they make excellent foundation materials for large structures.

In addition to cementation, *crystallization* can also contribute to the lithification of a rock. Crystallization usually refers to crystal growth within the void spaces in a rock. Under high pressures and temperatures, however, some of the original materials in the rock may chemically react to form crystals of new minerals that are more stable under the prevailing conditions. If temperatures and pressures reach sufficiently high levels, the alteration of sedimentary rocks is great enough to be called metamorphism. Because the transition between the two is gradual, it is sometimes hard to draw the line between sedimentary and metamorphic rocks. The boundary between sediments and sedimentary rocks is also difficult to define. Accordingly, some poorly lithified sedimentary rocks may be classified as soils for engineering purposes because their mechanical behavior may be closer to particle-aggregate systems than to well-cemented rock.

Characteristics of Sedimentary Rocks

Sedimentary rocks exhibit a variety of distinctive features. By carefully observing these characteristics, it is possible to assign the proper rock name to a specimen as well as to deduce the origin and depositional environment of the sediment that was lithified to form the sedimentary rock.

Texture and Classification

Perhaps the most important characteristic of a sedimentary rock is its texture, which, as is true for the igneous rocks, refers to the size and arrangement of the particles or grains that make up the rock. There are only two main types of texture with which we need to be concerned: *clastic* and *nonclastic*. Clastic rocks are composed of aggregates of individual mineral or rock fragments. When these fragments have been eroded, transported, and then deposited, the origin of the rock can be described as *detrital* (Table 5.2). Individual grains can be seen with the naked eye in conglomerates, breccias, and sandstones, which make up the coarse-grained clastic rocks. Hand-lens or microscopic observation is necessary for clasts in fine-grained sandstone and siltstone. Clay particles in shale and mudstone are so small that electron microscopes are used to study them.

If the texture of a rock is nonclastic, it means that the grains form an interlocking network similar to igneous rocks with crystalline texture. Nonclastic rocks are formed by chemical and biochemical precipitation from aqueous solutions. This type of texture can be observed in rocks formed by both inorganic and organic processes (Table 5.2). Organic precipitation is common to organisms that secrete shells composed of calcium carbonate or silica. However, if a rock is composed of an accumulation of shell fragments, its texture is considered to be clastic rather than nonclastic.

Once the texture of a rock is identified, it is necessary to measure the size of the grains. Geologists often use the Wentworth scale for classification of particle sizes (Table 5.3). Another commonly used measure of grain size is the phi unit (Table 5.3). This unit is defined by the formula

$$\phi = -\log_2 X \tag{5.1}$$

where X is the grain size in millimeters. Phi units are useful for statistical analysis because the grain-size distribution is often normally distributed when expressed in phi units.

Table 5.2	Classification of Sedimentary Rocks		
Origin	**Texture**	**Particle size or composition**	**Rock name**
Detrital	Clastic	Granule or larger	Conglomerate (round grains)
			Breccia (angular grains)
		Sand	Sandstone
		Silt	Siltstone
		Clay	Mudstone and shale
Chemical			
Inorganic	Clastic or nonclastic	Calcite, $CaCO_3$	Limestone
		Dolomite, $CaMg(CO_3)_2$	Dolomite
		Halite, NaCl	Salt
		Gypsum, $CaSO_4 \cdot 2H_2O$	Gypsum
Biochemical	Clastic or nonclastic	$CaCO_3$ (shells)	Limestone, chalk
			Coquina
		SiO_2 (diatoms)	Diatomite
		Plant remains	Coal

Source: From S. Judson, M. E. Kauffman, and L. D. Leet, *Physical Geology*, 7th ed., © 1987 by Prentice Hall, Inc., Englewood Cliffs, N.J.

Table 5.3   Wentworth Scale of Particle Sizes for Clastic Sediments

Size (mm)	Ø	Size name
		Boulder
256	−8	
		Cobble
64	−6	
		Pebble
4	−2	
		Granule
2	−1	
		Sand
1/16 (0.0625)	4	
		Silt
1/256 (0.0039)	8	
		Clay

Source: From C. K. Wentworth, A scale of grade and class terms for clastic sediments, *Journal of Geology*, 30:381, © 1922, University of Chicago Press.

The names of several particle sizes are used to form the rock name for the corresponding rocks. For example, sandstone is composed of particles between $\frac{1}{16}$ and 2 mm in diameter. Actually, there may be a wide range in grain sizes. Finer particles that fill the void spaces between larger grains are called *matrix*. Because of its resistance to erosion, sandstone often forms cliffs or upland areas (Figure 5.7). Detrital rocks with most particles larger than sand size are called *conglomerate* if the particles are rounded or *breccia* if the particles are angular (Figure 5.8). Fine-grained clastic rocks are called *shale* or *mudstone* (Figure 5.9). The term shale usually refers to a rock that tends to split into thin slabs parallel to the depositional layering of the sediment. These fine-grained rocks are typically weaker than sandstone and form low-angle, covered slopes or valleys in sedimentary rock terrains.

▲ FIGURE 5.7
Resistant sandstone (Entrada Formation) exposed in Arches National Park, Utah. Natural arches are formed by weathering processes and wind erosion. *Source:* Photo courtesy of R. A. Kehew.

(a)

(b)

▲ FIGURE 5.8
(a) Conglomerate is a detrital sedimentary rock composed of large, rounded particles (south-central Norway). (b) Breccia is similar in grain size to conglomerate but differs because the grains are angular (western Maryland). *Source:* Photos courtesy of the author.

The particle size of detrital rocks gives an important clue about the depositional environment. In the case of water-laid materials, the ability to transport particles is dependent upon the current velocity of the transporting fluid. Therefore, conglomerates represent sediment deposits of high-velocity streams, like those that rush down steep mountain slopes or form at the edge of a melting glacier. In contrast, silt and clay accumulate in low-energy environments, like lakes or seafloors.

The chemical and biochemical sedimentary rocks include a wide variety of materials. Chemical precipitation from water is responsible for deposits of limestone (Figure 5.10). Deposits of marine origin are the most common type of limestone. Most of Florida is underlain

(a)

(b)

▲ FIGURE 5.9
(a) An outcrop of shale showing characteristic layering (Mancos Shale near Aspen, Colorado).
Source: B. Bryant; photo courtesy of U.S. Geological Survey. (b) Under the microscope, the mineral composition of shale can be observed. Minerals in the light-colored bands are mostly quartz, whereas the dark-colored layers are made up of platy crystals of muscovite (Kuskokwim region, Alaska).
Source: W. M. Cady; photo courtesy of U.S. Geological Survey.

by fossiliferous limestone deposited under shallow marine conditions on a carbonate shelf. Classification of these rocks is based upon the types of grains and the amount of matrix making up the rock. Grain types include *bioclastic debris, oolites, intraclasts,* and *pellets*. Bioclastic debris includes the skeletal remains of organisms that secrete shells of calcium carbonate. Paleontologists, geologic specialists who study fossils and their evolution through geologic time, have identified thousands of these organisms. Examples include corals that

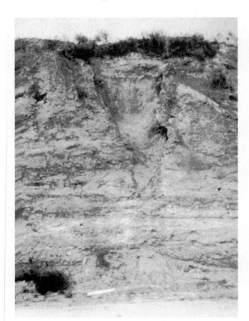

◄ FIGURE 5.10
An exposure of limestone in a quarry in Florida showing a solution cavity filled with soil that collapsed into
the void from above. *Source:* Photo courtesy of the author.

build extensive reefs composed of intergrown networks of skeletal calcite (Figure 5.11) and
solitary organisms like clams and snails. Ancient reefs buried in sedimentary basins are ea-
gerly sought by petroleum geologists because of their abundant void space, which may
serve as a reservoir for oil and gas. Oolites are spherical grains formed by precipitation of
concentric layers of calcium carbonate in a shallow-water marine environment where waves
or currents periodically roll grains across the bottom (Figure 5.12). Intraclasts are fragments
of precipitated calcium carbonate that are eroded and then redeposited, and pellets are composed of the fecal matter of organisms. These four grain types may or may not be surrounded by a matrix composed of calcium carbonate mud. Rock names describe the grain type and
matrix characteristics. For example, an *oomicrite* is a rock containing abundant oolites within
a muddy matrix. *Micrite* is an abbreviation for microcrystalline calcite.

▲ FIGURE 5.11
Brain coral (*Diploria*, species) and several species of Alcynonarian corals on a shallow reef in St. Croix,
U.S. Virgin Islands. *Source:* Photo courtesy of W. B. Harrison, III.

▶ FIGURE 5.12
Photomicrograph of oolitic limestone from core, ×30. Sand grains form the nuclei of many oolites (Hernando County, Florida). *Source:* P. L. Applin; photo courtesy of U.S. Geological Survey.

Areas underlain by carbonate rocks present a unique set of conditions for engineers and geologists involved in development and environmental problems. Although carbonate terrains frequently contain abundant groundwater resources, groundwater may easily become contaminated due to the ability of rain to dissolve calcium carbonate and create pathways for the rapid downward movement of contaminants. Florida and other limestone terrains are also plagued by rapid subsidence of the land surface to form *sinkholes*, as subsurface rocks are slowly dissolved by infiltrating rain or snowmelt (Chapter 11). Although dissolution of the rock may take thousands of years, collapse of the surficial soil into the underground cavity may be instantaneous. A solution cavity filled with younger soil materials is shown in Figure 5.10. The actual and potential damage caused by subsidence is a major threat to development in carbonate terrains. Groundwaters precipitate calcium carbonate under certain conditions as well as dissolve it; most commonly, this occurs in caves and around hot springs. This type of limestone is known as *travertine*. The spectacular travertine deposits at Mammoth Hot Springs in Yellowstone National Park (Figure 5.13) are a good example.

▲ FIGURE 5.13
These terraces at Mammoth Hot Springs at Yellowstone National Park were formed by precipitation of travertine from the emerging spring water. *Source:* Photo courtesy of the author.

Minerals precipitate from water because the concentration in the liquid phase is greater than the solubility of the solid phase. The concentration at the point of equilibrium between dissolution and precipitation is a function of temperature, pH, and other factors. Some minerals are so soluble that they rarely precipitate. For example, halite requires a very high concentration of sodium and chloride to initiate precipitation. The necessary concentrations are sometimes achieved by evaporation of water from a lake or marine basin. The Great Salt Lake in Utah (Figure 5.14), for example, is a remnant of a much larger lake, Lake Bonneville, that existed during a cooler and wetter climatic period during the Pleistocene Epoch. Climatic changes, including warmer temperatures and less precipitation, led to a gradual shrinking of the lake by evaporation. As the lake water became concentrated, precipitation of halite and other salts formed the Bonneville salt flats, the floor of ancient Lake Bonneville. Salt deposits of this type are called *evaporites*. Evaporites can also form in marine basins that are deeper than the surrounding parts of the sea. The isolation and depth of these basins limits normal oceanic circulation, and the necessary evaporative concentration can occur. Precipitation of evaporites follows a sequence controlled by the solubility of the evaporite minerals. As water begins to evaporate from seawater, calcite is the first mineral to precipitate. Gypsum follows as evaporation continues, and finally halite and other salts drop out of solution when the remaining water is nearly gone.

When beds of halite become buried by thousands of feet of other sedimentary rocks, a curious phenomenon sometimes occurs due to the mechanical behavior of salt under high pressures. Evaporite rocks are extremely weak, and under pressure, they behave as a plastic material and can be made to deform, or flow, for great distances. From an initially uniformly thick bed subjected to small differences in overburden pressure, salt begins to gradually flow toward a point in the bed that slowly thickens. Because of the low density of salt, thickened zones bulge upward, forming a *salt dome* (Figure 5.15). Sinking occurs adjacent to the dome because of the loss of salt there, due to the flow of salt toward the growing dome. More sediment is deposited in the low-lying areas during

Great Salt
Lake

Lake
Bonneville

Utah

◄ FIGURE 5.14
The Great Salt Lake is a small remnant of ancient Lake Bonneville. Evaporites were deposited as evaporation caused the lake to shrink in size.

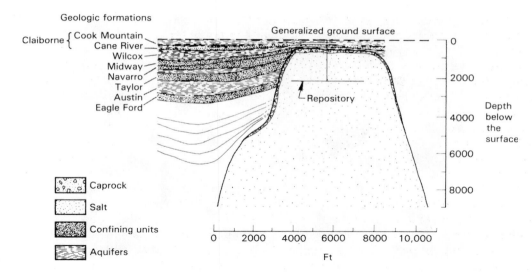

▲ FIGURE 5.15
Cross section of the Vacherie Salt Dome. Notice caprock above and on the sides of the dome. Sedimentary rock beds are thin above the dome and thicken away from the dome. Aquifers are rock units that contain economically significant groundwater supplies and confining units are beds that do not contain economically significant groundwater resources. *Source:* From U.S. Dept. of Energy, Draft Environmental Assessment—Vacherie Dome Site, Louisiana.

dome growth, so that rock formations thicken outward from the center of the dome. With continued dome expansion, salt may break through overlying rock units and rise hundreds of meters until a stable condition of equilibrium is reached and no further growth occurs. Dissolution of salt at the top and margins of the dome by groundwater flowing in the adjacent rock formations leaves behind less soluble material as a caprock. The formation of a caprock isolates the soluble salt to some extent from groundwater in adjoining rocks.

Salt domes sometimes form effective petroleum traps and have been exploited for petroleum production in the U.S. Gulf Coast region for many years. Because of its low permeability, some of our emergency supplies of oil are stored in large underground cavities excavated in salt. More recently, salt beds and salt domes have been suggested for another purpose: disposal of high-level radioactive wastes. Prior to selection of the Yucca Mountain site for further study, seven of the original nine potential repository sites chosen by the U.S. Department of Energy (DOE) were sedimentary rock sequences containing evaporite beds or domes. Studies of these materials have continued, and it is likely that these sites will eventually serve as repositories for some types of hazardous wastes.

Sorting

The range of particle sizes in a sedimentary rock is known as *sorting*. Rocks that contain a narrow range of sizes are described as *well sorted;* those that consist of a wide size range are termed *poorly sorted* (Figure 5.16). The value of sorting in interpretation of sedimentary rocks lies in its use as an environmental indicator. Poor sorting in a water-laid sedimentary rock suggests rapid fluctuations in current velocity and rapid deposition. Wind-deposited sediments, on the other hand, tend to be well sorted because wind is capable of transporting only small particles. Thus, the small particles are separated from the larger particles and accumulate in one location when the windstorm ceases. Sediment transported directly by a glacier can often be recognized by its very poor sorting. Although glaciers move

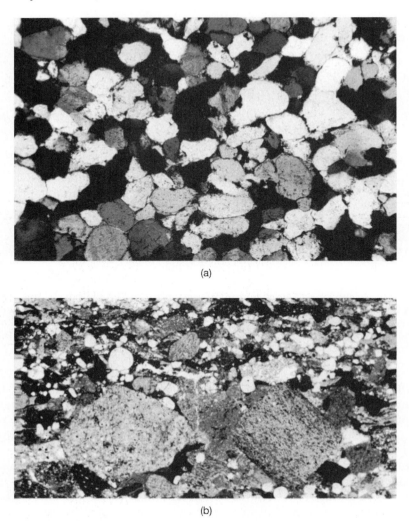

▲ FIGURE 5.16
(a) Photomicrograph of well-sorted quartz sandstone, Moffat County, Colorado. *Source:* W. R. Hansen; photo courtesy of U.S. Geological Survey. (b) A poorly sorted sandstone under the microscope. Particles range in size from silt to fine gravel, Valmy Formation, Nevada. *Source:* J. Gilluly; photo courtesy of U.S. Geological Survey.

very slowly, they are equally capable of picking up both huge boulders and small clay particles as they advance over an area. Very little segregation of the particles occurs during the movement of the glacier.

Sedimentary Structures

Sedimentary structures set sedimentary rocks apart from other rock types. The result of sediment transport and deposition, these structures are critical for understanding the conditions under which the sediments were deposited. Using sedimentary structures and other factors, geologists can reconstruct the environment of deposition of a sedimentary rock unit. When the depositional environment is known, variations in rock type and engineering characteristics can be predicted. If an engineering project involves subsurface data

► FIGURE 5.17
Bedding stands out even at a great distance in the well-exposed sedimentary rocks of the Grand Canyon. *Source:* J. R. Balsley; photo courtesy of U.S. Geological Survey.

that can be very expensive to obtain, reconstructions of sedimentary environments quickly become very relevant to the investigation.

The most fundamental sedimentary structure is *bedding*. Individual layers, or beds, of sediment stacked upon each other are the most recognizable indication of sedimentary rock (Figure 5.17). A bed is a unit that differs in some way from the beds above and below it. *Laminae* are layers similar to bedding that are less than 1 cm thick. Bed boundaries may be sharp or gradational. Bedding is the result of variation in the depositional process. A slight change in current velocity is all that is necessary to change the size of the material being transported and deposited. A river may deposit sand or gravel on the bed of its channel. Later, if that reach of the channel is abandoned in favor of a new channel, beds of silt and clay representing a lower-energy environment may be deposited over the coarser sediments. Thick sequences of sedimentary rocks like the Grand Canyon record much more drastic changes in depositional environments—alternating terrestrial and marine conditions, for example. Beds and other structures are the evidence of these changes.

Differences in the process of sedimentation can lead to different types of bedding. Three main types include *parallel bedding, cross bedding*, and *graded bedding* (Figure 5.18). Parallel bedding can be produced by transport of sediment by currents or by deposition in standing bodies of water where sediment particles fall through a column of water to their resting points on the bottom.

Cross bedding is the result of the deposition of sediment that moved in wavelike bedforms near the sediment-fluid interface. These conditions occur in streams and in windblown sand. Three major types of cross beds are recognized (Figure 5.19). Tabular and wedge-shaped cross beds are bounded by planar boundaries that are parallel for tabular and nonparallel for wedge-shaped cross beds. Trough cross beds have curved lower boundaries. The formation of cross beds is illustrated in (Figure 5.20). The wavelike

▲ FIGURE 5.26
Mud cracks indicate that this bed dried in the open air soon after deposition (Isle Royale National Park, Michigan). *Source:* R. B. Wolff; photo courtesy of U.S. Geological Survey.

oxygen is critical to the preservation of these materials because bacteria will decompose the organic matter if an oxygen supply exists.

Fossils

The presence of organic remains in sedimentary rocks is a major concern to the science of geology. It was primarily by the study of fossils that the geologic time scale was devised.

Living organisms consist of soft and hard parts. The soft parts, including cell tissue and internal organs, decay rapidly in the presence of oxygen when the organism dies. Hard parts—bones and shell, for example—can be preserved for long periods of time under favorable conditions in sedimentary deposits (Figure 5.27). Shells are usually

◀ FIGURE 5.27
Coquina, a rock composed almost entirely of fossils of marine organisms (Gila County, Arizona). *Source:* C. Teichert; photo courtesy of U.S. Geological Survey.

► **FIGURE 5.28**
Petrified stumps in western North Dakota preserved in their original growth position by the replacement of organic matter by dissolved silica carried by groundwater. The stumps have been exposed by erosion of the surrounding materials. *Source:* Photo courtesy of the author.

composed of calcium carbonate or silica, whereas bones consist of a phosphatic material. In anaerobic environments, soft organic remains can be preserved, although they may be greatly altered from their original state. These environments account for deposits of petroleum, which are derived from the remains of microscopic marine organisms, and coal, the product of terrestrial plants.

Fossils also include material that is not necessarily organic in origin. Petrified wood, for example, is composed of silica that has been precipitated by groundwater circulating through sedimentary materials. In this process, silica precipitation gradually replaces the organic material while retaining the original cellular structure (Figure 5.28).

Stratigraphy and Depositional Environments

When thick exposures of sedimentary rocks are examined—in the Grand Canyon, for example—one characteristic that is immediately apparent is that there are vertical changes in the rock units. The different appearance of adjacent rock units reflects changes in physical, or lithologic, properties of the rock. When rock units can be traced for a substantial distance, lateral changes in lithology within a particular unit can also be observed. The reason for these changes is simply that the boundaries of a specific depositional environment are finite. In fact, most depositional environments can be divided into subenvironments, or facies, each containing lithologically distinct sediments deposited contemporaneously with those in the adjacent subenvironment. For example, facies in a meandering stream environment are shown in Figure 5.29. A complete description of depositional environments is well beyond the scope of this book. The most important types of depositional environments are shown on Table 5.4. The major environments can be grouped into continental, marginal-marine, and marine types. The neritic environments occur on the continental shelves, which are actually part of the continents, but inundated by the oceans. These shallow-water environments are modern-day depositional analogies of periods in geologic time during which much larger areas of the continents were covered by shallow seas. The neritic environments are therefore very important natural laboratories to study processes similar to those that formed the rocks that cover large areas of the North American and other continents.

The branch of geology that deals with the horizontal and vertical changes and relationships between sedimentary rock units is known as *stratigraphy*. The use of only physical properties of the rock units would be an application of *lithostratigraphy*. But there are

▲ FIGURE 5.29
Generalized facies in a meandering stream environment. *Source:* From R. K. Matthews, *Dynamic Stratigraphy*, 2nd ed., © 1984 by Prentice Hall, Inc., Upper Saddle River, N.J.

Table 5.4 Generalized Classification of Depositional Environments		
Primary depositional setting	**Major environment**	**Subenvironment**
Continental	*Fluvial	*Alluvial fan *Braided stream *Meandering stream
	*Desert	
	Lacustrine	
	*Glacial	
	*Deltaic	*Delta plain *Delta front *Prodelta
Marginal-marine	*Beach/barrier island	
	*Estaurine/lagoonal	
	Tidal flat	
	Neritic	*Continental shelf **Organic reef
Marine	Oceanic	Continental slope Deep-ocean floor

*Dominantly siliciclastic deposition
**Dominantly carbonate deposition
Environments not marked by an asterisk(s) may be sites of siliciclastic, carbonate, evaporite, or mixed sediment deposition depending upon conditions. *Source:* S. Boggs, Jr., *Principles of Sedimentology and Stratigraphy*, 3rd ed., © 2001 by Prentice Hall, Inc., Upper Saddle River, N.J.

other types of stratigraphy as well. The application of fossils in sedimentary rocks to the interpretation of their lateral and vertical relationships is *biostratigraphy*. The changes in organisms over geologic time, the record of evolution, can be used to assign relative ages to rock units. For example, Ordovician fossils in North America are identical to those in the type area in Wales, whether or not specific rock units were ever physically continuous between the two regions. There are other subtypes of stratigraphy as well.

Since both vertical and lateral lithologic changes in sedimentary rocks are important, let's consider vertical changes first. The boundary between a lithologic unit and the unit immediately above or below it in a section is called a *contact*. Contacts can be either *conformable*, when the overlying unit was deposited very soon after the lower unit, or *unconformable*, when there is a significant break in sedimentation between the two units. The plane representing the break is an unconformity. The famous unconformity that helped James Hutton realize the great longevity of geologic time (Chapter 2) is one type of unconformity.

Some examples of different types of contacts are shown in Figure 5.30. The vertical change in lithology can be either abrupt or may occur over an interval of rock. Figure 5.31 shows an abrupt contact between the cross-bedded Permian Coconino Sandstone, a terrestrial deposit that was originally deposited as windblown sand dunes, and the marine Toroweap Formation, which is a limestone also of Permian age. When the contact occurs within an interval, the change can be a smooth transition from one lithology to another, or

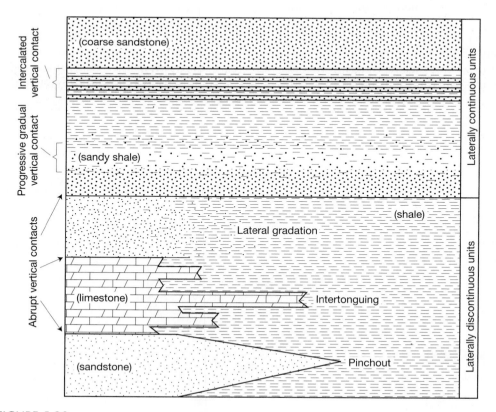

▲ FIGURE 5.30
Types of lateral and vertical contacts between lithologic units. *Source:* Sam Boggs, *Principles of Sedimentology and Stratigraphy*, 3rd ed., © 2001. Reprinted by permission of Pearson Educations, Inc., Upper Saddle River, N.J.

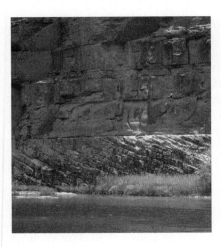

◀ FIGURE 5.31
An abrupt contact between the cross-bedded Coconino Sandstone and the overlying Toroweap Formation, Grand Canyon. *Source:* Photo courtesy of the author.

it can consist of alternating thin units of the upper and lower lithologies. This situation would imply that the depositional environment was fluctuating back and forth between two different facies for a period of time before the upper facies became dominant.

Lateral facies changes also produce characteristic relationships. These can also be gradational, but in a lateral rather than vertical sense, or other types, such as *pinchouts*, when one unit thins laterally until it terminates. *Intertonguing* contacts are similar to pinchouts, except that thin subdivisions of a thicker unit pinch out individually. It is easy to visualize how a pinchout contact could develop at a continental margin when sea level rises or falls (Figure 5.32). If sea level begins to rise, which stratigraphers call *transgression*, the new shoreline will move inland. The depositional facies of nearshore sand, offshore silty mud, and farther offshore mud will also shift laterally. The nearshore sand will be vertically overlain by

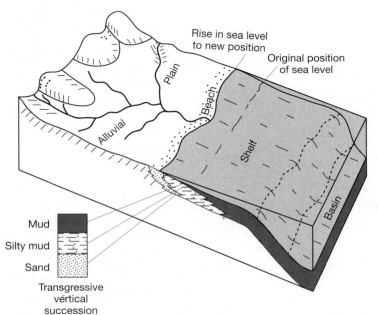

◀ FIGURE 5.32
Vertical and lateral relationships of facies during a marine transgression. *Source:* Sam Boggs, *Principles of Sedimentology and Stratigraphy*, 3rd ed., © 2001. Reprinted by permission of Pearson Education, Inc., Upper Saddle River, N.J.

silty mud after transgression and this sand will pinch out in the seaward direction. Multiple pinchouts could be created if sea level rises and falls over a period of geologic time.

The unconformity shown in Figure 2.31, which is known as an *angular unconformity*, is only one of several types that have been identified (Figure 5.33). Sometimes the unconformity is an undulating surface produced by erosion, but the rocks below have not been deformed as in the angular unconformity. Here the unconformity is quite obvious. There are also cases in which the overlying and underlying units are closely parallel

▶ FIGURE 5.33

Types of unconformities. The unconformities in all cases represent a significant period of geologic time during which the rocks below the unconformity were being folded, faulted, and/or eroded or during which erosion was minimal but no younger sediments were being deposited. *Source:* From S. Boggs, Jr., *Principles of Sedimentology and Stratigraphy*, 3rd ed., © 2001 by Prentice Hall, Inc., Upper Saddle River, N.J.

(a) Angular Unconformity

(b) Disconformity

(c) Paraconformity

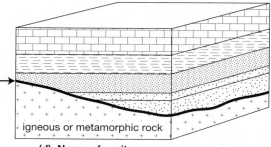

(d) Nonconformity

with no obvious surface of erosion. These unconformities are known as *paraconformities*. Finally, the flat-lying sedimentary rock sequence may overlay an erosion surface developed on igneous or metamorphic rocks. This is called a *nonconformity*.

The rocks exposed in the Grand Canyon contain several types of unconformities. The "Great Unconformity" is an angular unconformity between Precambrian sedimentary rocks and flat-lying Paleozoic rocks (Figure 5.34) in the eastern part of the Grand Canyon.

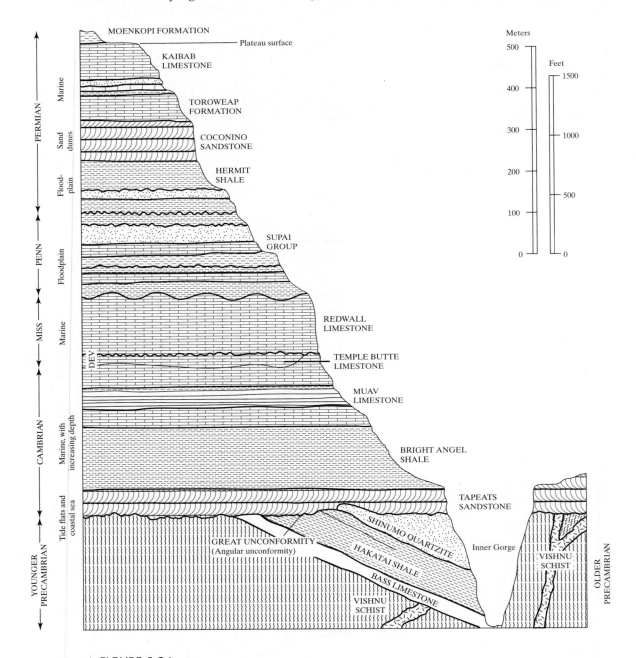

▲ FIGURE 5.34

Geologic column in the Grand Canyon. *Source:* © 1988. Illustration reprinted with permission of the publisher from *Pages of Stone: Geology of Western National Parks and Monuments* by Halka Chronic, The Mountaineers, Seattle.

▶ Figure 5.35

Unconformities between Precambrian and Cambrian rocks in the Grand Canyon. (a) Tilted and eroded Precambrian sedimentary rocks 700 to 1.5 billion years old form an angular unconformity with the overlying flat-lying Cambrian rocks. (b) The Inner Gorge, formed by the Precambrian Vishnu Schist (>2 billion years old). A nonconformity occurs at the contact between the schist and the flat-lying Cambrian rocks above. *Source:* Photos courtesy of the author.

(a)

(b)

Nearly a billion years of erosion and nondeposition separate these two units. In the western section of the canyon, Cambrian rocks lie directly on top of Precambrian metamorphic schist to form a nonconformity (Figure 5.35a, b). Disconformities in the section include the contact between the Cambrian Muav Limestone and the Devonian Temple Butte Limestone. A time interval of at least 100 million years, which encompasses the Ordovician and Silurian periods, is represented by this contact.

Six major unconformities have been discovered to be worldwide in extent. The rocks between them are largely conformable sequences of rocks separated by these major unconformities. In recent years, the recognition of these major rock sequences has led to the development of a new type of stratigraphy—*sequence stratigraphy*. Each major sequence is bounded above and below by an unconformity that represents a major erosional interval. A sequence would therefore involve a major trangression after the lower erosional interval, followed by a major regression and subaerial exposure. During the transgressive phase of the sequence, sea level rose dramatically and covered large parts of the continents with shallow seas. During this part of the sequence, nearshore sedimentary rocks, mainly sandstone,

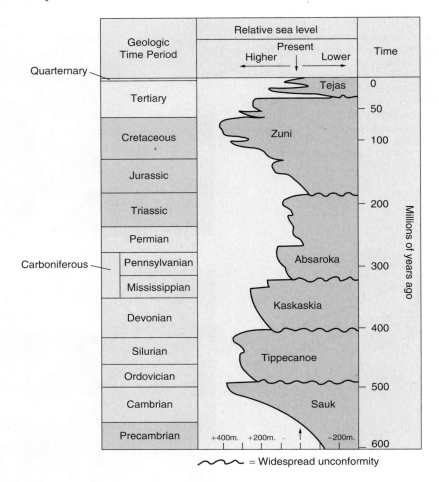

▲ FIGURE 5.36
The major sequences recognized in sequence stratigraphy. Each one represents a widespread sea-level rise and fall and is bounded by an unconformity. *Source:* Jon P. Davidson, Walter E. Reed, and Paul M. Davis, *Exploring Earth: An Introduction to Physical Geology*, 2nd ed., © 2002. Reprinted by permission of Pearson Education, Inc., Upper Saddle River, N.J.

would be overlain by offshore rocks, including shales and carbonate rocks. As the regression takes place, the shoreline migrates off the continent and nearshore formations are deposited on top of the offshore shales and carbonate rocks. Figure 5.36 shows the relationship of the six major sequences to the geologic time scale. The recognition of these sequences is a very important tool in correlating rock units in thick sedimentary rock strata, particularly in the subsurface. An obvious question is what caused such extensive sea-level changes? Three types of mechanisms have been suggested—expansion and retreat of large continental glaciers during an ice age, expansion or contraction of midocean ridge volumes during periods of rapid or slow seafloor spreading, and major tectonic uplift or subsidence of continental areas.

Geologic Maps and Cross Sections

Imagine that you are assigned to a highway design project in an area of sedimentary rocks. Since the route may cross several rock types with different engineering properties, where do you get preliminary information on the distribution of these rocks? With luck, there will be a published geologic map of the area available. For this reason, engineers should be familiar with the use of these maps.

The basic map unit on geologic maps is the *formation.* In a sedimentary rock terrain, a formation is a bed or group of related beds whose contacts can be traced where they are exposed or approximated where they are covered by other units or eroded away. Other units that may be shown on geologic maps include *members*, which are significant subdivisions of formations, and *groups*, which include two or more formations that appear to have a genetic relationship. These map units are called *rock units* because they are based strictly on lithologic characteristics without any consideration of age.

The field geologist constructs the map by measuring and describing vertical sequences of beds wherever they are well exposed. These measured sections are correlated to other sections across areas where the rocks are not exposed by using color, grain size, sedimentary structures, fossils, and any other characteristic that may be distinctive (Figure 5.37). Diagrams showing vertical sections and correlations are called *cross sections*.

▲ FIGURE 5.37
Correlation of rock units within a small area. *Source:* From W. L. Newman, *Geologic Time*, U.S. Geological Survey Pamphlet.

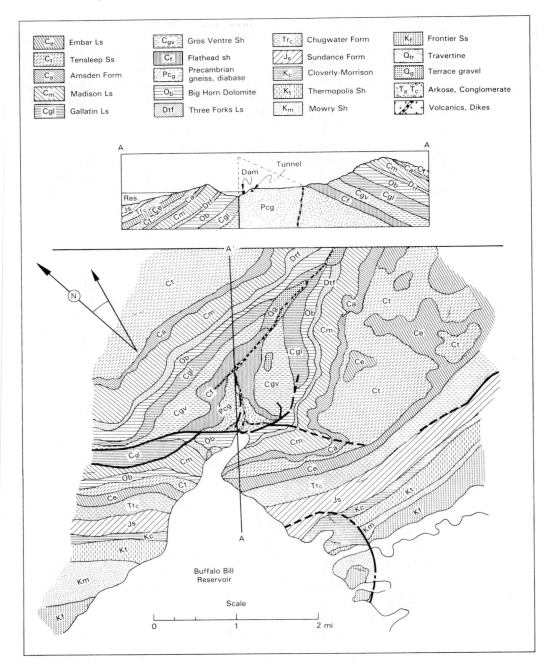

▲ FIGURE 5.38
Geologic map and cross section of the Cody Highway Tunnels area, Wyoming. *Source:* From W. F. Sherman, 1964, Engineering geology of the Cody Highway Tunnels, Park County, Wyoming, in *Engineering Geology Case Histories*, No. 4, P. D. Trask and G. A. Kiersch, eds.

Contacts are drawn on topographic or planimetric base maps to complete the geologic map, and the map and cross sections are often published together (Figure 5.38).

Correlations can be made at larger scales using other types of units. For example, units for which age can be assigned are known as *time-rock* units. A time-rock unit might consist

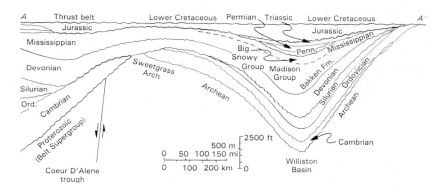

▲ FIGURE 5.39
Subsurface correlation of time-rock units across the Williston Basin. *Source:* After J. A. Peterson, 1988, Phanerozoic Stratigraphy of the northern Rocky Mountain region, in *The Geology of North America, vol. D-2: Sedimentary Cover North American Craton, U.S.,* Geological Society of America.

of the rocks that were deposited in the Cambrian period. Correlations across large distances and different depositional basins can be made (Figure 5.39). The unconformity-bounded units developed in sequence stratigraphy are particularly useful for this type of correlation.

Sedimentary Basins and Fossil Fuel Energy Resources

Sedimentary rocks contain many types of mineral resources. The Mississippi Valley lead-zinc ore deposits discussed in Chapter 3 are a good example. Energy resources, however, are particularly important because they are confined to sedimentary rocks. Petroleum and natural gas require very complex conditions for formation and concentration where they can be economically recovered. The first step in the sequence of events involves burial of abundant marine organisms in sedimentary basins that maintain low oxygen levels; in some cases, this has been caused by a restricted circulation between the basin and the surrounding ocean. Sedimentary basins can be formed in a variety of ways. Plate tectonic processes are associated with most of these types. For example, seafloor spreading creates oceanic basins on both sides of the midoceanic ridge. Deposition of sediments, however, is not the same in all parts of the basin. In the North Atlantic Ocean, sediment deposition is greatest near the North American continent because of the high sediment load carried to the coast by rivers draining the continent and decreases with distance away from the continent toward the center of the ocean. Thick sediment accumulations are also present near subducting plate margins, in which the source could be an island arc or a continental plate. Once sediment begins to accumulate in a basin, the weight of the sediment itself leads to subsidence of the basin, making it more likely to serve as a sediment trap.

During the Paleozoic era, much of North America consisted of relatively stable continental crust, a *craton*, covered by shallow seas during the major marine transgressions that took place. Sedimentary basins developed both near plate margins and in areas of crustal subsidence within the more stable craton (Figure 5.40). These areas are known as *intracratonic basins*, and as they gradually subsided, thicker sediments were deposited in them. The Williston Basin (Figure 5.39) is an intracratonic basin, as is the Michigan Basin (Figures 5.40, 5.41). Subsidence of the basin began in Cambrian time and continued intermittently through the Paleozoic. Over an interval of about 200 million years, the basin filled with about 4800 m of mostly marine sedimentary rock. A detailed diagram of the stratigraphy of the Michigan Basin is available at www.deq.state.mi.us/documents/deq-glm-rcim-geology-Stratigraphic_Nomenclature_2000.pdf. These sedimentary rocks include organic rich shales that served as *source rocks* for petroleum formation and more permeable rocks such as limestones, dolomites, and sandstones that constitute *reservoir rocks*.

The formation of petroleum from source rocks involves a complex series of events beginning with the deposition of the remains of marine organisms—mainly *phytoplankton* and *zooplankton*, which are microscopic plants and animals, respectively, that live in the upper part of the water column. When these organisms die and sink to the seafloor, they

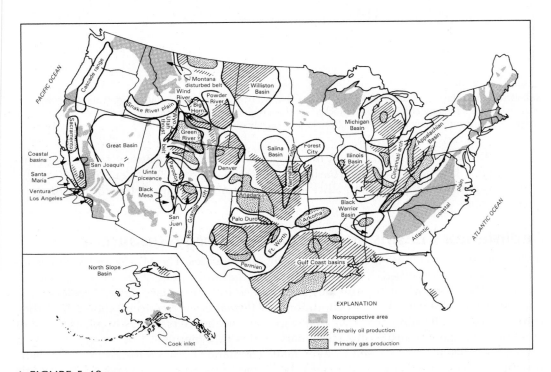

▲ FIGURE 5.40
Petroleum basin map of the United States. *Source:* From C. D. Masters and R. F. Mast, 1987, Geological setting of U.S. fossil fuels, *Episodes*, 10.

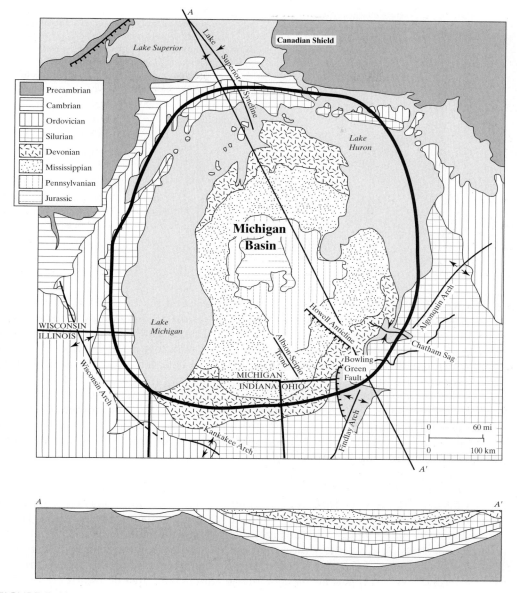

▲ FIGURE 5.41
Stratigraphic units exposed beneath glacial deposits in the Michigan Basin. Cross section shows bowl-shaped form of basin at depth. Heavy dark line shows approximate boundary of Michigan Basin.

can be partially preserved if they are buried by sediment and if oxygen levels are low. A very complex sequence of biochemical reactions begins to anaerobically degrade the organic matter. One of the first steps in this process is the release of gases such as carbon dioxide and methane. In subsiding basins, pressure is added through the weight of sediments deposited above the organic matter, and temperature increases with depth because of the geothermal gradient. The increases in temperature and pressure are critical to the formation of petroleum. One intermediate step is the formation of a substance called *kerogen,* which is a carbon-rich residue produced by expulsion of water and other elements

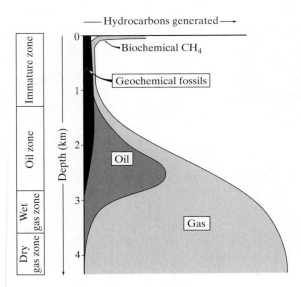

◀ FIGURE 5.42
Formation of petroleum and natural gas as a function of depth of burial. *Source:* Modified from Robert C. Laudon, *Principles of Petroleum Development Geology,* 1st ed., © 1996. Reprinted by permission of Pearson Education, Inc., Upper Saddle River, N.J.

that is insoluble in the organic acids that are also produced by the degradation of organic matter. Kerogen is a precursor for the formation of almost all oil and gas. With continued burial, the kerogen-rich sediment may enter a critical range of temperatures and pressures known as the *oil window*. The temperatures in the oil window range between 100°C and 200°C, which usually corresponds to a depth of burial between 1 and 3 km. Subjected to higher temperatures, the organic matter undergoes thermal *cracking,* or breakdown, of the organic compounds in petroleum to smaller molecules such as methane (CH_4), or natural gas. These relationships are summarized in Figure 5.42. Pressure has another effect during this processes. The rising fluid pressure in petroleum and natural gas produced from kerogen can exceed the strength of the fine-grained source rocks, creating microcracks, which allow the fluids to migrate into more *permeable* reservoir rocks. Permeability is discussed in more detail in Chapter 7.

The final component necessary to produce an economically significant accumulation of oil or gas in a sedimentary basin is a *trap* (Figure 5.43). The conditions necessary to form a trap include some kind of barrier that prevents the oil and gas from migrating. The barrier can be provided by a fault and the favorable configuration of a rock of low permeability, such as a shale or evaporite. Folds that are bowed upward (anticlines), illustrated in Figure 5.43c, were one of the first types of traps to be recognized by the petroleum industry. The oil and gas is trapped at the crest of the anticline against an overlying impermeable unit because oil and gas are less dense than the saltwater or brine that is also usually present in reservoir rocks. Fluids move toward regions of lower pressure; at the crest of an anticline, the hydraulic pressures acting on the oil and gas are greater in both directions and the hydrocarbons are therefore trapped in this location.

Many of the basins that contain oil and gas resources also contain coal. Coal differs from petroleum in that its primary source is land-based plants that accumulated in low, swampy regions during periods of subaerial exposure of the basins. The organic matter is preserved because of the high water table in these swamps and wetlands, which minimizes the amount of oxygen present to oxidize organic matter. Much like the formation of petroleum, the accumulating plant matter goes through a series of changes after burial and increase in temperature. During these changes, oxygen, hydrogen, and water are driven off and the original partly decomposed plant matter, known as *peat*, is converted to coals of increasing *rank* (Figure 5.44). High-rank coals, including bituminous and anthracite

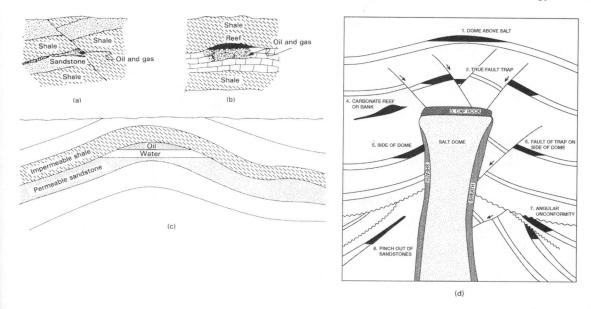

▲ FIGURE 5.43

Four types of petroleum traps. (a) Trap is caused by truncation of reservoir rocks by fault. (b) Oil accumulates in permeable reef structure surrounded by less permeable rocks. (c) Oil and gas accumulate at crest of anticline because they are less dense than surrounding fluids. (d) Various types of traps associated with salt domes. *Source*: Robert C. Laudon, *Principles of Petroleum Development Geology*, 1st ed., © 1996. Reprinted by permission of Pearson Education, Inc., Upper Saddle River, N.J.

	Rank stages	Characteristics	H_2O	Heat content

	Peat	Large pores
		Details of original
		  plant matter still
		  recognizable
		Free cellulose
				3000 kcal/kg
				(5400 Btul/lb)

	Soft brown coal	No free cellulose	~75	
	Dull brown coal	Marked compaction of plant structures	~35	4000 kcal/kg (7200 Btul/lb)
		Plant structures partly recognizable	~25	5500 kcal/kg (9900 Btul/lb)
	Bright brown coal		~10	7000 kcal/kg (12,000 Btul/lb)
	Bituminous	Plant structures no longer recognizable		8650 kcal/kg (15,500 Btul/lb)
	Anthracite			

Rank increasing (↓ along left margin)
Brown coal and lignite · *Hard brown coal* · *Hard coal* (left-side vertical labels)

▲ FIGURE 5.44

Characteristics of different ranks of coal.

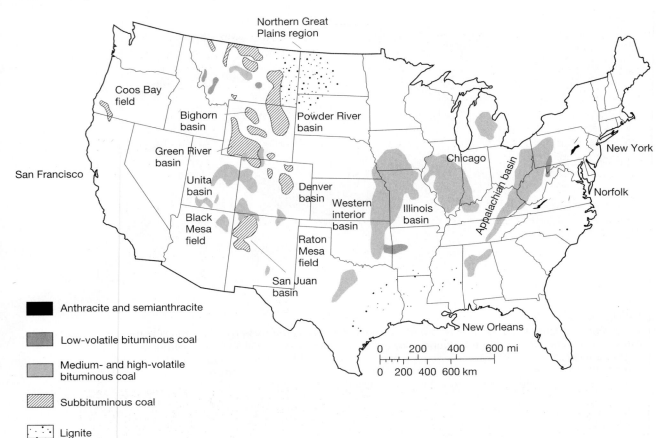

▲ **FIGURE 5.45**
Location and rank of major coal fields in the United States. *Source:* James R. Craig, David J. Vaughan, and Brian J. Skinner, *Resources of the Earth: Origin, Use, and Environmental Impact*, 3rd ed., © 2001. Reprinted by permission of Pearson Education, Inc., Upper Saddle River, N.J.

coals, are more valuable because they yield a greater heat content upon burning. Lignite, a form of low-rank coal, is abundant in the northern Great Plains of the United States, but its moisture content is too high to make transport to power plant sites economically feasible. Power plants are therefore constructed near the lignite mines and the electricity is transported to end uses through power grids. The most important coal-producing regions of the United States are shown on Figure 5.45.

Engineering in Sedimentary Rocks

Engineering conditions in sedimentary rocks are difficult to generalize because of the wide range in lithology, in degree of lithification, and in the orientation of bedding planes and other structures.

Stable vertical slopes can usually be excavated in well-cemented, horizontally bedded sandstones and limestones. Flatter slope angles must be cut for weaker rock types. Particularly important factors in the stability of sedimentary rock slopes are the direction and the amount of slope, or dip, of bedding. The most unfavorable situation occurs where bedding dips parallel or nearly parallel to the downslope angle of the slope. Bedding planes are zones of weakness in sedimentary rock masses and failure may occur. A huge landslide took place at the Vaiont Reservoir in Italy in 1963 partly because bedding dipped toward the center of the valley (Chapter 13).

Tunneling and underground mining in sedimentary rocks are influenced by lithology and structure (the orientation of the bedding). Where the stratigraphy of sedimentary rocks consists of horizontal or gently dipping beds, it is quite simple to predict the rock types to be encountered along the path of a tunnel. Difficulties arise in areas where the structure is more complex. Well-cemented sedimentary rocks are generally adequate for most types of building foundations. Special problems occur in limestones and evaporite deposits because these rocks are soluble under the action of flowing groundwater. The soils and rocks overlying underground cavities produced by chemical dissolution may collapse into the voids, damaging or destroying buildings constructed at the surface (Chapter 10). This phenomenon is also a problem above shallow underground mines. Dams and reservoir sites are subject to similar limitations. Undesirable leakage of water may occur along bedding planes or through solution cavities in the rock.

Like igneous rocks, the engineering properties of the sedimentary rocks are influenced by geologic events that take place long after deposition of the sediments. Strength can be increased by compaction and cementation. Alternatively, sedimentary rocks can be weakened by weathering. Shales are particularly susceptible to breakdown to clayey soils upon exposure to air and water. These relatively rapid changes can lead to problems in excavations and foundations.

Discontinuities in the rock caused by tectonic events are also important in determining the engineering behavior of a sedimentary rock mass. Faults and fractures can significantly weaken a body of rock, as well as allow the movement of water through the material.

Case in Point 5.1
Low-Level Radioactive Waste Disposal

One of the most controversial environmental problems facing nations that use nuclear power and possess nuclear weapons is the disposal of radioactive wastes. These wastes include high-level components that have long half-lives and must be isolated from the environment for thousands of years and low-level waste products that must be isolated for much shorter periods. Under current policy in the United States, a single repository is being developed for high-level commercial nuclear power plant wastes. In contrast, low-level waste, because of its lower hazard, will be disposed of in regional facilities licensed by the U.S. Department of Energy.

The history of radioactive waste disposal, like almost every other type of waste disposal, has progressed from little to no regard for health or the environment to a state approaching paranoia on the part of some individuals and groups in our society. As commercial nuclear reactors became common in the United States after World War II, low-level waste was commonly buried in shallow trenches at federally operated sites. After 1962, commercial sites were licensed to dispose of the waste from nuclear power plants (Figure 5.46). These sites were distributed across the country from arid to humid climatic settings. Burial methods changed little from earlier operations—wastes were buried in shallow trenches and capped with nearby soil materials (Figure 5.47). Materials of low permeability (a low ability to transmit fluids) were selected if possible to minimize migration of waste that could be leached by water infiltrating through the cap and into the waste. Unfortunately, most of these sites were not able to contain the wastes as anticipated and radionuclides migrated out of the trenches for various distances. Currently, only three commercial sites remain open until the new generation of engineered facilities can be developed. A brief history of two of the sites constructed in sedimentary rocks will be presented subsequently.

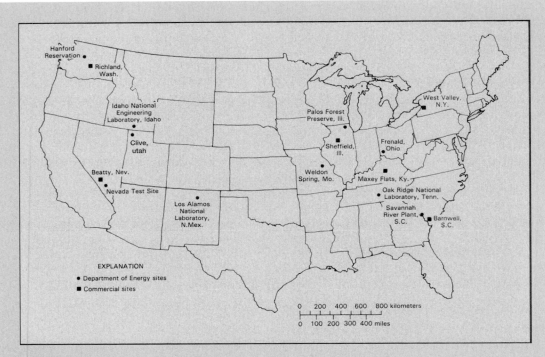

▲ FIGURE 5.46
Commercial and Department of Energy low-level radioactive-waste sites in the United States. *Source:* From M. S. Bedinger, 1989, *Geohydrologic Aspects for Siting and Design of Low-Level Radioactive-Waste Disposal,* U.S. Geological Survey Circular 1034.

▲ FIGURE 5.47
Active trenches containing low-level radioactive waste at the Barnwell, South Carolina, site. *Source:* Photo courtesy of the author.

The Oak Ridge National Laboratory has been used for burial of low-level wastes since 1944. Trenches were excavated in weathered material overlying Paleozoic sedimentary rocks that were mainly composed of shale and limestone (Figure 5.48). Water frequently accumulated in the trenches in this humid climate and was not initially considered to be a problem. Problems developed, however, and radionuclide migration occurred in several ways. In the carbonate rocks, leached radionuclides were transported through solution cavities and fractures. In the shale sites, a different process occurred. Because of the low permeability of these materials and the higher permeability of the trench caps, the filled trenches behaved according to the "bathtub" effect: Water infiltrated from the surface more rapidly than it drained through the trench walls and bottoms, and gradually waste liquids overflowed in seeps at the contact point between the cap material and the natural soils.

Maxey Flats, Kentucky, is located about 300 km north of the Oak Ridge site. Here, a commercial site was constructed in flat-lying, late Paleozoic shale and sandstone (Figure 5.49). Trenches were excavated in the soil developed upon the Nancy Member of the Borden Formation, a unit composed of interbedded sandstone and shale. Despite grading the trenches to a sump containing pipes to drain off accumulated water, radionuclides migrated out of the trenches. The primary pathway was found to be a fractured sandstone unit that occurred near the bottom of the trenches. Certain radionuclides were estimated to be migrating at a rate of 17 m per year, and seeps along hillsides near the trenches showed contamination by radioactive isotopes. In an attempt to reduce infiltration into the trenches, compacted clay caps were used to cover the trenches. When the clay caps failed to solve the problem, PVC (polyvinylchloride) sheets were used to cover the trenches. Although these covers have been effective, it is obvious that the site was poorly suited for the method of waste disposal chosen.

▲ FIGURE 5.48
Geologic cross section of sedimentary rock units at the Oak Ridge National Laboratory site in Tennessee. *Source:* From M. S. Bedinger, 1989, *Geohydrologic Aspects for Siting and Design of Low-Level Radioactive-Waste Disposal*, U.S. Geological Survey Circular 1034.

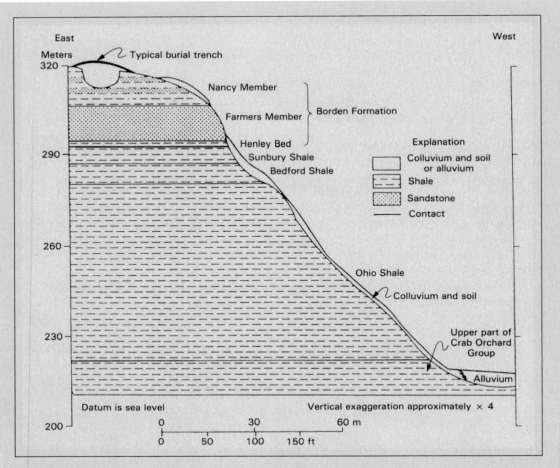

▲ FIGURE 5.49
Geologic cross section of sedimentary rock units at the Maxey Flats, Kentucky, disposal site. *Source:* From M. S. Bedinger, 1989, *Geohydrologic Aspects for Siting and Design of Low-Level Radioactive-Waste Disposal*, U.S. Geologic Survey Circular 1034.

Under the Low-Level Radioactive Waste Policy Act (1980) and Amendments Act (1985), states must take responsibility for their own waste. New facilities must meet very strict performance requirements. Siting will be based upon favorable geological and hydrological conditions as indicated by extensive geological investigations at prospective sites. In addition, the disposal facilities will be engineered structures, designed to provide multiple barriers to waste migration. Design concepts include both above- and belowground concrete vaults, with multiple permeability barriers and drainage systems to prevent any leachate from migrating into the surface or subsurface environment (Figure 5.50). Continuous monitoring of all components of the site will be required.

As part of the Low-Level Radioactive Waste Policy Act, regional waste compacts were established so that a group of states could band together to develop a site that would accept waste from all states in the compact (Figure 5.51). This strategy has been unsuccessful in establishing new sites for any of the compacts owing to public concerns about site selection and political disputes between states. Some states—Michigan, for example—have pulled out of regional compacts to avoid hosting a site that would accept waste from other states. Currently, only the 11 states of the Northwest and Rocky

▲ FIGURE 5.50
Potential design concepts for engineered low-level radioactive-waste repository. *Source:* From Michigan Low-Level Radioactive Waste Authority, 1989, *Siting Process Overview.*

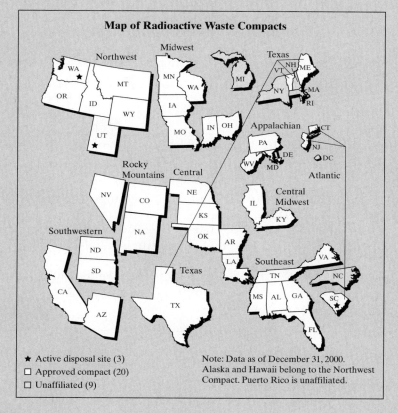

▲ FIGURE 5.51
Map of Low-Level Radioactive Waste compacts. Shaded states are not members of a compact. *Source:* Nuclear Regulatory Commission.

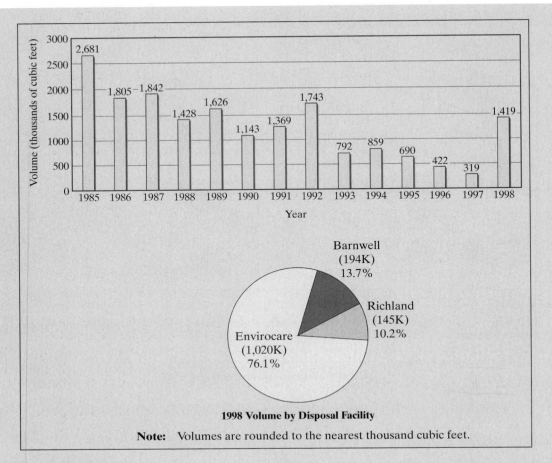

▲ FIGURE 5.52
Volumes of low-level radioactive waste disposed of in three existing commercial facilities, along with the proportions for each site. *Source:* U.S. Nuclear Regulatory Commission.

Mountain compacts, which use the Richland, Washington, site, and the 3 states of the Atlantic Compact, which use the Barnwell, South Carolina, site, have assured long-term access to disposal facilities. The other 36 states utilize Barnwell, which is phasing out its national acceptance of waste, and the Envirocare site in Clive, Utah. The amounts of waste and their division among these three sites is shown in Figure 5.52. Only Texas is close to permitting a new site. Ultimately, litigation may be the only way to break the logjam that currently exists and force compliance with the Low-Level Radioactive Waste Policy Act.

Summary and Conclusions

Sedimentary rocks originate from sedimentary processes that have been operating throughout geologic time. The first stage in the evolution of these rocks is the weathering and erosion of existing rocks. The products derived from this activity are then transported and deposited in a variety of terrestrial and marine environments. After deposition has occurred, then compaction, cementation, and crystallization cause lithification of the initially soft sediment.

Sedimentary rocks exhibit either clastic or nonclastic textures. Clastic textures are characteristic of the detrital rocks, those that have been transported to their point of deposition. The classification of the detrital rocks depends upon the size of the particles that make up the rock.

Although clastic textures may also be present in the chemical and biochemical sedimentary rocks, many of these rocks are composed of grains that have chemically precipitated at the site of deposition. Nonclastic textures, characterized by interlocking grain networks, result from this process.

The nature of each depositional environment imparts numerous characteristic features to the resulting rocks. Current strength, mode of transport, and water depth influence the sorting and bedding of the rock. Transport of sediment in ripples or dunes causes cross bedding to develop. Parallel bedding forms both by current action and by deposition in standing water. Turbidity currents transport high concentrations of suspended sediments that are deposited in graded beds.

The color and fossil content of sedimentary rocks are partially determined by geochemical factors. The presence or absence of oxygen is an extremely important variable in this regard.

Study of the stratigraphy of sedimentary rocks, using all the characteristics that have been described, has produced important information about the geologic past as well as practical data of great value to engineering projects. Fossil fuel energy resources are restricted to sedimentary rocks. Oil and gas form in subsiding sedimentary basins from the remains of microscopic marine organisms. Pressure and temperature drive the biochemical reactions to produce the organic compounds in petroleum and natural gas. The three conditions that are necessary for commercial accumulations are source rocks, reservoir rocks, and traps. Coal is the result of the conversion of organic matter derived from land plants under the action of burial and temperature increase.

Problems

1. How do sedimentary rocks fit into the geologic cycle?

2. What characteristics of the depositional environment influence the properties of sedimentary rocks?

3. How do postdepositional changes convert sediments to sedimentary rocks?

4. What chemical conditions influence the precipitation of nonclastic sedimentary rocks?

5. How do cross beds form?

6. What does color indicate about the geochemical conditions of the depositional environment?

7. What methods are available to gather data necessary to construct a stratigraphic cross section in an area? Why are cross sections useful in highway construction and other engineering projects?

8. What properties influence the strength or possible failure of slopes underlain by sedimentary rocks?

9. Why is petroleum production associated with ancient sedimentary basins? What factors must be present for an economically significant accumulation of oil or gas?

10. What do unconformities tell us about the history of a sedimentary rock sequence? Describe the different types and their implications.

11. What is sequence stratigraphy?

12. What types of sedimentary rocks are considered to be potentially suitable for high-level nuclear waste disposal? Why?

13. Describe the design features of low-level radioactive waste-disposal facilities. What problems have been encountered with respect to these sites in sedimentary rocks?

References and Suggestions for Further Reading

BEDINGER, M. S. 1989. *Geohydrologic Aspects for Siting and Design of Low-Level Radioactive-Waste Disposal.* U.S. Geological Survey Circular 1034.

BOGGS, S., Jr. 2001. *Principles of Sedimentology and Stratigraphy*, 3rd ed. Upper Saddle River, N.J.: Prentice Hall, Inc.

BLATT, H., G. MIDDLETON, and R. MURRAY. 1980. *Origin of Sedimentary Rocks*, 2nd ed. Englewood Cliffs, N.J.: Prentice Hall, Inc.

CHRONIC, H. 1988. *Pages of Stone*. Seattle: The Mountaineers.

CRAIG, J. R., D. J. VAUGHAN, and B. J. SKINNER. 2001. *Resources of the Earth*, 3rd ed. Upper Saddle River, N.J.: Prentice Hall, Inc.

GEOLOGICAL SOCIETY OF AMERICA. 1964. *Engineering Case Histories: Numbers 1–5*, P. D. Trask and G. A. Kiersch, eds. Boulder, Colo.

JUDSON, S., M. E. KAUFFMAN, and L. D. LEET. 1987. *Physical Geology*, 7th ed. Englewood Cliffs, N.J.: Prentice Hall, Inc.

LAUDON, R. C. 1996. *Principles of Petroleum Development Engineering*. Upper Saddle River, N.J.: Prentice Hall, Inc.

MASTERS, C. D., and R. F. MAST. 1987. Geological setting of U.S. fossil fuels. *Episodes*, 10(no. 4):308–313.

MICHIGAN LOW-LEVEL RADIOACTIVE WASTE AUTHORITY. 1989. *Siting Process Overview.*

PETERSON, J. A. 1988. *Phanerozoic Stratigraphy of the Northern Rocky Mountain Region: The Geology of North America, v. D–2, Sedimentary Cover North American Craton.* Boulder, Colo.: Geological Society of America.

U.S. DEPARTMENT OF ENERGY. 1984. Draft environmental assessment—Vacherie Dome site, Louisiana. (DOE/RW-0016). Washington, D.C.

Metamorphic Rocks and Processes

The igneous and sedimentary rocks that form at or beneath the earth's surface are not always the final result of the dynamic activity of the earth's crust. When exposed to the atmosphere, the minerals in these rocks may chemically alter to become new minerals that are more stable under the temperatures and pressures at the earth's surface. Similarly, when igneous and sedimentary rocks become buried deep within the crust because of the movements of lithospheric plates, they are subjected to elevated temperatures and pressures. Just as the rocks are unstable at low temperatures and pressures on the earth's surface, they are also unstable at the high temperatures and pressures existing at depths well below the surface. Under these conditions, minerals in the rock recrystallize to become new minerals, and new rock is formed. The processes that cause these changes are grouped under the term *metamorphism*. Study of these rocks yields valuable information about the effects of these conditions on rock and about the geologic history of a region.

Metamorphic Processes

The normal increases in temperature and pressure with increasing depth in the earth are sufficient to initiate metamorphic activity. In addition, tectonic stresses caused by the collision of two lithospheric plates, for example, can generate intense pressures oriented in a particular direction (Figure 6.1). The heat and pressure in deep crustal environments are two of the main components of metamorphic processes. A third component is the effect of liquid or gaseous solutions that move through the rocks and promote the chemical reactions that form new minerals.

Heat is perhaps the most important metamorphic agent. In Figure 6.2, the rate of increase of temperature with depth, or the *geothermal gradient*,

◄ FIGURE 6.1
Metamorphism at a convergent plate margin. Rocks metamorphosed may include sediments or other continental or oceanic crustal material.

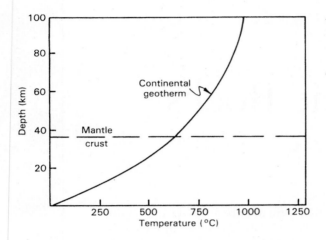

◄ FIGURE 6.2
Representative continental geotherm. *Source:* From G. C. Brown and A. E. Mussett, *The Inaccessible Earth,* © 1981 by George Allen & Unwin, Ltd., London.

suggests that the temperature in the crust at a depth of 15 km is approximately 300°C. This temperature is sufficient for the recrystallization of some minerals to begin. Because the geothermal gradient is highly variable from place to place, the depth associated with a particular temperature is not constant.

The effect of pressure varies at different depths in the crust. At shallow depths, rocks are relatively cold and brittle, so they can be fractured and crushed when subjected to high pressures. At greater depths, rocks are much softer because of the high temperatures. Under the action of pressure, they tend to deform by plastic flow, like modeling clay squeezed between the fingers. In the region of plastic deformation, pressure influences the types of new minerals formed. Typically, the atoms within the mineral structure are more closely packed together when the mineral crystallizes under high pressure.

The recrystallization of minerals during a metamorphic event is largely a solid-phase process. The rock does not actually melt and then recrystallize. The solid-phase reactions between minerals are greatly facilitated by the movement of small amounts of liquid or gaseous solutions through the rock. These solutions, which travel through the pores and cracks of the rock, add and remove various ions and molecules as the reactions occur. In this way, new chemical constituents can be brought into contact with mineral grains so that they may diffuse through the mineral structures during recrystallization.

Types of Metamorphism

Metamorphism is associated with several types of geologic events. The intrusion of a pluton, for example, brings magma into contact with existing crustal rocks (Figure 6.3). *Contact metamorphism* is the name given to the alteration of the surrounding *country rock* by the intruding magma. Heat is the most significant influence in contact metamorphism.

▶ FIGURE 6.3
The contact metamorphic aureole surrounding the Dartmoor Granite, Devon, England. *Source:* From R. Mason, *Petrology of the Metamorphic Rocks,* © 1978 by George Allen & Unwin, Ltd., London.

The effect of pressure is much less important. The extent of contact metamorphism, the zone of altered rock called the *metamorphic aureole,* rarely extends more than several hundred meters outward from the magmatic body. Thus, contact metamorphism is limited in areal extent to a thin shell around the pluton. More extensive effects may occur if solutions and vapors given off by the magma penetrate into the country rock along fractures. These *hydrothermal solutions* carry volatile components that separate from the cooling magma. Important vein-type ore deposits can be formed by hydrothermal solutions that migrate during contact metamorphic events.

A second type of metamorphism is characterized by much more extensive zones of rock alteration than found in the contact metamorphism zones. Its name, *regional metamorphism,* suggests a process that may transform rocks through huge portions of the crust. The conditions that produce regional metamorphism involve the effects of both temperature and pressure. Thus, regional metamorphism must occur deep within the crust, at depths of at least 10 km.

Geologic studies of metamorphic rocks indicate that the mineral content of rocks in regionally metamorphosed areas varies systematically. The specific group of minerals present in rock can be used to define a certain metamorphic *grade* in a particular area. High grade, for example, means that the rock was subjected to very high temperatures and pressures. Different metamorphic minerals can be produced from the same original rock under various metamorphic grades. Figure 6.4 shows the conditions of temperature and pressure under which the minerals kyanite, sillimanite, and andalusite form. All three metamorphic minerals have the same chemical composition (Al_2SiO_5) but different internal structures. Regional metamorphism produces areas in which metamorphic grade is very high. Grade decreases in all directions from the most intensely altered zones (Figure 6.5).

Regional metamorphism is often associated with the central cores of mountain ranges, whose rocks are exposed after the removal by erosion of thousands of meters of overlying

(a)

(b)

▲ FIGURE 6.16
St. Francis Dam (a) before and (b) after failure. *Source:* Photos courtesy of Los Angeles Department of Water and Power.

The geologic setting of the dam site is shown in Figure 6.17. On the east side of the valley, schists with foliation planes inclined toward the center of the valley formed the foundation material for the dam. The inherent weakness of the micaceous shists due to weak bonds along the foliation planes between adjacent mica grains was undoubtably a factor in the failure of the dam. Beneath the center of the dam, the schists were in fault contact with sedimentary rocks. The fault zone itself contained brecciated rocks produced during fault movement.

▲ FIGURE 6.17
Geologic cross section of the St. Francis Dam. Failure probably occurred in the schists beneath the east side of the dam. *Source:* Modified from C. F. Outland, *Man-made Disaster: The Story of St. Francis Dam*, © 1977 by the Arthur H. Clark Co., Glendale, Calif.

Although the precise cause of the dam collapse has never been proven, it is highly likely that water from the impounded reservoir seeped into the rocks beneath the dam, thus weakening the rock materials. On the east side, penetration of water into the schists may have weakened the resistance to failure along the inclined foliation planes in this abutment (Outland, 1977). At a critical point, failure of the rock abutment may have initiated the collapse of the dam.

Case in Point 6.2
Tunneling Problems in Metamorphic Rocks

Tunneling often presents some of the most complex and dangerous problems in civil engineering. These problems usually result from unknown or unexpected adverse geologic conditions encountered along the path of the tunnel.

The Harold D. Roberts Tunnel was constructed to divert water from the west slope of the Continental Divide, through the Front Range of the Rocky Mountains, for municipal use in the city of Denver, Colorado. Problems experienced during the construction of this 37.5-km concrete-lined tunnel were directly related to the physical properties of the rocks encountered along the centerline of the tunnel (Warner and Robinson, 1981). The geologic setting of this part of the Front Range consists of a highly complex series of igneous, metamorphic, and sedimentary rocks (Figure 6.18). The Precambrian metamorphic rocks, including schist, gneiss, quartzite, and migmatite, provided some of the most troublesome tunneling conditions along the route. Foliated and nonfoliated metamorphic rocks are present in various sections of the route, with biotite responsible for most of the foliation.

One of the major problems was the inflow of groundwater at high pressure from fractured rocks at certain locations. Flows as high as 32 L/s (500 gal/min) were measured. In order to test for zones of water inflow, which could not be predicted in advance, feeler holes were drilled ahead of the advancing tunnel face. When a flow was detected,

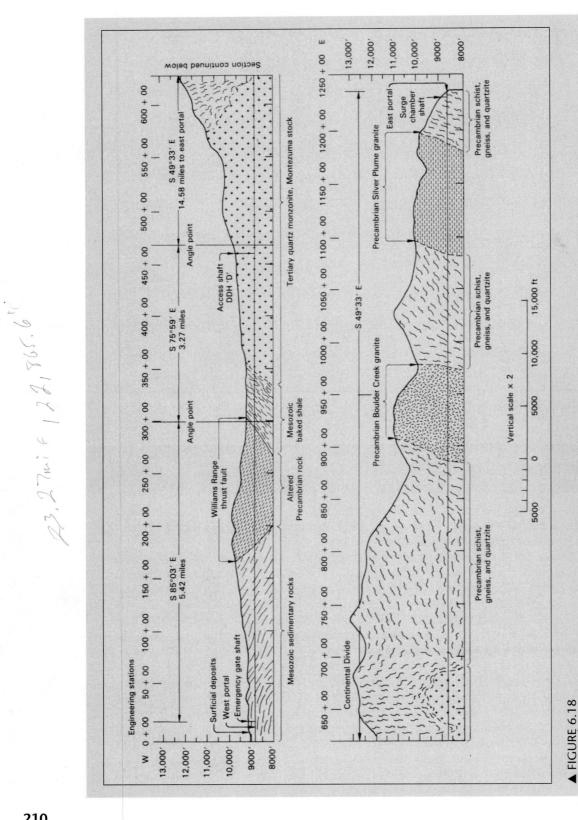

▲ FIGURE 6.18
Generalized geologic section along the route of the Harold D. Roberts Tunnel, Colorado. *Source:* From L. A. Warner and C. S. Robinson, 1981, *Geology of the Eastern Part of the Harold D. Roberts Tunnel, Colorado* (Stations 690 + 00 to 1238 + 58). U.S. Geological Survey Professional Paper 831-D. Diagram reprinted with permission of the Board of Water Commissioners, City and County of Denver, Colo.

grout (a cement slurry) was injected into the rock at high pressure to seal the fractures. Relatively strong gneiss frequently yielded high inflows of water through occasional fractures. Apparently, this rock behaved as a brittle material that could maintain open fractures, unlike the weaker foliated schist, in which fractures were most likely to seal themselves under the immense weight of the mountain range.

The cost of a tunnel is directly related to the number of steel supports that are needed to prevent collapse of the rock walls into the opening. These supports were installed as needed during construction at 0.6- to 0.18-m centers along the tunnel. Where the tunnel passed through strong, competent rock, steel supports were unnecessary. The percentage of a particular segment of tunnel that required steel supports ranged from 7.3 to 100, depending upon the rock type. When steel supports are needed at a section, the contractor must blast and excavate a larger cross-sectional area of rock than the estimated cross-sectional area, which is the area inside the *payline* (Figure 6.19). The extra area excavated is called *overbreak*. Overbreak increases the cost of a tunnel to the contractor because more concrete is needed to construct the finished tunnel lining. In Figure 6.20, mean overbreak, roughness factor, and eccentricity are plotted against percent support for a section of the Harold D. Roberts Tunnel. Roughness factor and eccentricity are parameters related, respectively, to irregularities and to differences in direction of the amount of overbreak. The graph clearly shows that, of the rock types compared, the foliated gneiss and schist had the highest values of the overbreak parameters and required the highest percentage of support. In this case, foliated metamorphic rocks were more costly for tunnel construction than other rock types.

▶ FIGURE 6.19
Cross section of a tunnel showing the overbreak, or area excavated beyond the payline to accommodate steel supports. The payline is the outer boundary of the area estimated for excavation prior to construction.

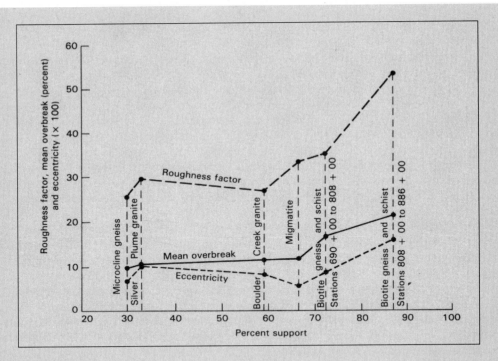

▲ FIGURE 6.20
Plot of various overbreak parameters versus percent support, showing that overbreak and therefore cost of tunneling increases in areas of foliated metamorphic rocks. *Source:* From L. A. Warner and C. S. Robinson, 1981, *Geology of the Eastern Part of the Harold D. Roberts Tunnel, Colorado* (Stations 690 + 00 to 1238 + 58), U.S. Geological Survey Professional Paper 831-D.

Summary and Conclusions

Metamorphic processes produce new rocks under the influences of heat, pressure, and hydrothermal solutions. Contact metamorphism is produced by the emplacement of magma bodies within the crust. Under the action of heat from the cooling magma, rocks are altered within a thin zone around the pluton. Nonfoliated metamorphic rocks are most commonly produced by contact metamorphism because of the minor effect of pressure in the process.

Pressure plays a much more important role in the formation of foliated rocks during regional metamorphism. Tectonic stress, the result of lithospheric plate interactions, combined with heat from deep burial, produces metamorphism on a regional scale. Metamorphic grade often decreases outward from a core of intensely altered rocks, where remelting may occur. This process may explain the huge granitic batholiths of the crust associated with migmatites.

Foliated metamorphic rocks develop by recrystallization of long, platy minerals oriented normal (perpendicular) to the direction of greatest pressure. Foliation is associated with various types of rock cleavage. With increasing metamorphic grade, slate, phyllite, schist, and gneiss can be formed. Nonfoliated rocks, such as quartzite and marble, involve recrystallization without the development of mineral-grain alignment.

The orientation of foliation is an important factor in construction in areas of metamorphosed rocks. The strength of a rock mass is much lower in the direction of the foliation than in other directions.

Problems

1. Where and why does metamorphism occur in the earth?

2. What are the differences between regional and contact metamorphism?

3. Explain how metamorphism can alter a rock chemically as opposed to mineralogically.

4. How is foliation produced in a metamorphic rock?

5. How do the characteristics of the original rock influence the properties of the metamorphic rock?

6. What are some of the engineering problems associated with metamorphic rocks?

References and Suggestions for Further Reading

DAVIDSON, J. P., W. E. REED, and P. M. DAVIS. 2002. *Exploring Earth*, 2nd ed. Upper Saddle River, N.J.: Prentice Hall, Inc.

MASON, R. 1978. *Petrology of the Metamorphic Rocks.* London: Allen & Unwin.

OUTLAND, C. F. 1977. *Man-made Disaster: The Story of St. Francis Dam.* Glendale, Calif.: Arthur H. Clark.

TURNER, F. J. 1981. *Metamorphic Petrology*, 2nd ed. New York: McGraw-Hill.

WARNER, L. A., and C. S. ROBINSON. 1981. *Geology of the Eastern Part of the Harold D. Roberts Tunnel, Colorado* (Stations 690 + 00 to 1238 + 58), U.S. Geological Survey Professional Paper 831-D.

CHAPTER 7

Mechanics of Rock Materials

It is not enough for engineers to have a basic understanding of the physical, chemical, and mineralogical characteristics of rocks, for in dealing with rock as an engineering material, engineers are involved with the *mechanics* of rock and other earth materials. Mechanics refers to the response of materials to applied loads. Although the same mechanical principles apply to all types of earth materials, two subdisciplines, *rock mechanics* and *soil mechanics*, have evolved to deal with the specific problems of rock and soil masses. The application of mechanics is also critical to an understanding of tectonic processes and the natural deformation of rock masses in the earth's lithosphere as well as in surficial processes such as mass wasting and land subsidence. We will address these topics in subsequent chapters.

General Types of Earth Materials
Rocks, Soils, and Fluids

From an engineering perspective, earth materials can be subdivided into three categories: rocks, soils, and fluids. Rocks are solid, dense aggregates of mineral grains. Igneous and metamorphic rocks are composed of interlocking mineral grains that were formed by crystallization from a magma or by recrystallization of an existing rock. The degree of interlocking of the mineral grains is one of several factors that determine how strong a rock will be. Chemically precipitated sedimentary rocks can also be interlocking aggregates of mineral grains. Clastic sedimentary rocks, on the other hand, are composed of particles that were derived from a preexisting rock. These particles were then transported by wind, water, or ice, deposited at a particular location, and bound together by various types and amounts of cementing agents. Even more important than the inherent properties of a rock are the properties of the rock mass, the field-scale volume of rock that includes fractures, joints, and other planes of weakness.

Soils, from the engineering standpoint, are similar to transported sedimentary rocks in that they consist of rock particles and minerals derived from preexisting rocks. Soils, however, lack strong cementing material between grains. For this reason, they usually can be excavated easily without heavy machinery or blasting. We will turn our attention to these materials in a later chapter.

The third major category of earth materials is composed of fluids. The earth's fluids include such liquids as water, magma, and petroleum, but they also include gases—for example, natural gas beneath the land surface and the gases of the atmosphere above the surface.

Phase Relationships

All rocks contain some void space between grains. This void space may be occupied by liquids and gases. For example, a sandstone rock unit several thousand meters below the surface may contain petroleum and natural gas. Near the surface, however, void spaces are usually filled or partially filled with water. Above the water table (Chapter 11), gases like nitrogen, oxygen, and carbon dioxide fill the remainder of the voids. Rocks must be considered as three-phase systems in many instances.

The volume of void space in an earth material can range from a very small percentage of the solid volume to several times the volume of the solid material. The relative amount of void space, along with the type and amount of fluid occupying the space, has an important influence on the mechanical behavior of the material. The relative amount of void space is quantified by the use of the parameters *porosity* and *void ratio* (Figure 7.1). The porosity, n, is defined as

$$n = \frac{V_v}{V_T} \tag{7.1}$$

where V_v is the volume of voids and V_T is the total volume of a representative volume of rock. The void ratio, e, is principally used by soils engineers. It can be defined as

$$e = \frac{V_v}{V_s} \tag{7.2}$$

where V_s is the volume of soil solids. Porosity is usually expressed as a percentage, whereas void ratio is expressed as a decimal.

▶ FIGURE 7.1
Definition of porosity and void ratio of an earth material. Void space (V_v) includes liquid and gas phases.

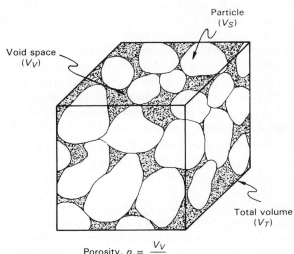

Particle
(V_S)

Void space
(V_V)

Total volume
(V_T)

Porosity, $n = \dfrac{V_V}{V_T}$

Void ratio, $e = \dfrac{V_V}{V_S}$

▲ **FIGURE 7.2**
Rocks with large, interconnected pore spaces allow the rapid migration of fluids like petroleum or water, but rocks with low permeability transmit fluids very slowly.

Although both porosity and void ratio describe the amount of void space in a soil or rock, they do not give any indication of the rate at which fluids will move through a saturated material. This property, called *permeability* (Figure 7.2), is determined by the size and degree of interconnection of the voids, as well as by the properties of the fluid, including its temperature, density, and viscosity. *Intrinsic permeability* measured in darcys or cm^2 (Table 7.1) refers to the material properties that influence fluid movement. The characteristics that determine intrinsic permeability include the size, shape, and packing of the grains as well as the degree of cementation and fracturing. Intrinsic permeability is of great importance to petroleum geologists and engineers. A rock may be saturated with oil, but if the permeability of the rock is low, oil would not move rapidly enough for an oil well to be considered economically feasible.

A composite parameter, *hydraulic conductivity* (m/s), is used by groundwater hydrologists and soils engineers to measure the ability of a rock or soil to transmit water (Table 7.1). Hydraulic conductivity, K, is defined as

$$K = k\frac{\rho g}{\mu} \tag{7.3}$$

where k is the intrinsic permeability, ρ and μ are the density and viscosity of water, respectively, and g is the acceleration of gravity. The properties of the material and the fluid are combined in hydraulic conductivity because water at relatively constant values of density and viscosity is usually the only liquid encountered when dealing with the upper part of the earth's crust. Contamination with petroleum fuels is an exception to this statement.

Table 7.1   Range of Values of Intrinsic Permeability, k (cm^2, darcys) and Hydraulic Conductivity, K (cm/s, m/s)

Rocks | Unconsolidated deposits

Rocks (top to bottom): Karst limestone, Permeable basalt, Fractured igneous and metamorphic rocks, Limestone and dolomite, Sandstone, Unfractured metamorphic and igneous rocks, Shale

Unconsolidated deposits (top to bottom): Gravel, Clean sand, Silty sand, Silt, loess, Glacial till, Unweathered marine clay

k (darcy)	k (cm^2)	K (cm/s)	K (m/s)
10^5	10^{-3}	10^2	1
10^4	10^{-4}	10	10^{-1}
10^3	10^{-5}	1	10^{-2}
10^2	10^{-6}	10^{-1}	10^{-3}
10	10^{-7}	10^{-2}	10^{-4}
1	10^{-8}	10^{-3}	10^{-5}
10^{-1}	10^{-9}	10^{-4}	10^{-6}
10^{-2}	10^{-10}	10^{-5}	10^{-7}
10^{-3}	10^{-11}	10^{-6}	10^{-8}
10^{-4}	10^{-12}	10^{-7}	10^{-9}
10^{-5}	10^{-13}	10^{-8}	10^{-10}
10^{-6}	10^{-14}	10^{-9}	10^{-11}
10^{-7}	10^{-15}	10^{-10}	10^{-12}
10^{-8}	10^{-16}	10^{-11}	10^{-13}

Source: From R. A. Freeze and J. A. Cherry, *Groundwater*, © 1979 by Prentice Hall, Inc., Englewood Cliffs, N.J.

Stress, Strain, and Deformational Characteristics

Stress

When a load is applied to a solid, it is transmitted throughout the material (Figure 7.3). The load subjects the material to pressure, which equals the amount of load divided by the surface area of the external face of the object over which it is applied. *Pressure*, therefore, is force per unit area. Within the material, the pressure transmitted from the external face to an internal location is called *stress*. Stress, at any point within the material, can also be defined as force per unit area.

Stresses can be classified according to their orientation within a body (Figure 7.4). Stresses of equal magnitude that act toward a point from opposite directions are called *compressive stresses*. When the stresses are directed away from each other, *tensile stress* is acting at the point. The third type of stress, *shear stress*, includes stresses that are offset from each other and act in opposite directions, as in a couple. On any plane passed through a

◀ FIGURE 7.3
A weight resting on a block causes pressure on the external surface of the block and stress on internal planes in the body.

◀ FIGURE 7.4
Classification of stress.

solid body, there are stresses acting normal to the plane, either compressional or tensional, as well as shear stresses acting parallel to the plane (Figure 7.5). In mechanics, it is possible to resolve the stresses acting at any point within an object into three mutually perpendicular *principal stresses* that are called the *maximum, intermediate,* and *minimum* principal stresses (Figure 7.5). The planes perpendicular to the directions of the three principal stresses, which are known as *principal planes,* have no shear stress acting upon them.

At shallow depths beneath the earth's surface, the vertical and lateral (horizontal) stresses present are due to the weight of the overlying rocks, soil, and air. Thus in Figure 7.6, the vertical stress, σ_v, acting on a horizontal plane at a depth h can be calculated as

$$\sigma_v = \gamma h + P_a \tag{7.4}$$

where γ is the unit weight of rock material and P_a is atmospheric pressure acting on the surface above the rock. Atmospheric pressure is usually neglected in calculating subsurface

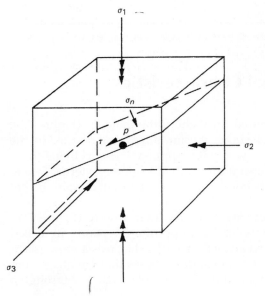

◀ FIGURE 7.5
Maximum (σ_1), intermediate (σ_2), and minimum (σ_3) principal stresses acting at point P within a body produce shear (τ) and normal (σ_n) stresses acting on a plane containing point P passed through the body.

► FIGURE 7.6
The vertical stress acting on shallow horizontal planes in the earth is the sum of the unit weight γ of the material times the depth (h) and the atmospheric pressure (P_a).

$$\sigma_v = \gamma h + P_a$$

► FIGURE 7.7
Vertical stress beneath a sequence of layers is the sum of the unit weight of each layer times its thickness. Atmospheric pressure is neglected.

$$\sigma_v = \gamma_a h_a + \gamma_b h_b + \gamma_c h_c$$

stresses because it is assumed to be a uniform component of all subsurface stresses. The vertical stress is calculated like hydrostatic stress in a liquid, although the lateral stress in the earth may or may not be equal to the vertical stress. In most subsurface applications, the vertical stress is considered to be the maximum principal stress. The minimum and intermediate principal stresses are assumed to be horizontal. When the horizon of interest is overlain by units of different unit weights, the stress components of each unit are summed to get the total vertical stress (Figure 7.7).

EXAMPLE 7.1

Calculate the vertical stress at a depth of 8 m at a location where a 5-m bed of sandstone with a unit weight of 25.1 kN/m^3 overlies a thick shale unit with a unit weight of 27.5 kN/m^3.

Solution

$h_a = 5$ m, $\gamma_a = 25.1$ kN/m^3

$h_b = 3$ m, $\gamma_b = 27.5$ kN/m^3

$$\sigma_v = \gamma_a h_a + \gamma_b h_b$$

$$\sigma_v = 25.1 \text{ kN/m}^3 \times 5 \text{ m} + 27.5 \text{ kN/m}^3 \times 3 \text{ m}$$

$$\sigma_v = 208 \text{ kN/m}^2$$

Mohr's Circle

Example 7.1 deals with the case in which we want to measure stresses upon the major principal plane. Suppose, however, that we now wish to measure stresses on a plane inclined at some angle, θ, measured counterclockwise from the major principal plane (Figure 7.8). It is possible to work in two dimensions with the assumption that, below ground surface, the intermediate principal stress, σ_2, is equal to the minimum principal stress, σ_3. The new plane is not a principal plane and therefore will be acted upon by both normal and shear stresses. The determination of stresses on planes inclined to the principal planes has applications in lab testing of rock strength, in investigating the stability of slopes, and in interpreting the geologic history of deformed rocks.

A graphical method for representing shear and normal stresses on inclined planes was devised by Otto Mohr (1835–1918). The plot, which is known as Mohr's circle, is shown in Figure 7.9. To construct the circle, values of σ_1 and σ_3 are plotted on the horizontal axis, which represents normal stress. By convention, compressive stresses are

◀ FIGURE 7.8
Position of a random plane oriented at angle θ measured counterclockwise from the major principal plane.

(a) (b) (c)

▲ FIGURE 7.9
(a) Stresses acting on an incremental element surrounding a point in the subsurface. (b) Mohr's circle plot for the stresses shown in (a). Deviator stress is the difference between σ_1 and σ_3. (c) Relationship of planes on the incremental element to points on Mohr's circle.

▶ FIGURE 7.10
General equations for shear
and normal stress derived from
Mohr's circle.

positive on the diagram. These points establish a circle of radius $(\sigma_1 - \sigma_3)/2$. The center is then plotted on the axis and the circle is drawn. Points on the circle represent the normal and shear stress on any plane oriented at angle θ from the principal plane, as measured from σ_3 (Figure 7.9). Because of the properties of a circle, the angle θ is equal to angle 2θ, measured from the center of the circle (Figure 7.9c). We now have a diagram representing the stress on any plane inclined relative to the principal plane. Normal and shear stress values can be determined graphically using the circle or by using the equations shown in Figure 7.10.

EXAMPLE 7.2

Vertical and horizontal principal stresses are 144 kN/m² and 36 kN/m², respectively. Determine the normal and shear stresses on a plane inclined at 45° to the principal plane. As shown, θ is measured counterclockwise from the major principal plane and angle 2θ, or 90°, is measured counterclockwise from the center of Mohr's circle. In this case, σ_n is equal to σ_3 plus the radius, and τ, the shear stress, is equal to the radius. It is evident from the diagram that a plane inclined at 45° from the major principal plane will have the highest value of shear stress of any plane.

Solution

$$\tau = \frac{\sigma_1 - \sigma_3}{2} = \frac{144 - 36}{2} = 54 \text{ kN/m}^2$$

$$\sigma_n = \sigma_3 + \text{radius} = 36 + 54 = 90 \text{ kN/m}^2$$

125.5
82.5

271 = 298.5

Deformation: Response to Stress

The application of stress to a body of rock or soil causes the material to yield or deform. The amount of deformation is called *strain*. The type and amount of strain that a particular material experiences depends on the type of stresses applied, as well as the depth and temperature. An understanding of deformation is important to geologists concerned with the behavior of earth materials during geologic processes and to engineers, who are more interested in deformation of material under the loads applied by engineering structures. Before considering natural materials, we will first review the deformation of so-called ideal materials.

Ideal Materials

Fundamental types of deformation can be demonstrated by the use of simple mechanical systems. As shown in Figure 7.11, the three basic types of material behavior are *elastic*,

◄ FIGURE 7.11
Types of idealized material deformation as illustrated by simple mechanical systems.

viscous, and *plastic.* Perfect elastic behavior, the type demonstrated by a spring, results in a linear plot of stress versus strain. The slope of the line relating stress and strain is an important material property called the *modulus of elasticity.* It can be stated as

$$E = \frac{\sigma}{\varepsilon} \tag{7.5}$$

where E is the modulus of elasticity, σ is the applied stress, and ε is the strain. The modulus of elasticity specifies how much strain will occur under a given stress. In this case, strain (ε) is measured as

$$\varepsilon = \frac{\Delta L}{L} \tag{7.6}$$

where ΔL is the change in length of the spring and L is the original length. Another characteristic of elastic materials is that they tend to return to their original condition when the stress is removed. Thus, elastic strain is recoverable. Some rocks approach ideal elastic behavior during deformations of small magnitudes. If the modulus of elasticity is known, it is therefore possible to predict the amount of deformation that will occur under an applied load.

Certain fluids display the type of behavior known as *viscous* (Figure 7.11). Stress is directly proportional to *strain rate* for these fluids. The constant of proportionality is the *viscosity* of the liquid, which is the slope of the stress versus strain rate plot. Even solid earth materials may exhibit viscous behavior under certain circumstances. The model for viscous behavior is a dashpot, in which a piston is moved through a cylinder containing a viscous liquid. The liquid flows through a space between the piston and the wall of the cylinder as the piston is pulled.

Plastic behavior, as demonstrated in Figure 7.11 by a block pulled along a surface, involves continuous deformation after some critical value of stress has been achieved. There is no deformation until the critical stress has been reached. Many rocks display plastic deformation under stress, but they differ from ideal plastic behavior in the other types of deformation that occur before plastic behavior begins.

Models that more closely approximate the behavior of real earth materials are provided by combining basic models of ideal behavior. The St. Venant model, for example, which combines elastic and plastic behavior, approximates the deformation of many rocks under stress. The Bingham model combines all three types of ideal behavior. This response has been used as a model for mud, lava, and similar flows. These processes will be described in Chapter 13.

Stress-Strain Behavior of Rocks

Although rocks can be compared with the idealized models just discussed, their actual behavior is more complex. One common method of testing rock behavior is the *unconfined compression test,* in which a cylindrical sample of rock is subjected to an axial load applied to the ends of the sample (Figure 7.12). As the axial stress is increased during the test, the changes in length of the sample can be measured. From this value, the strain at any instant can be determined using equation (7.6).

In Figure 7.13, a generalized stress-strain curve for rocks is shown. Unlike the ideal models, the curve is nonlinear and has three distinct segments. The first segment relates to the closing of microscopic pores and void spaces in the rocks as stress is applied at low levels. In the middle section of the curve, the nearly linear response approximates elastic behavior. The slope of this segment can be used to calculate a modulus of elasticity for the rock, even though the rock is not perfectly elastic (Figure 7.11). The final segment of the curve is most similar to plastic behavior. At the point of *failure,* where the sample crumbles and loses all resistance to stress, the curve terminates.

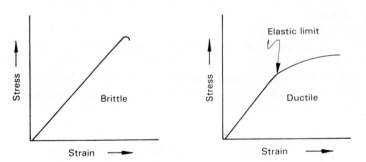

◄ FIGURE 7.12
Schematic diagram of the unconfined compression test. An axial load is applied to a rock or soil sample and the resulting strain is measured.

◄ FIGURE 7.13
A generalized stress-strain curve for rocks. The modulus of elasticity is measured by the slope of the tangent to segment 2 of the curve.

◄ FIGURE 7.14
Stress-strain curves for brittle and ductile rocks.

Different types of rocks vary considerably in their stress-strain behavior. In Figure 7.14, two types of response are shown. Rocks that display mostly elastic behavior until rupture are called *brittle*. Those that exhibit a significant amount of plastic response before failure are termed *ductile*. Ductile rocks usually deform elastically to a point called the *elastic limit*, where strain becomes more plastic with increasing stress.

Compressive Strength

We have described the failure of a brittle rock as the point when the rock loses all resistance to stress and crumbles. In a plastic material, a specific point of failure is harder to identify because deformation continues indefinitely at a constant level of stress. Failure in this case has to be defined as a certain amount, or percentage, of strain. For whatever criteria established for failure, *strength* is defined as the level of stress at failure.

When the stress at a certain point in the ground or in a laboratory test sample is resolved into its principal stresses, a *stress differential* is apparent between the three principal stresses. In an unconfined compression test, for example, the axial load becomes the greatest principal stress and the other two principal stresses are zero. When a stress differential exists, shear stresses develop on planes at all angles to the principal stress directions within the body (Figure 7.15). It is possible to resolve the principal stresses acting on the sample into shear and normal stresses acting on any plane within the sample (Figure 7.16). Mohr's circle is a handy way to calculate the value of these stresses. For an unconfined compression test, the σ_3 value would be plotted at the origin. For any plane, the shear stress, τ, tends to cause failure, and the normal stress, σ_n, tends to resist failure. On two critical planes within the rock, the combination of shear and normal stresses will produce the greatest tendency for failure. If the shear stress on these planes exceeds the *shear strength* of the rock in those directions, failure of the rock will occur. Therefore, the unconfined compression test indirectly determines the shear strength of

▶ FIGURE 7.15
Shear stresses on planes inclined to the principal planes produced by a stress differential.

▶ FIGURE 7.16
Shear and normal stresses acting on a plane within a test sample.

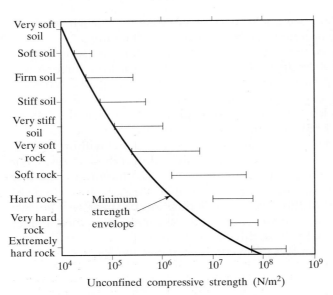

◄ FIGURE 7.17
Unconfined compressive strength plotted as a function as consistency or hardness of various types of earth materials. *Source:* From U.S. Department of Transportation, 1977. Rock slope engineering: workshops for planning, design, construction and maintenance of rock slopes for highways and railways. Federal Highway Administration, Washington, D.C.

the rock. A relationship between the consistency or hardness of various earth materials and the unconfined compressive strength is shown in Figure 7.17. The most useful aspect of this relation is the establishment of a minimum strength envelope. The large ranges of unconfined compressive strength for a particular consistency have to do with the type of joints, fractures, and other discontinuities in natural materials and other factors such as the orientation of the sample with respect to these features.

The relationship between shear and normal stresses during a strength test, and at failure, is critical to understanding the deformation behavior of the material. A test that can be used to determine this relationship directly is called, appropriately, the *direct shear* test (Figure 7.18). In the direct shear test, the failure plane is specified by the construction of the test cell so that variable shear and normal stresses can be applied to test for shear strength along the plane where the cell separates. Although the direct shear test is more commonly used for soils than for rocks, it provides an excellent illustration of strength, because shear strength is directly measured in contrast to the unconfined compression test. The test is usually conducted by applying a constant value of normal stress and then increasing the shear stress until failure occurs. Successive tests are then repeated at a higher normal stress. If the cell is filled with dry sand for a strength test, the results will be as shown in Figure 7.19. The line on the graph shows the relationship between shear stress and normal

◄ FIGURE 7.18
The direct shear test.

▶ FIGURE 7.19
Direct shear test results for dry sand.

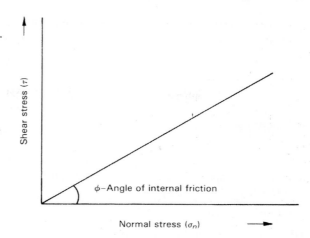

ϕ—Angle of internal friction

Normal stress (σ_n)

stress at failure. The constant slope indicates that shear strength is directly proportional to the normal stress applied to the failure plane. This result is obtained because the strength of the sand is controlled by frictional contact between sand grains, and the shear stress necessary to cause failure is directly dependent upon the normal stress applied to the failure plane. The angle that the plotted line makes with the horizontal axis (ϕ) is a basic property of the specific sand tested, and it is called the *angle of internal friction*. It depends upon the grain size, distribution, shape, and packing of the sand grains in the sample tested. The equation of the line relating shear strength, S, and normal stress, σ_n, is

$$S = \sigma_n \tan \phi \qquad (7.7)$$

The shear stress versus normal stress curve for dry sand passes through the origin because frictional contact between grains is the only component of strength in this material. For rocks, as well as for some soils, there are additional components of strength arising from the interlocking nature of grains in the rock, the cement in the pores of the rock, or attractive forces between grains or particles. These properties give the rock or soil a certain inherent strength that is independent of normal stress. In soils, this strength is called *cohesion*. These materials also have a strength component proportional to normal stress on the failure plane. The strength test results, then, for rocks and some cohesive soils, are of the type shown in Figure 7.20. The intercept on the vertical axis is a measure

▶ FIGURE 7.20
Generalized direct shear test results for rocks and some cohesive soils.

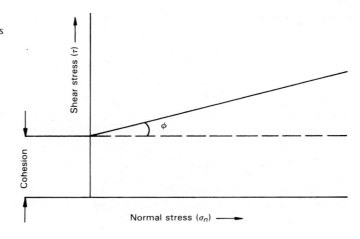

Normal stress (σ_n)

of the inherent strength or cohesion of the material. The formula for shear strength in this case can be written as

$$S = C + \sigma_n \tan \phi \tag{7.8}$$

where C is cohesion. This equation is called the Mohr-Coulomb equation.

EXAMPLE 7.3

A direct shear cell 25.4 cm by 25.4 cm in plan is filled with dry sand. A normal load of 6 kN is applied to the sample, and the shear force at failure is 3.5 kN. What is the angle of internal friction for the sand, based upon this one test?

Solution

The normal stress on the failure plane is 6 kN/0.065 m^2 = 92.3 kN/m^2, and the shear stress at failure (shear strength) is 53.8 kN/m^2. When plotted on a graph of shear stress versus normal stress, ϕ can be determined as shown.

$$\tan \phi = \frac{53.8}{92.3} = 0.58$$

$$\phi = 30.2°$$

Effects of Confining Pressure

The unconfined compression test is applicable to rock engineering where rock masses are exposed at the surface. However, in the design of a tunnel, an underground mine, or an underground waste repository, we would need to take into consideration the effects of deep burial of the rock upon its strength characteristics. Similarly, geologists must understand the deformation of rocks far below the earth's surface to interpret tectonic structures (Chapter 8) that were formed by deformation of the crust in the geologic past. Under these conditions, σ_3, the minimum principal stress, would no longer be equal to zero as in the unconfined compression test. The weight of the overlying column of rock translates into a pressure applied from all directions to any given element of rock at depth. This all-around stress is known as *confining pressure*. It is similar to hydrostatic pressure in that it increases with depth; however, this *lithostatic pressure* is not always

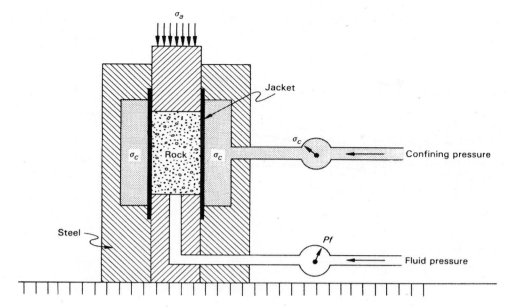

▲ FIGURE 7.21
Schematic diagram of triaxial test cell.

equal in all directions. The major principal stress, σ_1, may be applied from a vertical or lateral direction.

A laboratory test called the *triaxial test* (Figure 7.21) has been devised to more closely simulate the behavior of rocks or soil at depth. In this test, the cylindrical sample is enclosed in a jacket through which a radial confining pressure can be applied. Gases or liquids can be used to apply this all-around pressure. An axial load can then be applied, similar to an unconfined compression test, until failure occurs. With this apparatus, tests can be run repeatedly to study the effect of confining pressure upon deformation. The results of several tests can be used to establish a *failure envelope* (Figure 7.22) by drawing a line tangent to the Mohr's circle representing each test. The point of contact with each circle represents the shear strength for the corresponding values of confining and axial pressure. Notice that because the failure envelope is a sloping line, the point of contact with each circle does not occur at the highest value of shear stress, which would be at the top of the circle. Thus, failure does not necessarily take place on the plane where shear stress is the maximum, but on the plane where the combination of shear and normal stresses produce optimum failure conditions. When used for soils, the triaxial test data yield the angle of internal friction (the slope of the failure envelope) and the cohesion (the intercept of the shear stress axis).

Under increasing levels of confining pressure, as we might expect with increasing depth, rocks no longer deform as they did at shallow depth. Triaxial test results plotted on stress-strain diagrams illustrate several types of changes (Figure 7.23). First, as the confining

▶ FIGURE 7.22
Construction of failure envelope using Mohr's circle results from multiple tests.

▲ FIGURE 7.23

Effect of confining pressure on a rock sample (Solenhofen Limestone) in a triaxial test. Each curve represents a test conducted at a different confining pressure. *Source:* From M. P. Billings, *Structural Geology*, 3rd ed., © 1972 by Prentice Hall, Inc., Upper Saddle River, N.J.

pressure is increased, the rock passes through a transition from brittle to ductile behavior. Ductile response becomes dominant at confining pressures above 700 kg/cm². This result supports our expectation that rocks whose behavior may be brittle at the earth's surface become ductile under high confining pressures. Second, the tests run at higher confining pressures indicate that the strength of the rock increases with increasing confining pressure.

Confining pressure is not the only change in rock deformation with depth. Because the temperature increases steadily with depth, it is likely that rocks will behave differently at higher temperatures. Triaxial test cells can be constructed to investigate the effects of temperature on rock behavior under stress. Some results are shown in Figure 7.24. A particular rock type was tested at several temperatures under a constant confining pressure. Strength decreased at higher temperatures. In addition, the samples became ductile (reached their elastic limit) at a lower stress. Thus, plastic deformation will become more prevalent with increasing depth because of greater temperatures and pressures.

Another factor that affects rock deformation is time. Pressures are applied to rocks over millions of years in the earth. Although we obviously cannot duplicate such conditions in a lab test, it is possible to vary the rate of strain at which a rock sample approaches failure. In this way the effect of time can be investigated, although extrapolation to geologic time intervals is still difficult. When the strain rate of a triaxial test is decreased, two observations can be made. First, rock strength decreases with decreasing strain rate (increasing the time to failure); second, the rocks become more ductile at lower strain rates. These results hold true even at stresses below the elastic limit. The stresses are applied over long periods of time. In fact, the relationship between stress and strain rate can even be compared with that of a viscous material. Rocks, therefore, that are strong and brittle at the earth's surface may deform like a fluid deep within the earth over long periods of time. Lab values of unconfined compressive strength, along with other mechanical properties of rocks, are shown on Table 7.2.

▲ **FIGURE 7.24**
Effect of temperature on rock strength (Yule Marble). *Source:* From M. P. Billings, *Structural Geology*, 3d ed., © 1972 by Prentice Hall, Upper Saddle River, N.J.

Tensile Strengh

In addition to compressive strength, rocks also can mobilize significant resistance to tensile stresses prior to failure, a property which is known as *tensile strength*. As the values in Table 7.2 indicate, tensile strength is much less than compressive strength, in many cases in fact, only about 10% of compressive strength. The low tensile strength of rocks plays a role in some weathering processes that we will discuss in Chapter 9. There are other implications of this property as well. In the ancient civilizations of Egypt, Greece, and Rome, large temples were constructed using rock columns to support various types of roof structures. Resting on top of the columns were horizontal beams, or lintels, of rock (Figure 7.25). When a lintel spans two supporting columns, the beam tends to bow downward in the center and tensile stresses develop and increase with the distance between the columns. When the beams are composed of extremely hard, homogeneous rock, rock lintels can perform well and some have lasted several thousand years. However, when discontinuities or planes of weakness are present in the rock, the tensile strength may be even lower than the value for uniform, massive rock. For these reasons, rock beams had to be chosen very carefully to avoid any potential planes of failure. Very few rock masses even approach homogeneity, and construction with rock beams is a risky affair. The Romans solved this problem by extensive use of the arch in bridges, aqueducts (Figure 7.26), and buildings like the Coliseum, a design that transmits the load to the supporting columns by creating compressive stresses around the curved geometric form. With its high compressive strength, rock performs well in structures that utilize the arch. Modern building materials like steel, with a tensile strength as high as its compressive strength (Table 7.2) avoid problems of this type. The arch is still used in construction, especially in dams, in which the arch lies in a horizontal plane and transmits the stress to the rock abutments at the end of the dam.

Engineering Classification of Rock and Rock Masses

The performance of rock in engineering projects involves both the properties of rock as determined on small, intact samples tested in the laboratory and the properties of the rock mass in the field. The behavior of the rock mass is almost always weaker than lab tests

Table 7.2 Mechanical Properties of Rock and Other Materials

Rock type	Locality	ρ Density (g/cm^3)	E Modulus of Elasticity ($\times 10^9$ N/m^2)	Ultimate Strength σ_c Compressive Strength ($\times 10^6$ N/m^2)	σ_t Tensile Strength ($\times 10^6$ N/m^2)
Amphibolite	California	2.94	92.4	278.0	22.8
Andesite	Nevada	2.37	37.0	103.0	7.2
Basalt	Michigan	2.70	41.0	120.0	14.6
Basalt	Colorado	2.62	32.4	58.0	3.2
Basalt	Nevada	2.83	33.9	148.0	18.1
Concrete	—	2.7–3.2	2.1–1.0	0.41–0.21	0.04–0.02
Conglomerate	Utah	2.54	14.1	88.0	3.0
Diabase	New York	2.94	95.8	321.0	55.1
Diorite	Arizona	2.71	46.9	119.0	8.2
Dolomite	Illinois	2.58	51.0	90.0	3.0
Gabbro	New York	3.03	55.3	186.0	13.8
Gneiss	Idaho	2.79	53.6	162.0	6.9
Gneiss	New Jersey	2.71	55.16	223.0	15.5
Granite	Georgia	2.64	39.0	193.0	2.8
Granite	Maryland	2.65	25.4	251.0	20.7
Granite	Colorado	2.64	70.6	226.0	11.9
Graywacke	Alaska	2.77	68.4	221.0	5.5
Gypsum	Canada	2.32	—	22.0	2.4
Limestone	Germany	2.62	63.8	63.8	4.0
Limestone	Indiana	2.30	26.96	53.1	4.07
Marble	New York	2.72	54.0	126.9	11.7
Marble	Tennessee	2.70	48.3	106.0	6.5
Phyllite	Michigan	3.24	76.5	126.0	22.8
Quartzite	Minnesota	2.75	84.8	629.0	23.4
Quartzite	Utah	2.55	22.06	148.0	3.5
Salt	Canada	2.20	4.64	35.5	2.5
Sandstone	Ohio	2.17	10.52	38.9	5.17
Sandstone	Utah	2.20	21.37	107.0	11.0
Schist	Colorado	2.47	8.96	15.0	—
Schist	Alaska	2.89	39.3	129.6	5.5
Shale	Utah	2.81	58.19	215.8	17.2
Shale	Pennsylvania	2.72	31.2	101.4	1.38
Siltstone	Pennsylvania	2.76	30.6	113.0	2.76
Slate	Michigan	2.93	75.85	180.0	25.5
Steel	—	7.85	200.00	365.0	365.0
Tuff	Nevada	2.39	3.65	11.3	1.17
Tuff	Japan	1.91	76.0	36.0	4.31

Source: P. H. Rahn, *Engineering Geology: An Environmental Approach*, 2nd ed., © 1996 by Prentice Hall, Inc., Upper Saddle River, N.J.

would suggest because of the natural variations in composition and the planes or zones of weakness that are present in almost all rock masses.

Classification of Intact Rock

Laboratory properties of many types of rocks are shown on Table 7.2. Values such as these relating to stress-strain behavior are utilized in the engineering classification of

► FIGURE 7.25
Horizontal beam used to support the roof of Karnak Temple, Luxor, Egypt. The beam shows a crack at the center formed in response to tensile stresses. *Source:* Photo courtesy of the author.

▲ FIGURE 7.26
Roman aqueduct in Segovia, Spain, which uses the arch in construction. Stones were fitted so carefully that mortar was not necessary. *Source:* Photo courtesy of the author.

rock developed by D. U. Deere and R. F. Miller (Deere, 1968). One component of the classification is the unconfined compressive strength (Table 7.3). Five categories are established for an intact sample of rock. The modulus of elasticity forms the other aspect of the classification. This classification applies only to rock that is internally continuous, or intact. Most masses of rock contain discontinuities that strongly influence engineering behavior. Rock-mass properties will be discussed in a subsequent section.

Table 7.3	Strength Classification of Intact Rock	
Class	Description	Unconfined Compressive Strength (kg/cm^2)
A	Very high strength	>2250
B	High strength	1125–2250
C	Medium strength	562–1125
D	Low strength	281–562
E	Very low strength	<281

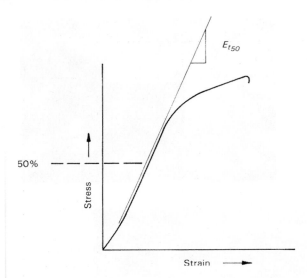

Definition of the tangent modulus $(E_{t_{50}})$ used in the Deere and Miller classification. The tangent is drawn at 50% of unconfined compressive strength.

Table 7.4 Modulus of Elasticity Classification of Rocks	
Description	$E_{t_{50}}\,(\mathbf{kg/cm^2 \times 10^5})$
Very stiff	8–16
Stiff	4–8
Medium stiffness	2–4
Low stiffness	1–2
Yielding	0.5–1
Highly yielding	0.25–0.5

In Figure 7.27, the definition of the particular modulus value used is illustrated. $E_{t_{50}}$ is the modulus obtained by taking the slope of a line tangent to the stress-strain curve at 50% of the unconfined compressive strength. The classes of modulus of elasticity are listed in Table 7.4. Together, the modulus of elasticity $(E_{t_{50}})$ and the unconfined compressive strength (σ_a) can be used to calculate a modulus ratio, which is defined as

$$M_R = \frac{E_{t_{50}}}{\sigma_a} \qquad (7.9)$$

Modulus-ratio values are plotted on a diagram (Figure 7.28) divided into fields of high modulus ratio ($>500:1$), medium modulus ratio ($500:1-200:1$), and low modulus ratio ($<200:1$). The position of plotted points on the modulus-ratio diagram gives a visual comparison of rocks and indicates the strength and modulus values, the rock properties that control engineering behavior.

Igneous Rocks

The strength of igneous rocks is high when the rock is composed of a dense network of interlocking crystals. This condition is usually present in the intrusive rocks, which had sufficient time for crystallization to develop a three-dimensional network. As shown in Figure 7.29, intrusive rocks generally have a high modulus of elasticity and a medium modulus ratio.

The extrusive igneous rocks have a much wider range in strength and modulus. Lava flows may develop strength and modulus properties that are similar to the intrusive rocks,

▶ FIGURE 7.28
Classification diagram for the Deere and Miller classification of intact rock. *Source:* Modified from D. U. Deere, *Rock Mechanics in Engineering Practice*, K. G. Stagg and O. C. Zienkiewicz, eds., © 1968 by John Wiley & Sons, Ltd., London.

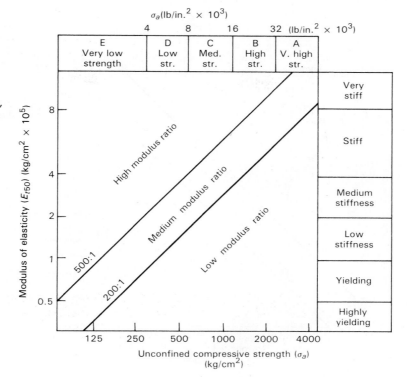

▶ FIGURE 7.29
Engineering classification of intact igneous rock. *Source:* Modified from D. U. Deere, *Rock Mechanics in Engineering Practice*, K. G. Stagg and O. C. Zienkiewicz, eds., © 1968 by John Wiley & Sons, Ltd., London.

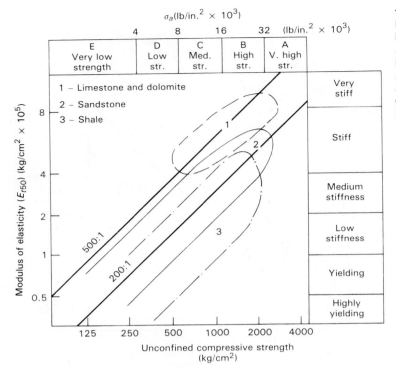

◀ FIGURE 7.30
Engineering classification of intact sedimentary rock. *Source:* Modified from D. U. Deere, *Rock Mechanics in Engineering Practice*, K. G. Stagg and O. C. Zienkiewicz, eds., © 1968 by John Wiley & Sons, Ltd., London.

but vesicular extrusive rocks are usually weaker. The pyroclastic rocks extend the field of the extrusive rocks into the area of low strength and modulus because of their low density and high porosity caused by their formation from airfall or flow processes. Most extrusive rocks, however, fall into the medium modulus ratio category.

Sedimentary Rocks

Sedimentary rocks have extremely variable strength and modulus properties (Figure 7.30). For the clastic rocks, these values depend on the rock characteristics acquired in the depositional environment, as well as on all changes that have affected the rock in the lithification process. Important depositional factors include grain-size distribution, sorting, rounding, and mineral composition. Lithification processes, such as compaction and cementation, usually increase the strength of the rock. Clastic rocks of medium and coarse grain size usually fall into the moderate modulus-ratio class over a wide range of strength and modulus values. Shale exhibits a strong tendency for plastic deformation. As a result, the modulus of elasticity is low and the modulus ratio overlaps into the low zone.

Nonclastic rocks differ in engineering properties according to the composition of the rocks. Limestone and dolomite generally have medium to high strength and modulus ratios. Evaporite rocks exhibit weak, plastic behavior. The tendency for plastic deformation at low strength in the evaporite rocks explains why these rocks are considered to be a possible option for high-level nuclear waste repositories. Because of their stress-strain behavior, any cavities or cracks in the rocks are sealed by plastic flow. This property is important in preventing migration of waste products from the disposal site.

Metamorphic Rocks

Metamorphism increases the strength of some sedimentary rocks by compaction and recrystallization. Thus (Figure 7.31), the modulus ratio of quartzite is limited to the high-strength

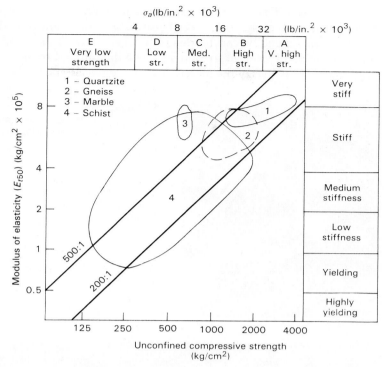

► FIGURE 7.31
Engineering classification of
intact metamorphic rock.
Source: Modified from D. U.
Deere, *Rock Mechanics in Engineering Practice*, K. G. Stagg
and O. C. Zienkiewicz, eds.,
© 1968 by John Wiley & Sons,
Ltd., London.

side of the diagram relative to the corresponding sedimentary rock sandstone. Marble is an exception to the trend of increased strength following metamorphism. When limestone or dolomite recrystallizes, the crystals in the resulting marble are large. The strength of a rock generally is proportional to the surface area of contact between grains, and a fine-grained rock has a higher contact area than a coarse-grained rock. Therefore, limestone and dolomite lose strength when they are metamorphosed. Because only the grain size changes during metamorphism, and the mineral composition remains the same, the modulus-ratio field of marble falls within the high modulus-ratio section of the chart.

Schists have a wide variation in strength and modulus because of their strongly oriented foliation, which produces planes of weakness parallel to the foliation. The strength of a schist depends upon the direction from which stresses are applied. If the stress configuration tends to produce failure in the direction of the foliation, the strength is minimized. A much greater strength develops if the stresses tend to cause failure perpendicular to the foliation planes. Gneisses are limited to the high-strength and modulus regions of the chart because the coarse bands of alternating light- and dark-colored minerals in gneiss do not result in highly developed planes of weakness in the rock.

Rock-Mass Properties and Classification

The strength and deformational properties of rocks that we have discussed usually are determined by laboratory tests on rock samples. These samples are necessarily small, intact specimens taken from large bodies of rock at a field location. Although the test results obtained from intact samples are useful for comparison of properties between various rock types, the strength values cannot be directly applied to the overall rock mass in the field situation. The reason for this apparent discrepancy is that the behavior of a rock mass under load in the field is partially controlled by the strength developed along discontinuities in the rock and by the weathering characteristics, rather than by the strength of the intact portions of the rock itself. Discontinuities are present in almost every type of rock and they act to lower the strength of the rock mass (Figure 7.32). For example, in the direct shear and triaxial tests, shear strengths can be determined under variable conditions of confining stress (σ_3). The angle of internal friction φ can be determined by this procedure. Table 7.5 illustrates that the φ values of the materials filling joints are much lower than values determined

◀ **FIGURE 7.32**
Highly jointed Precambrian igneous rock mass, south Sinai Peninsula, Egypt. *Source:* Photo courtesy of the author.

Table 7.5   Appoximate Friction Angles and Cohesion Values for Rocks and Joint Infilling Materials

Rock	$\phi°$ (intact rock)	$\phi°$ (discontinuity)	$\phi°$ (ultimate)	c (massive rock) $kN\ m^{-2}$
Andesite	45	31–35	28–30	
Basalt	48–50	47		
Chalk		35–41		
Diorite	53–55			
Granite	50–64		31–33	100–300
Greywacke	45–50			
Limestone	30–60		33–37	50–150
Monzonite	48–65		28–32	
Porphyry		40	30–34	100–300
Quartzite	64	44	26–34	
Sandstone	45–50	27–38	25–34	50–150
Schist	26–70			
Shale	45–64	37	27–32	25–100
Siltstone	50	43		
Slate	45–60	25–34		

Infilling material	$\phi°$ (approximate)
Remolded clay gouge	10–20
Calcitic shear zone material	20–27
Shale fault material	14–22
Hard rock breccia	22–30
Compacted hard rock aggregate	40
Hard rock fill	38

Source: P. H. Rahn, *Engineering Geology: An Environmental Approach*, 2nd ed., © 1996 by Prentice Hall, Inc., Upper Saddle River, N.J.

▲ FIGURE 7.33
Rock bolts used to prevent failure of sandstone slabs bounded by discontinuities in the Colorado River canyon downstream from the Glen Canyon Dam. *Source:* Photo courtesy of the author.

for intact rocks. Under stress, therefore, the rock will fail along existing planes of weakness rather than develop new fracture planes within intact, solid rock (Figure 7.33). Therefore, it is very important to determine the properties of the rock mass as well as the properties of intact rock within the rock mass.

Some of the types of rock discontinuities are shown in Table 7.6. Depositional discontinuities are acquired by sedimentary rocks as their constituent particles are deposited as sediment. These discontinuities include bedding planes between beds and laminations of different lithology (Chapter 5). Sedimentary structures, including mud cracks and ripple marks, also represent discontinuities in the rock. In addition to knowing the types of discontinuities, engineers must also determine their spacing, orientation, and roughness. The orientation is particularly important in considering the ability of a rock slope to support itself under the load of the rocks and soils that make up the slope. Failure in this situation occurs when part of the rock mass breaks away along a discontinuity and moves downslope (Figure 7.34). Slope movements, which will be discussed in more detail later, constitute a very significant category of geologic hazards.

Table 7.6 Rock Discontinuities	
Sedimentary	**Metamorphic**
Bedding planes	Foliation
Sedimentary structures (mud cracks, ripple marks, cross beds, and so on)	Cleavage
Unconformities	**Igneous**
	Cooling joints
Structural	Flow contacts
Faults	Intrusive contacts
Joints	Dikes
Fissures	Sills
	Veins

▲ FIGURE 7.34
Rockfall on eastbound lane of Interstate 70, Jefferson County, Colorado. Failure occurred along joints and foliation planes in Precambrian gneiss. *Source:* W. R. Hansen; photo courtesy of U.S. Geological Survey.

Case in Point 7.1
The Ancient Egyptians and Rock-Mass Properties

As living gods, the pharaohs of ancient Egypt constructed massive temples to glorify themselves and the large cadre of nonliving deities worshiped in their religion. The obelisk was one of the most impressive symbols of their power and importance. The obelisks at Karnak Temple, near present day Luxor, were erected about 3500 years ago (Figure 7.35). These narrow rock spires, which are as much as 30 m high and covered with hieroglyphics, are chiseled out of beautiful pink, coarse-grained granite that crops out at Aswan, about 150 km up the Nile from Luxor. To construct these monuments, workers had to locate a suitable exposure of unweathered rock with no significant fractures, excavate the obelisk to its final size in place from the outcrop with primitive hand tools, move it from the quarry to a boat on the Nile, float it downriver for 150 km, and raise it at the temple without damage. That this challenging series of steps did not always succeed is illustrated by an unfinished obelisk in an Aswan quarry (Figure 7.36). The obelisk was excavated and shaped to its final form on three sides but then abandoned in place when a large fracture developed in the rock. This obelisk, at 42 m in length, would have been the largest ever quarried and the largest object sculpted from a single stone in history. The Aswan granite is remarkable for its uniformity and lack of rock-mass discontinuities. The unfinished obelisk, however, is an indication of the limits of rock-mass homogeneity.

▲ FIGURE 7.35
Obelisks as high as 30 m rise above the walls of Karnak Temple, Luxor, Egypt. *Source:* Photo courtesy of the author.

▶ FIGURE 7.36
An unfinished 42-m-long obelisk abandoned in a quarry near Aswan, Egypt. *Source:* Photo courtesy of the author.

Table 7.7	Rock Quality Designation (RQD)
RQD (%)	**Description of Rock Quality**
0–25	Very poor
25–50	Poor
50–75	Fair
75–90	Good
90–100	Excellent

Quantification of rock-mass properties is very difficult because of the number of variables involved. One index that is used is called the *rock quality designation* (RQD). During site investigation for an engineering project, test holes are drilled to determine subsurface rock formations. One sampling technique is to obtain a *core*, which is a small cylindrical sample of the entire vertical interval drilled. The core can then be used for laboratory tests of the rocks penetrated. The length of an individual piece of core is dependent upon the degree and orientation of fractures in the rock. Cores of highly fractured rock will consist of long, unbroken segments. RQD is defined as the percentage of core recovered in pieces 10 cm or longer in length. The index is expressed as a percentage of the total depth drilled (Table 7.7). Thus, in 10 m of drilling, recovery of 9.2 m in pieces 10 cm or longer in length would have an RQD of 92% and be described as excellent quality.

Several rock-mass classifications have been devised for specific applications. Few of these, if any, are suitable for all purposes. A rock-mass classification devised to predict the type of excavation needed for various types of materials is shown in Figure 7.37. Numerical values can be assigned for each variable and summed to obtain an overall rock-mass classification. Problems arise in assigning numerical values to largely descriptive variables such as weathering characteristics. Rock is divided into five classes in this classification based on the uniaxial compressive strength of intact rock and the joint spacing in the rock mass. The resulting rock-mass categories can be related to methods of excavation, types of tunnel support systems, and other applications.

Rock-mass classification for underground excavations, including mines and tunnels, is particularly important because of the potential loss of life that could result from collaspse of the uncompleted excavation. One of the most widely used classifications is the Gemechanics Classification of Bieniawski (1989), which is illustrated in Table 7.8. Also known as the Rock Mass Rating (RMR) system, the classification is based on six individual factors: the uniaxial compressive strength of the rock material in a particular section of the tunnel, the RQD value, the spacing of discontinuities, the condition of discontinuities, the groundwater conditions, and the orientation of discontinuities. Each factor is rated as shown, and the ratings are summed to give a total rating between 1 and 100. Five rock-mass classes are defined by ranges in the total ratings. The sum of factors 1 through 5 gives an initial RMR score, which then can be modified by factor 6, the orientation of discontinuities. The five descriptive classes of orientations—very favorable, favorable, etc.—are based upon the strike, or compass heading, and dip (angle of inclination) of the discontinuities with respect to the orientation of the tunnel axis. Table 7.9 presents the criteria. A fracture running parallel to the tunnel axis with a high dip angle would be the worst possible case, as it would create the most likely condition for blocks of rock to collapse into the tunnel. The nonzero values of factor 6 are negative and are subtracted from the previous total to yield the total rating. Part D of Table 7.8 summarizes some applications of the RMR classes.

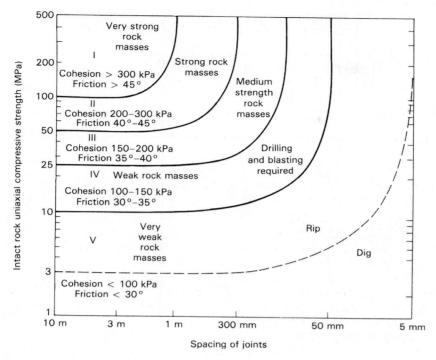

▲ FIGURE 7.37

Strength diagram for jointed rock masses showing types of excavation needed. *Source:* From Z. T. Bieniawski, 1974, Geomechanics classification of rock masses and its application to tunnelling, *Proc. 3rd Cong. Int. Soc. Rock Mech.*, 1:27–32. Reprinted with permission of Butterworth-Heinemann Limited.

Knowledge of the stand-up time, the time that a rock-cut opening will stand without support, is essential to tunnel construction. The method of excavation and support are related to the RMR value in Table 7.10. The easiest way to advance a tunnel is by blasting or cutting the full face, but this is only possible for very good or good rock. For less favorable conditions, the tunnel cross section is often excavated in sections, or *drifts*, because smaller openings are easier to control and support can be installed gradually as the cross section is enlarged. Several examples of drift sequences are shown in Figure 7.38. In Figure 7.38a, the first excavation is known as a top heading. Completion of side drifts (2) will create a bench allowing roof supports to be installed, *arch ribs* in this case. The bottom can then be excavated using a bottom drift (3), followed by lateral drifts to install *wall ribs* below the arch ribs. In Figure 7.38b, the lower side drifts are first in the sequence, followed by the top drifts (2), the top side drifts (3), and finally the central core (4) of the tunnel. In very weak rock masses, drifts surrounding the final tunnel cross section are first drilled and filled with concrete (Figure 7.38c). The capping beam (4) is constructed of reinforced concrete for further support. In Figure 7.38d, the central drift is excavated first and then expanded laterally to install supports. The use of steel for tunnel supports, with its great strength, revolutionized tunnel construction as it did building construction. Timbers were the dominant form of support in the centuries prior to the advent of steel construction. In many cases, timbers were used temporarily until masonry walls or arches made of building stone were constructed to provide permanent support.

Table 7.8 The Geomechanics Classification of Rock Masses (Rock Mass Rating System) Rating is determined by evaluating factors 1 through 5 and summing. Final value is determined by subtraction of the value of factor 6, if necessary.

A. CLASSIFICATION PARAMETERS AND THEIR RATINGS

	Parameter		Range of Values				For this low range, uniaxial compressive test is preferred		
1	Strength of intact rock mineral	Point-load strength index (MPa)	>10	4–10	2–4	1–2			
		Uniaxial compressive strength (MPa)	>250	100–250	50–100	25–50	5–25	1–5	<1
		Rating	15	12	7	4	2	1	0
2	Drill core quality RQD (%)		90–100	75–90	50–75	25–50	<25		
	Rating		20	17	13	8	3		
3	Spacing of discontinuities		>2 m	0.6–2 m	200–600 mm	60–200 mm	<60 mm		
	Rating		20	15	10	8	5		
4	Condition of discontinuities		Very rough surfaces Not continuous No separation Unweathered wall rock	Slightly rough surfaces Separation <1 mm Slightly weathered walls	Slightly rough surfaces Separation <1 mm Highly weathered walls	Slickensided surfaces or Gouge <5 mm thick or Separation 1–5 mm Continuous	Soft gouge >5 mm thick or Separation >5 mm Continuous		
	Rating		30	25	20	10	0		
5	Groundwater	Inflow per 10 m tunnel length (L/min)	None	<10	10–25	25–125	>125		
		Ratio $\dfrac{\text{Joint water pressure}}{\text{Major principal stress}}$	0	<0.1	0.1–0.2	0.2–0.5	>0.5		
		General conditions	Completely dry	Damp	Wet	Dripping	Flowing		
		Rating	15	10	7	4	0		

244

B. RATING ADJUSTMENT FOR DISCONTINUITY ORIENTATIONS

Parameter			Range of Values				
			Very Favorable	Favorable	Fair	Unfavorable	Very Unfavorable
Strike and dip orientations of discontinuities	Ratings	Tunnels and mines	0	−2	−5	−10	−12
		Foundations	0	−2	−7	−15	−25
		Slopes	0	−5	−25	−50	−60

C. ROCK MASS CLASSES DETERMINED FROM TOTAL RATINGS

Rating	100–81	80–61	60–41	40–21	<20
Class no.	I	II	III	IV	V
Description	Very good rock	Good rock	Fair rock	Poor rock	Very poor rock

D. MEANING OF ROCK MASS CLASSES

Class no.	I	II	III	IV	V
Average stand-up time	20 yr for 15-m span	1 yr for 10-m span	1 wk for 5-m span	10 h for 2.5-m span	30 min for 1-m span
Cohesion of the rock mass (kPa)	>400	300–400	200–300	100–200	<100
Friction angle of the rock mass (deg)	>45	35–45	25–35	15–25	<15

Source: Z. T. Bieniawski, *Engineering Rock Mass Classifications*, New York, Wiley Interscience, © 1989.

Table 7.9 Effect of Discontinuity Strike and Dip Orientations in Tunneling

Strike Perpendicular to Tunnel Axis Drive with Dip		Drive against Dip	
Dip 45–90	Dip 20–45	Dip 45–90	Dip 20–45
Very favorable	Favorable	Fair	Unfavorable

Strike Parallel to Tunnel Axis		Irrespective of Strike
Dip 20–45	Dip 45–90	Dip 0–20
Fair	Very unfavorable	Fair

Source: Z. T. Bieniawski, *Engineering Rock Mass Classifications*, New York, Wiley Interscience, © 1989.

Table 7.10 Guidelines for Excavation and Support of Rock Tunnels in Accordance with the Rock Mass Rating System

Rock-mass Class	Excavation	Support Rock Bolts (20-mm dia, Fully Grouted)	Shotcrete	Steel Sets
Very good rock (I) (RMR: 81–100)	Full face 3-m advance	Generally, no support required except for occasional spot bolting		
Good rock (II) (RMR: 61–80)	Full face 1.0–1.5-m advance Complete support 20 m from face	Locally, bolts in crown 3 m long, spaced 2.5 m, with occasional wire mesh	50 mm in crown where required	None
Fair rock (III) (RMR: 41–60)	Top heading and bench 1.5–3-m advance in top heading Commence support after each blast Complete support 10 m from face	Systematic bolts 4 m long, spaced 1.5–2 m in crown and walls with wire mesh in crown	50–100 mm in crown and 30 mm in sides	None
Poor rock (IV) (RMR: 21–40)	Top heading and bench 1.0–1.5-m advance in top heading. Install support concurrently with excavation 10 m from face.	Systematic bolts 4–5 m long, spaced 1–1.5 m in crown and walls with wire mesh	100–150 mm in crown and 100 mm in sides	Light to medium ribs spaced 1.5 m where required

(continued)

Table 7.10   *(Continued)*

Rock-mass Class	Excavation	Support		
		Rock Bolts (20-mm dia, Fully Grouted)	Shotcrete	Steel Sets
Very poor rock (V) (RMR: <20)	Multiple drifts 0.5–1.5-m advance in top heading. Install support concurrently with excavation. Shotcrete as soon as possible after blasting.	Systematic bolts 5–6 m long, spaced 1–1.5 m in crown and walls with wire mesh. Bolt invert.	150–200 mm in crown, 150 mm in sides, and 50 mm on face	Medium to heavy ribs spaced 0.75 m with steel lagging and forepoling if required. Close invert.

Source: Z. T. Bieniawski, *Engineering Rock Mass Classifications*, New York, Wiley Interscience, © 1989.

▶ FIGURE 7.38
Methods of advancing a tunnel heading in incompetent rock. Drifts are excavated in numbered sequence. (a) Top heading and bench. (b) Side drifts and supports installed first. (c) Drifts driven outside cross section of tunnel and filled with concrete before excavation of tunnel. (d) Center drift excavated first followed by side and lower drifts to install support.

Case in Point 7.2
The Gros Ventre Slide: Role of Discontinuities in Rock-Mass Stability

A good example of the effect of discontinuity orientation upon slope movements is provided by the Gros Ventre slide, the largest historic rock slide in the United States (Figures 7.39 and 7.40). Sedimentary rock formations are oriented as shown in the Gros Ventre Valley near Jackson, Wyoming (Alden, 1928). On the south slope of the valley, bedding planes are

▲ FIGURE 8.12
Aerial view of a monocline, Arizona. *Source:* Photo courtesy of the author.

Joints

Joints (Figure 8.15) are fractures in which there has been no movement parallel to the failure plane. Although joints are produced by unloading (Chapter 9) and cooling of lava (Chapter 4), our main concern here is for joints caused by tectonic stresses. These fractures commonly occur in sets consisting of numerous parallel planes spaced throughout a rock mass at regular intervals. One or more joint sets, each having a distinctive orientation, are often present.

Joints can be produced by tensional, shearing, and compressional stresses. If structures other than joints are found in an area, the joints may be related to the stresses that are responsible for those structures. For example, Figure 8.16 shows two possible orientations of intersecting joint sets associated with folds. Joint sets that intersect at high angles are known as *conjugate joint sets*. The engineering properties of rock masses are established by the type, origin, and spacing of joints and other discontinuities (Chapter 7). Strength is controlled by these fractures because failure usually occurs along planes of weakness rather than within intact rock. The orientation of joints may impart a degree of *anisotropy* to the rock mass; in other words, strength may vary with the direction from which the stress is applied because of the spatial distribution of joints. Rock-slope stability is also controlled by the spacing and orientation of joints. These factors determine the size and direction of failure of blocks that may be unstable, thus affecting the engineering of highway cuts, open-pit mine slopes, and other excavations.

Another rock-mass property that is influenced by jointing is permeability. In many rock types, most of the movement of groundwater, petroleum, and also contaminants is through fractures rather than through pores in the intact rock. In rocks of low permeability, exploration for groundwater involves identifying major joint traces on the ground surface. The point of intersection of two or more joints is a prime location for high well yields in such areas. Planning for waste-disposal facilities must also consider fracture patterns. Contaminated groundwater can migrate rapidly along joints and faults; these fractures must therefore be identified and monitored.

108°   104°   103°

Wyoming | South Dakota

South Dakota
Wyoming

Location map

44°

Cenozoic igneous centers

Cenozoic sedimentary rocks

Mesozoic rocks from Newcastle Sandstone
(L. Cret.) to top of U. Cretaceous

Mesozoic rocks below Newcastle Sandstone
(L. Cret.). Includes Spearfish Formation,
which is partly Permian

Paleozoic rocks

Precambrian granite and metamorphic rocks

N

Scale
40   0   40 km

▲ FIGURE 8.13

Generalized geology of the Black Hills of South Dakota and Wyoming. Erosion has truncated the original surface of the dome, and now the oldest rocks (Precambrian metamorphics) are exposed at the center of the structure surrounded by successively younger rock units. *Source:* From F. R. Karner and D. L. Halvorson, 1987, The Devils Tower, Bear Lodge Mountains, Cenozoic igneous complex, Northeastern Wyoming, in *Geological Society of America Centennial Field Guide—Rocky Mountain Section*, Geological Society of America.

Faults

Faults are fractures along which movement of rock masses has occurred parallel to the fault plane. Faults are classified according to the type of movement, or slip, that has taken place. The two categories established on this basis include dip-slip faults and strike-slip faults.

Dip-slip faults, in which the relative movement of rock masses is in the direction of the dip of the fault plane, include *normal* faults, *reverse* faults, and *thrust* faults (Figure 8.17). Normal and reverse faults can be differentiated if the relative displacement of a bed or other lithologic feature can be determined. The blocks on opposite sides of the fault plane are defined as the *hanging wall* and the *foot wall*. The hanging wall is always the block above the fault plane, and the foot wall is always the block below the plane. Therefore, by

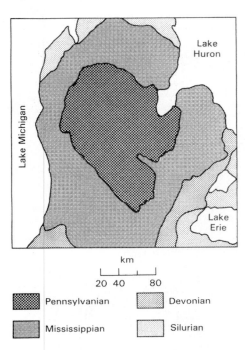

◀ FIGURE 8.14
Map of the Michigan Basin showing progressively younger rock units toward the center of the structure.

Lake Huron

Lake Michigan

Lake Erie

km

20 40 80

Pennsylvanian   Devonian

Mississippian   Silurian

▲ FIGURE 8.15
Closely spaced joints in an igneous rock body (Whatcom County, Washington). *Source:* M. H. Staatz; photo courtesy of U.S. Geological Survey.

definition, if the hanging wall has moved downward with respect to the footwall, the fault is a normal fault (Figure 8.18), and relative movement in the opposite sense is characteristic of reverse faults.

The amount of slip on a dip-slip fault can range from fractions of a centimeter to thousands of meters. The Teton fault in Wyoming has a combined displacement of 11,000 m

▶ FIGURE 8.16
Conjugate joint sets caused by stress that produced folding: (a) fold with vertical diagonal joints; (b) fold with strike joints dipping about 30°. *Source:* From M. P. Billings, *Structural Geology,* 3d ed., © 1972 by Prentice Hall, Inc., Englewood Cliffs, N.J.

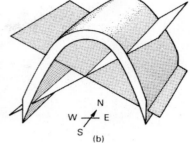

(a)

(b)

▶ FIGURE 8.17
Types of dip-slip faults: (a) normal fault, hanging wall is displaced downward with respect to foot wall; (b) reverse fault, hanging wall is displaced upward with respect to foot wall; (c) thrust fault, similar to reverse fault except that fault plane is inclined at a lower angle.

Foot wall

Hanging wall

(a)

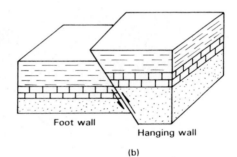

Foot wall

Hanging wall

(b)

(c)

(Love, 1987). Figure 8.19 shows a block diagram and cross section of this impressive structure. Paleozoic through Quaternary sedimentary rocks are preserved in the down-dropped hanging wall, but these formations have been eroded away on the foot-wall block. Of the total displacement on the fault, only about 1500 to 2000 m are evident at land surface in the rugged Teton mountain front (Figure 2.55), which is the steepest and youngest mountain front in the Rocky Mountains. Most of the displacement has occurred in the last 9 million years. The exposed and eroded plane of a fault, in this case the Teton mountain front, is called a *fault scarp*. The Teton fault scarp owes its justifiably

◀ FIGURE 8.18
Normal fault, northern Arizona. Displacement is defined by the offset of light-colored beds. *Source:* Photo courtesy of the author.

(a)

(b)

▲ FIGURE 8.19
Block diagram of the Teton Range and surrounding area, Wyoming. The Teton mountain front is a fault scarp produced by erosion of resistant Precambrian rocks on the upthrown block of the Teton fault. (a) Dimensions of the block shown are approximately 85 km by 70 km. (b) Cross section showing displacement across the Teton fault and associated rock units. *Source:* From J. D. Love, 1987, Teton mountain front, Wyoming, in *Geological Society of America Centennial Field Guide—Rocky Mountain Section,* Geological Society of America.

famous topography to the youth of the fault and the resistance to erosion of the Precambrian metamorphic and igneous rocks in the foot wall.

It is not always possible to identify the type of fault by the slope of the topographic scarp. Although the topography of the Colorado Front Ranges is similar to the Tetons, the faults that produced these mountain fronts were reverse faults. Notice in Figure 8.20 that the topographic scarp formed by the mountain front is almost at right angles to the dip of the reverse faults at the base of the slope. This occurs because erosion has produced a more stable topographic scarp that is no longer parallel to the fault scarp. The scarp in this case is known as a *fault-line* scarp.

Thrust faults are very similar to reverse faults except that the angle of the fault plane with respect to the horizontal is quite low. They involve intense crustal compression and can result in large displacements. The McConnell thrust in the Canadian Rockies of Alberta, for example (Figure 8.21), has a displacement of about 40 km. The Lewis overthrust (Figure 8.22), exposed in Glacier National Park, has a similar scale of displacement. In isolated exposures, thrusts produce relationships that can be confusing. Notice from the cross section of the McConnell thrust that Cambrian and Ordovician sedimentary rocks are thrust over Cretaceous rocks. Without knowing that a thrust fault was present, you might conclude that an outcrop of this section violates the Principle of Superposition because

▲ FIGURE 8.20
Schematic cross section through the eastern flank of the Colorado Front Range. The primary structure, the Rampart Range fault, is a high-angle reverse fault. Precambrain granite in the hanging wall was faulted upward relative to Paleozoic sedimentary rocks to the east. Width of section approximately 25 km. *Source:* From J. B. Noblett, A. S. Cohen, E. M. Leonard, B. M. Loefller, and D. A. Gevirtzman, 1987, The Garden of the Gods and basal Phanerozoic nonconformity on and near Colorado Springs, Colorado, in *Geological Society of America Centennial Field Guide—Rocky Mountain Section*, Geological Society of America.

▲ FIGURE 8.21
Multiple thrusts in the Front Ranges, Canadian Rockies. Cambrian and Devonian rocks are thrust over Cretaceous rocks along the McConnell thrust. *Source:* From J. Suppe, *Principles of Structural Geology,* © 1985 by Prentice Hall, Inc., Englewood Cliffs, N.J. Reproduced with permission from Prentice Hall, Inc., and from the Geological Survey of Canada, GSC Map 1272A.

▲ FIGURE 8.22
Photo of Lewis overthrust (dark line along exposure of fault plane), Glacier National Park. Precambrian metasedimentary rocks are thrust over Cretaceous rocks. *Source:* Photo courtesy of the author.

older rocks overlie younger rocks! Careful mapping and identification of the fault would be necessary to resolve the problem.

Dip-slip faults sometimes occur in pairs (Figure 8.23). A down-dropped block bounded by two normal faults is called a *graben*. Rift zones, in which the lithosphere is being pulled apart by plates moving away from each other, usually contain large grabens at the surface (Figure 8.24). The structure formed when the central block between two dip-slip faults is up-thrown is known as a *horst* (Figure 8.23). *Strike-slip* faults are characterized by the lateral movement of fault blocks along the strike of the fault plane (Figure 8.25). These

(a)

◀ FIGURE 8.23
Pairs of normal faults forming (a) a graben and (b) a horst.

(b)

▶ FIGURE 8.24
(a) Right-lateral and (b) left-lateral strike-slip faults. The fence on the surface can be used to determine amount of displacement.

(a)

(b)

faults are either *right-lateral* or *left-lateral* types, depending on the relative movement of fault blocks. The two varieties can be distinguished by facing the fault from either side and noting the direction, either left or right, toward the continuation across the fault of a linear feature such as a stream, fence, or road that passed straight across the fault prior to displacement (Figure 8.25). The San Andreas fault is an excellent example of a right-lateral, strike-slip fault (Figure 8.26).

Dip-slip and strike-slip faults are merely end members of fault types that may display any combination of dip-slip and strike-slip displacement. Faults that contain components of both types of movements are called *oblique-slip* faults. The actual type of movement that takes place along any individual fault is the result of the type and orientation of the principal stresses imposed upon a crustal block. As shown in Figure 8.27, dip-slip faults are associated with horizontal stresses. When these stresses are compressional, reverse and thrust faults will occur. When the horizontal stresses are tensional, normal faults develop. Strike-slip faults are caused by shearing stresses acting as a couple on a crustal block. When fault movement takes place, the type and amount of displacement is readily observable. Fault scarps can be used to measure the amount of displacement on a dip-slip fault. These ruptures displace any surface features present. An active strike-slip fault can often be recognized by a characteristic set of topographic features (Figure 8.28). These include bends in stream courses or offsets of other linear topographic features, as well as depressions, or *sag ponds*, along the trend of the fault.

Fault scarps and other types of topography associated with active faults are rapidly modified by erosion. Because of this, there may be no scarps marking the presence of

▲ FIGURE 8.25
Right-lateral displacement of 1.45 m can be measured by the offset of a paint strip along a road near Kobe, Japan. The rupture occurred during the 1995 Hanshin-Awaji (Kobe) earthquake. *Source:* Carol Prentice; photo courtesy of U.S. Geological Survey.

◀ FIGURE 8.26
The San Andreas fault in the Carrizo Plains of California. *Source:* F. E. Wallace; photo courtesy of U.S. Geological Survey.

inactive faults, and as a result these structures can only be located by careful geologic mapping. Clues may be provided by linear stream courses or other topographic features. Stream courses are readily established along fault zones because of the low resistance to erosion caused by the crushing and grinding of rock along the fault plane.

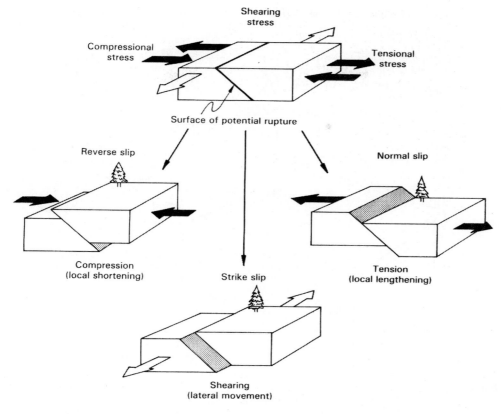

▲ FIGURE 8.27
Relationship of stress orientation to the style of faulting. *Source:* From E. A. Hay and A. L. McAlester, *Physical Geology: Principles and Perspectives*, 2nd ed., © 1984 by Prentice Hall, Inc., Englewood Cliffs, N.J.

The identification of active or potentially active faults is critical to the siting of such projects as nuclear power plants, schools, hospitals, and other buildings. Fault zones should be avoided for these types of structures. Inactive faults are also relevant to engineering design because they may bring rocks of different engineering properties into contact, or they may control the permeability and groundwater movement through a rock mass.

Geologic Maps and Cross Sections

Geologic maps are among the best sources of information for preliminary site location and design. For this reason, engineers need to become familiar with the construction and use of these maps. Geologic maps are drawn after extensive field work, during which the geologist observes as many outcrops as possible in the area being mapped. Lithology, fossils, contacts, and any other information are systematically depicted. Rock units are divided into formations, which are traced through the map area by plotting their contacts with other formations on a base map, which is commonly a topographic map. When the area contains folded sedimentary rocks, strike and dip are measured frequently and plotted on the base map at the location of the outcrop. These measurements form the basis for interpretation of these structures. Vertical aerial photographs are commonly used before and during field work.

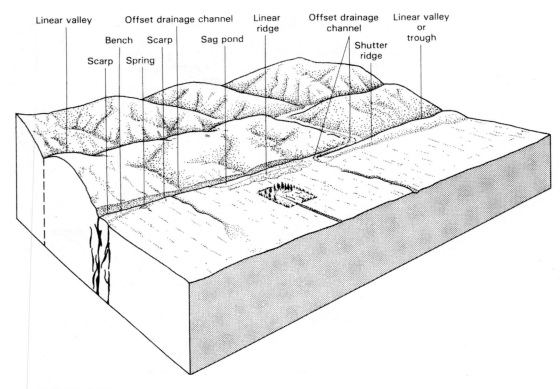

▲ FIGURE 8.28
Topography developed along an active strike-slip fault. *Source:* From R. D. Borcherdt, ed., 1975, *Studies for Seismic Zonation in the San Francisco Bay Region*, U.S. Geological Survey Professional Paper 941-A.

The geologic map is the final result of this effort (Figure 8.29). Rock units are shown on the map by colors or patterns. Because contacts are rarely exposed continuously across an area, they are sometimes drawn by using indirect evidence. The geologic map, therefore, is actually the geologist's interpretation of the geology. A certain amount of uncertainty and error in placing contacts and representing structural relationships is unavoidable. For this reason, contacts should not be transferred from a geologic map drawn at a scale of 1:24,000 or smaller to an engineering site map at a much larger scale and be expected to be precisely accurate. More detailed mapping is usually required when the exact position of a contact is required.

Several types of geologic maps are made. When bedrock formations are exposed in an area, these may be the only units shown on the map, even though there may be surficial materials, like stream or windblown sediments, overlying the bedrock units in some parts of the map area. Even though these materials are not mapped, they may be very important for foundation design or other engineering reasons. Some geologic maps show the surficial geologic materials and some show both surficial and bedrock units. It is important to examine the map and read the report to determine what types of units were mapped by the geologist.

Cross sections are drawn by projecting contacts and other information from the map onto a vertical plane oriented along a line that is shown on the geologic map (Figure 8.29). These diagrams are the geologist's interpretation of the distribution of map units in the subsurface. Cross sections are based on outcrop data or direct subsurface data from boreholes or well records, if available. In an area of complex geology, there is almost never as much subsurface control as the geologist would like to have. For this reason, cross sections are subject to more uncertainty than geologic maps and should be used with caution.

▲ FIGURE 8.29
Geologic map and cross section. *Source:* From S. Judson, M. E. Kauffman, and L. D. Leet, *Physical Geology,*
7th ed., © 1987 by Prentice Hall, Inc., Englewood Cliffs, N.J.

Case in Point 8.1
Effects of Rock Structure on Dam Construction

Careful geologic mapping was necessary prior to construction of the Arbuckle Dam near
Ardmore, Oklahoma, which was completed in 1966 by the U.S. Bureau of Reclamation.
The complex geologic setting of this dam consists of a series of tightly folded and faulted
Paleozoic sedimentary rocks and beautifully illustrates the role of geology in a major engineering project (Jackson, 1969). The Arbuckle Dam is located in a region of rolling hills
known as the Arbuckle Mountains. Rocks of the Arbuckle Mountains consist primarily of
a sequence of shales and limestones deposited in a Paleozoic sea (Table 8.1). In late Paleozoic time, the rocks were deformed by compression from the southwest into a series of
broad anticlines and synclines. Some of the folds were overturned and some rocks in the
anticlines were thrust over younger rocks in the adjacent synclines. A later sequence of
events included uplift, high-angle normal faulting, and erosion. Sediments eroded from

Table 8.1   Generalized Stratigraphic Section of Arbuckle Dam and Reservoir

	Formation	Thickness (feet)
Quaternary	ALLUVIUM	0–30
Pennsylvanian	VANOSS FM.—Poorly to moderately cemented conglomerate, shale, and sandstone of outwash debris on north flank of mountains.	650–900
	DEESE FM.—Varicolored shale; well-cemented conglomerate; thin beds of hard limestone and sandstone	up to 2000
	SPRINGER FM., GODDARD MBR.—Black shale, noncemented but very compact, bituminous, appears graphitic.	1000–3000
Mississippian	CANEY FM.—Dark-colored shale, moderately hard, limy, fissile.	150
	SYCAMORE FM.—Blue to weathered limestone, hard, tough.	0–50
	WOODFORD FM.—Dark-colored shale, siliceous and cherty, platy and brittle	400–500
Devonian-Silurian	HUNTON GR.—Thick to medium bedded limestone, marlstone	250
Cambrian-Ordovician	SYLVAN FM.—Gray-green fissile shale.	200
	VIOLA FM.—Bituminous limestone.	450
	ARBUCKLE GR.—Limestone, dolomite, shale, sandstone.	5000–8000

Source: From J. L. Jackson, 1969, Geologic studies at Arbuckle Dam, Murray County, Oklahoma, *Bulletin of the Association of Engineering Geologists*, 5:79–100.

the rising mountain mass were deposited around the flank of the uplift. During Mesozoic time, marine rocks were deposited over the Paleozoic sequence and then removed by erosion. Continued erosion during the Cenozoic era has produced the present landscape. Major structural elements of the region are shown in Figure 8.30.

Geologic conditions at the dam site were defined by careful geologic mapping and test drilling (Figure 8.31). The stream valley is established in a sequence of steeply dipping, faulted shales and limestones. The center of the valley is a graben formed by several normal faults, including the highly deformed Reagan fault zone at the base of the left abutment (the section of the dam that meets the valley wall). Rocks are fractured, crushed, and altered along the fault and shear zones.

One of the major concerns in dam construction is to prevent excessive leakage from the reservoir through the foundation rocks beneath the base of the dam. Leakage can result in the loss of reservoir storage as well as in the possibility of failure and collapse of the dam. A common procedure to prevent such leakage involves sealing dam foundations with a grout curtain. The grout curtain is produced by injecting grout (cement slurry) under high pressure into holes drilled in a line across the valley floor. Grout is forced into pores, cavities, and fractures within the rock and then hardens to form a low-permeability seal for the dam. At the Arbuckle Dam, 135 grout holes were used, ranging in spacing from 25 to 110 ft (7.6 to 33.5 m). The profile of the grout curtain shown in Figure 8.31 illustrates the amount of grout needed to seal the foundation. The

▶ FIGURE 8.30
Geologic setting of the Arbuckle Mountains, showing major faults and folds. *Source:* From J. L. Jackson, *Bulletin of the Association of Engineering Geologists,* 5: 79–100, © 1969 by the Association of Engineering Geologists.

▶ FIGURE 8.31
Cross section of the Arbuckle Dam site and profile of the grout curtain emplaced beneath the dam. *Source:* From J. L. Jackson, *Bulletin of the Association of Engineering Geologists,* 5: 79–100, © 1969 by the Association of Engineering Geologists.

most permeable area was the Reagan fault zone at the base of the left abutment. Relatively small amounts of grout were needed across the valley, except at a fault and unconformity in the vicinity of the right abutment.

Construction of the Arbuckle Dam required a thorough understanding of the geology of the site. At this project, the distribution of folds, faults, and other structures was critical to the design of the dam.

Earthquakes

At 5:12 A.M. on April 18, 1906, a great earthquake struck San Francisco. During the next 60 seconds, violent ground shaking devastated portions of the city and caused panic and terror. The ground was actually observed to be moving in waves in some places; train tracks were bent like wire, and whole buildings or chimneys collapsed to the ground (Figure 8.32). The aftermath of the earthquake was even more destructive because fires broke out throughout the city and firefighters were helpless because the city water main had been ruptured by the shaking. In all, at least 700 people were killed and 250,000 were left homeless.

The 1906 earthquake was the result of a sudden rupture along a 400-km-long segment of the San Andreas fault, the boundary between two lithospheric plates. Since that time, the United States has experienced only one other great earthquake, the Alaska earthquake of 1964. Fortunately, it occurred in a sparsely populated area. In recent years, building codes have been continually modified to minimize damage from earthquakes. In nearly every metropolitan area, however, many buildings predate modern building practices.

Eighty-three years after the 1906 earthquake, many Americans were settling in on the evening of October 17, 1989, to watch the third game of the World Series to be played in San Francisco, California. What they saw instead, was the gradual unfolding of the effects of another severe earthquake in the San Francisco Bay region, this one centered under the Loma Prieta mountain near the town of Santa Cruz. This earthquake, while not as strong as the 1906 tremblor, still wreaked havoc in the Bay area (Figure 8.33). Sixty-two people were killed, 3757 injured, and 12,053 displaced, and property damage amounted to a staggering

◀ FIGURE 8.32
Damage to the San Francisco City Hall by the 1906 earthquake. *Source:* W. C. Mendenhall; photo courtesy of U.S. Geological Survey.

▶ FIGURE 8.33
Rubble from collapsed building on Jefferson Street in the Marina District of San Francisco resulting from the Loma Prieta earthquake. *Source:* D. Perkins; photo courtesy of U.S. Geological Survey.

$6 billion. Perhaps even more frightening because an anxious nation watched on television as the drama of rescue and human suffering played out the Loma Prieta event, along with an even more destructive earthquake near Los Angeles in January 1994, were ominous warnings of things to come. While geologists work to develop the ability to predict earthquakes, the role of engineers to design and build structures to resist strong ground motion is crucial.

Occurrence

Earthquakes are caused primarily by the rupture of rocks in the earth's crust along faults. Most earthquakes take place along faults in the vicinity of plate boundaries (Figure 8.34). Earthquakes also occur far from plate boundaries. Often, the faults associated with these intraplate earthquakes are deeply buried and not visible at land surface.

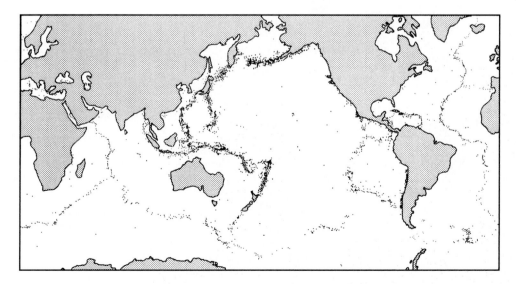

▲ FIGURE 8.34
Locations of shallow earthquakes during a 7-year period. The epicenters are mainly concentrated on tectonic plate margins. *Source:* Data from ESSA, U.S. Coast and Geodetic Survey.

Elastic Rebound Theory

During earthquakes, bodies of rock on opposite sides of faults move relative to each other. Both dip-slip and strike-slip displacements are recognized, as well as combinations of the two. During the Loma Prieta earthquake, for example, the relative strike-slip displacement between the Pacific and North American plates was 1.9 m. In addition, there was 1.3 m of uplift of the Pacific plate relative to the North American plate (Figure 8.35), despite the common perception of the San Andreas fault as a strike-slip fault. Vertical displacement may be due to a bend in the trend of this section of the fault.

A longstanding theory in geology, formulated by the analysis of surface rupture after the 1906 earthquake, holds that rocks deform in an elastic manner until brittle failure takes place (Figure 8.36). Elastic strain that had been accumulating on either side of

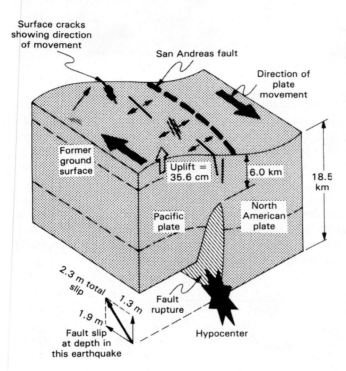

◀ **FIGURE 8.35**
Relative horizontal and vertical movement of plates during the Loma Prieta earthquake. *Source:* From P. L. Ward and R. A. Page, *The Loma Prieta Earthquake of October 17, 1989*, U.S. Geological Survey Pamphlet.

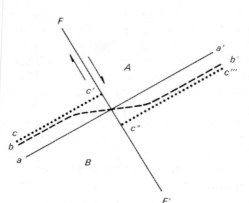

◀ **FIGURE 8.36**
Elastic rebound theory. Line *a–a'* is a real or imaginary linear feature that crosses a fault (*F–F'*) such as a road or fence. Elastic strain produces distortion of the line to the form shown as line *b–b'*. At the time of rupture, block *A* slides rapidly past block *B* and the elastic strain energy is now represented by segments *c–c'* and *c"–c'''*. *Source:* From M. P. Billings, *Structural Geology*, 3rd ed., © 1972 by Prentice Hall, Inc., Englewood Cliffs, N.J.

the fault is suddenly released, causing the rocks on either side of the fault to abruptly slide past each other. Elastic strain energy is released in the form of *seismic waves*, which radiate outward through the rocks in all directions from the *hypocenter*, or point of rupture. The energy that can be obtained from elastic strain can be painfully experienced by holding a rubber band against someone's arm, stretching it outward, and then releasing it suddenly from the opposite end. The propagation of seismic waves during fault movement is a consequence of elastic rebound, and the hypothesis is known as the *elastic rebound theory*. It is significant that seven damaging earthquakes occurred in the San Francisco Bay region during the 83 years prior to the great earthquake of 1906, but only two have occurred in the 83 years since. Release of strain energy during the 1906 earthquake may thus have relieved the stress on the fault for a period of decades. Some seismologists believe that elastic strain energy has now accumulated to the point where ruptures are beginning to occur. This may suggest that in the coming decades more frequent earthquakes similar to the Loma Prieta event may precede the next "big one" in the San Francisco region.

Although the elastic rebound theory satisfactorily explains earthquakes produced by the rupture of the brittle rocks near the earth's surface, the theory becomes more difficult to apply to those earthquakes generated at depths greater than a few 10s of kilometers beneath the surface. At these depths, the pressure exerted by the overlying rocks makes elastic behavior less likely. A more comprehensive theory for the origin of earthquakes in regions of plastic deformation is thus needed. Intensive research underway at present may lead to an improved explanation for deep earthquakes.

Detection and Recording

During rupture at the hypocenter of an earthquake, the elastic strain energy stored in the rocks is suddenly released in the form of *seismic waves*. These waves are propagated outward from the hypocenter through the rocks in all directions. This seismic energy causes rapid vibrations in the rocks in the subsurface as well as on the land surface itself. The instrument designed to measure and record the characteristics of seismic waves is called a *seismograph* (Figure 8.37). It consists of a weight suspended on a string or wire that acts as a pendulum. The inertia of the mass resists its motion relative to the rigid frame to which it is attached. When the pendulum is connected either mechanically or electrically to a pen in contact with a rotating drum and chart, a record of the wave vibrations, called a *seismogram*, is produced. The pendulum of a seismograph is constrained to move in only one direction. Therefore, a complete seismological station must contain three seismographs. One measures the vertical component of motion, and the other two measure horizontal motion in directions at right angles to each other.

Ground motion from an earthquake usually is most severe near the *epicenter*, the point on the earth's surface directly above the hypocenter. Instruments called *accelerographs* are used to measure the strong motion of an earthquake in the vicinity of the epicenter. Accelerographs differ from seismographs in that they begin recording only during strong earth movements. Seismographs record seismic waves continuously. Accelerographs measure ground acceleration relative to g, the acceleration of gravity, which has a value of 9.8 m/s^2. The highest acceleration recorded was 1.25 g during the 1971 San Fernando, California, earthquake. Accelerations are measured in three directions at right angles to each other (Figure 8.38). Commonly, accelerographs are installed at different levels in a building or in different geologic settings in order to evaluate the response to seismic waves of different types of structures, soils, or rocks. Seismograph stations are now integrated into regional and worldwide networks. These networks use similar equipment and process data uniformly in order to standardize data gathering and analysis.

◀ FIGURE 8.37
A seismograph is the instrument used to measure seismic waves. *Source:* Photo courtesy of U.S. Geological Survey.

Magnitude and Intensity

The strengths and effects of earthquakes are determined in two ways. *Magnitude* is a quantitative measure of the energy released by an earthquake, as obtained from seismograph records. A method for determining magnitude developed by seismologist Charles F. Richter defines magnitude as the logarithm of the largest seismic wave amplitude produced by the trace of a standard seismograph located at a distance of 100 km from the epicenter. Magnitudes (M) obtained by this method are commonly referred to as the magnitude on the *Richter scale*. The highest magnitude values measured have been approximately 8.9. The logarithmic aspect of magnitude means that an earthquake of magnitude 7, for example, produces 10 times the wave amplitude of an earthquake of magnitude 6. Great earthquakes are classified as those with a magnitude of 8 or more. A new magnitude measure, the *moment magnitude* (M_w) is now commonly used in place of the Richter magnitude. The moment magnitude is based on the *seismic moment* (M_0), of the earthquake, which is defined as

$$M_0 = SA\mu \tag{8.1}$$

where S is the amount of slip (m) on the fault during the earthquake, A is the area (m^2) of the fault plane that ruptured, and μ is the modulus of rigidity of the rocks at the

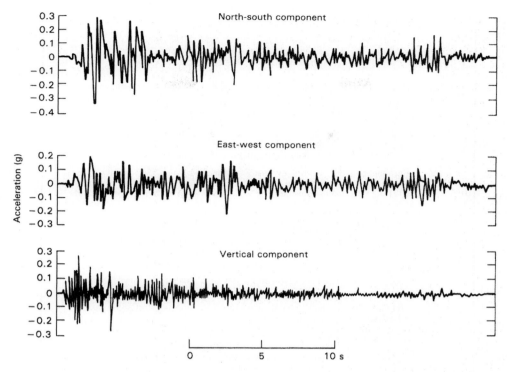

▲ FIGURE 8.38
Accelerogram of the 1940 Imperial Valley, California, earthquake recorded at El Centro, California.
Source: From W. W. Hays, 1980, *Procedures for Estimating Earthquake Ground Motions*, U.S. Geological Survey Professional Paper 1114.

hypocenter (Nm^{-2}). The modulus of rigidity measures the resistance of the rocks to shearing distortion by the seismic waves. The moment magnitude is calculated from the seismic moment as

$$M_w = 2/3 \log M_o - 6.0 \tag{8.2}$$

The SI units of moment magnitude are Nm. The moment magnitude is considered to be more quantitative than the Richter magnitude and better correlated with the amount of energy released by the earthquake. The amount of seismic energy released in an earthquake is a function of the volume of rock releasing strain energy during the event. Thus, the amount of energy liberated is maximized in great earthquakes. In fact, the small number of high-magnitude earthquakes that occur every year release more energy than the combined energy of the hundreds of thousands of smaller earthquakes that occur annually.

Unlike earthquake magnitude, *intensity* is a subjective measure that is determined by observing the effects of the earthquake on structures and by interviewing people who experienced the event. The most commonly used intensity scale is the *Modified Mercalli* scale, which has 12 divisions (Figure 8.39). Each division describes a particular level of human response and building performance. Intensity generally decreases with distance from the epicenter (Figure 8.40); local geologic conditions, however, also influence the amount of ground shaking. We will discuss these factors in more detail later.

Earthquakes cannot be predicted with accuracy. On a global basis, however, there is a consistent relationship between magnitude and *frequency* (Figure 8.41). The inverse relationship illustrated in Figure 8.41 shows that whereas on an average there are about

MODIFIED MERCALLI INTENSITY SCALE OF 1931

(Abridged)

I. Not felt except by a very few under especially favorable circumstances.

II. Felt only by a few persons at rest, especially on upper floors of buildings. Delicately suspended objects may swing.

III. Felt quite noticeably indoors, especially on upper floors of buildings, but many people do not recognize it as an earthquake. Standing motor cars may rock slightly. Vibration like passing of truck. Duration estimated.

IV. During the day felt indoors by many, outdoors by few. At night some awakened. Dishes, windows, doors disturbed; walls made cracking sound. Sensation like heavy truck striking building. Standing motor cars rocked noticeably.

V. Felt by nearly everyone; many awakened. Some dishes, windows, etc., broken; a few instances of cracked plaster; unstable objects overturned. Disturbance of trees, poles and other tall objects sometimes noticed. Pendulum clocks may stop.

VI. Felt by all; many frightened and run outdoors. Some heavy furniture moved; a few instances of fallen plaster or damaged chimneys. Damage slight.

VII. Everybody runs outdoors. Damage negligible in buildings of good design and construction; slight to moderate in well-built ordinary structures; considerable in poorly built or badly designed structures; some chimneys broken. Noticed by persons driving motor cars.

VIII. Damage slight in specially designed structures; considerable in ordinary substantial buildings with partial collapse; great in poorly built structures. Panel walls thrown out of frame structures. Fall of chimneys, factory stacks, columns, monuments, walls. Heavy furniture overturned. Sand and mud ejected in small amounts. Changes in well water. Disturbed persons driving motor cars.

IX. Damage considerable in specially designed structures; well designed frame structures thrown out of plumb; great in substantial buildings, with partial collapse. Buildings shifted off foundations. Ground cracked conspicuously. Underground pipes broken.

X. Some well-built wooden structures destroyed; most masonry and frame structures destroyed with foundations; ground badly cracked. Rails bent. Landslides considerable from river banks and steep slopes. Shifted sand and mud. Water splashed (slopped) over banks.

XI. Few, if any (masonry), structures remain standing. Bridges destroyed. Broad fissures in ground. Underground pipe lines completely out of service. Earth slumps and land slips in soft ground. Rails bent greatly.

XII. Damage total. Waves seen on ground surfaces. Lines of sight and level distorted. Objects thrown upward into the air.

▲ FIGURE 8.39
The Modified Mercalli scale (abridged) for determining earthquake intensity.

10,000 earthquakes a year of magnitude 4, there are fewer than 10 earthquakes every year above magnitude 7.

Seismic Waves
Wave Types

Seismic waves generated during an earthquake fall into two basic categories: *body waves* that travel through the earth's interior from the earthquake's hypocenter, and *surface*

▶ FIGURE 8.40
Seismic intensity map of the 1971 San Fernando, California, earthquake. *Source:* From W. W. Hays, 1980, *Procedures for Estimating Earthquake Ground Motions*, U.S. Geological Survey Professional Paper 1114.

▶ FIGURE 8.41
Average relationship between magnitude and frequency on a global basis. *Source:* From B. A. Bolt, *Nuclear Explosions and Earthquakes: The Parted Veil,* © 1976 by W. H. Freeman and Co., San Francisco.

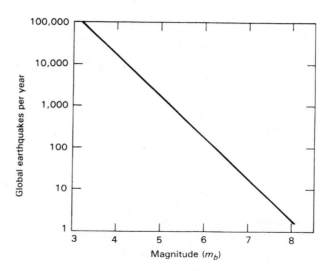

waves that move along the earth's surface from the epicenter. Body waves can be subdivided further into *P* (*primary*) and *S* (*secondary*) waves. *P* waves are also known as compressional waves because they travel through rocks as alternate compressions and expansions of the material (Figure 8.42). The movement of individual rock particles is a back-and-forth motion in the direction of propagation. *P* waves are the fastest-moving seismic waves. Their velocity (α), in m/s is expressed as

$$\alpha = \sqrt{\left(\kappa + \tfrac{4}{3}\mu\right)\rho} \tag{8.3}$$

where (κ) is the bulk modulus of the rock (resistance to volume change), μ is the modulus of rigidity (resistance to change in shape), and ρ is the density (g/cm^3).

 S waves are also known as shear waves because of their mode of travel (Figure 8.43). The travel of *S* waves is analogous to the movement of a rope anchored to a wall at one

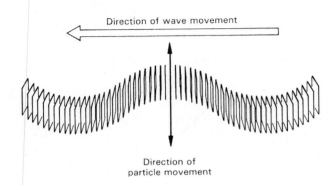

▲ FIGURE 8.42

The motion of *P* waves. The wave moves through rock as a sequence of compressions and expansions parallel to the direction of wave movement. Particles vibrate in back-and-forth fashion in the direction of wave movement. *Source:* After E. A. Hay and A. L. McAlester, *Physical Geology: Principles and Perspectives,* 2nd ed., © 1984 by Prentice Hall, Inc., Englewood Cliffs, N.J.

◀ FIGURE 8.43

Transmission of *S* waves. Particle movement is perpendicular to direction of wave movement through rock. *Source:* After E. A. Hay and A. L. McAlester, *Physical Geology: Principles and Perspectives,* 2nd ed., © 1984 by Prentice Hall, Inc., Englewood Cliffs, N.J.

end and shaken up and down with your hand at the other end. Although the wave motion moves from your hand toward the wall, the movement of the rope is up and down as the wave passes. This shear deformation, which is transverse to the direction of wave propagation, explains why *S* waves cannot be transmitted through a liquid: A liquid has no resistance to shear stresses. The discovery that *S* waves do not travel through the earth's outer core is the main evidence for the hypothesis that this part of the core is liquid-like. The velocity of *S* waves (β), which is lower than that of *P* waves, can be expressed as

$$\beta = \sqrt{\frac{\mu}{\rho}} \tag{8.4}$$

The modulus of rigidity, μ, is equal to zero in liquids; therefore, the *S*-wave velocity is zero.

Surface waves, which arrive at seismograph stations after both *P* and *S* waves, include *Love* and *Rayleigh* waves. Surface waves are characterized by long periods (10 to 20 s) and wavelengths of 20 to 80 km. The period measures the amount of time that elapses between the arrival of two successive wave crests, and wavelength measures the distance separating identical points on successive waves. Love waves originate from *S* waves that reach the surface at the epicenter. As they move outward along the surface of the earth, they produce only horizontal ground motion. Rayleigh waves resemble the motion of ocean waves, with predominantly vertical displacement. A seismogram showing the arrival of *P* waves, *S* waves, and surface waves is shown in Figure 8.44.

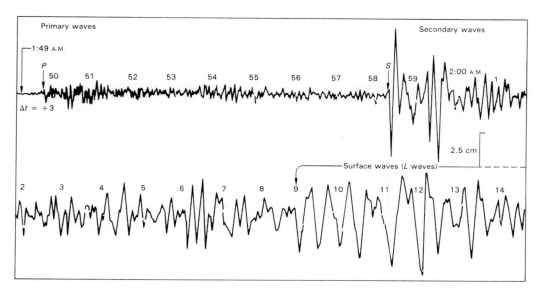

▲ FIGURE 8.44
Portion of a seismogram showing the arrival of P, S, and surface waves. *Source:* From S. Judson, M. E. Kauffman, and L. D. Leet, *Physical Geology*, 7th ed., © 1987 by Prentice Hall, Inc., Englewood Cliffs, N.J.

Table 8.2   Travel Time and Time Interval for Arrival of Seismic Waves

Distance from source, km	Travel Time				Interval between P and S (S–P)	
	P		S			
	min	s	min	s	min	s
2,000	4	06	7	25	3	19
4,000	6	58	12	36	5	38
6,000	9	21	16	56	7	35
8,000	11	23	20	45	9	22
10,000	12	57	23	56	10	59
11,000	13	39	25	18	11	39

Distance	Time (S–P)
km 500	50 sec
	40
300	30
200	20
100	10
75	8
50	6
	4
25	
20	2
10	
5	
0	0

Source: From S. Judson, M. E. Kauffman, and L. D. Leet, *Physical Geology*, 7th ed., © 1987 by Prentice Hall, Inc., Englewood Cliffs, N.J.

Time-Distance Relations

The difference in velocity between P and S waves means that the time interval between the first arrival of P waves and S waves increases with distance from the earthquake source (Table 8.2). Thus, the time interval between P-wave and S-wave arrivals measured at a seismograph station is an indication of the distance of the station from the epicenter. Figure 8.45 illustrates how S–P time differences vary with distance. The S–P time difference can be used for locating the epicenter of a given earthquake if at least three

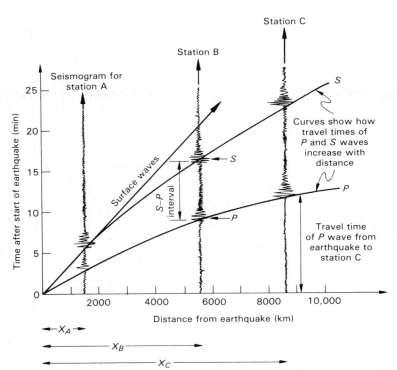

▲ FIGURE 8.45

Variation of the S–P time interval with distance from the epicenter. As the distance from the epicenter increases from seismograph station X_A to X_C, the S–P time interval increases. *Source:* From F. Press and R. Siever, *Earth*, 3rd ed., © 1982 by W. H. Freeman and Co., San Francisco.

◀ FIGURE 8.46

The epicenter of an earthquake can be located by plotting three circular arcs with radii determined from the S–P time interval as shown in Figure 8.45. The epicenter is located at the intersection of the three arcs. *Source:* From F. Press and R. Siever, *Earth*, 3rd ed., © 1982 by W. H. Freeman and Co., San Francisco.

seismograms from different stations are available for the earthquake. The distance from the epicenter corresponding to the S–P time difference is used as the radius of a circle to plot a circular arc in the vicinity of the epicenter (Figure 8.46). The point where three circular arcs intersect is the epicenter.

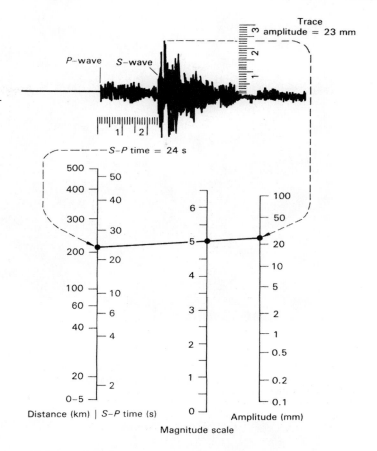

► FIGURE 8.47
Determination of the Richter magnitude from the *S–P* time interval and the maximum trace amplitude measured from a seismogram. *Source:* From W. W. Hays, 1980, *Procedures for Estimating Earthquake Ground Motions*, U.S. Geological Survey Professional Paper 1114.

The *S–P* time interval, along with maximum trace amplitude, is also useful in determining the magnitude (Figure 8.47). If the plotted points representing the trace amplitude and *S–P* time interval are connected with a straight line on a nomogram, the magnitude of the earthquake is indicated where the line crosses the center scale.

Earthquake Hazards

Most earthquakes accompany major movements of crustal blocks along faults. Crustal blocks can also move without generating seismic waves. Such movement, called *aseismic creep*, is desirable from the standpoint that it relieves accumulating strain gradually, thus preventing a buildup of strain leading to a major earthquake. Aseismic creep also has detrimental effects because of the associated displacement that takes place along faults. This displacement can rupture utility lines and wells and cause damage to streets, sidewalks, and other structures. A particularly dramatic example of aseismic creep is the 1963 failure of the Baldwin Hills Reservoir in southern California, an impoundment that stored water for the City of Los Angeles (Figure 8.48). Aseismic displacement along faults located beneath the reservoir caused fracturing of the reservoir's underdrain system and impermeable asphalt membrane. Water leakage through the reservoir floor and dam eventually led to the complete failure of the dam and the draining of the impoundment.

Hazards associated with earthquakes include surface faulting, ground shaking, ground failure, and destructive ocean waves called *tsunamis*. Rupture of the ground surface by fault

▲ FIGURE 8.48
Aerial view of the Baldwin Hills reservoir after failure. Aseismic creep along faults beneath the impoundment ruptured the impermeable seal of the bottom. Water then migrated through the dam at an increasing rate until a section of the dam collapsed and the reservoir drained rapidly. *Source:* Photo courtesy of Los Angeles Dept. of Water and Power.

▲ FIGURE 8.49
Fault scarp formed during the 1959 Hebgen Lake earthquake in Montana. *Source:* I. J. Witkind; photo courtesy of U.S. Geological Survey.

displacement includes both vertical and lateral motion. Fault scarps are the result of surface rupture formed by dip-slip displacement during an earthquake (Figure 8.49). Obviously, any structure built directly across a fault that sustains surface rupture during an earthquake is likely to be damaged or destroyed (Figure 8.50). By mapping faults known to be active (those in which movement has occurred within recent geologic time), engineers can avoid these hazardous areas for some or all types of structures. Unfortunately, development has occurred over active faults in some areas (Figure 8.51). In addition, surface displacement frequently has occurred along faults that were thought to be inactive. Ground shaking is usually the greatest threat to buildings and human life associated with earthquakes. The peak acceleration attained by a structure multiplied by its mass defines the dynamic force applied to the structure during an earthquake. Ground acceleration depends on the distance from the epicenter, the depth of the hypocenter, and the characteristics of the soils and rock beneath the foundation. The response of the structure depends upon its size and design. We will examine these factors more closely in the following sections.

(a) Thrust displacement

(b) Vertical displacement

(c) Left-lateral displacement

(d) Extensional displacement

(e) Localized differential displacements

▲ FIGURE 8.50

Types of structural damage caused by ground rupture during the San Fernando earthquake. *Source:* From U.S. Geological Survey, 1971, *The San Fernando, California, Earthquake of February 9, 1971*, U.S. Geological Survey Professional Paper 733.

◄ FIGURE 8.51
Urban development along the San Andreas fault, San Mateo County, California. Arrows show location of fault. *Source:* R. W. Wallace; photo courtesy of U.S. Geological Survey.

Case in Point 8.2
Effects and Damage from the Great New Madrid Earthquakes

Although the interior of the United States is generally considered to be seismically benign, a devastating series of earthquakes centered around New Madrid, Missouri, struck the central United States over a 5-month period in 1811 and 1812. The magnitudes of several of these earthquakes are estimated to have reached the mid 7s in the Richter magnitude scale, or over 8.0 in terms of moment magnitude; the seismic intensity map of the earthquakes (Figure 8.58) indicates that the earthquakes were felt over most of the eastern United States.

The first of the large earthquakes occurred in the early morning hours of December 16, 1811. The five largest earthquakes occurred between Blytheville, Arkansas and New Madrid, Missouri, coincident with the epicenters of frequent smaller earthquakes that occur to this day in an area known as the New Madrid seismic zone (Figure 8.52). The proximity of the epicenters of the large quakes to the Mississippi River was responsible for the level of damage that was sustained. Most of the population of the day lived in farms or small villages along the river. The intense ground motion within and along the river generated great waves that washed onto the banks, throwing boats with them. The weak saturated soils in the river banks failed and slumped into the channel. Several towns totally disappeared as a result. Saturated sand liquefied and flowed under the stresses of seismic shaking. Where the liquefying sand was overlain by clayey sediment, the sand burst through cracks or fissures and flowed onto the land surface. The remnants of hundreds of these sand boils are still visible throughout the region. Aftershocks were felt daily after the December 16 event. Large earthquakes struck again on January 7 and February 12, 1812. These earthquakes were widely felt as far away as the eastern seaboard. The February 12

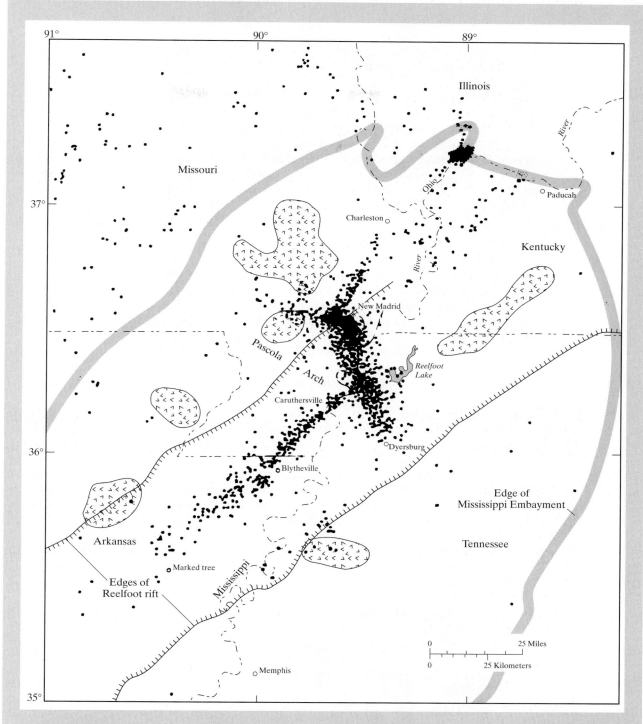

▲ FIGURE 8.52
Earthquake epicenters and other features of the New Madrid seismic zone. Most of the epicenters are concentrated along linear zones, interpreted to be faults that failed during the 1811–1812 earthquakes. The faults are concentrated within the boundaries of the Reelfoot rift, a deeply buried rift zone that was intruded by magma near the end of Precambrian time. *Source:* From W. Spence, R. B. Herrmann, A. C. Johnston, and G. Reagor, *Responses to Iben Browning's Prediction of a 1990 New Madrid, Missouri, Earthquake,* U.S. Geological Survey Circular 1083, 1993.

event was perhaps the strongest. During the event, a block of crust along the Mississippi River within the fault zone was upthrown to form a horst. This caused the river to reverse its course and flow northward for a short time. At least two low waterfalls formed in the river channel as fault scarps were instantaneously created. Although the great erosional force of the river began to wear down the scarps almost immediately, dozens of boats were capsized as they were swept over the falls, drowning most of their occupants. Eyewitness accounts of the river paint a terrifying picture of giant waves and raging, turbulent waters crashing into the banks. In all, more than 2000 shocks occurred between December and April or May. Loss of life was minimized by the sparse population and the low, wooden buildings. A comparable event today, with the greatly increased population and infrastructure, would undoubtedly cause the greatest loss of life and property from a natural disaster in the history of the United States.

Ground failure includes a variety of processes. During intense shaking, soil structure and behavior can be drastically altered. Seismic stresses often cause soils to densify and compact. Certain saturated soils lose strength and liquefy almost instantaneously during seismic shaking. In this state, the soil can no longer support buildings, which then may sink and rotate into the soil as complete units (Chapter 10). This process, called *liquefaction*, was responsible for extensive damage to buildings in both the Loma Prieta earthquake and the June 21, 1990, earthquake in northern Iran (Figure 8.53). This 7.7 magnitude earthquake left 40,000 dead, 60,000 injured, and 500,000 homeless. A major factor in the loss of life was the poor performance of buildings during ground shaking. Downslope mass movements, including slumps, slides, flows, and avalanches can be triggered by earthquakes. Often, these movements involve huge masses of material and are responsible for great damage and loss of life. The Lower Van Norman Dam (Figure 8.54) nearly failed during the 1971 San Fernando, California, earthquake because of a large slump that developed on the upstream side. If the reservoir level had not been maintained at a lower elevation than normal design limits called for (because of concerns for the safety of the dam), it most likely would have totally failed during the earthquake. The imminent failure during the earthquake necessitated the evacuation of 80,000 people downstream. Other examples of mass movements caused by earthquakes are discussed in Chapter 13.

An earthquake hazard that is greatly feared in coastal areas is the tsunami, or seismic sea wave. These are high-energy waves generated by earthquakes whose epicenters lie

◄ FIGURE 8.53
Differential settlement due to liquefaction during the 1990 northern Iran earthquake. The building to the left sank relative to the small attached structure at center, which has a separate foundation. *Source:* M. Mehrain.

► FIGURE 8.54
Lower Van Norman Dam near San Fernando, California, after damage by the 1971 earthquake. Slumping of parts of the dam into the reservoir is evident. *Source:* Photo courtesy of U.S. Geological Survey.

beneath the seabed. Tsunamis, as exemplified by the December, 2004 Indian Ocean tsunami, travel rapidly across the ocean basins and cause great damage when they impact coastlines. Tsunamis are described more fully in Chapter 15.

In addition to the hazards just mentioned, other problems are encountered during earthquakes. These include fires, disruption of water supplies, and disease. As the population of seismically active areas continues to increase, the potential for damage and loss of life grows accordingly.

Seismic Risks and Land-Use Planning

Assessment of seismic hazard for an area is a complex procedure that must take into account the history of seismic events in the area, the location of active faults, the regional and local responses to seismic events, and the seismic characteristics of the surficial geologic materials. If the evaluation of all these factors indicates that the area may be subjected to a potentially damaging earthquake, measures can be taken to reduce the risks associated with the event. These measures can include land-use planning and zoning as well as the adoption of stringent building codes.

Location of Earthquakes

Locations of earthquake epicenters recorded during a particular period of time provide a good indication of seismically active areas. Such maps (Figure 8.34) show a high concentration of earthquakes along lithospheric plate boundaries. Earthquakes are not limited to plate boundaries, however. Figure 8.55 indicates that damaging earthquakes have struck many areas in the United States during the past 200 years. Table 8.3 provides information on the magnitude and effects of these events.

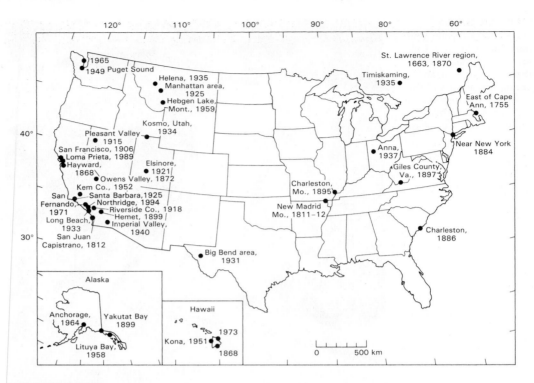

▲ FIGURE 8.55

Location of past destructive earthquakes in the United States. *Source:* From W. W. Hays, 1980, *Procedures for Estimating Earthquake Ground Motions*, U.S. Geological Survey Professional Paper 1114.

In addition to determining the location of earthquake epicenters, it is also important to map and identify active faults. A number of specific criteria are used for recognition of an active fault. For example, the U.S. Nuclear Regulatory Commission (NRC) defines a *capable* fault as one that has experienced displacement at least once within the past 35,000 years. A map of major faults in California is shown in Figure 8.56. If a particular fault segment has not undergone rupture within a certain period of time, this does not necessarily mean that the segment is inactive. Instead, it may indicate that that section of the fault is *locked*, a condition in which large amounts of elastic strain accumulate prior to infrequent, but large-magnitude, earthquakes. Locked sections of faults are called *seismic gaps* because plots of earthquake hypocenters show a lack of events in these areas. A cross section of the San Andreas fault from northern to central California (Figure 8.57) indicates three seismic gaps: the San Francisco Peninsula gap, the southern Santa Cruz Mountains gap, and the Parkfield gap. The lower cross section shows that the southern Santa Cruz Mountains gap was filled by the Loma Prieta earthquake and its aftershocks. Seismic gaps are closely monitored because the longer a gap exists, the stronger the eventual earthquake may be in that area.

Ground Motion

The response of the earth's crust to an earthquake varies both on a regional and local scale. The decrease in amplitude of seismic waves as they move outward from the epicenter is known as *seismic attenuation*. The seismic attenuation characteristics of the San Francisco earthquake and the New Madrid, Missouri, earthquake are shown in Figure 8.58. The contours define zones of equal seismic intensity as measured by the Modified Mercalli scale. Even though the San Francisco earthquake was much larger in magnitude than

Table 8.3   Property Damage and Lives Lost in Notable U.S. Earthquakes

Year	Locality	Magnitude	Damage (million dollars)	Lives Lost
1811–12	New Madrid, Mo.	7.5 (est.)	—	—
1865	San Francisco, Calif.	8.3 (est.)	0.4	—
1868	Hayward, Calif.	—	0.4	30
1872	Owens Valley, Calif.	8.3 (est.)	0.3	27
1886	Charleston, S.C.	—	23.0	60
1892	Vacaville, Calif.	—	0.2	—
1898	Mare Island, Calif.	—	1.4	—
1906	San Francisco, Calif.	8.3 (est.)	500.0	700
1915	Imperial Valley, Calif.	—	6.0	6
1925	Santa Barbara, Calif.	—	8.0	13
1933	Long Beach, Calif.	6.3	40.0	115
1935	Helena, Mont.	6.0	4.0	4
1940	Imperial Valley, Calif.	7.0	6.0	9
1946	Hawaii (tsunami)	—	25.0	173
1949	Puget Sound, Wash.	7.1	25.0	8
1952	Kern County, Calif.	7.7	60.0	8
1954	Eureka, Calif.	—	2.1	1
1954	Wilkes-Barre, Pa.	—	1.0	—
1955	Oakland, Calif.	—	1.0	1
1957	Hawaii (tsunami)	—	3.0	—
1957	San Francisco, Calif.	5.3	1.0	—
1958	Khantaak Island and Lituya Bay, Alaska	—	—	5
1959	Hebgen Lake, Mont.	—	11.0	28
1960	Hilo, Hawaii (tsunami)	—	25.0	61
1964	Prince William Sound, Alaska	8.4	500.0	131
1965	Puget Sound, Wash.	—	12.5	7
1971	San Fernando, Calif.	6.5	553.0	65

Source: From Office of Emergency Preparedness, 1972, *Disaster Preparedness*, Report to Congress of the United States, Vol. 3.

the New Madrid event (Table 8.3), it was felt over a much smaller area. The regional crustal properties of the eastern and western parts of the United States are therefore inferred to be very different.

On a local scale, the intensity of ground motion is largely a function of the surficial geology. Soft soils tend to vibrate much more strongly than stiff soils or rock. This relationship is clearly illustrated by the response to earthquakes in San Francisco. The geologic setting of the city consists of deposits of bay mud and alluvium (river-deposited soils) in low-lying areas (Figure 8.59) and areas of bedrock at shallow depths throughout the rest of the city. Figure 8.60 shows the distribution of intensity during the 1906 earthquake. The areas of high intensity correspond closely to the areas underlain by bay mud and alluvium. Damage from the Loma Prieta earthquake closely followed the pattern of damage from the 1906 tremblor. The Marina district (Figure 8.60) was the most heavily damaged sector of the city, with structural damage to 150 buildings. This area was a lagoon underlain by bay mud at the time of the 1906 earthquake. It was later filled by rubble from destroyed buildings to create a fairground for the 1915 Panama-Pacific International Exposition. In 1989, land use in the Marina district was dominated

▲ FIGURE 8.56

Major faults and locations of past ruptures in California and Nevada. *Source:* From W. W. Hays, 1980, *Procedures for Estimating Earthquake Ground Motions*, U.S. Geological Survey Professional Paper 1114.

by expensive residential dwellings. During the earthquake, ground motion caused liquefaction of the bay muds and damage to many buildings (Figure 8.61).

The relationship between surficial material and ground motion is further illustrated by the recordings of accelerographs in San Francisco of a distant underground nuclear explosion (Figure 8.62). Horizontal ground motion was much greater for bay mud and alluvium than for bedrock when subjected to the same seismic waves. Similar results were

▲ **FIGURE 8.57**
Cross sections along the San Andreas fault from north of San Francisco to Parkfield. The upper section shows locations of earthquakes from January 1969 through July 1989, and the lower section shows the Loma Prieta earthquake and its aftershocks, which filled one of three gaps in the cross section. *Source:* From P. L. Ward and R. A. Page, *The Loma Prieta Earthquake of October 17, 1989*, U.S. Geological Survey Pamphlet.

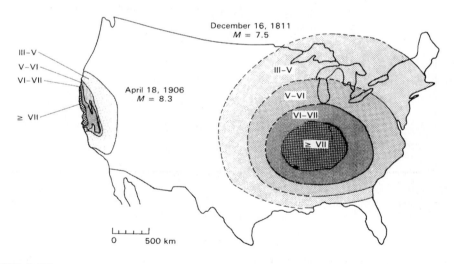

▲ **FIGURE 8.58**
Seismic intensity plotted for the 1906 San Francisco earthquake and the 1811 New Madrid, Missouri, event. Even though the San Francisco earthquake had a larger magnitude, it affected a smaller area because of differences in regional seismic attenuation. *Source:* From W. W. Hays, 1980, *Procedures for Estimating Earthquake Ground Motions*, U.S. Geological Survey Professional Paper 1114.

obtained by accelerographs that recorded some of the larger aftershocks of the Loma Prieta earthquake. Soft soils actually amplify incoming seismic waves, a condition that may have contributed to the collapse of part of the Nimitz Freeway (Interstate 880) in Oakland, California. The section that failed was a two-tiered roadway known as the Cypress structure (Figure 8.63). The mode of failure involved collapse of the upper roadway onto the lower section. Vehicles on the lower section were crushed and 41 people were killed.

▲ FIGURE 8.59

Generalized geology of the San Francisco area. *Source:* From R. D. Borcherdt, ed., *Studies for Seismic Zonation of the San Francisco Bay Region*, U.S. Geological Survey Professional Paper 941-A.

Although the entire Cypress structure was damaged, only that portion constructed on bay mud collapsed (Figure 8.63).

The common practice of using artificial fill to expand urban areas in low-lying coastal cities creates a perilous risk for large populations throughout the world. The January 17, 1995 Hanshin-Awaji earthquake in Kobe, Japan ($M_w = 6.9$), became the most expensive earthquake in terms of property damage to date. In addition to nearly 5400 deaths, the earthquake damaged or destroyed 152,000 buildings along with many other types of structures and utilities, generating an estimated cost of $100 billion. Damage was concentrated in reclaimed (filled) land and in adjacent areas underlain by unconsolidated materials of various types (Figure 8.64). Ground shaking in the filled areas caused extensive liquefaction and lateral spreading. Expulsion of water and soil during the spreading, much like the New Madrid earthquakes, led to subsequent settlement of the land surface. Like the Nimitz Freeway, elevated expressways in Kobe collapsed or were

▲ FIGURE 8.60

Relative intensity of the 1906 earthquake in San Francisco. The areas of highest intensity correspond to areas of bay mud and alluvium, as shown in Figure 8.59. *Source:* From R. D. Borcherdt, ed., *Studies for Seismic Zonation of the San Francisco Bay Region,* U.S. Geological Survey Professional Paper 941-A.

extensively damaged. Building collapse was responsible for about 90% of the loss of life, but only about 60% of the property loss. The remaining 10% of deaths were caused by fires ignited by natural gas leaks and/or electrical problems. About 7000 buildings were destroyed by fire.

The damage to Kobe and San Francisco is a grave warning to coastal cities with similar geologic settings. For example, Oakland, California, also has large areas of artificial fill overlying weak estuarine sediments, with an active fault (the Hayward fault) that has produced historic earthquakes (Figure 8.65). These cities must take all possible measures to minimize risk from moderate earthquakes that reoccur on a fairly frequent basis. One of the most effective ways to reduce future losses in areas of high seismic risk is land-use planning. Zoning ordinances of various types can prevent or restrict development in critical areas. An obvious restriction is to prevent development over active faults. Although such a restriction seems to be only common sense, past development has largely ignored

▲ FIGURE 8.61
The third story of a three-story building in the Marina district of San Francisco. The first and second stories collapsed due to ground shaking and liquefaction. *Source:* D. Perkins; U.S. Geological Survey.

▲ FIGURE 8.62
Recordings of horizontal ground motion generated by an underground nuclear explosion. The ground motion is much stronger for stations located on bay mud and alluvium. *Source:* From R. D. Borcherdt, ed., *Studies for Seismic Zonation of the San Francisco Bay Region*, U.S. Geological Professional Paper 941-A.

▶ **FIGURE 8.63**
Surficial geology of the Oakland, California, area. Segment AB of the Nimitz Freeway, constructed on bay mud, collapsed. Segment BC was damaged but did not collapse. *Source: From S. E. Hough, P. A. Fregerg, R. Busby, E. F. Field, and K. H. Jacob, Eos, Transactions of the American Geophysical Union, Nov. 21, 1989.*

▲ **FIGURE 8.64**
The geological setting of Kobe, Japan, showing the location of reclaimed ground and other surficial deposits. The stippled area is the Japanese Meterological Agency (JMA) intensity 7 area, where damage was greatest. *Source: From T. L. Holzer, 1995, U.S. Geological Survey. The 1995 Hanshin-Awaji (Kobe), Japan, Earthquake, GSA Today, 5(8):154–167.*

▲ FIGURE 8.65
Similarity of Kobe, Japan, and Oakland, California. The two cities are built along an active fault next to a bay and contain large areas of reclaimed land subject to widespread liquefaction. *Source:* From T. L. Holzer, 1995, U.S. Geological Survey.

the presence of known active faults (Figure 8.51). Further zoning options can establish minimum distances, or *setbacks*, from active faults for buildings such as schools and hospitals, where damage could result in numerous casualties. Similar restrictions could be applied to areas of potential landsliding or severe ground shaking. The most stringent standards for site location and construction should be established for engineering projects such as dams and nuclear power plants, whose satisfactory performance in an earthquake is necessary to prevent massive losses in human life and property. In most parts of the United States, these types of structures now require the most detailed site investigations and seismic risk-assessment procedures possible.

Earthquake Engineering

The effect of an earthquake upon a building is determined by the relationship and interaction between the characteristics of the seismic waves that reach the site, the response of the materials beneath the building, and the design and construction of the building itself.

Horizontal ground motions often present the most severe test of a building's ability to withstand an earthquake without failure. The resulting lateral accelerations impose inertial forces on a structure. In the absence of strict building codes, structures rarely are designed to resist lateral forces of the magnitude imposed by strong ground motion. Figure 8.66 illustrates the nature of these inertial forces. For low, rigid structures (Figure 8.66, left) the peak lateral force can be calculated as the product of the mass of the building and the peak lateral acceleration. The lateral force for flexible buildings subjected to short-duration seismic shaking can be less than the product of mass times acceleration. This force decrease is achieved because part of the energy is absorbed by the bending action of the frame (Figure 8.66, center). In tall, very flexible buildings, the lateral force can even exceed $M \times A$ because of the increase in inertial force caused by the oscillation of the structure by repeated vibrations (Figure 8.66, right).

One of the most important factors in earthquake damage is the relationship between the vibrational period of a structure and the period of the material upon which it is built. When set into motion, geologic materials, as well as buildings, tend to vibrate at a certain rate. This characteristic rate is defined by the *fundamental period*. The period of a wave or

Mass of structure is *M*

Ground moves with acceleration *A*

Rigid structure

$F = MA$

Flexible structure

For very short duration of motion

$F < MA$

Flexible structure

For prolonged duration of motion whose frequency is near that of structure

$F > MA$

▲ **FIGURE 8.66**
Inertial forces developed by different types of buildings when subjected to earthquake shaking. *Source:* From H. J. Degenkolb, 1977, *Earthquake Forces on Tall Structures*, Booklet 2717A, Bethlehem Steel Corp., Bethlehem, Pa.

vibration is the time interval between the passage of two corresponding points on the waveform, the time between two successive crests, for example. Fundamental periods of geologic materials range from well below 1 s for bedrock and stiff soils to several seconds or more for deep, soft soils. The fundamental period of a building varies with the height of the structure. Tall buildings have long periods (several seconds), and low buildings have short periods. The effect of building height upon acceleration is illustrated in Figure 8.67. Accelerographs were installed in the basement and penthouse of a 10-story warehouse that was shaken by the 1952 Tehachapi, California, earthquake. Accelerations were much greater in the penthouse than in the basement. In addition, the

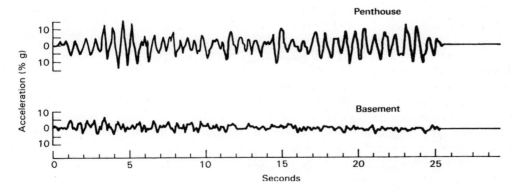

▲ **FIGURE 8.67**
Differences in acceleration between the basement and penthouse of a 10-story reinforced-concrete building. *Source:* From F. W. Housner, Design spectrum, in R. L. Wiegel (ed.), *Earthquake Engineering*, © 1970 by Prentice Hall, Inc., Englewood Cliffs, N.J.

penthouse accelerations clearly show the effect of the fundamental period of the building, with maximum accelerations occurring every few seconds. No such pattern is evident in the basement accelerogram.

When the period of a soil and a building are similar, the most hazardous situation occurs. Resonance between the building and the soil leads to more violent shaking and greater damage is usually the result. Tall buildings sustain greater damage when constructed on deep, soft soils because of their similarity in vibrational period. Similarly, small, rigid buildings may perform more poorly when sited upon short-period materials such as bedrock. Figure 8.68 illustrates the effect of building height upon relative seismic damage. The damage intensity for tall structures rises dramatically when the fundamental period of the foundation soil is longer than 1 s.

Construction materials and design often play a critical role in the performance of a building during an earthquake (Table 8.4). Small, wood-frame structures are generally the safest type as long as they are anchored securely to their foundations. Improperly anchored houses have been observed to shear apart from their foundations during lateral ground accelerations. Steel-frame or reinforced-concrete construction methods are least hazardous for multistory buildings. Buildings that provide the least degree of protection include those of nonreinforced masonry, brick and mortar, and adobe construction. Structures of this type are literally deathtraps because of their heavy weight and brittle behavior. The Guatemalan earthquake of 1976 provides a good example of the high correlation between adobe construction and nearly total destruction of villages in areas of moderate- to high-intensity earthquakes. (Figure 8.69). Likewise, in the 1990 northern

◄ FIGURE 8.68
Structural damage intensity for different height buildings related to depth of soil and fundamental period of soil. Tall buildings sustain maximum damage when constructed on long-period soil deposits. *Source:* From H. B. Seed et al., 1972, Soil conditions and building damage in 1967 Caracas earthquake, *Journal of the Soil Mechanics and Foundations Division,* American Society of Civil Engineers, SM-8: 787–806.

Table 8.4 Earthquake Ratings for Common Building Types

Simplified Description of Structural Types	Relative Damageability (in order of increasing susceptibility to damage)
Small wood-frame structures, i.e., dwellings not over 3000 sq ft, and not over 3 stories	1
Single or multistory steel-frame buildings with concrete exterior walls, concrete floors, and concrete roof. Moderate wall openings	1.5
Single or multistory reinforced-concrete buildings with concrete exterior walls, concrete floors, and concrete roof. Moderate wall openings	2
Large-area wood-frame buildings and other wood-frame buildings	3 to 4
Single or multistory steel-frame buildings with unreinforced masonry exterior wall panels; concrete floors and concrete roof	4
Single or multistory reinforced-concrete frame buildings with unreinforced masonry exterior wall panels, concrete floors, and concrete roof	5
Reinforced-concrete bearing walls with supported floors and roof of any materials (usually wood)	5
Buildings with unreinforced brick masonry having sandlime mortar and with supported floors and roof of any materials (usually wood)	7 up
Bearing walls of unreinforced adobe, unreinforced hollow concrete block, or unreinforced hollow clay tile	Collapse hazards in moderate shocks

Source: From D. Armstrong, 1973, *The Seismic Safety Study for the General Plan*, Sacramento, Calif.: California Council on Intergovernmental Relations.

Note: This table is not complete. Additional considerations would include parapets, building interiors, utilities, building orientation, and frequency response.

▲ FIGURE 8.69
Devastation of adobe houses in the town of Tecpan, Guatemala, by the 1976 earthquake. *Source:* Photo courtesy of U.S. Geological Survey.

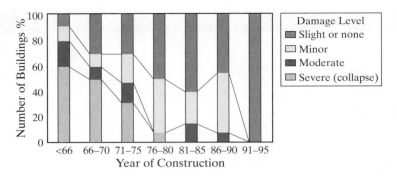

▲ FIGURE 8.70

Levels of damage to reinforced concrete buildings in Kobe, Japan, plotted by time of construction. Performance increased significantly in buildings constructed after new building codes were adopted in 1981. *Source:* From T. L. Holzer, 1995, U.S. Geological Survey. The 1995 Hanshin-Awaji (Kobe), Japan, Earthquake, *GSA Today*, 5(8):154–167.

Iran earthquake, destruction was overwhelmingly linked to structures with unreinforced masonry walls and roofs.

Stringent building codes are among the most effective measures to minimize earthquake damage. Structural bracing to increase the lateral resistance of steel and reinforced concrete buildings to ground motion are particularly effective. The performance of reinforced-concrete buildings in the Kobe earthquake was highly dependent on age (Figure 8.70), with newer buildings faring much better than older ones. New building codes were adopted in Kobe in 1981 and damage was greatly reduced in buildings constructed under these codes.

Although building performance is certainly a major concern, the effects of earthquakes upon *lifeline* systems are also of vital importance. Lifeline systems include electricity and fuel supplies, water and wastewater facilities, transportation lines, and communication networks. Loss of these facilities in a large urban area can lead to greater problems than the earthquake itself. Emergency rescue and relief efforts are severely hampered by damage to energy, transportation, and communication systems. The breakdown of water and sanitary networks can lead to disease and the inability to combat fires that may break out. Eighty-three years after uncontrollable fires destroyed the city of San Francisco after the earthquake of 1906, fires in the Marina district could not be fought with city water due to the rupture of water mains during the 1989 earthquake.

Earthquake Prediction

Prediction of earthquakes is a scientific goal that, if achieved, could prevent widespread loss of life in a major earthquake. Unfortunately, accurate prediction is not currently possible, although intensive research is proceeding in many areas.

Two types of earthquake prediction are theoretically possible. The first type is long-term forecasting, in which the probability of an earthquake along a particular fault segment within a certain time interval is calculated by studying seismic gaps and historical records of earthquakes that have occurred along that fault segment. Table 8.5 presents this type of analysis for major faults in California, and Figure 8.71 graphically illustrates the probabilities. That the Parkfield segment of the San Andreas fault was locked is quite clearly illustrated by the delineation of the Parkfield gap in Figure 8.57. The high probability of an earthquake in the Parkfield area is discussed below. By plotting the numbers of earthquakes within specific time intervals against their magnitudes, diagrams similar to Figure 8.41 can be constructed for a local area. From this plot it is possible to determine

Table 8.5	Probabilities of Major Earthquakes on California's Primary Fault Segments				
Fault Segment	**Date of Most Recent Event**	**Expected Magnitude**	**Estimated Recurrence Interval (yr)**	**Probability of Occurrence between 1988 and 2018 (percent)**	**Level of Reliability (A, Most Reliable)**
		San Andreas fault			
North Coast	1906	8	303	<10	B
San Francisco Peninsula	1906	7	169	20	C
Southern Santa Cruz Mountains	1906	6.5	136	30	E
Central creeping	—	—	—	<10	A
Parkfield	1966	6	21	>90	A
Cholame	1857	7	159	30	E
Carrizo	1857	8	296	10	B
Mojave	1957	7.5	162	30	B
San Bernardino Mountains	1812(?)	7.5	198	20	E
Coachella Valley	1680 ± 20	7.5	256	40	C
		Hayward fault			
Northern East Bay	1836(?)	7	209	20	D
Southern East Bay	1868	7	209	20	C
		San Jacinto fault			
San Bernardino Valley	1890(?)	7	203	20	E
San Jacinto Valley	1918	7	184	10	C
Anza	1892(?)	7	142	30	D
Borrego Mountain	1968	6.5	189	<10	B
		Imperial fault			
Imperial	1979	6.5	44	50	C

Source: From P. L. Ward and R. A. Page, 1989, *The Loma Prieta Earthquake of October 17, 1989*, U.S. Geological Survey Pamphlet.

the *recurrence interval,* or the average time interval between earthquakes of a specific magnitude. Predictions can then be made that an earthquake of that magnitude has a high probability of occurrence within a specified time interval, if the date of the last earthquake is known.

Research leading to short-term predictive ability, involving a time interval small enough for evacuation of an area, for example, has focused on *precursors* that have been observed prior to previous earthquakes. Precursors are physical or chemical phenomena that occur in a typical pattern before an earthquake. These phenomena include changes in the velocity of seismic waves, the electrical resistivity of rocks, the frequency of preliminary earthquakes (foreshocks), the deformation of the land surface, and the water level or water chemistry of wells in the area. Many of these precursors can be explained by a theory called the *dilatancy model.* Under this hypothesis, rocks in the process of strain along a fault show significant dilation or swelling before rupture. This volume increase is caused by the opening of *microcracks,* which are minute failure zones in weaker mineral grains in the rock and along grain boundaries. As the porosity increases during the formation of microcracks,

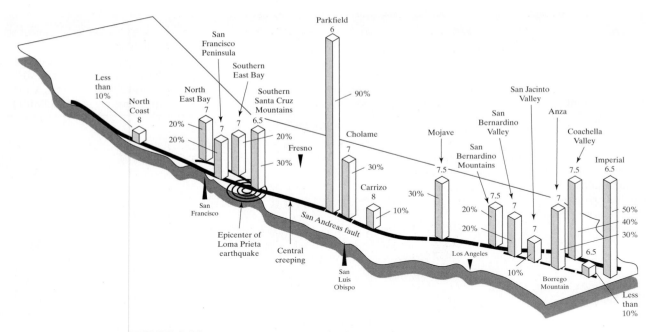

▲ FIGURE 8.71
Probabilities of earthquakes between 1988 and 2018 for segments of major faults in California. Box size is proportional to probability and expected magnitude is shown above each box. Probabilities were determined one year before the Loma Prieta earthquake. *Source:* From P. L. Ward and R. A. Page, 1989, *The Loma Prieta Earthquake of October 17, 1989*, U.S. Geological Survey Pamphlet.

groundwater flows into the highly stressed areas. These changes in density and water content affect the ability of the rock to transmit seismic waves and conduct electricity. Therefore, seismic-wave velocity and electrical resistivity progressively change as the overall rupture along the fault draws near. Localized changes in land-surface elevation are also related to volume changes at depth. An area of recent uplift along the San Andreas fault near Los Angeles, which has been named the Palmdale Bulge (Figure 8.72), is being monitored in great detail as a possible indicator of a future earthquake.

Volume changes and groundwater movement may be reflected by changes in water levels in wells as well as by changes in the chemical composition of groundwater. Radon gas has been observed to increase in wells prior to earthquakes. These increases are perhaps related to the release of radon gas from rocks during the formation of microcracks. The pattern of seismic activity in the vicinity of an area of imminent fault rupture is also significant. This pattern consists of an initial rise in the number of small events, followed by a decline in foreshocks just prior to the major earthquake. The decline may represent a temporary increase in rock strength before the newly formed microcracks are filled with water.

The precursor phenomena can be grouped into stages according to the dilatancy model (Figure 8.73). Stage I consists of a gradual stress buildup along the fault. Stages II and III are correlated with dilatancy and water influx. Stage IV is the major earthquake, and Stage V is the aftermath of the event. If every earthquake followed the sequence shown in Figure 8.73, with uniform stage duration, earthquake prediction would be a simple matter. Instead, each earthquake is unique in terms of specific precursor behavior patterns and length of precursor stages. Two magnitude-5 earthquakes preceded the Loma

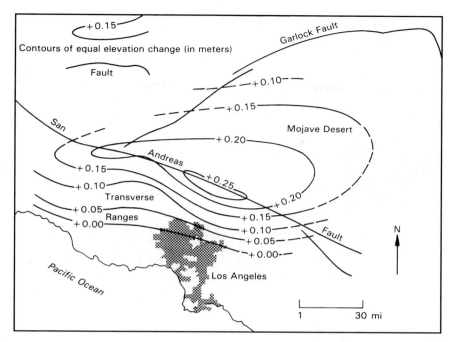

▲ FIGURE 8.72

Uplift along the San Andreas fault in the vicinity of Palmdale, California. This activity may be a precursor to a future earthquake. *Source:* From F. Press and R. Siever, *Earth*, 3rd ed., © 1982 by W. H. Freeman and Co., San Francisco.

Prieta event by 15 and 2 months. In each case, a public advisory was issued stating that the earthquakes could be foreshocks to a stronger earthquake within 5 days. However, the fault did not cooperate and prediction was not successful. Continued research and study of future earthquakes will certainly lead to refinement of the dilatancy model, or to a replacement model with more accurate predictive capabilities.

The strange behavior of animals before an earthquake has been recognized for centuries and is now being studied scientifically for use in earthquake prediction. The range of unnatural behavior patterns is extremely diverse. Many animals seem restless and frightened. Snakes have come out of hibernation during the winter and frozen to death. Normal eating and sleeping patterns are abandoned. Although the exact means by which animals sense an impending earthquake are unknown, suggestions have included changes in subsurface water content, acoustic emissions, and electrostatic effects.

As a result of past earthquake disasters, the People's Republic of China has developed the world's most advanced earthquake prediction program. Since the 1970s, a number of successful predictions have been made. In 1975, prediction of the Haicheng earthquake (magnitude 7.3) was responsible for saving tens of thousands of lives by evacuation from hazardous buildings. A notable failure, however, was the inability to predict the 1976 Tangshan event (magnitude 7.6), in which perhaps as many as three-quarters of a million people were killed. Two recent examples illustrate the uncertainties and dangers of earthquake predictions. The high probability of an earthquake in the Parkfield segment of the San Andreas fault was mentioned earlier. Concern about the Parkfield area was based on the historical earthquake record (Figure 8.74), in which significant earthquakes occurred in 1857, 1881, 1901, 1922, 1934, and 1966. The

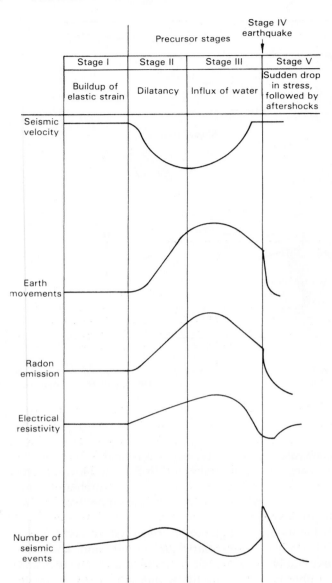

◄ FIGURE 8.73
Trends in precursor activity that have been related to major earthquakes by use of the dilatancy model. *Source:* From F. Press, 1975, Earthquake prediction, *Scientific American*, May 1975, with permission.

average recurrence interval of these events was 22 years, with a maximum time between events of 32 years. Because of the historical regularity of the events, and because the previous six events had ruptured the same area on the fault, the U.S. Geological Survey issued a prediction in 1985 that there was a 95% probablility that a magnitude 5.5–6 earthquake would occur between 1985 and 1993. In anticipation of the upcoming earthquake, a highly detailed monitoring network was set up at Parkfield, including seismometers, creep meters, borehole strain meters, and other instruments, many of which are monitored in real time and can be accessed from the web site, quake.wr.usgs. gov/research/parkfield/index.html. In addition to the monitoring network, the U.S. Geological Survey established a four-tiered system of alerts: D, C, B, and A, with increasing level of concern for the magnitude-6 event. With a status A alert, the public is

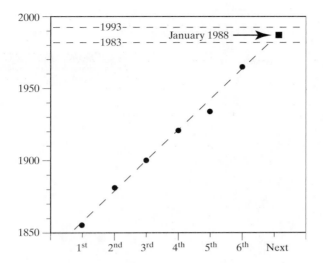

▲ FIGURE 8.74
Past earthquakes on the Parkfield section of the San Andreas fault occurred in 1857, 1881, 1901, 1922, 1934, and 1966. The extrapolation of the trend of these events led to the prediction of a moderate earthquake in 1988, which did not actually occur until 2004. *Source:* National Earthquake Prediction Evaluation Council Working Group, 1994, *Earthquake Research at Parkfield, California, 1993 and Beyond—Report of the NEPEC Working Group to Evaluate the Parkfield Earthquake Prediction Experiment*, U.S. Geological Survey Circular 1116.

notified and recommended to follow earthquake preparedness procedures that have been widely circulated to residents. If the earthquake does not occur within 72 h, the alert is considered to be a false alarm. As of 2001, two level-A alerts had been issued, triggered by M_w-4.7 and 4.8 earthquakes that occurred in 1992 and 1993. Both proved to be false alarms. The predicted earthquake finally arrived on Sept. 28, 2004, a M_w-6.0 event. The arrival of the main event 11 years after the prediction window shows that precise earthquake prediction continues to elude our best efforts. The monitoring system, however, has produced much new information about the fault and its seismicity.

A much more troubling example of earthquake prediction was the forecast of a New Madrid earthquake on December 3, 1990. The prediction, a 50% chance of an earthquake within a 5-day period, was made by Dr. Iben Browning, an independent scientist, albeit one who had no training in seismology. Following closely on the heels of the widely publicized damage from the 1989 Loma Prieta earthquake, Dr. Browning's prediction attracted local and national media coverage. Despite the disavowal of Browning's methods and prediction by earthquake experts, New Madrid became the center of a media frenzy. As the day drew near, tourists and media crews descended on the town. Schools and factories were closed. Twenty-two million dollars worth of earthquake insurance was purchased by homeowners. As the period came and went, nothing happened. Although the publicity of a potential earthquake probably motivated many people in the region to become more knowledgeable about earthquake preparedness, the public's confusion between real and "junk" science illustrates a real problem in creating an effective warning system for natural disasters.

With our ever-growing knowledge of earthquake processes, prediction is bound to become more accurate. In 2003, the U.S. Geological Survey issued another prediction, this time for the San Francisco Bay regions (USGS, 2003). This one predicts a magnitude 6.7 or greater earthquake within the 30-year period 2003–2032, with a 62% probability. This prediction considers the numerous faults that potentially could rupture to produce

▲ FIGURE 8.75
30-year probabilities of ruptures for faults in the San Francisco Bay area. The probability for an earth-quake on any of the faults at a magnitude of 6.7 or greater is 62%. *Source:* U.S. Geological Survey, 2003.

this event (Figure 8.75), each of which has its own, smaller probability of producing the earthquake (Table 8.6). The worst-case scenario for this earthquake, a 7.9 event similar to the 1906 earthquake, is estimated to produce a loss of life of 5800 people if it strikes during working hours in the rapidly growing Bay area. This prediction really highlights the need to prepare for the event with all of the tools available.

Table 8.6   The 10 Most Damaging Earthquake Scenarios for the San Francisco Bay Area and Their 30-Year Probabilities

	30-year Probability	Magnitude
Rodgers Creek	15.2%	7.0
Northern Calaveras	12.4%	6.8
Southern Hayward (possible repeat of 1868 earthquake)	11.3%	6.7
Northern + Southern Hayward	8.5%	6.9
Mt. Diablo	7.5%	6.7
Green Valley-Concord	6.0%	6.7
San Andreas: Entire N. CA segment (possible repeat of 1906 earthquake)	4.7%	7.9
San Andreas: Peninsula segment (possible repeat of 1838 earthquake)	4.4%	7.2
Northern San Gregorio segment	3.9%	7.2
San Andreas: Peninsula + Santa Cruz segment	3.5%	7.4

Source: USGS, 2003.

Case in Point 8.3
The San Fernando and Northridge Earthquakes: (Seismic) History Repeats Itself

Among recent earthquakes in the United States, the February 9, 1971, earthquake in San Fernando, California, and the nearby January 17, 1994, Northridge earthquake stand out for several reasons. First, both events struck the same outlying vicinity of a major metropolitan area, Los Angeles. Thus, despite the relatively moderate magnitude of these events (Table 8.7), great damage was caused to the densely populated area. Second, although the earthquakes occurred in a region known to be seismically active, displacement took place along faults that either were not known or considered to be active prior to the earthquakes. Third, the network of seismograph stations that recorded the 1971 event provided the most detailed and abundant data ever gathered for an earthquake. These data provide many interesting facts and interpretations, including the highest ground accelerations ever recorded. These accelerations commonly ranged from $0.5\ g$ to $0.75\ g$, with peak recorded values of more than $1\ g$. Accelerations of this magnitude are particularly significant because they exceed by several times the design values used for earthquake-resistant structures. Finally, the earthquakes provided a performance test for many modern buildings constructed under recent building codes. Unfortunately, many of the structures did not perform well (Table 8.8). Even worse, some critical highway bridges that were destroyed and rebuilt after the 1971 earthquake collapsed again in 1994.

The San Fernando earthquake was generated by movement along a number of minor faults near a bend in the San Andreas fault. The locations of the main shock and major aftershocks are shown in Figure 8.76. Fortunately, the earthquake took place about 6:00 A.M.; at this hour schools were unoccupied and streets and highways were nearly deserted. Had the earthquake struck just 1 h or 2 h later, the loss of life would have been much greater. Ground shaking from the main shock lasted about 1 min; the duration of strong motion was only about 10 s. For a moderate event, the destruction inflicted by the

Problems

1. How can the deformation of rocks deep within the earth's crust be simulated in laboratory tests of rock samples?

2. What evidence leads us to believe that rocks exposed at the surface were deformed at great depths?

3. What field measurements are made in the study of rock structures? What do they indicate about the structures?

4. In what type of geologic setting or province would you be likely to find highly folded rocks?

5. What is the difference between faults and joints?

6. Describe the relationship between fault type and stress orientation.

7. Summarize the ideas that led up to the theory of plate tectonics.

8. List as many ways as you can how folds and faults would influence the construction of tunnels and dams.

9. Describe the major structural features of your state, province, or region.

10. Why doesn't the elastic rebound theory satisfactorily explain all earthquakes?

11. Describe the instruments used to measure earthquakes.

12. Contrast earthquake magnitude and intensity.

13. Why can't S waves travel through the earth's core?

14. A city near the epicenter of an earthquake sustains major damage to poorly constructed buildings, although hospitals and schools designed to resist damage from ground motion sustain only minor damage. Tall, thin structures such as chimneys and towers topple throughout the city, and newly deposited mounds of sand are found on the ground surface near the river. What was the intensity of the earthquake?

15. Based on Figure 8.40, what were the probable effects of the San Fernando earthquake in San Diego, California?

16. Using Table 8.2, estimate the distance between the earthquake epicenter and the point at which the seismogram shown in Figure 8.44 was recorded.

17. If the $S–P$ time interval measured from a seismogram is 32 s and the maximum amplitude is 50 mm, what was the Richter magnitude of the earthquake?

18. How is a tsunami generated?

19. How do the geologic materials beneath a site influence its response to an earthquake?

20. A 12-story building is constructed in an area in which the soils are 600 ft deep. What percentage of structural damage can be expected?

21. Approximately how many earthquakes of magnitude 7 occur worldwide every year?

22. Suggest two likely locations for future earthquakes along the San Andreas fault south of San Francisco. How are these areas identified as possible earthquake epicenters?

23. What are the major areas of research concerned with earthquake predictions?

24. How might the effects of the San Fernando earthquake have differed if the magnitude were 8.0 instead of 6.5?

References and Suggestions for Further Reading

BAROSH, P. J. 1969. *Use of Seismic Intensity Data to Predict the Effects of Earthquakes and Underground Nuclear Explosions in Various Geologic Settings.* U.S. Geological Survey Bulletin 1279.

BERLIN, G. L. 1980. *Earthquakes and the Urban Environment*, vols. 1, 2, 3. Boca Raton, Fla.: CRC Press.

BILLINGS, M. P. 1972. *Structural Geology*, 3rd ed. Englewood Cliffs, N.J.: Prentice Hall, Inc.

BLAIR, M. L., and W. E. SPANGLE. 1979. *Seismic Safety and Land-use Planning—Selected Examples for California.* U.S. Geological Survey Professional Paper 941-B.

BOLT, B. A. 1976. *Nuclear Explosions and Earthquakes: The Parted Veil.* San Francisco: W. H. Freeman.

BORCHERDT, R. D., ed. 1975. *Studies for Seismic Zonation of the San Francisco Bay Region.* U.S. Geological Survey Professional Paper 941-A.

COSTA, J. E., and V. R. BAKER. 1981. *Surficial Geology: Building with the Earth.* New York: John Wiley.

DEGENKOLB, J. J. 1977. *Earthquake Forces on Tall Structures*, Booklet 2717A. Bethlehem, Pa.: Bethlehem Steel Corp.

ECKEL, E. B. 1970. *The Alaska Earthquake March 27, 1964: Lessons and Conclusions.* U.S. Geological Survey Professional Paper 546.

ESPINOSA A. F., ed. 1976. *The Guatemalan Earthquake of February 4, 1976: A Preliminary Report.* U.S. Geological Survey Professional Paper 1002.

HAY, E. A., and A. L. MCALESTER. 1984. *Physical Geology: Principles and Perspectives*, 2nd ed. Englewood Cliffs, N.J.: Prentice Hall, Inc.

HAYS, W. W., ed. 1980. *Procedures for Estimating Earthquake Ground Motions*. U.S. Geological Survey Professional Paper 1114.

HOLZER, T. L. 1995. The 1995 Hanshin-Awaji (Kobe), Japan, Earthquake: *GSA Today*, 5(8):154–167.

HOUGH, S. E., P. A. FREGERG, R. BUSBY, E. F. FIELD, and K. H. JACOB. 1989. Did mud cause freeway collapse? *Eos, Transactions of the American Geophysical Union*, 70(47):1497.

JACKSON, J. L. 1969. Geologic studies at Arbuckle Dam, Murray County, Oklahoma. *Bulletin of the Association of Engineering Geologists*, 5:79–100.

JUDSON, S., M. E. KAUFFMAN, and L. D. LEET. 1987. *Physical Geology*, 7th ed. Englewood Cliffs, N.J.: Prentice Hall, Inc.

KARNER, F. R., and D. L. HALVORSON. 1987. The Devils Tower, Bear Lodge Mountains, Cenozoic igneous complex, northeastern Wyoming, in *Geological Society of America Centennial Field Guide—Rocky Mountain Section*. Boulder, Colo.: Geological Society of America, 161–164.

LOVE, J. D. 1987. Teton mountain front, Wyoming, in *Geological Society of America Centennial Field Guide—Rocky Mountain Section*. Boulder, Colo.: Geological Society of America, 173–176.

NATIONAL EARTHQUAKE PREDICTION EVALUATION COUNCIL WORKING GROUP. 1994. *Earthquake Research at Parkfield, California, 1993 and Beyond–Report of the NEPEC Working Group to Evaluate the Parkfield Earthquake*. U.S. Geological Survey Circular 1116.

NICHOLS, D. R., and J. M. BUCHANAN-BANKS. 1974. *Seismic Hazards and Land-use Planning*. U.S. Geological Survey Circular 690.

NOBLETT, J. B., A. S. COHEN, E. M. LEONARD, B. M. LOEFFLER, and D. A. GEVIRTZMAN. 1987. The Garden of the Gods and basal Phanerozoic nonconformity in and near Colorado Springs, Colorado, in *Geological Society of America Centennial Field Guide—Rocky Mountain Section*. Boulder, Colo.: Geological Society of America, 335–342.

OFFICE OF EMERGENCY PREPAREDNESS. 1972. *Disaster Preparedness*. Report to Congress of the United States, vol. 3.

PRESS, F. 1975. Earthquake prediction. *Scientific American*, 232(5):14–23.

PRESS, F., and R. SIEVER. 1982. *Earth*, 3rd ed. San Francisco: W. H. Freeman.

ROBINSON, G. D., A. A. WANEK, S. H. HAYS, and M. E. MCCALLUM. 1964. *Philmont Country: The Rocks and Landscape of a Famous New Mexico Ranch*, U.S. Geological Survey Professional Paper 505.

SEED, H. B., R. U. WHITMAN, H., DEZFULIAN, R. DOBRY, and I. M. IDRISS. 1972. Soil conditions and building damage in the 1967 Caracas earthquake. *Journal of the Soil Mechanics and Foundations Division*. American Society of Civil Engineers, SM-8:787–806.

SPENCE, W., R. B. HERRMANN, A. C. JOHNSTON, and G. REAGOR. 1993. *Responses to Iben Browning's Prediction of a 1990 New Madrid, Missouri, Earthquake*. U.S. Geological Survey Circular 1083.

SUPPE, J. 1985. *Principles of Structural Geology*. Englewood Cliffs N.J.: Prentice Hall, Inc.

U.S. GEOLOGICAL SURVEY. 1976. *Earthquake Prediction: Opportunity to Avert Disaster*. U.S. Geological Survey Circular 729.

U.S. GEOLOGICAL SURVEY. 1982. *The Imperial Valley, California, Earthquake of October 15, 1979*. U.S. Geological Survey Professional Paper 1253.

U.S. GEOLOGICAL SURVEY, and NATIONAL OCEANIC AND ATMOSPHERIC ADMINISTRATION. 1971. *The San Fernando, California, Earthquake of February 9, 1971*. U.S. Geological Survey Professional Paper 733.

U.S. Geological Survey. 2003. *Is a powerful quake likely to strike in the next 30 years?* U.S. Geological Survey Fact Sheet 039-03.

WALLACE, R. E. 1974. *Goals, Strategy, and Tasks of the Earthquake Hazard Reduction Program*. U.S. Geological Survey Circular 701.

WALLACE, R. E. 1984. *Faulting Related to the 1915 Earthquake in Pleasant Valley, Nevada*. U.S. Geological Survey Professional Paper 1274-A.

WALLACE, R. E., ed. 1990. *The San Andreas Fault System, California*. U.S. Geological Survey Professional Paper 1515.

WARD, P. L., and R. A. PAGE. 1989. *The Loma Prieta Earthquake of October 17, 1989*. U.S. Geological Survey Pamphlet.

WESSON, R. L., E. J. HELLEY, K. R. LAJOIE, and C. M. WENTWORTH. 1975. Faults and future earthquakes, in *Studies for Seismic Zonation in the San Francisco Bay Region*, R. D. Borcherdt, ed., U.S. Geological Survey Professional Paper 941-A:A5–A30.

WIEGEL, R. L., ed. 1970. *Earthquake Engineering*. Englewood Cliffs, N.J.: Prentice Hall, Inc.

YOUD, T. L., and S. N. HOOSE. 1978. *Historic Ground Failures in Northern California Triggered by Earthquakes*. U.S. Geological Survey Professional Paper 993.

CHAPTER 9

Weathering and Erosion

We have followed the various igneous, metamorphic, and sedimentary processes leading to the formation of rocks on the surface and below the surface of the earth's crust. We have seen how tectonic forces driven by the earth's internal energy elevate deeply formed rocks to the earth's surface. Progression through the geologic cycle now brings us to the processes acting upon, or just below, the surface of the earth that are associated with the action of water, ice, wind, and gravity. These processes, fueled by external energy derived from the sun, are components of the hydrologic cycle. Together, they lead to the disintegration and decomposition of continental rocks, followed by the transportation of debris and chemical elements to the ocean basins, where sedimentation precedes the ultimate formation of new continental rocks.

Weathering is the mechanical and chemical breakdown of rocks exposed at the earth's surface into smaller particles that may differ in composition from the original substance. Although the mechanical and chemical aspects of weathering will be discussed separately, they occur simultaneously and are related in many ways.

Weathering processes are of interest and importance to a broad range of scientists and engineers. Soil scientists study weathering processes in relation to the genesis of agriculturally productive soils from rocks and unconsolidated deposits. Hydrogeologists investigate the chemical constituents that are released by weathering reactions in the upper few meters of the soil zone and are then carried downward to influence the chemical composition of groundwater. Geologists and engineers have demonstrated the role of weathering in landslides and other types of slope failure. In the construction industry, the effect of weathering on building stone is of great concern, an effect obvious to anyone who has observed the faded and indistinct inscriptions on old tombstones. The deterioration of building stone, road aggregate, and other construction materials is a serious and expensive consequence of weathering phenomena. Perhaps the most important influence of weathering is its relevance to the design and performance of structures built upon weathered rock. Weathering gradually weakens rock; in some climates the end result is a deep mantle of residual soil with vastly different engineering properties from the parent rock. The effects of

► FIGURE 9.1
Climatic influences on types of weathering processes. *Source:* From L. Peltier, 1950, The geographical cycle in periglacial regions as it is related to climatic geomorphology; reproduced by permission from the *Annals of the Association of American Geographers*, 40:219, Fig. 3.

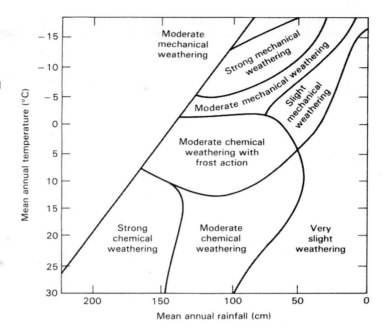

weathering must be assessed at every foundation site where design requires the presence of predominantly unaltered rock.

The types and intensity of weathering processes that occur in a particular area are primarily the result of climate. Temperature controls the mechanical processes caused by the freezing and thawing of water, as well as influences the rates of chemical weathering reactions. The other main climatic variable is precipitation, because water plays a part in both physical and chemical weathering processes. The dominant type of weathering that can be expected in a particular climatic region is indicated in Figure 9.1, which utilizes mean annual rainfall and mean annual temperature to characterize climate.

Mechanical Weathering

Mechanical weathering includes processes that fragment rocks into smaller particles by exerting forces greater than the strength of the rock. Usually, these forces act within rock masses to overcome the tensile strength of the rock, which is much lower than its compressive strength. The principal mechanical and chemical weathering types are shown in Table 9.1.

Frost action is one of the most effective mechanical weathering processes. When water freezes within a rock mass, its volume expansion (Figure 9.2) exerts pressure potentially greater than the tensile strength of most rocks (Chapter 7). The increase in volume is caused by the formation of a hexagonal structure in which the water molecules are farther apart than in their more densely packed condition in liquid water. The maximum pressure exerted by this process occurs at a temperature of −22°C. However, experimental field studies have shown that temperatures rarely reach this point. In addition, the freezing temperature of water declines if dissolved solutes are present and as the pressure increases. Because of these and other arguments it is not considered likely that in-place freezing of water in rocks can account for the large amount of shattering that is observed (Figure 9.3). A current alternative hypothesis involves the growth of ice in rock pores and fractures by migration of water to the freezing front (Figure 9.4). In this model, a thin layer of liquid water adsorbed to the rock surfaces resists freezing because of its more ordered structure. Water will migrate from unfrozen portions of the rock to the ice/water

Table 9.1 Mechanical and Chemical Weathering Processes

Mechanical
    Frost action
    Salt weathering
    Temperature changes
    Moisture changes
    Unloading
    Biogenic processes

Chemical Solution:

$$CaCO_3 \ + \ H_2CO_3 \ \longrightarrow \ Ca^{2+} \ + \ 2HCO_3^-$$

  (Calcite)   (Carbonic   (Calcium   (Bicarbonate
             acid)   ion)   ion)

Hydrolysis:

$$2KAlSi_3O_8 \ + \ 2H^+ \ + \ 9H_2O \ \longrightarrow \ Al_2Si_2O_5(OH)_4 \ + \ 4H_4SiO_4 \ + \ 2K^+$$

  (Orthoclase)   (Kaolinite)   (Dissolved   (Dissolved
                                              silica)   potassium)

Hydration:

$$CaSO_4 \ + \ 2H_2O \ \longrightarrow \ CaSO_4 \cdot 2H_2O$$

  (Anhydrite)   (Gypsum)

Oxidation:

$$4FeSiO_3 \ + \ O_2 \ + \ H_2O \ \longrightarrow \ 4FeO(OH) \ + \ 4SiO_2$$

  (Pyroxene)   (Limonite)   (Dissolved
                                           silica)

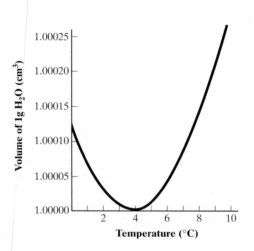

◄ FIGURE 9.2
The change in volume of water with temperature.

interface in the pore under an energy gradient that exists because unfrozen water has more free energy (the type of energy that drives chemical reactions) than water in its frozen form. The migration of water to the freezing front is a function of the porosity and permeability of the rock, the dissolved solids content of the pore water, and other factors. As the incoming water freezes in the pore, the resulting pressure causes the pore to propagate by the formation of microfractures at the pore boundaries. This model is consistent

▲ FIGURE 9.3
Angular blocks of rock in the Madison Range, Montana, produced from an outcrop by frost action. *Source:* Photo courtesy of the author.

▶ FIGURE 9.4
Ice forming within a rock pore or fracture. A thin layer of ordered water lines the rock surface. Pore water from unfrozen areas of the rock migrates to the pore under an energy gradient and freezes in the pore, adding to the volume of ice and causing microfractures to form and enlarge the cavity.

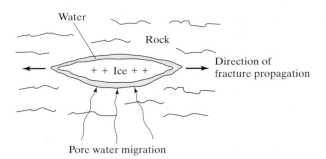

with field conditions, as indicated by experimental studies and field measurements. The size and shape of the angular rock fragments that are produced by frost action depend upon the joint spacing and orientation in the rock (Figure 9.3). On slopes, rock fragments produced in this manner may fall, slide, or roll to lower elevations, where they form a continuous sloping mantle of loose rock called a *talus* slope (Figure 9.5).

The effects of crystallization of salt within a rock are somewhat similar to the effects of freezing water. The expansion and weakening of the rock caused by salt crystallization is called *salt weathering*, and it presents one of the most serious threats to the durability and appearance of stone buildings. The salt that crystallizes from solution in the rock includes not only sodium chloride but also a number of chloride, sulfate, and carbonate salts that can enter the rock in several ways. Sedimentary rocks may retain salts from their original deposition. Other dissolved constituents may be contributed by chemical weathering processes. The salt may also be introduced after the rock is quarried. Polluted urban air and rainfall, for example, are excellent sources of salt.

Salt can crystallize from an aqueous solution as a crust on the rock surface or within rock pores and voids. Often, a concentration of salt forms in a thin layer just below the

▲ FIGURE 9.5
Accumulation of loose rock on a talus slope, southeastern Iceland. *Source:* Photo courtesy of the author.

rock surface. In this case, the surficial layer gradually deteriorates until it splits, or spalls, off from the rock face, at which time the process begins again. Inside the rock, salt crystals can weaken the rock in at least three ways. The first way is the pressure and expansion caused by the initial growth of the salt crystal. Second, upon heating, the salt crystals expand more than the surrounding minerals of the host rock, so that pressure is created. And third, salt crystals tend to absorb water into their structure to form hydrated crystals. This entails a volume expansion and an increase in pressure upon the adjacent grains, which can reach extremely high values (Figure 9.6). These hydration pressures can at

▲ FIGURE 9.6
Hydration pressures produced by hydration of $CaSO_4 \cdot \frac{1}{2}H_2O$. *Source:* From E. M. Winkler and E. J. Wilheim, 1970, Salt burst by hydration pressures in architectural stone in urban atmosphere, *Geological Society of America Bulletin*, 81:567–572.

times exceed the tensional strength of the rock, causing spalling of thin, exterior layers. Unfortunate demonstrations of salt weathering were initiated by the removal of carved stone monuments from Egypt and other arid regions and their relocation to more humid climatic regimes in Europe and the United States. Inscriptions that were clearly readable after thousands of years in their original climate are destroyed in a hundred years of weathering in cool, humid climates.

Case in Point 9.1
Salt Weathering and Ancient Egyptian Monuments

The ancient Egyptians built most of their temples and monuments along the flood plain of the Nile River. In the 1960s, the historical pattern of yearly Nile floods followed by dry periods was broken by the construction of the Aswan High Dam. The storage of Nile water in Lake Nasser, upstream from the dam, allowed the Egyptians to regulate their water usage and, for the first time, grow two or sometimes three crops per year irrigated by water released from the dam during the dry season. Although the enhanced crop yields were a great benefit to the country, an unintended consequence of the increased irrigation was a rise in water tables beneath the flood plain. Groundwater gradually began to saturate the foundations of stone monuments and temples. The water moved upward in the stone by capillary action, carrying its dissolved salts along with it. As the water evaporates in the dry air, salt precipitates, causing damage to the rock by salt weathering. Figure 9.7 shows the Sacred Lake of Dendara Temple, a 2000-year-old temple in Upper Egypt near Qena. Cleopatra visited this temple with her lover Marc Antony and may well have bathed in this pool. Today, the impoundment is dry because of pumping wells installed around the perimeter designed to lower the water

▲ FIGURE 9.7
View of the Sacred Lake of Dendara Temple, Egypt. Photo taken from the roof of the temple. The water table is kept artificially low by pumping groundwater and piping it away from the site.
Source: Photo courtesy of the author.

◀ FIGURE 9.8
Salt precipitate on the lower walls of the lake. Groundwater rises from the water table into the sandstone blocks by capillary action. When the water evaporates into the air, salt crystals grow in the rock, causing decomposition and weakening of the stones. *Source:* Photo courtesy of the author.

table. Figure 9.8 shows the lower part of the walls of the lake. The white precipitate is salt precipitated from groundwater drawn up into the stone from the high water table. In some temples, dewatering, excavation, and repair have been necessary to preserve ancient foundations from the rising water tables.

Temperature changes have long been suspect as a weathering mechanism because of the observation of cracked boulders on desert surfaces with little evidence of chemical weathering. Early lab experiments involving multiple cycles of heating and cooling did not appear to cause fracturing in rock samples. More recent work, however, suggests that the experiments did not fully duplicate natural conditions and that the mechanism may be significant. Part of the problem is that there are several variables potentially involved in the process. First, the thermal conductivity of the rock, or the ability to conduct heat into the rock from the surface, is important. If the thermal conductivity is low, heating and cooling will be concentrated in the uppermost layers of rock, producing more expansion and contraction than in rocks of high thermal conductivity. The coefficient of thermal expansion is also important. This coefficient describes the amount of expansion of a specific mineral related to a specific rise in temperature. Rocks composed of minerals with different coefficients of thermal expansion will experience differential stresses because some minerals tend to expand more than others and therefore exert different amounts of pressure. Finally, the range in temperature fluctuations and the rate of change of temperature within the rock are also important. The combination of all these factors may produce the results that are observed in the field.

Moisture changes, in the form of alternate wetting and drying, have experimentally been shown to cause expansion and contraction that may lead to weakening of the overlying rock. This mechanism may be even more effective when combined with temperature changes. Only a small amount of water in a rock may be sufficient to accentuate the effects of heating and cooling.

A peculiar type of weathering develops on natural rock faces composed of rock that is susceptible to spalling. In a process that is not as yet totally understood, rounded cavities or hollows form and gradually enlarge. The hollows may be found individually or in a dense network of closely spaced holes (Figure 9.9). A variety of names have been applied to these features, but *honeycomb weathering* is a useful descriptive term for examples such

▲ FIGURE 9.9
Honeycomb weathering, Meteor Crater, Arizona. (Camera lens cap for scale.) *Source:* Photo courtesy of
the author.

as that seen in Figure 9.9. The origin of the pits and hollows probably involves the breakdown of a protective mineral coating composed of iron and manganese oxides and the
spalling off of thin layers of rock. Moisture and wind may then be involved in removal of
detached grains. The action of lichens and other organisms has also been invoked in some
theories. Although honeycomb weathering is not restricted to dry climates, the fact that it
is common in arid regions suggests that mechanical weathering plays a significant role in
its formation.

Rock, like any other material, deforms under the application of stress. Deeply buried
rock experiences both elastic and plastic compression under the weight of the overlying
rocks. Upon removal of the overburden, which could occur during stream erosion of a valley or by deep excavation for a construction project, the elastic component of the deformation is recovered and the rock expands. The expansion caused by *unloading* is often
sufficient to fracture the rock along planes parallel to the surface on which stress has been
released. This type of weathering, also called *exfoliation*, produces thin slabs of rock
bounded by joints oriented parallel to the ground surface as the overlying material is removed and stress is released (Figure 9.10). The jointed rock slabs generated by exfoliation
are susceptible to further weathering and erosional processes. As the rock slabs are gradually removed from the slope, the unloading action continues and new joints and slabs
form. A process that produces a somewhat similar result is called *spheroidal weathering*, in
which thin concentric layers form and gradually split off the outside of weathering boulders. Unlike exfoliation, the forces that cause the curved slabs to deteriorate and break
away from the central core are due to chemical weathering reactions. Exposures of deeply
weathered granitic rock commonly contain spheroidal cores of unweathered rock surrounded by weathered rock (Figure 9.11). Over time, the cores are gradually decreased in
size by spheroidal weathering.

◄ FIGURE 9.10
Exfoliation results in the separation of thin slabs of rock from a rock mass along planes parallel to the surface; Sierra Nevada Mountains, California. *Source:* N. K. Huber; photo courtesy of U.S. Geological Survey.

◄ FIGURE 9.11
Rounded boulders (upper surface) produced by spheroidal weathering. The roadcut below exposes rounded corestones surrounded by weathered residual soil; near Lake Tahoe, California. *Source:* Photo courtesy of the author.

Organisms participate in both mechanical and chemical weathering processes. The mechanical effects are dominated by plant roots that exploit thin joints or cracks in rock as they grow. The pressures exerted by the growing roots can wedge apart blocks of rock, leading to acceleration of other weathering processes in the larger openings.

Chemical Weathering

Most rocks originally formed under conditions very different from those that exist at the earth's surface. In particular, the igneous and metamorphic rocks crystallized at very high temperatures and pressures. When these rocks are exposed to the lower temperatures and

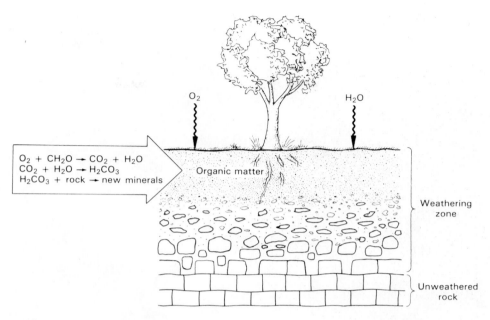

▲ FIGURE 9.12
Production of acidic weathering solutions by formation of CO_2 in the soil zone.

pressures present at the surface, they are unstable and tend to react chemically with components of the atmosphere to form new minerals that are more stable under those conditions. The most important atmospheric reactants are oxygen, carbon dioxide, and water. In polluted air, however, other reactants are available.

Figure 9.12 illustrates conditions in the zone of plant growth—the upper meter or so of the soil zone—as rainfall or snowmelt percolates downward from the surface. Decaying organic matter in this zone, under aerobic conditions, generates carbon dioxide. For example, the reaction

$$O_2 + CH_2O = CO_2 + H_2O \tag{9.1}$$

indicates that a carbohydrate (CH_2O) will react with oxygen to produce carbon dioxide. Carbon dioxide in the soil zone can reach levels of several orders of magnitude greater than its concentration in the atmosphere; it then reacts with infiltrating rainfall or snowmelt by the reaction

$$H_2O + CO_2 \longleftrightarrow H_2CO_3 \tag{9.2}$$

to produce carbonic acid. Production of carbonic acid lowers the pH by partial dissociation according to the reaction

$$H_2CO_3 \longleftrightarrow H^+ + HCO_3^- \tag{9.3}$$

The resulting increase in hydrogen-ion concentration (lower pH) causes a more intense attack on minerals by the solution. Decomposition of organic matter can also produce *humic* acids, which, like carbonic acid, yield hydrogen ions to lower the pH.

The specific types of weathering reactions that have been recognized are listed in Table 9.1 along with example chemical reactions. When a mineral completely dissolves during weathering, the reaction is known as *solution*. The tendency of a mineral to dissolve

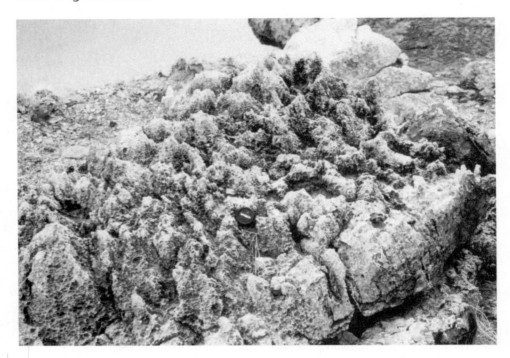

▲ FIGURE 9.13
Solution of exposed limestone outcrop to produce etched, pitted surface; Lee's Ferry, Arizona.
Source: Photo courtesy of the author.

in weathering solutions is described as its *solubility*. Evaporite minerals dissolve readily in water, whereas carbonate minerals are somewhat less soluble. Even so, solution of limestone and dolomite is extremely important. When limestone is exposed at the surface, pitting and etching of the rock is obvious (Figure 9.13). Even more significant is the subsurface solution of carbonate rocks to form caves and a type of landscape known as *karst* topography. Karst differs from other landscapes in that surface streams are rare and drainage takes place underground. The characteristics of karst areas will be described in Chapter 11. Some silicate minerals weather by solution, but their solubilities are very low. Quartz, for example, usually dissolves appreciably only under the intense weathering conditions found in the tropics. Quartz differs from the carbonate minerals in that its solubility increases as the pH increases.

Hydrolysis is the reaction between acidic weathering solutions and many of the silicate minerals, including the feldspars. The reaction illustrated in Table 9.1 indicates that in hydrolysis a feldspar mineral reacts with hydrogen ions to form several dissolved products as well as a solid product, the clay mineral kaolinite. Other clay minerals are produced by similar weathering reactions. This reaction is an example of the breakdown of a mineral that is stable at high temperature and pressure to form a new mineral that is stable under conditions near the earth's surface. Notice that hydrolysis produces dissolved silica and potassium in the orthoclase example. These constituents are carried downward out of the weathering zone in the water in which they are dissolved toward the water table, where they become part of the groundwater flow system. Hydrolysis reactions are responsible for producing deposits of clay that are mined for use in many industrial processes.

The adsorption of water into the lattice structure of minerals is called *hydration*. The formation of gypsum from anhydrite by hydration is illustrated in Table 9.1. Clay minerals are also susceptible to hydration. The volume expansion that accompanies hydration is an important contributor to the physical weakening and breakdown of a rock. The role of hydration in salt weathering was previously described.

The reaction of free oxygen with metallic elements is familiar to everyone as rust. This process, an example of *oxidation*, affects rocks containing iron and other elements. In an oxidation reaction, iron atoms contained in minerals lose one or more electrons each and then precipitate as different minerals or amorphous substances. For example, in the weathering of pyroxene by oxidation, as illustrated in Table 9.1, iron is oxidized and hydrated to form the mineral limonite. The presence of limonite in rocks or soils is indicated by brownish or reddish staining.

Stability

The stability of minerals under the action of chemical weathering agents at the earth's surface depends upon the difference between the conditions at the surface and the conditions under which the mineral originally crystallized. Minerals occurring in igneous rocks can be grouped into a sequence of relative weathering stability that is the exact opposite of their order of crystallization from the magma or lava (Figure 9.14). This relationship is known as *Goldich's stability series*. For example, olivine is the first mineral to crystallize from a silicate melt and is therefore the most unstable mineral in a weathering environment. Olivine will weather rapidly when exposed to the atmosphere. Quartz, the last mineral to crystallize from a magma, is very stable in most weathering environments. It dissolves so slowly that, for all practical purposes, it is considered to be insoluble except under humid, tropical conditions. The stability of quartz explains why it is so common in sedimentary rocks such as sandstone. Feldspars and ferromagnesian minerals from the original rocks weather relatively rapidly to clay minerals, leaving quartz to be transported and deposited in depositional environments. The clay minerals that form by the weathering of feldspars and other

▲ FIGURE 9.14
Goldich's stability series, which shows the relative stability under weathering conditions of primary minerals in igneous rocks.

Climate	Hot Arid	Humid Tropical	Humid Temperature
Table 9.2 Mineral Composition of Weathered Rocks as a Function of Climate			
Mineral		Percentages	
Quartz	28	8	50
Illite	32	—	24
Chlorite	15	—	—
Montmorillonite	12	—	7
Kaolinite	—	12	14
Gibbsite	—	68	—
Geothite	5	10	—
Gypsum	5	—	—
Calcite	2	—	—
Others	1	2	5

Source: From F. C. Beavis, *Engineering Geology*, © 1985 by Blackwell Scientific Publications, Inc., Melbourne.

minerals are stable weathering products that are deposited as muds in low-energy depositional environments and that later are consolidated to form shales.

The weathering products of rock-forming minerals vary as a function of climate and other factors. Table 9.2 shows the secondary mineral composition of residual soils weathered from rocks of initially similar composition in three different types of climate in Australia. The intense leaching of the humid tropical climate left a weathering residue dominated by iron and aluminum-hydroxide secondary minerals (goethite and gibbsite, respectively). Most of the quartz was removed from the original rocks. Weathering products in the hot arid and humid temperate climates contain greater percentages of clay minerals and quartz. Determination of the mineralogical weathering products is important because these minerals influence the strength and other engineering properties of residual soils.

The importance of climate in chemical weathering cannot be overemphasized. Climate directly controls temperature and the amount and timing of precipitation and indirectly influences other variables that play a role in weathering, such as vegetation. The changes in these variables are responsible both for the types of weathering products, as indicated in Table 9.2, and also the depth of weathering. Broad variations in rock weathering on the earth's surface reflect the gradual changes in climate and vegetation occurring between the poles and equator (Figure 9.15). Chemical weathering is minimal under the cold conditions at the poles. In the midlatitude, cool moist conditions, with forest vegetation, the depth of weathering increases and the weathering products are dominated by kaolinite. In the semi-arid to arid regions centered around 30° north and south of the equator, temperature rises, but precipitation declines, reducing the amount of chemical weathering. The most intense weathering occurs near the equator, where temperature and precipitation are both high. The depth of weathering penetrates tens to hundreds of meters into near-surface rock masses. Silicate minerals, even including quartz, are removed by the intense conditions (Table 9.2), leaving weathering products enriched in iron and aluminum oxides, sometimes to the degree that the soils become ore deposits for these metals. In climatic zones at the edges of the tropic known as savannas, most of the rain occurs in a distinct wet season as opposed to a more even distribution closer to the equator. The drying and oxidation of the soils during the dry season can lead to the formation of a hardened crust of iron and/or aluminum oxides. The soils, known as *laterites* when they are composed of iron oxide and *bauxite* when they are composed of aluminum oxides, are unsuitable for agriculture. Some tropical soils harden soon after deforestation and thus must be abandoned for agriculture. The weathering zones shown in Figure 9.15 are generalized and will be modified by topography and other variables.

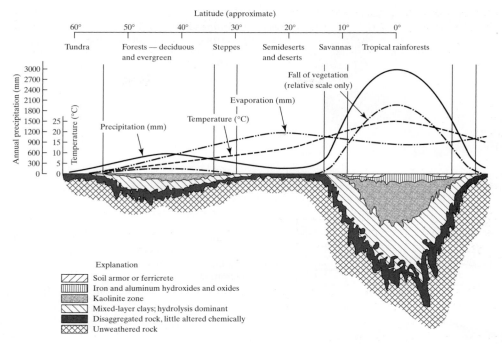

▲ FIGURE 9.15
Generalized weathering zones along a transect from the poles to the equator. *Source:* From A. L. Bloom, *Geomorphology: A Systematic Analysis of Late Cenozoic Landforms,* 3rd ed., © 1998 by Waveland Press, Long Grove, IL.

Weathering and Landforms

The significance of weathering, as can be surmised from the preceding discussion, depends upon the climate and lithology of the rocks being weathered. When conditions are favorable, and sufficient time is available, weathering can produce distinctive landscapes. To illustrate, we will focus upon igneous rock terrains that have been exposed to active chemical weathering for long periods of geological time. Under these circumstances, the effects of weathering can penetrate to great depths. Alteration of the granitic rocks decreases with depth until a boundary called the *weathering front* is reached. This boundary, which can be sharp or transitional, separates weathered from unweathered rock (Figure 9.16). When the spacing of joints is irregular, the depth of the weathering front can be highly variable, penetrating to great depths where joints occur close together and extending to much shallower depths where joints are spaced far apart. If such an area is subjected to increased erosion, perhaps due to uplift or climate change, the weathered residual materials are removed, leaving plains of low relief broken only by isolated hills of granite, which represent the former zones of shallow

▶ FIGURE 9.16
Weathering front developed on granite. The material changes from residual sandy clay at the top of the section to an increasing proportion of unaltered rock in corestones with increasing depth. The jointed rock at the base of the section is only slightly weathered. The depth to the weathering front may be tens of meters. *Source:* After A. J. Gerrard, 1988, *Rocks and Landforms,* Unwin Hyman.

◀ FIGURE 9.17
Inselberg composed of granitic rocks projecting above a plain; the formation is illustrated in Figure 9.18; Joshua Tree National Monument, California. *Source:* Photo courtesy of the author.

(a)

◀ FIGURE 9.18
Formation of inselbergs by a two-stage process: (a) vertical section through granitic rocks with varied spacing of joints; (b) deep weathering under humid or subhumid climate; (c) erosion of residual soil under arid conditions leaving isolated remnants of unweathered rock. *Source:* Modified from D. D. Trent, 1984, Geology of the Joshua Tree National Monument, *California Geology*, 37:75–86.

(b)

(c)

subsurface weathering. These striking residual hills are called *inselbergs* (Figure 9.17). Inselbergs can be composed of any lithology, but they are particularly common in arid, granitic terrains. Figure 9.18 illustrates the presumed steps in the formation of such a landscape.

Rock Weathering and Engineering

The degree and pattern of weathering are among the most important factors to be determined in an on-site engineering investigation. The effects of weathering on the engineering properties of rocks include a decrease in strength, a loss of elasticity, a decrease in density, and increases in moisture content and porosity. These detrimental changes can be critical to the suitability of a site for structures such as arch dams, which require maximum strength and elasticity.

The subsurface distribution of joints and faults is particularly important to determine in on-site engineering investigation because of the control these discontinuities exert upon the depth of the weathering front. Rock masses may be deeply weathered along fractures and unaltered in intact rock at shallow depth between the fractures. Weathering conditions

▲ FIGURE 9.19
Degree of rock weathering at the site of an underground power station in Australia. Weathering zones include fresh rock (F), slightly weathered rock (SW), highly weathered rock (HW), and completely weathered rock (CW). *Source:* From F. C. Beavis, *Engineering Geology,* © 1985 by Blackwell Scientific Publication, Inc., Melbourne.

at an underground hydroelectric power plant in Australia are shown in Figure 9.19. Weathering zones are classified as fresh (F), slightly weathered (SW), highly weathered (HW), and completely weathered (CW). Tunnel and underground excavations for the project were sited in order to avoid faults and associated deep-weathering zones.

Erosion

The chemical and physical breakdown of rocks by weathering processes leads to the formation of materials that can be easily transported by processes operating at the earth's surface. *Erosion* is the detachment and transportation of surface particles under the action of one or more of the forces and agents listed in Table 9.3. The materials that are susceptible to erosion include not only weathered residual products developed from rocks but any unconsolidated surficial deposit. We will refer to these materials as soil in the engineering sense, that is, any unconsolidated material regardless of thickness or origin.

Erosion by Water

Throughout the world, erosion by running water is the most important type of erosion in terms of the amount of sediment removed from the land surface. Sediment eroded from slopes is subsequently transported by streams to the oceans. The amount of slope erosion is a complex function of rock type, climate, vegetation, and topography. In its natural state, the amount of soil erosion from a landscape, as measured by the sediment load carried by streams draining it, can be related to the climate and vegetation supported by the climate (Figure 9.20). The highest sediment yield occurs under semi-arid conditions, where rainfall occurs as infrequent severe storms with high runoff and vegetation is too sparse to protect the land surface. Sediment yield decreases in desert environments because there is just too little rainfall to produce runoff; what runoff there is cannot sustain sufficient streamflow to carry the sediment. Sediment yield also decreases in more humid

Table 9.3   Erosional Agents
Gravity (mass wasting)
Water (fluvial erosion)
Wind (aeolian erosion)
Glacial ice
Waves

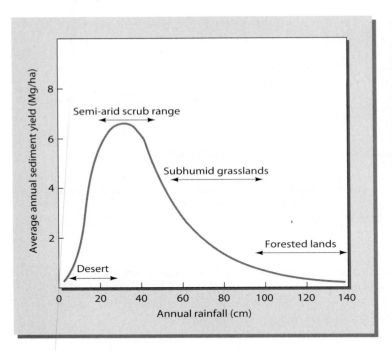

◀ FIGURE 9.20
The relationship between sediment yield and annual precipitation for landscapes in their natural states. Maximum sediment yield occurs in semi-arid climates. *Source:* From N. C. Brady and R. R. Weil, 2002, *The Nature and Properties of Soils*, 13th ed., Prentice Hall, Inc., Upper Saddle River, N.J.

▲ FIGURE 9.21
Erosional retreat of slope composed of shale with resistant conglomerate at top. Conglomerate blocks protect shale from weathering to form pedestals; Lee's Ferry, Arizona. *Source:* Photo courtesy of the author.

environments supporting grassland and forest vegetation because a dense cover of grasses and/or trees protects the soil from the type of rapid runoff that leads to erosion.

A dramatic example of erosion is shown in Figure 9.21. The slope in the background is composed of a bed of resistant conglomerate at the top of the slope underlain by nonresistant shale. In the semi-arid to arid climate of the area, erosion of the shale is rapid and the slope is continually retreating. Blocks of conglomerate break off and roll or slide downslope as the underlying shale is eroded. When these blocks come to rest, they protect the shale beneath from erosion. As the shale continues to be eroded outside the protected area, *pedestals* are formed, which consist of conglomerate blocks supported by thin columns of shale.

Erosion of land altered by human activity is much more rapid than erosion of land in its natural state. Unfortunately, agricultural lands are the source of much of the sediment removed by erosion. The loss of soil productivity caused by the erosion of topsoil is one of the most serious problems facing the human population. Erosion in stream and river channels is also a significant problem. Bridge piers, levees, and other structures can be damaged or destroyed by river erosion.

Processes

Soil erosion by water occurs in a variety of ways. Initiation of erosion commonly includes detachment of soil particles from the surface by *raindrop erosion*. The energy transferred to the surface by the impact of raindrops is sufficient to dislodge individual soil particles from particle aggregates at the soil surface. Downslope movement of particles also begins by the splash of the raindrop.

In the early stages of a rainstorm, most of the precipitation infiltrates into the ground. If the duration and intensity of the precipitation are sufficient, a saturated zone develops just below the soil surface and water begins to pond on the surface. We will discuss these relationships in more detail in Chapter 11. When ponded water reaches a depth greater than the height of surface irregularities, water flows downslope in a thin, wide sheet as *overland flow*. You may have noticed this type of flow on gently sloping, paved parking lots. On soil surfaces, this shallow flow of water can transport detached soil particles and is given the name *sheet erosion*.

Sheet erosion is a rather rare and inefficient erosional process and is soon replaced by *rill erosion* (Figure 9.22). The channelized flow of water in rills has a higher velocity

▲ FIGURE 9.22
Severe rill erosion on a steeply sloping crop field. Notice the sediment deposition at the base of the slope. *Source:* Photo courtesy of USDA Natural Resources Conservation Service.

▲ FIGURE 9.23
Progressive loss of agricultural or grazing land is caused by development and enlargement of gullies.
Source: Photo courtesy of USDA Natural Resources Conservation Service.

and a greater ability to transport particles downslope. With time, rills enlarge into gullies (Figure 9.23), which gradually extend themselves *headward* (upstream) and may render the land unfit for agriculture.

Not all erosion takes place at the surface. When sediments rich in swelling clays (Chapter 10) are exposed in valley slopes, vigorous erosion can take place below the surface, a process called *piping*. Although piping has been observed in many climatic settings, it is particularly prevalent on the steep, dry slopes of arid regions. The combination of clay-rich soils and active erosion both above and below the land surface leads to a distinctive type of topography called *badlands*. The term refers to a deeply eroded landscape riddled with steep, bare slopes that is exceptionally unpleasant to cross on foot or horseback if you have to go anywhere in a hurry. The exposed clay surfaces expand when wet and crack as they shrink and dry. Over time, cracks are enlarged into a network of tunnels just under the surface that convey water and sediment to exit holes near the base of the slopes (Figure 9.24). Piping can lead to costly damage when it develops under roads and other structures. Enlargement of the pipes eventually leads to roof collapse and surface subsidence, affecting any structures that may lie above the pipes (Figure 9.25).

Erosional Rates

Attempts have been made to predict the amount of erosion that will occur within a given amount of time for various types of land settings. These methods yield only rough estimates because of the large number of variables in the soil erosion process. Some of the main variables are listed in Table 9.4. Some of these factors have been combined into the *Universal Soil Loss Equation*, which is stated

$$A = RKLSCP \tag{9.4}$$

where

A = average annual soil loss (tons/acre)
R = rainfall factor

$$K = \text{soil-erodibility factor}$$
$$LS = \text{slope length—steepness factor}$$
$$C = \text{cropping factor}$$
$$P = \text{conservation factor}$$

The rainfall factor in this equation is obtained from the product of the kinetic energy of raindrops and the maximum 30-min intensity of the storm. Values for this index in the eastern United States are shown in Figure 9.26. It is apparent from this map that the most

▲ FIGURE 9.24
Piping in clayey slopes in an arid climate; exit holes are visible at the base of the slope. Petrified Forest National Park, Arizona. *Source:* Photo courtesy of the author.

1 Shale and sandstone of Cretaceous Mancos Shale

2 Tan silt and clay, sandy in places, of Quaternary age

3 Flood plain of Aztec Wash

4 Pipe system

5 Block left as natural bridge

6 Debris blocks undermined and sapped by pipes

7 Culvert

8 Flow of ephemeral drainage

9 Plunge pool

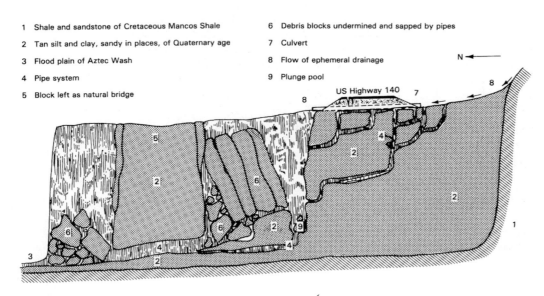

▲ FIGURE 9.25
Schematic diagram of piping beneath U.S. Route 140, southwestern Colorado. *Source:* From G. G. Parker and E. A. Jenne, 1967, 46th Annual Meeting, Highway Research Board, Washington, D.C., U.S. Geological Survey, Water Resources Division, p. 27.

Table 9.4 Variables Affecting Soil Erosion
Rainfall
Intensity
Duration
Soil Characteristics
Porosity
Permeability
Moisture content
Grain size and shape
Topography
Orientation of slope
Slope angle
Length of slope
Vegetation
Type and distribution on slope

▲ FIGURE 9.26
Average values of the rainfall-erosion index for the eastern United States. *Source:* From D. D. Smith and W. H. Wischmeier, 1962, *Advances in Agronomy,* 14:109–148.

intense rainfall events occur in the Gulf Coast region. The factor K, relating to the characteristics of the soil, is a function of the grain-size distribution. Soils composed of silt and fine sand are the most erodible types. The topography of the area is evaluated by the product of L (slope length) and S (slope angle expressed as a percent). Each factor is calculated as the ratio of the actual value to standardized plots of land 72.6 feet in length and 6 feet in width, with 9% slope. For example, a slope 50 feet in length would have an L value of $50/72.6 = 0.69$. The cropping factor ranges from high values (severe erosion) for barren land and row crops, to low values for natural grasses and woodlands. Vegetated land that is cleared for construction will have a substantial increase in this factor. Crop management techniques are incorporated into the conservation factor. Erosion-control measures such as contour plowing and terracing reduce the value assigned to this factor from its maximum, in the case where no erosion control methods are implemented. Tables and nomographs are available for obtaining values for the factors in the Universal Soil Loss Equation.

The Universal Soil Loss Equation was developed mainly for agricultural lands. Other methods of estimating erosion rates include measuring the amount of sediment carried by rivers downstream from the eroded area and measuring the amount of sediment trapped in reservoirs constructed along river courses. These methods indicate that erosion is greatly accelerated by certain changes in land use. An excellent example of these changes is illustrated in Figure 9.27. When Europeans arrived in the United States, they initiated a series of land-use changes. One of the first changes to occur in arable areas was the conversion of land to crop production. This process involved clearing forested land or breaking sod in prairie regions. Both activities caused a decrease in infiltration, an increase in runoff, and an increase in soil erosion. Marginal agricultural areas often reverted back to forest or grazing as more productive farmland was developed. The erosion was correspondingly decreased in those areas.

The most drastic increases in soil erosion and sediment yield in streams occur during urbanization of drainage basins. Erosion is most severe during and immediately after construction (Figure 9.28). After the construction phase, erosion decreases because much of the

▶ FIGURE 9.27
Sediment yield from lands undergoing various land-use changes. *Source:* From M. G. Wolman, 1967, *Geografiska Annaler,* 49-A:385–395. Used by permission of the Swedish Society for Anthropology and Geography.

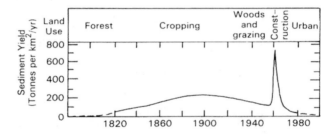

▶ FIGURE 9.28
Soil erosion from cleared, unprotected slopes during housing construction. *Source:* Photo courtesy of USDA Natural Resources Conservation Service.

land surface is covered by buildings and paved areas. In addition, runoff is collected and conveyed to stream channels through an artificial network of drains and storm sewers.

Erosion Control

The use of erosion-control methods must be increased throughout the world in an attempt to combat excessive soil erosion. In agricultural lands, these methods include *contour farming, strip cropping, terracing,* and *conservation tillage.* Their use becomes more important as the slope of the land increases. Contour farming is the orientation of crop rows parallel to the elevation contours of the land. This practice disrupts overland flow of water and thereby inhibits sheet and rill erosion. Strip cropping is the alternation of strips of erosion-inhibiting crops, such as grasses, with more erodible row crops. Grasses trap sediment eroded from crop rows and also increase moisture infiltration. The optimum width of the strips is determined by the amount of land slope. Terracing is a more intensive method of erosion control requiring construction of alternating embankments and terraces on the slope. Conservation tillage refers to a variety of practices that reduce the amount of disturbance and exposure of soil during cropping. Historical plowing practices involved inversion of the soil to bury plant residue and produce a bare soil surface, which was extremely susceptible to erosion. The advent of herbicides that could kill weeds without overturning the soil, and new equipment such as chisel plows that could plant seeds through a cover of plant residue created a revolution in farming in the United States and other countries (Figure 9.29). Not only do these new techniques help minimize soil erosion but they also save time, fuel, and money.

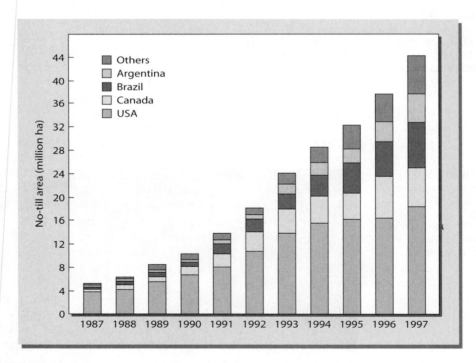

▲ FIGURE 9.29
Increases in the adoption of conservation tillage. No-till is actually one of several procedures included in conservation tillage. *Source:* From N. C. Brady and R. R. Weil, 2002, *The Nature and Properties of Soils,* 13th ed., Prentice Hall, Inc., Upper Saddle River, N.J.

▲ FIGURE 9.30
Newly constructed street (notice curb and partially buried drain in foreground) filled with sediment eroded from adjacent housing lots. *Source:* Photo of USDA Natural Resources Conservation Service.

Although the area converted to urban use is small in comparison with the amount of farmland, the sediment yield from urbanizing areas can be greater by a factor of 100 or more than the sediment yield of natural or agricultural areas nearby. The problem of controlling erosion in land undergoing urbanization is directly related to the problem of excessive sedimentation, particularly in undesirable locations. Excessive sedimentation can clog highways, ditches, and drains, causing expensive cleanup operations (Figure 9.30). Increased sediment yield to stream systems kills fish, increases flooding, and decreases the capacity of reservoirs. Chemical pollutants, including pesticides and herbicides, are often adsorbed on soil particles and carried by erosion to streams and lakes.

In urban areas, new construction is likely to pose the most serious erosion and sedimentation problems. Sediment yields from areas that annually produce from 70 to 280 tonnes/km^2 can increase to as much as 28,000 tonnes/km^2 during certain types of construction. Guy (1976) has documented many examples of effective and ineffective erosion and sediment-control measures and has also suggested methods to reduce erosion by applying sound hydrologic and geomorphic principles to the design of these systems. Soil erosion can often be drastically reduced by planning construction during periods of low expected rainfall, dividing the construction site so that only small areas are cleared of vegetation at any one particular time, and utilizing temporary vegetation cover on soil stockpiles and other exposed areas whenever possible.

When additional erosion control measures are necessary, *diversions, bench terraces*, and *detention basins* can be constructed. Diversions consist of channels and ridges designed to collect overland flow and carry it away from areas that must be protected from erosion or sedimentation. Bench terraces reduce the length of the slope, and, as implied in equation (9.4), reduce erosion. Diversions are often routed to detention basins (Figure 9.31), which are designed to trap sediment carried by runoff from the construction site. Temporary sediment-control measures are often installed until vegetation or paving reduces the area of unprotected ground at a site. Straw or hay bales, for example, form effective sediment

▲ FIGURE 9.31
A sediment detention basin used to trap sediment in an urban area. *Source:* Photo courtesy of USDA Natural Resources Conservation Service.

▲ FIGURE 9.32
Failure of a straw-bale filter designed to trap sediment in an unvegetated ditch along a newly constructed highway. Flow in the ditch overtopped the straw-bale dam and eroded a new channel adjacent to the straw bales. *Source:* Photo courtesy of the author.

traps around storm-drain inlets or culverts, when flow velocities are not extremely high. These temporary measures fail, however, when the amount and velocity of overland flow overwhelm the straw-bale structure (Figure 9.32).

Erosion by Wind

The impact of wind erosion has never been more clearly demonstrated than in the Great Plains of the United States during the 1930s (Figure 9.33). Dry conditions coupled with steadily increasing destruction of the natural grass cover devastated the agricultural economy of the region.

▲ FIGURE 9.33
A dust storm in Colorado during the 1930s. Notice the drifts of sand around fences and buildings.
Source: Photo courtesy of USDA Natural Resources Conservation Service.

The erosion of soil by wind is similar to erosion by water. Wind, however, is not as effective an erosional agent as water because of its lower density. Variables in the wind-erosion system can be divided into characteristics of the wind and characteristics of the soil and land surface. Wind variables include wind velocity and duration and the length of the open area without obstacles over which the wind blows. Important properties of the soil and land surface include particle size and size distribution, moisture content, and vegetation. Particles of typical density within the size range of 0.1 to 0.15 mm in diameter (very fine to fine sand) are most susceptible to wind erosion. The distribution of grains upon the surface is also very important. Soils containing large particles gradually become protected as wind erosion proceeds. As the fine particles are removed, coarser grains become concentrated at the surface, preventing further erosion of the soil. Very fine particles in the soil, including silt and clay, promote the formation of soil aggregates. These clumps composed of smaller particles are held together by the cohesion of the silt and clay. In order for soil aggregates to be eroded by wind, they must first be broken down by raindrop impact, abrasion by wind-transported sediment, and other processes.

Moisture is one of the most important soil properties. Moist soil is cohesive and resistant to wind erosion. The cohesive forces provided by the soil water are lost when the soil dries. This explains why droughts and wind erosion go hand in hand.

Vegetation is the final soil and land-surface variable. Vegetation provides physical protection for the soil; it also holds moisture, increases the roughness of the surface, and adds organic binding agents to the soil. The removal of vegetation during plowing and other activities is one of the primary factors in increased wind erosion.

Wind erosion can be greatly reduced by proper farming and land-use practices. *Shelterbelts* are one of the most common attempts to control wind erosion (Figure 9.34). These linear bands of trees or other plants decrease wind velocity and the unobstructed distance over which the wind blows. Farming methods for wind-erosion control include strip cropping and the planting of temporary cover crops, rather than allowing the soil to remain bare between periods of crop production. The processes of wind erosion are discussed further in Chapter 16.

◄ FIGURE 9.34
A shelterbelt planted to minimize erosion of soil by wind. *Source:* Photo courtesy of USDA Natural Resources Conservation Service.

Case in Point 9.2
Erosion, Sedimentation, and Reservoir Capacity

The interrelationship between erosion and deposition is well illustrated by the accompanying photos of Lake Ballinger Dam in Texas. Figure 9.35a shows the dam and reservoir prior to its abandonment in 1952. Lake Ballinger Dam was built in 1920 to provide a municipal water supply. Sedimentation gradually reduced the maximum depth of the reservoir from about 11 m to slightly more than 1 m by 1952. The excessive sedimentation was caused by runoff from cultivated fields in the small drainage basin of the reservoir. Figure 9.35b, taken after the final drainage of the reservoir, shows the greatly decreased capacity caused by excessive sedimentation. Better soil conservation practices in the drainage basin could have alleviated the problem. Thus the problem of erosion in one area is compounded by undesirable sedimentation in another area.

▲ FIGURE 9.35a
Lake Ballinger Dam, Texas, before abandonment. *Source:* Photo courtesy of USDA Natural Resources Conservation Service.

▲ FIGURE 9.35b
The dam after it was abandoned in 1952 because of excessive sedimentation. *Source:* Photo courtesy of USDA Natural Resources Conservation Service.

Summary and Conclusions

Rocks in contact with the atmosphere at the earth's surface are subjected to a wide variety of physical and chemical weathering processes that lead to their gradual decay and disintegration. Expansive pressures within rocks are caused by frost action, salt weathering, moisture-content changes, temperature changes, and the removal of overburden load. These physical processes are complemented by chemical reactions between atmospheric constituents and unstable minerals to form new compounds that are more stable under the surface conditions of low temperature and pressure. Chemical weathering reactions include complete dissolution of the original mineral (solution), formation of new minerals (hydrolysis), addition of water to the mineral structure (hydration), and reaction with oxygen to form new substances (oxidation). Minerals derived from igneous rocks weather in reverse order from their order of crystallization. The resistance of quartz to weathering explains its abundance in sedimentary rocks produced by the weathering and erosion of previously existing rocks.

Erosion by water and wind is most severe when land is altered from its natural state. Prediction of rates of fluvial erosion by such methods as the Universal Soil Loss Equation is difficult because of the large number of variables involved in the process. Erosion damage is often initiated by raindrop impact and sheet erosion. Sheet erosion progresses to rill and gully erosion during the gradual destruction of arable land by fluvial action. Land cleared for construction presents the most vulnerable condition for soil erosion. Various farming practices can be adopted to minimize erosion. Similarly, erosion of lands under development can be greatly limited by construction planning and utilization of erosion-control structures designed with an understanding of the hydrologic and geomorphic processes involved.

Problems

1. Describe the specific climatic conditions in which the following types of weathering are dominant:
   (a) Strong chemical weathering
   (b) Moderate chemical weathering with frost action
   (c) Strong mechanical weathering

2. How does salt weathering cause the deterioration of building stone?

3. What is the relationship between temperature and hydration pressure of gypsum?

4. Write a chemical equation for the solution of dolomite by carbonic acid.

5. Write a chemical raction for the hydrolysis of albite ($NaAlSi_3O_8$).

6. Why do minerals fit into their respective positions in Goldich's stability series?

7. Summarize the formation and function of carbon dioxide in weathering processes.

8. What engineering properties of rock are affected by weathering, and what changes take place?

9. During construction of a highway, a wooded slope is cleared and excavated to increase the slope from 6% to 14%. The C factor increases from 0.1 (mature woodland) to 1.0 (bare ground) prior to stabilization of the cut with vegetation. Assuming that R, L, and P do not change, how much of an increase in erosion can be expected? (*Hint:* Calculate the erosion before and after road construction, and from these values determine the factor by which erosion has increased.)

10. What is the range in the rainfall erosion index along the Mississippi Valley of the eastern United States?

11. What methods are used to minimize erosion during construction?

12. What are the consequences of excessive soil erosion? Which one(s) would you expect to be most severe in your area?

References and Suggestions for Further Reading

BEAVIS, F. C. 1985. *Engineering Geology*. Melbourne: Blackwell Scientific.

BLAND, W., and D. ROLLS. 1998. *Weathering: An Introduction to the Scientific Principles*. London: Arnold.

BLOOM, A. L. 1998. *Geomorphology: A Systematic Analysis of Late Cenozoic Landforms*, 3rd ed., Long Grove, IL: Waveland Press.

BRADY, N. C., and R. R. WEIL. 2002. *The Nature and Properties of Soils*, 13th ed. Upper Saddle River, N.J.: Prentice Hall, Inc.

COOKE, R. V., and J. C. DOORNKAMP. 1974. *Geomorphology in Environmental Management*. Oxford: Clarendon Press.

COSTA, J. E., and V. R. BAKER. 1981. *Surficial Geology: Building with the Earth*. New York: John Wiley.

GERRARD, A. J. 1988. *Rocks and Landforms*. London: Unwin Hyman.

GUY, H. P. 1976. Sediment-control methods in urban development: Some examples and implications, in *Urban Geomorphology*, D. R. Coates, ed. Geological Society of America Special Paper 174, pp. 21–35.

PARKER, G. G., and E. A. JENNE. 1967. Structural failure of western U.S. highways caused by piping. 46th Annual Meeting, Highway Research Board, Washington, D.C., U.S. Geological Survey, Water Resources Division, p. 27.

PELTIER, L. 1950. The geographical cycle in periglacial regions as it is related to climatic geomorphology. *Annals of the Association of American Geographers*, 40: 214–236.

SMITH, D. D., and W. H. WISCHMEIER. 1962. Rainfall erosion. *Advances in Agronomy*, 14:109–148.

TRENT, D. D. 1984. Geology of the Joshua Tree National Monument, *California Geology*, 37:75–86.

WINKLER, E. M., and E. J. WILHELM. 1970. Salt burst by hydration pressures in architectural stone in urban atmosphere. *Geological Society of America Bulletin*, 81: 567–572.

WOLMAN, M. G. 1967. A cycle of sedimentation and erosion in urban river channels. *Geografiska Annaler*, 49-A: 385–395.

Soils, Soil Hazards, and Land Subsidence

Perhaps no term causes more confusion in communication between various specialized groups of earth scientists and engineers than the word *soil*. The problem arises in the reasons for which different groups of professionals study soils. Soil scientists, or *pedologists*, constitute a group interested in soils as a medium for plant growth. For this reason, pedologists focus most of their attention on the organic-rich, weathered zone that supports plant growth—the upper meter or so beneath the land surface—and refer to the rocks or sediments below the weathering zone as *parent material*. Soil scientists have developed a complex system of classification for soils that is based on the physical, chemical, and biological properties that can be observed and measured in the soil.

Soils engineers, the corresponding group of engineering soil specialists, take an entirely different approach to the study of soil. To soils engineers, the word *soil* connotes any material that can be excavated with a shovel. This definition places no limitations on depth, origin, or ability to support plant growth. The engineering classification of soil is based on the particle size, the particle-size distribution, and the plasticity of the material. These characteristics relate closely to the behavior of soil under the application of load.

Most geologists fall somewhere between pedologists and soils engineers in their approach to soils. Geologists are interested in soils and weathering processes as indicators of past climatic conditions and in relation to the geologic formation of useful materials ranging from clay deposits to metallic ores. The pedological soil classification is complicated for nonspecialists and, therefore, geologists commonly use textural classifications of soil. Geologists usually refer to any loose material below the plant growth zone as sediment or *unconsolidated material*. The term *unconsolidated* may be confusing to engineers because *consolidation* specifically refers to the compression of saturated soils in soils engineering, in contrast to its more general usage in geology.

An important aspect of soil with respect to engineering applications is whether the material can be classified as *transported* or *residual*. Transported soils are deposits of rivers, glaciers, and other surficial geologic processes. Residual soils are those that developed in place by the weathering processes discussed in Chapter 9. This distinction is important because the origin of the material is relevant to its thickness, continuity or discontinuity over a landscape, and the nature of the lower contact with unweathered bedrock.

The Soil Profile

In the first part of this chapter we shall investigate soil from the viewpoint of soil scientists. *Pedogenic,* or soil-forming, processes include the chemical weathering processes that we discussed in Chapter 9, as well as certain additional physical, chemical, and biological processes. The result of this activity is a sequence of recognizable layers, or *horizons,* that constitute the *soil profile.*

A number of possible horizons can be described in the field. Natural exposures can be used or pits can be dug to examine the soil profile (Figure 10.1). Profile descriptions are based on recognition of *master horizons* and subdivisions of master horizons. The criteria for master horizons are given in Table 10.1. A surficial horizon, the O, is used to designate fresh or partly decomposed organic matter. The A horizon, recognized by its dark color, is a zone of mixed mineral material and partly decomposed organic matter called *humus.* The E horizon is a zone of *eluviation,* which describes the leaching of metal oxides and clays. As a result, it is often light colored in comparison to the A and B horizons. Substances leached from the E horizon accumulate in the B horizon, which is known as an *illuvial* zone. A horizon that is strongly cemented by calcium carbonate is labeled a K horizon. Below the B or K horizons, there are several possible horizons that may be somewhat weathered but do not have the properties of the previously described horizons. The C horizon generally consists of the parent material from which the soil was formed. A horizon composed of consolidated bedrock is described as R.

◀ FIGURE 10.1
A soil profile exposed in a pit dug for soil classification. *Source:* Photo courtesy of USDA Natural Resources Conservation Service.

Table 10.1   General Description of Soil Horizons

Soil Horizon Nomenclature

Master Horizons

O horizon	Surface accumulation of mainly organic matter overlying mineral soil.
A horizon	Accumulation of humified organic matter mixed with mineral fraction; the latter is dominant. Occurs at the surface or below an O horizon; Ap is used for those horizons disturbed by cultivation.
E horizon	Usually underlies an O or A horizon, characterized by less organic matter and/or fewer sesquioxides (compounds of iron and aluminum) and/or less clay than the underlying horizon. Horizon is light colored due mainly to the color of the primary mineral grains because secondary coatings on the grains are absent.
B horizon	Underlies an O, A, or E horizon; shows little or no evidence of the original sediment or rock structure. Several kinds of B horizons are recognized, some based on the kinds of materials illuviated into them, others on residual concentrations of materials. Selected subdivisions are:

Bh horizon Illuvial accumulation of organic matter–sesquioxide complexes that either coat grains, form pellets, or form sufficient coatings and pore fillings to cement the horizon.

Bk horizon Illuvial accumulation of alkaline earth carbonates, mainly calcium carbonate; the properties do not meet those for the K horizon.

Bo horizon Residual concentration of sesquioxides, the more soluble materials having been removed.

Bq horizon Accumulation of secondary silica.

Bt horizon Accumulation of silicate clay that has either formed *in situ* or is illuvial; hence it will have more clay than the assumed parent material and/or the overlying horizon. Illuvial clay can be recognized as grain coatings; bridges between grains; coatings on ped surfaces or in pores; or thin, single, or multiple near-horizontal discrete accumulation layers of pedogenic origin (clay bands or lamellae). In places, subsequent pedogenesis can destroy evidence of illuviation.

By horizon Accumulation of gypsum.

Bz horizon Accumulation of salts more soluble than gypsum.

K horizon	A subsurface horizon so impregnated with carbonate that its morphology is determined by the carbonate. Carbonate coats or engulfs all primary grains in a continuous medium to make up 50% or more by volume of the horizon. The uppermost part of a strongly developed horizon commonly is laminated. The cemented horizon corresponds to some caliches and calcretes.
C horizon	A subsurface horizon, excluding R, like or unlike material from which the soil formed or is presumed to have formed. Lacks properties of A and B horizons but includes materials in various stages of weathering.

Cox and Cu horizons In many unconsolidated Quaternary deposits, the C horizon consists of oxidized C overlying seemingly unweathered C. It is suggested the Cox be used for oxidized C horizons and Cu for unweathered C horizons.

Cr horizon In soils formed on bedrock, there commonly will be a zone of weathered rock between the soil and the underlying rock. If it can be shown that the weathered rock has formed in place, and has not been transported, it is designated Cr.

R horizon	Consolidated bedrock underlying soil. It is not unusual for this and the Cr horizon to have illuvial clay in cracks; the latter would be designated Crt.

(Continued)

Flocculated

Dispersed

◀ FIGURE 10.13
Clay particles displaying flocculated and dispersed fabrics.

the load of structures, called *settlement*, can be very damaging when it occurs in excessive or variable amounts beneath a building.

The index properties of fine-grained, or cohesive, soils are somewhat more complicated than the index properties of cohesionless soils because of the influence of clay minerals. The type and amount of clay minerals are, therefore, very significant. With respect to engineering behavior, *consistency* is the most important characteristic of cohesive soils. It refers to the strength and resistance to penetration of the soil in its in-place condition. Consistency is determined by the arrangement of soil particles, particularly clay particles, in the soil, which is called the soil's *fabric*. Although many types of fabric are possible, soils in which edge-to-face contact of clay particles exists (Figure 10.13), or soils that have *flocculated* fabrics, are much stronger than soils that have the parallel arrangement of particles found in *dispersed* fabrics. Flocculated fabrics change to dispersed fabrics during *remolding*, which involves disturbance and alteration of the soil by natural processes or during various lab tests. The consistency of a soil can be determined by field tests in which the soil is evaluated in place or by lab tests on samples that have been carefully handled to avoid remolding. The unconfined compression test is often used as an indication of consistency. In practice, the relative terms *soft, medium, stiff, very stiff,* and *hard* are applied to describe consistency. The relationship between these terms, unconfined compressive strength values, and the results of simple tests made with the hand and other objects are shown in Table 10.6.

The ratio of unconfined compressive strength in the undistured state to strength in the remolded state defines the index property called *sensitivity*. The sensitivity, S_t, can be expressed as

$$S_t = \frac{\text{strength in undisturbed condition}}{\text{strength in remolded condition}} \tag{10.14}$$

Soils with high sensitivity (Table 10.7) are highly unstable and can be converted to a remolded or dispersed state very rapidly, with an accompanying drastic loss of strength. The disastrous slope movements that follow these transformations are discussed in Chapter 13.

The water content, defined in the previous section, is an important influence upon the bulk properties and the behavior of a soil. In the remolded state, the consistency of the soil is defined by the water content. Four consistency states are separated by the water content at which the soil passes from one state to another (Figure 10.14). These water content values are known as the *Atterberg limits*. For example, the *liquid limit* is the water content at which the soil-water mixture changes from a liquid to a plastic state. As the water content decreases, the soil passes into a semisolid state at the *plastic limit*, and a solid state at the *shrinkage limit*. The shrinkage limit defines the point at which the volume of the soil becomes nearly constant with further decreases in water content. The Atterberg limits can be determined with simple laboratory tests. The use of the Atterberg limits in predicting the behavior of natural, in-place soil is limited by the fact that they are conducted on

Table 10.6   Consistency and Strength for Cohesive Soils

Consistency	Shear Strength (kg/cm^2)	Unconfined Compressive Strength (kg/cm^2)	Feel or Touch
Soft	<0.25	<0.5	Blunt end of a pencil-size item makes deep penetration easily
Medium (medium stiff or medium firm)	0.25–0.50	0.50–1.0	Blunt end of a pencil-size object makes 1.25-cm penetration with moderate effort
Stiff (firm)	0.50–1.0	1.0–2.0	Blunt end of a pencil-size object can make moderate penetration (about 0.6 cm)
Very stiff (very firm)	1.0–2.0	2.0–4.0	Blunt end of a pencil-size object makes slight indentation; fingernail easily penetrates
Hard	>2.0	>4.0	Blunt end of a pencil-size object makes no indentation; fingernail barely penetrates

Source: From D. F. McCarthy, *Essentials of Soil Mechanics and Foundations,* 4th ed., © 1993 by Prentice Hall, Inc., Englewood Cliffs, N.J.

Table 10.7   Sensitivity of Clays

Type	Sensitivity Value
Nonsensitive	2–4
Sensitive	4–8
Highly sensitive	8–16
Quick	>16

remolded soils. The relationship between moisture content and consistency defined by the Atterberg limits may not be the same in soils in the undisturbed state. Therefore, the Atterberg limits are mainly used for classification rather than for the prediction of soil behavior under field conditions.

The most useful engineering classification of soils is the *Unified Soil Classification System* (Figure 10.15). This classification gives each soil type a two-letter designation. For coarse-grained soils, the first letter, either *G* for gravel or *S* for sand, refers to the dominant particle size in the soil. The second letter is either *W*, for well graded, *P*, for poorly graded, or *M* or *C*, for coarse-grained soils that contain more than 12% of silt or clay, respectively. The first letter of the designation for fine-grained soils is *M* or *C*, for silt or clay, respectively. The second letter, either *H* (high) or *L* (low), refers to the plasticity of the soil as defined in Figure 10.16. The chart is a plot of *plasticity index (PI)* against liquid limit *(LL)*. The plasticity index is defined as

$$PI = LL - PL \qquad \qquad (10.15)$$

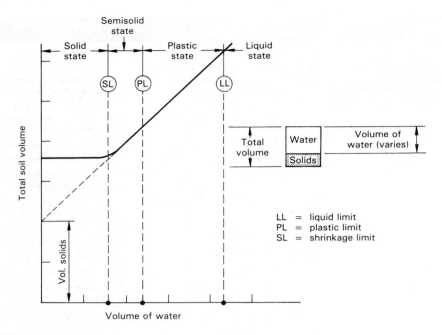

▲ FIGURE 10.14
Atterberg limits. As the volume of water (water content) in a volume of soil decreases, the soil passes through four states, separated by the Atterberg limits. *Source:* From D. F. McCarthy, *Essentials of Soil Mechanics and Foundations*, 4th ed., © 1993 by Prentice Hall, Inc., Upper Saddle River, N.J.

The difference between the liquid and plastic limits (*PI*) is a measure of the range in water contents over which the soil remains in a plastic state. The plotted position of a soil with respect to the *A-line* on the plasticity chart determines whether the soil receives the letter *H* for high plasticity or the letter *L* for low plasticity. Highly organic soils, which form a final category in the classification, are also subdivided into high- and low-plasticity types. Once a soil has been classified by the Unified system, predictions can be made of the soil's permeability, strength, compressibility, and other properties.

Shear Strength

The shear strength of a soil determines its ability to support the load of a structure or remain stable upon a hillslope. Engineers must therefore incorporate soil strength into the design of embankments, roadcuts, buildings, and other projects. The division of soils into cohesionless and cohesive types for classification is also useful for discussing strength.

In the Mohr-Coulomb theory of failure, shear strength has two components—one for inherent strength due to bonds or attractive forces between particles, and the other produced by frictional resistance to shearing movement equation (7.7). The shear strength of cohesionless soils is limited to the frictional component. When the direct shear test is used to investigate a cohesionless soil, successive tests with increasing normal stress will establish a straight line that passes through the origin (Figure 10.17). The angle of inclination of the line with respect to the horizontal axis is the angle of internal friction. Example 7.3 illustrates the determination of angle of internal friction from the results of direct shear tests. Values of the angle of internal friction are given in Table 10.8. If the soil is dense when tested, initially higher values for the angle of internal friction will be measured, but with increasing amounts of strain, the angle will decline to the approximate ranges seen in Table 10.8.

The shear strength of a cohesive soil is more complicated than that of a cohesionless material. The differences are due to the role of pore water in a cohesive soil. Most cohesive

► FIGURE 10.15
Engineering classification of soils by the Unified Soil Classification System. *Source:* Modified from D. F. McCarthy, *Essentials of Soil Mechanics and Foundations,* 4th ed., © 1993 by Prentice Hall, Inc., Upper Saddle River, N.J.

Unified Soil Classification System
(ASTM Designation D-2487)

Major division			Group symbols	Typical names
Coarse-grained soils — More than 50% retained on No. 200 sieve	Gravels — 50% or more of coarse fraction retained on No. 4 sieve	Clean gravels	GW	Well-graded gravels and gravel-sand mixtures, little or no fines
			GP	Poorly graded gravels and gravel-sand mixtures, little or no fines
		Gravels with fines	GM	Silty gravels, gravel-sand-silt mixture
			GC	Clayey gravels, gravel-sand-clay mixtures
	Sands — More than 50% of coarse fraction passes No. 4 sieve	Clean sands	SW	Well-graded sands and gravelly sands, little or no fines
			SP	Poorly graded gravels and gravel-sand mixtures, little or no fines
		Sands with fines	SM	Silty sands, sand-silt mixtures
			SC	Clayey sands, sand-clay mixtures
Fine-grained soils — 50% or more passes No. 200 sieve	Silts and clays — Liquid limit 50% or less		ML	Inorganic silts, very fine sands, rock flour, silty or clayey fine sands
			CL	Inorganic clays of low to medium plasticity, gravelly clays, sandy clays, silty clays, lean clays
			OL	Organic silts and organic silty clays of low plasticity
	Silts and clays — Liquid limit greater than 50%		MH	Inorganic silts, micaceous, or diatomaceous fine sand or silts, elastic silts
			CH	Inorganic clays of high plasticity, fat clays
			OH	Organic clays of medium to high plasticity
Highly organic soils			Pt	Peat, muck, and other highly organic soils

soils in field conditions are at or near saturation because of their tendency to hold moisture and their low permeability. When load is applied to a soil of this type, the load is supported by an increase in the pore-water pressure until pore water can drain into regions of lower pressure. At that point, soil particles are forced closer together and the strength increases, just like a cohesionless soil. Time is an important factor, however, because it takes longer for water to move out of a low-permeability material.

Strength tests for cohesive soils are usually made in triaxial cells in which the drainage of the sample can be controlled. Test conditions can allow (1) no drainage of the soil during loading; (2) drainage during an initial phase of loading, followed by failure in an undrained condition; and (3) complete drainage during very slow loading to failure. The response of the sample is different in each case. For the most basic case, in which the sample is undrained throughout the test, the results will yield a straight line on a plot of normal stress versus shear stress (Figure 10.18). The reason for this consequence is that the soil

◀ FIGURE 10.16
Chart for determining classification of fine-grained soils by the Unified Soil Classification System. *Source:* From D. F. McCarthy, *Essentials of Soil Mechanics and Foundations*, 4th ed., © 1993 by Prentice Hall, Inc., Upper Saddle River, N.J.

◀ FIGURE 10.17
Determination of shear strength from direct shear data for cohesionless soil.

Table 10.8 Values of ϕ for Cohesionless Soils	
Soil Type	**Angle ϕ, Degrees**
Sand and gravel mixture	33–36
Well-graded sand	32–35
Fine to medium sand	29–32
Silty sand	27–32
Silt (nonplastic)	26–30

Source: From D. F. McCarthy, *Essentials of Soil Mechanics and Foundations*, 4th ed., © 1993 by Prentice Hall, Inc., Upper Saddle River, N.J.

particles cannot be forced closer together without drainage of pore water and thus cannot develop greater resistance to shear failure. The shear-strength value determined in this type of test is termed *cohesion* (Chapter 7). The property of soils in the field that controls cohesion is consistency. Stiff or hard soils are compacted in their natural state and therefore have more shear strength than soft soils. The values of shear strength given in Table 10.6 are comparable to cohesion values obtained from an undrained triaxial test. Strength tests of soils that are mixtures of cohesive and cohesionless material yield failure envelopes similar to Figure 7.19.

$N_c > N_b > N_a$ (S_c, S_b, S_a represent ultimate shearing forces)

Shear intercept is the value of cohesion, c

Failure envelope

Area; A, is cross-sectional area of sample upon which the normal load, N, acts.

Shear stress, τ $\left[\tau = \dfrac{S}{A}\right]$

Normal stress, σ $\left[\sigma = \dfrac{N}{A}\right]$

▲ FIGURE 10.18
Determination of cohesion from undrained triaxial test results for cohesive soil.

EXAMPLE 10.2

A saturated, cohesive soil is tested in a triaxial cell. During the test, no drainage is allowed from the sample. At failure, the major and minor principal stresses are 2600 and 1000 psf, respectively. What is the cohesion of the sample?

Solution

Because the sample is undrained, increases in the normal stress will have no effect on the strength, and the failure envelope will be a straight line. σ_1 and σ_3 are plotted on the horizontal axis to form a diameter of a Mohr's circle plot. The failure envelope will be tangent to the top of the circle at a distance from the origin on the vertical axis equal to the radius of the circle. Thus,

$$c = r = \frac{\sigma_1 - \sigma_3}{2}$$

$$c = \frac{2600 \text{ psf} - 1000 \text{ psf}}{2} = 800 \text{ psf}$$

$\sigma_1 = 2600$ psf

$\sigma_3 = 1000$ psf

$c = 800$ psf

$r = \dfrac{\sigma_1 - \sigma_3}{2}$

$90°$

σ_3 (1000 psf) σ_1 (2600 psf)

It is evident from Table 10.6 that unconfined compressive strength is equal to twice the cohesion (shear strength) of cohesive soil. This can be shown on a plot similar to that in Example 10.2. The unconfined compressive test is one in which σ_3 is zero and σ_1 at failure is the unconfined compressive strength. When the Mohr's circle is plotted with σ_3 at the origin, it is apparent that σ_1 is the diameter of the circle and the cohesion is the radius.

When a landslide occurs, it is an indication that the stress within the soil mass exceeded the shear strength of the soil at the time of failure. The load imposed by buildings and other structures on most soils is rarely great enough to overcome the ultimate shear strength or *bearing capacity* of the soil. Case in Point 10.1 describes an example of a building that actually did cause failure of the soil upon which it was constructed.

Settlement and Consolidation

Despite the lack of large-scale bearing-capacity failures caused by the load upon the soil exerted by structures, the soil does deform under the load applied. The usual response is a volume decrease in the soil beneath the foundation. *Settlement* is the vertical subsidence of the building as the soil is compressed. Excessive settlement, particularly when it is unevenly distributed beneath the foundation, can result in serious damage to the structure (Figure 10.19).

The tendency of a soil to decrease in volume under load is called *compressibility*. This property is evaluated in the *consolidation test*, in which a soil sample is subjected to an increasing load. The change in thickness is measured after the application of each load increment

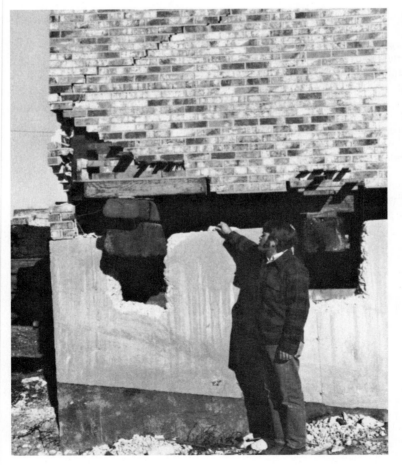

◄ FIGURE 10.19
Damage to a building by settlement. The house has been raised to its original elevation. The gap between the wall and the foundation shows the amount of settlement that has occurred. *Source:* Photo courtesy of USDA Soil Conservation Service.

▶ FIGURE 10.20

Schematic diagram of a consolidation test to determine soil compressibility. *Source:* From D. F. McCarthy, *Essentials of Soil Mechanics and Foundations,* 4th ed., © 1993 by Prentice Hall, Inc., Upper Saddle River, N.J.

(Figure 10.20). During the consolidation test, soil particles are forced closer together as in strength tests. Thus, compression of the soil causes a decrease in void ratio. Data from the consolidation test are plotted as void ratio against the log of vertical pressure applied to the sample (Figure 10.21). The slope of the resulting straight line is called the *compression index,* C_c. The compression index is an important property of a soil because it can be used to predict the amount of settlement that will occur from the load of a building or other structure. Clay-rich soils and soils high in organic content have the highest compressibility.

The compression of a saturated clay soil is a slightly different process from the compression of an unsaturated material. As we mentioned earlier, when the pores of a soil are filled with water, compression is impossible because the void volume is totally occupied. The increase in load upon the saturated clay during a consolidation test or actual field situation is initially balanced by an increase in fluid pressure in the pores of the soil (Figure 10.22). Higher fluid pressure in the clay causes the pore water to flow out of the clay toward any direction where fluid pressure is lower. Only after pore fluid is removed can the soil compress to a lower void ratio. This process is called *consolidation.* It is an important phenomenon to consider when constructing a building on a saturated clay because consolidation is very slow. The movement of pore water out of the soil is a function of the permeability, and clay has the lowest permeability of any soil. Therefore, it may take years for the soil to reach equilibrium under the load imposed. The settlement occurring during this period can be dangerous to the building. A famous example of consolidation is the Leaning Tower of Pisa (Figure 10.23). Tilting of the tower is the result of nonuniform consolidation in a clay layer beneath the structure. This ongoing process may eventually lead to failure of the tower.

▶ FIGURE 10.21

Consolidation test data plotted as decrease in void ratio versus log of vertical pressure. *Source:* From D. F. McCarthy, *Essentials of Soil Mechanics and Foundations,* 4th ed., © 1993 by Prentice Hall, Inc., Upper Saddle River, N.J.

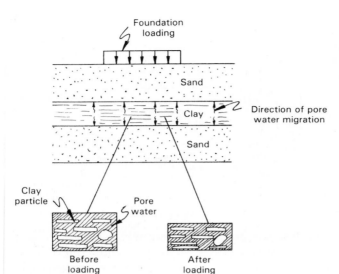

◀ FIGURE 10.22
Consolidation of saturated clay under the load exerted by a foundation. Clay particles are forced closer together after pore water is forced out of the unit.

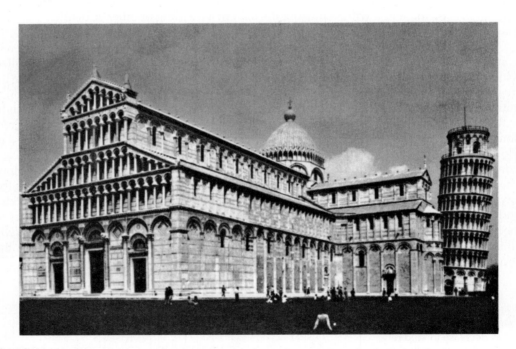

▲ FIGURE 10.23
The Leaning Tower of Pisa (tower at right), whose tilt is the result of nonuniform consolidation beneath the structure. *Source:* Photo courtesy of the Italian Government Travel Office, E.N.I.T.

Clay Minerals

Clay minerals are so important to soils engineering that we must take a closer look at their properties. Clays are good examples of *colloids*, particles so small that their surface energy controls their behavior. In particular, the electrical charge on the particle surface influences the interaction between adjacent clay particles. In Figure 10.24, the effect of the clay-surface

▶ FIGURE 10.24

Attraction of cations and water molecules to the charged surface of a clay particle.

charge upon the pore-water solution surrounding it is shown. Surface charges on clay particles are negative under most conditions. In order to balance these charges, cations from the pore-water solution and water molecules are attracted to the particle surface. The attraction of water is possible because of the polar nature of the water molecule. The positive side of the water molecule is, therefore, attracted to the negative clay surface. The negative charges on the surface of the clay particle in combination with the attracted cations and water molecules are called the *diffuse double layer*. Water molecules in diffuse double layers behave somewhat differently from water that is beyond the double layer in pore spaces. In addition, the positively charged sides of the water molecules cause repulsive forces between the double layers of adjacent clay particles.

When clays are initially deposited by settling to the bottom in a lake or the ocean, for example, both attractive and repulsive forces between clay particles are present. The relative strength of these forces determines which of the fabrics shown in Figure 10.13 will develop during sedimentation. Repulsive forces are stronger when the electrolyte content (ionic concentration) of the solution is weak. Therefore, *dispersed* fabrics are more common in clays deposited in freshwater and *flocculated* fabrics characterize seawater deposition.

The electrostatic properties of clays explain many aspects of their engineering behavior. Because of the tendency of clays to absorb water, the water content of clay soils is frequently near or above the plastic limit. In the plastic state, the strength of clay soils is relatively low. When load is applied to clay soils, clay particles can be forced closer together or forced to reorient themselves, causing a new fabric to develop. The large decreases in void ratio that accompany these adjustments to load are responsible for the high compressibility of clay. As we have seen, when pore water is expelled during consolidation, the decrease in void ratio may be a very slow process. Because of these properties, clay soils can create troublesome engineering problems. Unstable slopes and settlement of foundations are examples. Site investigations for heavy structures must not neglect the presence of clay soils at depth; even though surficial soils may have low compressibility, a clay bed in the subsurface may lead to settlement of the structure (Figure 10.22).

Case in Point 10.1
Bearing Capacity Failure in Weak Soil

A rare example of a true bearing-capacity failure of a soil beneath a foundation is provided by the Transcona, Manitoba, grain elevator (White, 1953). This structure (Figure 10.25) was built in 1913 by the Canadian Pacific Railway. During the initial filling of the elevator, the underlying soils failed along a nearly circular surface, causing a rotation of the elevator to an angle of 27° from the vertical (Figure 10.26). Despite the sinking and rotational

▲ FIGURE 10.25
The Transcona grain elevator, located near Winnipeg, Manitoba, as it appears today. *Source:* Photo courtesy of the author.

▲ FIGURE 10.26
Rotation of the grain elevator by bearing-capacity failure of the weak foundation soil. *Source:* Sketched from a photo by L. S. White.

movement, the elevator was largely undamaged. By excavating beneath the side that was rotated upward and by underpinning the structure with piles, engineers were able to restore the elevator to an upright position, where it now rests at a depth of 7 m lower than the depth of the original foundation. Years later, when testing and analysis of the soils were done, it was determined that the pressure of the fully loaded elevator exceeded the shear strength of the soil. The soils in the area are clay-rich sediments deposited in a Pleistocene glacial lake. These materials have a very low strength and high compressibility. Failures similar to the Transcona grain elevator are less likely today because of great progress in the field of soil mechanics with respect to testing and analysis of soils.

Soil Hazards

The basic engineering principles of soil behavior provide the theory for understanding the interactions between engineering structures and soils. In practice, however, the performance of soil under imposed engineering load can be quite complex. In order to identify and predict potentially hazardous soil processes, the engineer must determine the relationships and interactions between the soil and the rock units, the groundwater flow system, the climate, and the vegetation of an area. All these factors can influence soil processes that can damage existing structures or require special designs for new projects.

Expansive Soil

Surprisingly, damage to structures caused by soils that expand by absorbing water is the most costly natural hazard in the United States on an average annual basis. The annual bill for repair and replacement of highways and buildings affected by expansive (swelling) soils runs into billions of dollars. The portions of the United States underlain by expansive soils are shown in Figure 10.27.

The tendency for soils to swell can be explained by the characteristics of clay particles that we have already considered. The unbalanced electrostatic charges on clay-particle surfaces draw water molecules into the area between silicate sheets, thus forcing them apart. The cations attracted to the clay surfaces provide another factor in swelling behavior. Because of the attraction of the negatively charged clay-particle surfaces for cations, small spaces within or between clay particles may contain a higher concentration of cations than larger pores within the soil. These conditions (Figure 10.28) create an *osmotic potential* between the pore fluids and the clay-mineral surfaces. Normally, cations diffuse from a higher concentration to a lower concentration in order to evenly distribute the ions throughout the solution. In expansive soils, because ions are held by the clay particles,

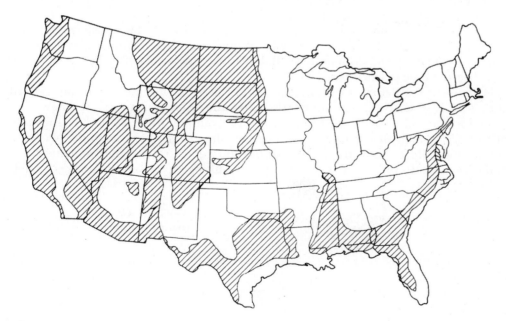

▲ FIGURE 10.27

Distribution of expansive soils in the United States. *Source:* From H. A. Tourtelot, 1974, Geologic origin and distribution of swelling clays, *Bulletin of the Association of Engineering Geologists*, 11. Used by permission of the Association of Engineering Geologists.

Clay particle

Soil pore

⊕ — Water molecule
⊕ — Cation

◄ FIGURE 10.28
Swelling of clay-rich soils. Water molecules diffuse into clay particles in order to equalize the ionic concentration between the particle and the soil pore.

water moves from areas of low ionic concentration (high concentration of water) to areas of high ionic concentration (low concentration of water) within clay particles or aggregates. This influx of water exerts pressure, which causes the clay to swell.

If a clay soil is subjected to drying conditions—for example, when evaporation is removing water from the soil near the land surface—a suction effect is exerted on the soil that causes water molecules that are not held tightly to clay particles to be drawn out into the large pores of the soil and to move upward to replace the evaporated water. This loss of water from the clay leads to shrinkage, the reversal of the swelling process.

Swelling and shrinkage of a soil can occur only with changes in water content. Therefore, the potential for damage from expansive soil is limited to the upper zone of the soil in which seasonal changes in moisture content take place. This zone is called the *soil-moisture active zone* (Figure 10.29). Below this zone, even though the soil may have the potential to shrink and swell, volume changes will not take place because the water content of the soil is constant.

Prediction of swelling behavior can be accomplished by lab tests designed to measure the pressure generated by soil being wetted. A quick estimate of swelling behavior can be made using the Atterberg limits of the soil. Table 10.9 shows a classification of swelling potential based on the plasticity index.

Damage to a structure is possible when as little as 3% volume expansion takes place (Figures 10.30 and 10.31). Failure results when the volume changes are unevenly distributed in the soil beneath the foundation. For example, water-content changes in the soil around the edge of a building can cause swelling pressure beneath the perimeter of the building, while the water content of the soil beneath the center remains constant (Mathewson et al., 1975). The type of failure resulting from this situation is called *end*

Foundation

Soil
moisture
active
zone

Water table

◄ FIGURE 10.29
The soil-moisture active zone.

Table 10.9 Correlation of Swelling Behavior with Plasticity Index	
Degree of Expansion	**Plasticity Index**
Very high	>35
High	25–41
Medium	15–28
Low	<18

Source: From R. D. Holtz and W. D. Kovacs, *An Introduction to Geotechnical Engineering,*
© 1981 by Prentice Hall, Inc., Upper Saddle River, N.J.

▲ FIGURE 10.30
Structural damage to a group of apartment buildings in San Antonio, Texas, caused by shrink-swell
behavior of expansive soils. *Source:* Photo courtesy of USDA Soil Conservation Service.

lift (Figure 10.32). The factors that could cause end lift include excessive lawn or shrub
watering beside the building and insufficiently long drain spouts that do not carry
drainage from the roof far enough away from the structure. If a house lot is not graded
properly, the land may slope toward the house. This would concentrate drainage at the
edge of the building and high infiltration could occur. The opposite of end lift is *center
lift* (Figure 10.32), where swelling is focused beneath the center of the structure or
where shrinkage takes place under the edges. Either process causes the foundation to
bulge beneath the center. A possible cause of soil shrinkage near the edges of a structure
is the close proximity of trees. Certain trees remove large quantities of water from the
soil by the process of *transpiration*; soil shrinkage in the vicinity of the roots is the result.

If highly expansive soils are recognized before construction, special foundation designs can be used. One method involves support of the building by using a grade beam

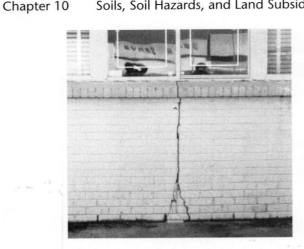

◄ FIGURE 10.31
Crack in a wall of an office building caused by swelling clays, San Antonio, Texas. The office building and adjacent shopping plaza were demolished because of structural damage by swelling soils.
Source: Photo courtesy of the author.

End lift

◄ FIGURE 10.32
Two types of structural failure caused by swelling soils. *Source:* From C. C. Mathewson, J. J Castleberry, and R. T. Lytton, 1975, Analysis and modeling of home foundations of expansive soils in central Texas, *Bulletin of the Association of Engineering Geologists*, 12: 275–302. Used by permission of the Association of Engineering Geologists.

Center lift

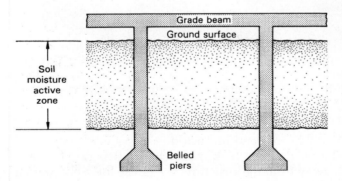

Grade beam
Ground surface
Soil moisture active zone
Belled piers

◄ FIGURE 10.33
Belled pier and grade beam foundation utilized to prevent damage from expansive soils. *Source:* From B. M. Das, *Principles of Foundation Engineering*, © 1984 by Wadsworth, Inc., Belmont, Calif.

resting upon belled piers that extend below the base of the soil-moisture active zone (Figure 10.33).

Hydrocompaction

A different type of soil problem is presented by certain soils in arid regions. When water and load are applied to these soils, their structure collapses to a more dense state. This type of behavior, called *hydrocompaction*, is the result of the origin and history of the soils. Sediments

► FIGURE 10.34

Loose, unsaturated soil structure maintained by pore
water and clay bonds.

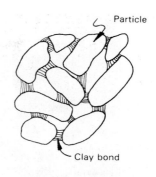

deposited by wind or water usually develop a very loose, open structure. Soils susceptible to hydrocompaction, which are predominantly composed of silt and sand grains rather than clay particles, exist in an unsaturated state. The low-density structure is maintained by weak bonds formed by water, clay, or soluble precipitants that bridge pores between particles (Figure 10.34). When the pores are only partially filled with water, the water exerts tensional forces that tend to hold adjacent particles together. In addition, small amounts of clay or soluble precipitants (such as gypsum) between particles act as a temporary cementing agent. When water saturates the soil, both the water and clay bonds are destroyed. Pore water loses its tensional effect when the pores become saturated, and clay bonds are suspended in the solution upon wetting. The soil will then collapse to a denser structure under its own weight or the weight of a structure built upon the soil. Saturation is often caused by lawn watering around a foundation and leakage of water lines, storm sewers, and canals. When the collapse of the soil occurs, settlement of foundations or rupture of utility lines is usually the result.

Liquefaction

The term *liquefaction* is commonly applied to the conversion of saturated sand or silt to a liquid state under rapid or cyclic stresses of the type that might be caused by earthquakes, vibrations, or explosions. Normally, a sand would respond to increased stress or loading by the expulsion of pore water associated with a decrease in void ratio. Water can be driven out readily because of the high hydraulic conductivity of sand. Under rapid or cyclic loading, however, the increased stress is transferred to the pore water as the particles are forced closer together. If the pore-water pressure is great enough to suspend the particles within the pore fluid, total loss of shear strength occurs and fluid behavior is the result. An excellent example of liquefaction occurred in the Niigata, Japan, earthquake of 1964 (Figure 10.35). Liquefaction of a saturated sand beneath the foundations of the apartment buildings was responsible for a temporary loss of shear strength in the soils supporting the buildings. When this happened, the buildings rotated or sank into the fluidized soil. Little structural damage resulted from these movements, and the buildings were later jacked upright and new foundations were constructed.

A chilling example of liquefaction has only recently been deduced from historical records of the effects of an earthquake on June 7, 1692 in the town of Port Royal, Jamaica. Although Port Royal was more civilized in 1692, it still had not lived down its reputation from several decades earlier as the bawdy and boisterous home port of pirates and buccaneers, including the famous Henry Morgan. The trouble with Port Royal was its construction upon a narrow sandy spit of land bordering the mainland (Figure 10.36). The saturated, low-density sands beneath the town were ideal for liquefaction. During the earthquake, the town was almost completely devastated, including a substantial portion of the town that

▲ FIGURE 10.35
Rotation of buildings due to liquefaction of saturated sand during the 1964 Niigata, Japan, earthquake. *Source:* Photo courtesy of the U.S. Geological Survey.

▲ FIGURE 10.36
Approximate area of Port Royal, Jamaica, showing the portion of the city lost by liquefaction and subsidence below sea level during the earthquake of 1692. *Source:* David S. Brumbaugh, *Earthquakes: Science and Society*, 1st ed., © 1999, p.145. Reprinted by Permission of Pearson Education, Inc., Upper Saddle River, N.J.

simply sank into the liquefying sands and disappeared below sea level. A 17th-century newspaper account of the tragedy includes a graphic illustration of the effects and eyewitness descriptions of the fate of some of the 2500 people lost in the earthquake (Figure 10.37). People, along with buildings and other structures, sank into the liquefying sand. Some

▲ FIGURE 10.37

A 17th-century newspaper account of the earthquake and some of the grisly effects of liquefaction. The descriptions of events at the locations of the letters on the figure include the following: "A. The Houses Falling ... H. The Ground rolling under the Minister's Feet ... T. The Sea washing the dead Carkasses out of their Graves and Tombs, and dashed to pieces by the Earthquake ... V. People swallow'd up in the Earth, several as high as their Necks, with their heads above Ground. W. The Dogs eating of Dead Men's Heads."

were lost below, but others were able to paddle their way upward and survived. Some of the most unfortunate were partially trapped in the sand when the earthquake stopped. The sand immediately regained its strength, crushing the partially buried citizens, some of them with only their heads projecting above land surface, later to be eaten by dogs scavaging through the remains of the town.

Land Subsidence

Subsidence describes several related localized or areally widespread phenomena associated with sinking of the land surface. The vertical movements involved range from sudden collapse to slow, gradual declines in surface elevation. In a general way, subsidence can be subdivided into types caused by (1) removal of subsurface fluids, (2) drainage or oxidation of organic soil, and (3) surface collapse into natural or excavated subsurface cavities.

Subsidence attributed to withdrawal of fluid is a widely distributed occurrence in the United States and other countries. Although it is most commonly heavy pumping of groundwater that leads to this type of subsidence, depletion of oil and gas reservoirs has been responsible for subsidence in some areas. A prominent example is the Wilmington oil field in Long Beach, California, in which maximum elevation decreases of about 9 m were attributed to the withdrawal of petroleum. Direct damage to structures and indirect damage because of increased flooding proved very costly to the city. Groundwater withdrawals have accounted for subsidence over thousands of square kilometers of land in the western United States and other areas. The problem occurs in agricultural areas like the San Joaquin Valley in California, where the water is used for irrigation, as well as in urban areas worldwide, including Las Vegas, Nevada; Houston, Texas; Mexico City, Mexico; and Venice, Italy, where the groundwater is withdrawn for municipal water supplies. The subsidence bowl (area undergoing subsidence) in Houston is one of the largest in the United States. The subsidence is correlated with accelerated groundwater pumping since the 1930s, although petroleum withdrawals may also be significant, at least on a local basis.

Subsidence is most prevalent in geologically young, poorly lithified sequences of saturated sands and clays. Groundwater is pumped from the sandy units, but these materials have low compressibilities. According to the *Principle of Effective Stress*, which is further discussed in Chapter 11, pumping of fluids from sandy, permeable (aquifer) material causes some compaction because when the hydraulic pressure is lowered by pumping water from the pore spaces, the pressure on the sand grains increases and they tend to be forced closer together. This effect causes compaction of the aquifer and potential land subsidence, but if water levels are allowed to recover, the compaction is reversible, or elastic (Figure 10.38). However, the situation is different in the fine-grained units interbedded with the aquifers.

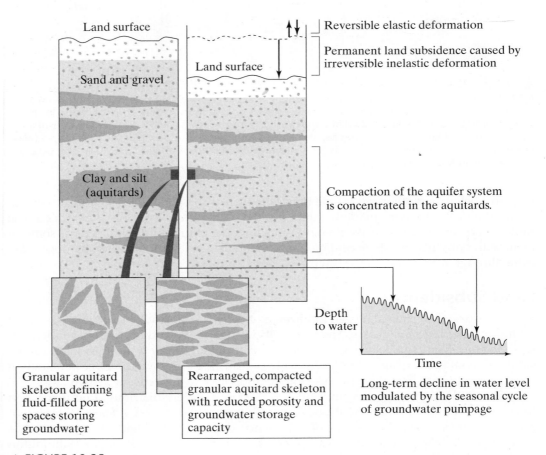

▲ FIGURE 10.38
Land subsidence caused by withdrawal of fluids. As water is pumped from the sand beds, they compact slightly, but recover elastically upon cessation of pumping and the rise of groundwater levels. Irreversible compaction of compressible fine-grained beds in the sequence accounts for most of the land subsidence. *Source:* From U.S. Geological Survey, *Land Subsidence in the United States*, Circular 1182.

The lowered hydraulic pressure in the sands causes the fine-grained beds to slowly compact as pore water is drained into the more permeable beds above and below. The process is closely analogous to a consolidation test. The fine-grained units compact irreversibly when fluids are removed, and because they are more compressible, most of the ultimate subsidence can be accounted for by consolidation of silt and clay beds. This irreversible process is slow because of the low permeability of the materials.

Subsidence caused by groundwater pumpage for irrigation has been especially problematic in the semi-arid agricultural basins of California. The Santa Clara Valley, better known in recent years as "Silicon Valley," is a classic case. The valley, which extends southward from San Francisco Bay, developed as a center for production of irrigated fruits and vegetables in the early part of the 20th century. Subsidence, which reached its peak in the late 1960s (Figure 10.39), became a major problem. The elevation of downtown San Jose declined from 98 feet in 1910 to 84 feet in 1995. The causes of this subsidence are obviously related to groundwater pumpage (Figure 10.40). As pumpage increased in the first half of the century, water levels declined and land subsidence accelerated. The response by water managers was to import surface water through an elaborate system of reservoirs and aqueducts that could be used for artificial recharge. These measures raised the water levels in basin aquifers and slowed the rate of subsidence to a very small amount. The effects of land subsidence included the inundation of 117 acres of previously dry land by San Francisco Bay and the necessity to maintain a gradient on stream channels so that water would actually flow to the bay. The cost of the subsidence, including the construction of levees around the south side of San Francisco Bay, raising railroad and road grades, construction of sewage pumping stations, replacing wells that collapsed due to subsidence, and many other measures, has been estimated at $300 million in 1998 dollars. Maintenance costs will continue into the future.

A similar scenario, but even more severe, occurred in the San Joaquin Valley, which lies east of the Sierra Nevada Mountains and south of the Santa Clara Valley. This much larger

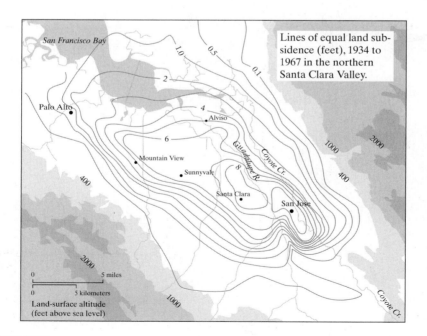

▲ FIGURE 10.39
Contours of equal amount of subsidence of the land surface in the Santa Clara Valley, northern California. *Source:* From U.S. Geological Survey, *Land Subsidence in the United States,* Circular 1182.

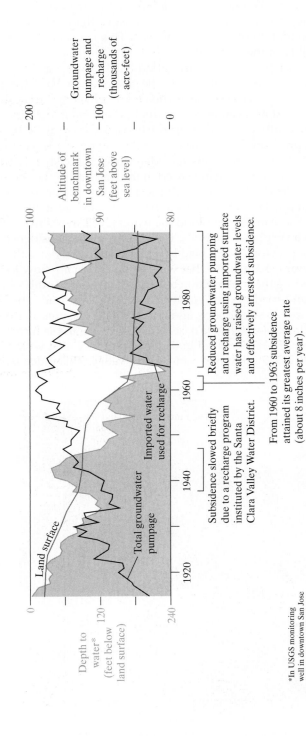

▲ FIGURE 10.40

Graph of total groundwater pumpage, depth of water table levels below land surface, and land surface elevation in San Jose, California. Subsidence reached its maximum rates in the 1960s, which necessitated artificial recharge to the affected aquifers. As groundwater levels rose, the rate of subsidence became much more gradual. *Source:* From U.S. Geological Survey, *Land Subsidence in the United States*, Circular 1182.

▶ FIGURE 10.41
Subsidence in the San Joaquin Valley, California, caused by withdrawal of groundwater. *Source:* Photo courtesy of Thomas Holzer.

valley experienced subsidence that reached 28 feet at its greatest point (Figure 10.41). The subsidence in this valley has been described as the largest human alteration of the earth's surface.

Accompanying subsidence in many locations are ruptures of the land surface. These ground failures take the form of fissures as long as several kilometers, in which the surface cracks or separates laterally (Figure 10.42), or faults, in which vertical displacement across the rupture occurs. More than 160 faults in the Houston-Galveston area have been mapped and many have also been discovered in the Las Vegas Valley. Leveling surveys have shown that recent movement across the faults corresponds to the long-term trends of groundwater pumpage. Surface fissures often occur along fault zones because different amounts of subsidence occur on opposite sides of the fault (Figure 10.43). The differential subsidence may

▲ FIGURE 10.42
Fissures formed during land subsidence in southern Arizona. *Source:* Photo courtesy of Thomas Holzer.

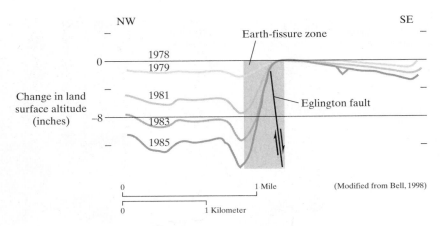

▲ FIGURE 10.43
Cross section of a fault zone in Las Vegas Valley. Subsidence caused by groundwater withdrawal is greater on one side than the other and the resulting tensional stresses lead to fissures at the ground surface. *Source:* From U.S. Geological Survey, *Land Subsidence in the United States,* Circular 1182.

come about because the fault acts as a partial barrier for groundwater flow. Thus, one side of the fault is affected more by groundwater withdrawals than the other side and tensional stresses are created between the two sides. Surface fissuring in urban areas is an expensive hazard because the cracking and tilting of the land surface can damage foundations, roads, and utilities.

Subsidence by a different mechanism occurs when highly organic soils are drained during land-use conversion to agriculture or construction. Soils of this type, frequently termed *peat* or *muck*, are formed in marshes, bogs, and other poorly drained settings with water tables near the land surface. The high water table inhibits oxidation of organic matter, so this material accumulates in the soil. Characteristics of the soils include dark color, low density, and high compressibility. When drained, organic soils are exposed to oxidizing conditions as the water table drops. The resulting decomposition of the organic content, in combination with some physical compaction due to the removal of pore fluids, leads to the gradual subsidence of the land surface. Damage to buildings can be caused by settlement of the structure if it bears upon the organic soils. If the building is constructed on foundations that bear upon stronger materials below the organic soils, subsidence of the soil around the structure can leave a gap between the floor slab and the ground surface (Figure 10.44 and Case in Point 10.2).

The Everglades, South Florida's great freshwater wetland system, has been irreparably damaged by drainage and subsidence of the exposed peat soils. About 50% of the original wetland has been drained with a system of canals so the land could be used for agriculture or development. Prior to drainage, the wetlands was essentially a huge, slow-moving river of surface water draining to the Gulf of Mexico. Subsidence in the agricultural areas (Figure 10.45) has reversed the gradient of the land surface. Current attempts to restore portions of the original wetlands by the federal government will never be able to reestablish southerly flow through the wetlands. Instead of the original "saw grass prairie" ecosystem, lakes will develop because the land no longer slopes uniformly to the south as it did before subsidence. Continuing oxidation of the drained wetland soils will eventually render them unfit for agriculture as the soils gradually shrink and disappear.

The third major type of subsidence is associated with the collapse of overlying materials into large underground cavities. Although the processes that cause this type of subsidence generally occur in bedrock below the soil zone, collapse of the surface soils

► FIGURE 10.44
Subsidence of peat and muck soils around a house built on pilings. The land surface around the house has subsided about 0.6 m in 2 years. The porch and steps have fallen off. *Source:* Photo courtesy of USDA Soil Conservation Service.

► FIGURE 10.45
Land subsidence in the vast Everglades wetlands complex of South Florida has made true restoration of the original surface-water flow system and ecosystem impossible. *Source:* From U.S. Geological Survey, *Land Subsidence in the United States,* Circular 1182.

into the voids below is the usual manifestation. The cavities may be excavated or natural. Subsidence over underground coal mines constitutes the most common example of subsidence over excavated cavities. This process has affected about 8000 km^2 in the United States, mostly in the eastern part of the country. The effects of subsidence depend upon the method of mining and the thickness of the overburden. When the coal is entirely removed from a seam during mining, surface subsidence is usually contemporaneous with mining. A basin-shaped depression commonly develops over the mined-out area if the coal seam is deeper than about 30 meters. Alternatively, columns or *pillars* of coal may be left in place to provide support for the mine roof. Subsidence above mines of this type may take place many years later after the mine is abandoned. Unfortunately, buildings, or even towns, have been built unknowingly over abandoned underground mines. The modes of subsidence are shown in Figure 10.46. Two of the most common surficial impacts of subsidence over abandoned mines are *sinkholes* or *troughs*. Sinkholes (Figure 10.47) are circular pits formed by collapse of the surface soil into the mine voids when the coal seam is fairly shallow. Troughs are larger subsidence features that develop over deeper mines.

Natural cavities form in rock primarily by dissolution of the rock by circulating groundwater. Limestone caves are the most common example of solution cavities, although other soluble rock types behave similarly. In limestone terrains, also known as *karst* topography, surface depressions, called *sinkholes*, or *dolines*, can form in several different ways (Figure 10.48). If limestone is exposed at or near the surface, depressions can form gradually by dissolution of the near-surface rock and transport of the residual soil into cavities below the surface. These depressions are known as *solution* sinkholes. Alternatively, surface limestone can collapse into large voids dissolved into the rocks below, producing a *collapse* sinkhole. Similar types of sinkholes also form in areas in which the limestone is overlain by sandy or clayey sediment. *Cover subsidence* sinkholes develop more gradually

▲ **FIGURE 10.46**
Mechanisms of subsidence associated with subsidence over abandoned underground coal mines.
Source: From R. E. Gray and R. W. Bruhn, 1984, Coal mine subsidence—eastern United States, in *Man-Induced Land Subsidence*, T. L. Holzer, ed., Geological Society of America, Reviews in *Engineering Geology*, vol. 6.

▲ FIGURE 10.47
Sinkholes formed by subsidence over abandoned coal mines, western North Dakota. *Source:* Photo courtesy of North Dakota Public Services Commission.

▶ FIGURE 10.48
Four types of sinkholes that form by dissolution of limestone. The cover subsidence and cover collapse sinkholes develop by the gradual or sudden transport of sediment overlying the limestone into solution cavities below. (a) solution sinkhole, (b) collapse sinkhole, (c) cover subsidence sinkhole, and (d) cover collapse sinkhole.

by washing of the overlying soil into solution cavities in the limestone below. *Cover collapse* sinkholes often form suddenly by the failure and sinking of the overlying soil into large voids or caverns below.

Surface subsidence, leading to the development of sinkholes, often occurs in dry periods when water tables are declining or when water tables are lowered artificially by pumping. Groundwater below the water table provides a measure of support for the sediment above a limestone cavern by the hydrostatic pressure exerted by the fluid. When the water table drops, this support is lost and surface collapse is initiated (Figure 10.49). The Winter Park, Florida, sinkhole (Figure 10.49), which formed in 1981, provides a spectacular example of urban disruption by subsidence over a cavity in soluble rock. The collapse occurred over a period of several days with no prior warning.

◄ FIGURE 10.49
The Winter Park, Florida, sinkhole as it appeared just after formation in 1981. *Source:* Photo courtesy of Jammal & Associates.

Case in Point 10.2
Land Subsidence in New Orleans

With respect to construction and development, the geologic setting of New Orleans is one of the most challenging in the world. Its major problems stem from its location on the low-lying Mississippi Delta. Despite being 75 km from the mouth of the river, the average elevation of the city is only 0.4 m above the elevation of the Gulf of Mexico. It is of more than slight concern that periodic floods of the river rise to as much as 6.5 m above the Gulf of Mexico through the heavily diked urban area. As tragically illustrated by Hurricane Katrina in September, 2005, New Orleans is highly vulnerable to hurricanes travelling across the Gulf of Mexico.

As if those problems were not enough, the foundation conditions of the city are extremely poor, particularly throughout the extensive areas of swamp and marshland occupied by urban development. The general subsurface conditions are shown in Figure 10.50. The Holocene deposits are composed for the most part of fine-grained deltaic and fluvial sediments. In cypress swamps and other low-lying areas, these materials are rich in organics and highly compressible. Large, heavy structures are built on piles that are driven into

▲ FIGURE 10.50
Block diagram showing surface and subsurface geological conditions of New Orleans, La.
Source: From C. R. Kolb and R. T. Saucier, 1982, Engineering geology of New Orleans, in *Geology Under Cities*, R. F. Legget, ed., Geological Society of America, Reviews in *Engineering Geology*, vol. 5.

the more competent Pleistocene sediments that usually lie within 25 m of the surface. Smaller structures built in former swamp areas rely on shorter timber piles limited to the Holocene sediments.

To proceed with expansion of the city into marsh areas, drainage of the surface soils and lowering of the water table was carried out. This led to rapid subsidence because the organic rich soils occupy one-half or less the volume in a dry condition compared with their natural state. Lowering of the water table also initiates oxidation and decomposition of the soils, with further, slow subsidence. For single-family homes and other lightly loaded structures, the consequence is a sinking land surface relative to the house, which rests on piles. Left untouched, it is possible to see under the entire foundation slab. The efforts to deal with the problem at one home are shown in Figure 10.51. As the years went by, the homeowners simply added steps in increments to reach the gradually sinking front yard. Total subsidence at the house is 0.81 m. Subsidence up to 2 m has been documented in other areas.

▲ FIGURE 10.51
Subsidence of organic soils in New Orleans after drainage. The photo on the left shows the house built on pilings just after construction in 1941. The photo on the right shows the same house in 1992. Front-porch steps were added over the years as the ground surface subsided. *Source:* Photo courtesy of J. O. Snowden.

Summary and Conclusions

The term *soil* has different meanings to various groups of earth scientists and engineers. Soil scientists are concerned with material produced by pedogenic processes that can sustain plant growth. Soils engineers consider soil to be any transported sediment or residual material that can be excavated with hand tools.

Pedogenic processes acting upon rock or sediment produce recognizable horizons. The main horizons include the A horizon, the zone of high organic-matter content; the E horizon, a zone of eluviation; the B horizon, a zone of illuviation of clay and metal oxides; and the C horizon, a transition zone between highly altered and nonaltered parent material.

The development of horizons in the soil profile is controlled by the five soil-forming factors: *Parent material* and *topographic position* (relief) set the stage for the action of *climate* and *organisms* over a period of *time*. Of these factors, climate is dominant. The possible variations and interactions between the soil-forming factors can produce a vast range of soil types. As a result, pedogenic classification of soils has been difficult. A modern classification system based entirely on measurable soil properties has replaced an older, inadequate system. Because of the complexity of the new system, some terms and concepts from the older system are still in use.

The objective of engineering soils classification is to predict soil behavior in an engineering project. The index properties, which are used to classify soils, can be correlated with the strength and compressibility of the soil. For cohesionless soils, the relevant index properties are grain-size distribution, grading, in-place density, and relative density. Cohesive soils are more difficult to characterize because of the interaction between clay minerals and water. The index properties used include clay content, consistency, water content, sensitivity, and the Atterberg limits. The index properties are used to classify the soil in the Unified Soil Classification System.

The performance of a soil as a foundation support material is dependent upon its shear strength and compressibility. The bearing capacity of the soil is rarely exceeded by a structure, as it was beneath the Transcona grain elevator. Even though a soil may not fail, damage to structures will be experienced if settlement takes place. Settlement is the subsidence of a foundation because of compression of the underlying soil. Consolidation is the time-dependent compression of clay soil as pore water is slowly expelled from the loaded soil.

Many of the engineering problems associated with clay can be explained by its colloidal properties. The surface charge of clay particles attracts water and cations from the pore solution. The attraction or repulsion of adjacent particles during deposition determines the initial fabric. During consolidation, reorientation of particles often occurs. The low strength of clay is indicative of high Atterberg limits and high water contents.

Engineering projects must often be built in areas where hazardous soil processes are active. The tendency of clay soils to absorb water leads to volume expansion that can be very damaging to structures. Within the soil-moisture active zone, changes in water content of the soil will lead to swelling or shrinking. Land-use changes are often responsible for this behavior. Construction and landscaping practices that promote swelling include improper lot grading and excessive lawn watering. If highly expansive soils are present, special foundation designs may be necessary.

Hydrocompaction is a type of soil collapse that takes place when arid-region soils are saturated and loaded. Saturation destroys soil moisture or clay bonds that impart temporary strength to the soils. When these bonds are removed, the soil collapses to a more dense state.

Subsidence causes numerous engineering problems due to the collapse or sinking of the land surface. These phenomena originate from the withdrawal of fluids from the soil, oxidation, or drainage of organic soils and the subsequent collapse of surface material into subsurface cavities.

Problems

1. Why is there confusion or disagreement over the use of the term *soil*?

2. Summarize the processes that lead to the formation of soil horizons.

3. Under what conditions do soils become mineral deposits?

4. Explain the process of soil swelling.

5. Under what conditions are cohesionless soils susceptible to hydrocompaction?

6. Using the volumetric relationships in soils presented in Figure 10.9, show that

$$e = \frac{n}{1 - n}$$

7. A sample of undisturbed soil has a dry weight of 110 lb and a volume of 1 ft^3. If the specific gravity of the soil solids is 2.65, determine the void ratio and the porosity.

8. A container with a volume of 0.0084 m^3 is filled with 13.62 kg of dry sand. The container is then filled carefully with water so that the condition of the sand is not changed. When filled, the total weight (mass) of the soil and water is 16.98 kg. What is the void ratio of the sand, and what is the specific gravity of the soil particles?

9. An undisturbed sample of clay has a volume 0.52 ft^3. The wet weight is 64 lb and the dry weight is 50 lb. If the specific gravity of the clay particles is 2.66, find the water content, void ratio, and degree of saturation.

10. A sample of saturated clay taken from an excavation has a volume of 500 cm^3. If the sample has a wet weight of 810 g and a dry weight of 500 g, what is the specific gravity of soil solids?

11. How does the engineering classification of soils differ from the engineering classification of rocks?

12. What are the Atterberg limits, and what is their usefulness?

13. Give the USCS symbol for the following soils.
    (a) 40% of a sample passes the No. 200 sieve; 30% of the coarse fraction passes the No. 4 sieve; contains significant silt fraction.
    (b) 4% passes the No. 200 sieve; 60% of the coarse fraction passes the No. 4 sieve; little or no fines; poorly graded.
    (c) 96% passes the No. 200 sieve; PI—40%; LL—65%.

14. Give some examples of the significance of the properties of clay minerals in soils engineering.

15. What soil characteristics influence compressibility?

16. Why is consolidation a time-dependent process?

17. A saturated cohesive soil is tested in an undrained strength test in a triaxial cell. The major and minor principal stresses at failure are 1.5 kg/cm^2 and 0.44 kg/cm^2, respectively. Determine the cohesion and the unconfined compressive strength of the material. What is the consistency classification of the soil?

18. What geological conditions are likely to lead to land subsidence upon groundwater withdrawals?

References and Suggestions for Further Reading

BIRKELAND, P. W. 1999. *Soils and Geomorphology,* 3rd ed. New York: Oxford University Press.

BRADY, N. C., and R. R. WEIL. 2002. *The Nature and Property of Soils,* 13th ed. Upper Saddle River, N.J.: Prentice Hall, Inc.

BLOOM, A. L. 1998. *Geomorphology: A Systematic Analysis of Late Cenozoic Landforms,* 3rd ed. Upper Saddle River, N.J.: Prentice Hall, Inc.

BRUMBAUGH, D. S. 1999. *Earthquakes, Science and Society.* Upper Saddle River, N.J.: Prentice Hall, Inc.

CLARK, G. R., II. 1995. Swallowed up. *Earth,* April 1995, 34–41.

DAS, B. M. 1984. *Principles of Foundation Engineering.* Monterey, Calif: Brooks/Cole Engineering Division.

GRAY, R. E., and R. W. BRUHN. 1984. Coal mine subsidence—eastern United States, in *Man-Induced Land Subsidence,* T. L. Holzer, ed. Geological Society of America, Reviews in *Engineering Geology,* 6:123–149.

HOLTZ, R. D., and W. D. KOVACS. 1981. *Introduction to Geotechnical Engineering.* Upper Saddle River, N.J.: Prentice Hall, Inc.

HOLZER, T. L. 1984. Ground failure by ground-water withdrawal from unconsolidated sediment, in *Man-Induced Land Subsidence,* T. L. Holzer, ed. Geological Society of America, Reviews in *Engineering Geology,* 6:67–101.

JUDSON, S., M. E. KAUFFMAN, and L. D. LEET. 1987. *Physical Geology,* 7th ed. Upper Saddle River, N.J.: Prentice Hall, Inc.

KOLB, C. R., and R. T. SAUCIER. 1982. Engineering geology of New Orleans, in *Geology Under Cities,* R. F. Legget, ed. Geological Society of America, Reviews in *Engineering Geology,* 5:75–93.

MATHEWSON, C. C., J. P. CASTLEBERRY, and R. T. LYTTON. 1975. Analysis and modeling of the performance of home foundations of expansive soils in central Texas. *Bulletin of the Association of Engineering Geologists,* 12:275–302.

MCCARTHY, D. F. 1993. *Essentials of Soil Mechanics and Foundations,* 4th ed. Upper Saddle River, N.J.: Prentice Hall, Inc.

PECK, R. B., and F. G. BRYANT. 1953. The bearing capacity failure of the Transcona elevator. *Geotechnique,* 3:201–208.

SOIL SURVEY STAFF. 1960. *Soil Classification, A Comprehensive System—7th Approximation.* Washington, D.C.: U.S. Department of Agriculture, Soil Conservation Service.

TOURTELOT, H. A. 1974. Geologic origin and distribution of swelling clays. *Bulletin of the Association of Engineering Geologists,* 11:259–275.

U.S. GEOLOGICAL SURVEY. 1999. *Land Subsidence in the United States,* Circular 1182.

WHITE, L. S. 1953. Transcona elevator failure: eyewitness account. *Geotechnique,* 3:209–214.

11 CHAPTER

Groundwater

The importance of groundwater to our society has never been greater than it is today. There is no more fundamental natural resource than freshwater; we depend on it daily for drinking, sanitation, agriculture, industry, and recreation. In the past, most of our needs could be met by such easily accessible surface-water sources as rivers, lakes, and reservoirs. Those days have passed however; the potential for expansion of surface-water sources is limited. Groundwater is our only alternative for obtaining large amounts of freshwater at a reasonable cost. Currently, nearly one-half the population of the United States uses groundwater as a drinking water source; soon, one-half the total water usage will come from groundwater.

The steadily increasing use of groundwater has raised important concerns about this resource. First, we must learn to evaluate and manage groundwater reservoirs so that they are not mined to exhaustion like ore deposits or other nonrenewable resources. Second, we have realized that groundwater is not a pristine substance that exists in total isolation from activities taking place on the land's surface. Instead, it is now known that past waste-disposal practices as well as the storage and handling of hazardous materials have contaminated groundwater supplies in many locations. Extensive governmental and private efforts are now underway to clean up the most severe contaminant sites and to protect the vast majority of unaffected groundwater reservoirs. Contamination and remediation in the subsurface environment is the subject of Chapter 12.

The occurrence and movement of groundwater were once thought to be mysterious and unpredictable. Even today, misconceptions about groundwater are held by the general public, the media, and even some geologists and engineers. A thorough understanding of groundwater is critical for engineers because subsurface fluids are of major importance in civil, environmental, mining, and petroleum engineering.

Groundwater Flow

Darcy's Law

A 19th-century French engineer named Henri Darcy laid the groundwork for the modern study of groundwater with experiments involving the flow of water through a column filled with sand. These experiments established that the volumetric flow rate of water through saturated sand was proportional to the energy gradient, or the loss of energy per unit length of flow path. A schematic of Darcy's apparatus is illustrated in Figure 11.1. To express *Darcy's Law* in a commonly used manner, we will define the *specific discharge*, *v*, as the volumetric flow rate, *Q*, measured in cubic meters per second or similar units, divided by the cross-sectional area of the flow tube *A*. Thus,

$$v = \frac{Q}{A} \tag{11.1}$$

The form of energy involved in the flow process is the mechanical energy possessed by the fluid at each point in its flow path. This energy is the sum of three components: elevation, or position (potential energy); movement (kinetic energy); and pressure. When dealing with groundwater, we can neglect the kinetic energy contribution because of the very slow velocities of groundwater flow. The remaining components, elevation and pressure, need not be evaluated individually, but instead can be measured together by the level to which water at any point in the flow system will rise above an arbitrary datum. This parameter, as indicated in Figure 11.1, is called *hydraulic head*. The thin tubes through which water rises to measure the head are called *manometers*.

Also shown in the diagram are the two components of hydraulic head. Elevation head is the distance above the datum to the level of the manometer intake, and pressure head is the height of the rise of water in the tube above the intake. Notice that the elevation above the datum, or head, in the manometers decreases in the direction of flow. This is a consequence of the loss of mechanical energy along the flow path as it is converted to heat

◀ FIGURE 11.1
Schematic diagram of a lab apparatus that illustrates the parameters involved in Darcy's Law.

through friction between the fluid and the sand grains and also because of friction between water molecules in the fluid. Darcy's Law requires evaluation of the *hydraulic* (energy) *gradient*, and this is expressed as

$$I = \frac{h_1 - h_2}{L} \tag{11.2}$$

where

I = is the hydraulic gradient
h_1 and h_2 = are head values at the points where the manometers are inserted into the flow system
L = is the distance between the manometers measured in the flow direction

A final parameter is required for a complete expression of Darcy's Law, and that is the constant of proportionality between the specific discharge and the hydraulic gradient. With this term, K, Darcy's Law can be stated as

$$\frac{Q}{A} = v = -K\frac{h_2 - h_1}{L} \tag{11.3}$$

or, in differential form,

$$v = -K\frac{dh}{dl} \tag{11.4}$$

The minus sign simply indicates that groundwater flow is a mechanical process with an irreversible loss of mechanical energy, or head, in the direction of flow.

The constant K, known as the *hydraulic conductivity*, is a very important parameter in groundwater hydrology. It is related to intrinsic permeability, the ability of a porous medium, like the sand in the flow tube, to transmit a fluid under a given hydraulic gradient equation (7.3). A distinction is often made between the terms *hydraulic conductivity* and *permeability* because hydraulic conductivity is defined to include the properties of the fluid as well as the properties of the medium, whereas permeability is restricted to the properties of the medium. It is easy to visualize that more water would flow through a tube containing gravel than one filled with silt or clay under the same hydraulic gradient. It is also true, however, that more water than molasses would flow through sand, again with a constant hydraulic gradient. Therefore, it is important to remember that the proportionality constant in Darcy's Law, K, encompasses both the characteristics of the fluid and the properties of the medium. The combination of the two into one constant is convenient in groundwater work because the density and viscosity of water in most near-surface groundwater reservoirs do not vary greatly. Hydraulic conductivity can then be assumed to represent the permeability of the units through which the groundwater flows.

Darcy's Law under Field Conditions

One of the reasons Darcy's Law is so useful is that it can be directly applied to field situations. The device used to measure head in the field is called a *piezometer* (Figure 11.2); as in a manometer, the head measurement depends on the level to which water in a tube rises above a datum. In this case, the datum used is sea level and the tube is a pipe inserted into a hole drilled into the ground, which is much like a normal water well. A piezometer differs from a well in several respects, however. A well usually contains a water intake section, or *screen*, that is as long as possible so that the yield of the well can be maximized. A piezometer, on the other hand, has a very short screened interval in

◄ FIGURE 11.2
Design of a piezometer, the instrument used for measuring hydraulic head in the field.

order to obtain a head measurement at one specific point in a flow system. Also, piezometers are usually smaller in diameter than wells and should be carefully sealed with cement or clay above the screen to isolate the point of measurement from other parts of the groundwater flow system.

With head measurements obtained from several piezometers, it is possible to calculate the hydraulic gradient. Head measurements in a piezometer are made by determining the depth to water from the top of the pipe with a tape and then subtracting the depth from the elevation of the top of the pipe. Figure 11.3 illustrates the similarity between groundwater flow through a subsurface rock unit and flow through the Darcy apparatus (Figure 11.1). The hydraulic gradient indicates the direction of groundwater flow. In many places, groundwater has a vertical as well as a horizontal component of flow. In these areas, it is necessary to install a network of piezometers at various depths to determine the three-dimensional distribution of head in the flow system.

◄ FIGURE 11.3
Groundwater flow through a bed of rock or sediment. The flow direction is indicated by a decrease in head from left to right. Hydraulic gradient can be determined in the same way as in the Darcy experiment.

EXAMPLE 11.1

Two piezometers are completed in a sand bed 3 m thick with a hydraulic conductivity of 10^{-4} m/s. Piezometer A, at an elevation of 162 m above sea level, has a depth to water of 12 m. Piezometer B, located at an elevation of 153 m, has a depth to water level of 16 m. The direction of groundwater flow is in the same direction as a straight line drawn between A and B, which are separated by a distance of 1500 m. What is the rate of flow (in m^3/s) through a cross section of the bed perpendicular to the flow direction that is equal to its thickness and 1 m wide?

Solution

In piezometer A, the hydraulic head, h_A = 162 m − 12 m = 150 m. In piezometer B, the head, h_B = 153 m − 16 m = 137 m. Since the two piezometers are installed along the direction of flow, we can apply Darcy's Law directly:

$$v = -K \frac{h_A - h_B}{L} = \frac{-(10^{-4}\text{ m/s})\ 150\text{ m} - 137\text{ m}}{1500\text{ m}}$$

$$= 8.7 \times 10^{-7}\text{ m/s}$$

We can neglect the negative sign because it merely tells us that flow is in the direction from higher head to lower head (from A to B). Finally, because we need to determine the total flow through a column of sand 1 m wide and equal to the thickness of the bed (3 m), and because $v = Q/A$,

$$Q = v \times A = 8.7 \times 10^{-7}\text{ m/s} \times 3\text{ m}^2 = 2.6 \times 10^{-6}\text{ m}^3/\text{s}$$

The Water Table

During drilling for installation of a well or piezometer, water is encountered under different physical conditions. The uppermost zone contains water in the pores of rock and soil along with a gas phase. The incomplete filling of void spaces by water gives rise to the name *unsaturated zone* for this part of the subsurface (Figure 11.4a). The unsaturated zone is also referred to as the *vadose zone*. Below the unsaturated zone, at a depth determined by many factors (including climate, topography, and geologic setting), void spaces in the material are filled with water; this region is called the *saturated zone*. The saturated zone can be subdivided into the *phreatic zone*, below a surface known as the *water table*, and the *capillary fringe*, which lies above the water table. The capillary fringe and phreatic zones can be separated by the value of fluid pressure in the pore water at a particular point (Figure 11.5). Just as fluid pressure in a lake increases from zero at the surface (that is, normal atmospheric pressure) to greater values with depth, the fluid pressure in the saturated zone increases with depth from the point at which it is equal to atmospheric pressure, which is the horizon that is defined as the *water table*. The water table can be located by measuring the water-level elevation in a shallow well that only extends several

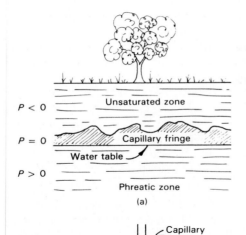

◀ FIGURE 11.4
(a) Zones of subsurface water. The relative value of fluid pressure is shown to the left of each zone. Zero represents atmospheric pressure. (b) Rise of water in a capillary tube—the same phenomenon that produces the capillary fringe.

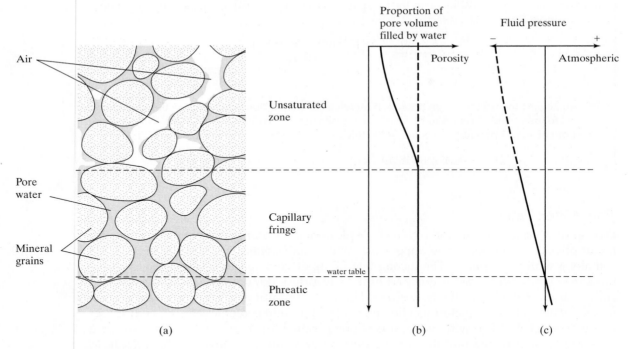

▲ FIGURE 11.5
(a) Distribution of pore water and air in pores in the subsurface. Pores are completely filled with water below the top of the capillary fringe. In the unsaturated zone above, the pores contain variable amounts of water between and surrounding grains, as well as air. (b) The fraction of void space that is filled with water, which increases from the surface downward (unless a heavy rainfall event has saturated the surface) to the top of the capillary fringe, where saturation begins. At that point and below, the volume of pore water is equal to the porosity. (c) Pore pressure changes with depth. In the unsaturated and capillary zones, the fluid pressure is less than atmospheric and water is held against gravity by adhesion to the soil particles and surface tension at the interface between water and air. The fluid pressure increases to atmospheric at the water table and continues to increase with depth in the phreatic zone.

meters into the saturated zone. The water level in such a well corresponds to the water table at that point.

The saturated material above the water table contains fluid in which the fluid pressure is less than atmospheric. This zone, the *capillary fringe*, is analogous to the rise of water above the free-water surface in a capillary tube (Figure 11.4b, Figure 11.5). The capillary rise is caused by forces of adhesion between the water and the capillary tube and of *surface tension* at the curved surface, or *meniscus*, of the water at the top of the capillary tube. Surface tension is an upward-directed force that results from the tendency of the water surface to assume a shape of minimum area when in contact with another fluid (in this case, air) with which it does not mix. If the pores in the soil are small, as in clay and silt, capillary rise above the water table occurs just as in a single capillary tube. The capillary effect varies with soil type: A small-to-nonexistent rise occurs in gravel, whereas rises of several meters or more are possible in fine-grained soils with high silt and clay contents. The fluid pressure, then, will be positive (greater than atmospheric) below the water table in the phreatic zone, equal to atmospheric at the water table, and negative (less than atmospheric) in the capillary fringe and in the unsaturated zone.

Groundwater Flow Systems and Flow Nets

Figure 11.6a shows a cross section through an area of hilly topography. The position of the water table, high beneath the hills and near the surface in the valley, gives a good indication of the distribution of head. Groundwater will flow from the areas where head is highest, called *recharge areas*, to areas where head is lowest, called *discharge* areas. The discharge area in Figure 11.6a is a river channel. Groundwater emerges as springs and seeps along the sides and bottoms of the channel, thereby sustaining the flow of the stream during dry periods. The component of stream flow derived from groundwater is the *base flow*. Rivers that flow all year because of groundwater discharge are *perennial streams*; streams that flow only during periods of rainfall that generate surface runoff are *ephemeral streams*.

Aside from recognizing that groundwater will somehow flow from the recharge areas to the discharge areas in Figure 11.6a, we would find it useful to know exactly what paths the groundwater will follow in this flow system. If we make some simplifying assumptions, the problem can be solved mathematically and graphically. A solution requires specified boundary conditions for the region of flow. For this flow system we can specify *no-flow* boundaries,

▶ FIGURE 11.6
(a) A cross section showing the relationship of the water table to topography. (b) A flow net showing the position of flow lines and equipotential lines in the flow system.

that is, boundaries normal to which there is no flow, at the bottom and sides of the flow region in Figure 11.6b. The left and right no-flow boundaries are considered to be imaginary because if the hills are symmetrical, a *groundwater divide* will develop near the crest of the hill and serve as a plane of symmetry. Groundwater will tend to flow away from this vertical plane as it moves toward discharge points in valleys on opposite sides of the hill. The lower no-flow boundary could represent an impermeable bed of rock at this depth in the flow system.

With the boundary conditions that we have established, we can determine the head at any point by developing and solving equations that describe flow in this system, using numerical methods that require computers to calculate the distribution of head or by using graphical techniques. Once we know the distribution of head, we can draw contour lines called *equipotential lines* on the diagram, as in Figure 11.6b. Equipotential lines connect points of equal head. The configuration of the equipotential lines will allow us to draw one other group of lines, *flow lines*. These lines, which are drawn perpendicular to equipotentials, indicate the paths that groundwater will follow in this flow system, the objective of this exercise. Thus, the resulting *flow net* indicates that under these conditions, groundwater travels along long, curving paths from a recharge point to a discharge point. A flow net drawn for an area of glacial topography in western Michigan is shown in Figure 11.7. A flow net such as this is only an approximation of groundwater flow because of the complexity of the subsurface geology. Most flow systems involve materials that are *heterogeneous* to some extent, that is, materials that vary in hydraulic conductivity. If these variations are relatively simple and continuous, as, for example, in a sequence of beds of known hydraulic conductivity extending through the entire region of flow, the flow net can represent their effects upon groundwater movement. If the hydraulic conductivity variations are discontinuous or if the subsurface geology is not well known, however, the flow net is only a generalized depiction of groundwater flow.

▲ FIGURE 11.7
Conceptual groundwater flow net through a series of ridges and valleys formed by glacial advances in western Michigan. *Source:* From W. T. Straw, R. N. Passero, and A. E. Kehew, 1993, Conceptual hydrologic glacial facies models, in *Environmental Impacts of Agricultural Activities*, Y. Eclestein and A. Zaporozec, eds., Water Environment Federation.

▶ FIGURE 11.8

(a) Plan view of an area in which two streams flow toward the bottom of the page and join at the bottom of the diagram. The area between the streams is a recharge area and groundwater flows in the direction of flow lines, which are perpendicular to equipotential lines. The equipotential lines are simply contour lines of water-table elevation. (b) Cross section of flow system parallel to $A–A'$.

(a)

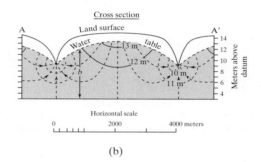

(b)

A flow net can also be drawn on a map or plan view. The contours of head used in this case are simply contours of the water-table elevation (Figure 11.8). The flow lines, which are still perpendicular to equipotentials in plan view, are very useful in predicting the movement of pollutants that may be introduced into the flow system from waste-disposal sites.

Recharge and Discharge Processes

The input to the dynamic, circulating groundwater flow systems below the earth's surface is recharge in the form of rainfall and snowmelt that percolates downward from the surface to the water table. Without recharge, the water table would steadily drop, wells would go dry, and the great groundwater reservoirs of the earth would be depleted.

If enough information is available, groundwater flow systems can sometimes be divided into *local, intermediate,* and *regional* types, based upon the relationship between recharge and discharge areas. Figure 11.9 shows a cross section of a hypothetical drainage basin that slopes from right to left. The land surface, and also the water table, has an undulating configuration with highs and lows superimposed on the overall slope (dashed line). Any groundwater flow, as defined by flow lines, that begins as recharge at highs on the water table and curves to discharge points in the adjacent low areas is part of a local flow system. Some of the groundwater recharges from a high point near the headwaters of the drainage basin and discharges at a low point near, but to the right of, the ultimate discharge point on the left side of the cross section, passing beneath several local flow systems in the process. Groundwater in this region of the cross section is called an *intermediate* flow system. The remainder of the groundwater

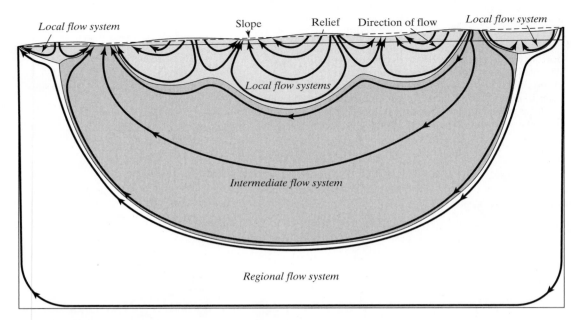

▲ FIGURE 11.9
Cross section showing the development of local, intermediate, and regional groundwater flow systems in a basin that has local undulations superimposed on a regional slope. The width of the flow system could be many kilometers and the depth to the base of the regional flow system could be hundreds to thousands of meters. *Source:* From J. A. Tóth, 1963, A theoretical analysis of regional groundwater flow in small drainage basins, *Journal of Geophysical Research*, 68: 4795–4811.

recharges at the far right of the section, descends below both the local and intermediate flow systems, and discharges at the far left side of the cross section. This is known as a *regional* flow system because it originates as recharge at the highest point in the flow system and discharges at the lowest point. Groundwater in these three types of flow systems can vary significantly in many characteristics, including temperature, age, and chemical composition.

When rainfall hits the ground surface or when snow melts, several possible pathways are available for water movement (Figure 11.10). If the water sinks into the soil, *infiltration* has occurred. Soils differ greatly in the rate at which water is absorbed, which is called the *infiltration capacity*. Coarse-grained soils usually have higher infiltration capacities than fine-grained soil. The infiltration capacity is also influenced by such other factors as slope, type of vegetation, and the existing soil-moisture condition. As infiltration continues, the infiltration capacity decreases in a manner shown in Figure 11.11. The limiting value of the infiltration capacity is the hydraulic conductivity of the soil, which is reached when the soil becomes saturated. After the infiltration capacity drops below the rainfall rate, water ponds on the surface and soon after begins to flow across the surface to lower elevations if the land is sloping. This process, *overland flow*, is more common in dry regions with fine-grained soils that have low infiltration capacities.

The infiltration of water into the ground does not necessarily mean that groundwater recharge will result, because there are two processes that can recycle water back to the atmosphere. Direct *evaporation* of the water from the soil is one mechanism. The other, *transpiration* (Figure 11.10), is the utilization of soil water and subsequent release of water vapor to the atmosphere by plants. Some plants, including certain species of trees, have the ability to take up large amounts of water through their roots and then return a high percentage of it to the atmosphere through their leaves. Evaporation and transpiration are so

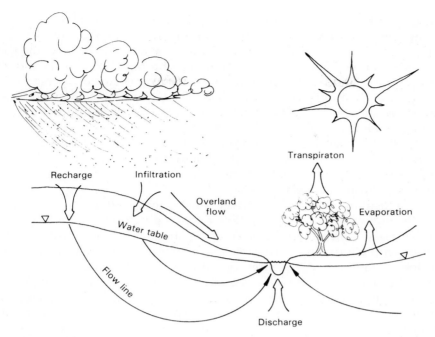

▲ FIGURE 11.10
The interaction between atmospheric water and groundwater.

▶ FIGURE 11.11
The decrease in infiltration capacity that occurs after a zone of saturation develops near the soil surface. The minimum value of infiltration capacity is hydraulic conductivity. *Source:* From R. A. Freeze and J. A. Cherry, *Groundwater*, © 1979 by Prentice Hall, Inc., Upper Saddle River, N.J.

difficult to separate quantitatively that they are frequently discussed together under the term *evapotranspiration*.

The amount of water available for groundwater recharge is, therefore, the fraction not lost by overland flow or evapotranspiration. Even part of the remaining water may be held in the unsaturated zone and not actually reach the water table. Some water must move downward to the water table for recharge to occur. In many areas, recharge may take place only for short periods of the year. The rise in the water table during the spring and early summer, when precipitation is high and evapotranspiration low, is the most common indication of groundwater recharge. In some arid and semi-arid areas, this may be the only period of recharge in a typical year, aside from occasional heavy thunderstorms.

Groundwater discharge is the opposite of recharge; in discharge areas water moves from the saturated zone into the unsaturated zone and perhaps to land surface. If the near-surface sediments are homogeneous and the water table is high, discharge areas will occur in topographic lows where the water table intersects the land surface, as in lakes and streams. The elevation of the stream or lake actually represents the position of the water table at that point. A surficial body of water is not required, however, sometimes discharge areas can be recognized by indications of a high water table and abundant evapotranspiration. Evidence for this type of discharge area includes persistent swampy conditions during all or most of the year, plants like willow and cottonwood that extend their roots below the water table to obtain moisture (*phreatophytes*), and saline or alkaline soil conditions. Saline and alkaline soils indicate evaporation of groundwater from the soil, with the dissolved salts carried by groundwater left behind in the soil. In some areas, particularly where the water table is deep, groundwater recharge, rather than discharge, occurs from topographic low points. In these situations, most of the recharge is derived from ephemeral stream channels or lakes.

Springs, which are localized discharge points, represent a more obvious type of discharge. Springs can vary greatly in flow rate, ranging from barely a trickle to more than 10 m^3/s. Some general types of springs are shown in Figure 11.12. In stratigraphic sequences with alternating permeable and nonpermeable beds, *contact springs* can develop. Joints, faults, and fractures commonly provide conduits for the upward movement of groundwater to the surface. The high temperature and highly mineralized chemical composition of some springs attest to the depths to which some of these conduits extend. Some of the most spectacular springs are located in areas of soluble bedrock like limestone. Groundwater chemically reacts with these rocks and, over thousands of years, can dissolve large amounts of rock, forming subsurface caves and passages. Tremendous amounts of water may issue from springs in limestone terrains (Figure 11.13).

◀ FIGURE 11.12
Some geologic settings in which springs occur.

▶ FIGURE 11.13
Vasey's Paradise, a series of springs emanating from limestone rock units in the Grand Canyon. *Source:* Photo courtesy of the author.

Groundwater Resources

The most obvious reason for studying groundwater is its value as a resource. In some parts of the United States, particularly in rural areas, groundwater constitutes the only water supply. Surface water, the alternative source, when available, must be extensively treated before drinking and is quite often limited in quantity. Increases in demand for water in the United States will have to be met predominantly by expanding utilization of groundwater.

Aquifers

An *aquifer* is a saturated body of rock or soil that transmits economically significant quantities of groundwater. The most important property of an aquifer is its hydraulic conductivity, for it is the rate at which water can move to the screen of a pumping well to replace the water pumped out that determines the yield of the well. This rate of movement, under natural or pumping conditions, is controlled by the hydraulic conductivity. Geologic materials that constitute productive aquifers include sand and gravel, highly fractured rocks, and soluble rocks containing large subsurface openings for groundwater movement and storage.

Materials that do not transmit economically significant quantities of groundwater are known as *aquitards*. Aquitards are usually composed of dense, nonfractured rock units or sedimentary beds of silt and clay. Sedimentary rock terrains often consist of alternating aquifers and aquitards. Exploration for groundwater in these areas involves location of the permeable rock horizons. Some materials are referred to as *aquicludes*, units that transmit no groundwater. It is unlikey, however, that true aquicludes exist in nature, because all materials seem to convey some groundwater, however small the amount.

Aquifer type is determined by the presence or absence of an overlying aquitard. Aquifers that lack overlying aquitards and have the water table as their upper boundary are *unconfined aquifers*. These often occur in surficial deposits of sand and gravel. Because the water table is the upper boundary, contour lines of water-table elevation drawn on a map indicate the direction of flow of groundwater in an unconfined aquifer. As shown in

◀ FIGURE 11.14
An unconfined aquifer showing water-table contours, which are elevations above sea level of the water table, projected onto the surface.

Figure 11.14, the slope of the water table is a close approximation of the hydraulic gradient, providing the flow in the aquifer is nearly horizontal. Similarly, the elevation of the water table is closely equivalent to the hydraulic head in an unconfined aquifer.

Confined aquifers are bounded above and below by aquitards. As a result of this confinement, there are important differences between the flow of groundwater in confined aquifers and the flow in unconfined aquifers. The water level, or head, in a well in a confined aquifer may be above or below the water table; in fact, there is no specified relationship between the head in a confined aquifer and the water table. This distinction between the two aquifer types is shown in Figure 11.15. If the heads in a confined aquifer are contoured on a map, as previously shown for the water table in unconfined aquifers, the contours will define an imaginary surface called the *potentiometric surface* (Figure 11.15). The potentiometric-surface map is similar to the water-table contour map in that it indicates the distribution of head and therefore the direction of water movement in a confined aquifer. The major difference

▲ FIGURE 11.15
A confined aquifer, in which the water level in a well rises to the potentiometric surface. Contours of the potentiometric surface are shown on the land surface.

▶ FIGURE 11.16
A geologic and topographic configuration that produces flowing wells whenever the potentiometric surface is above the land surface.

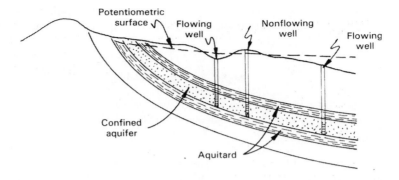

between the potentiometric surface and the water table is that the potentiometric surface is not an actual surface in the ground analogous to the water table; it is a hypothetical level defined by the elevations of water levels in wells in confined aquifers. Confined aquifers are also known as *artesian aquifers*. The fluid pressure in some artesian aquifers may be so high that the head, or potentiometric surface, is above the land surface. A well completed in this type of aquifer will flow without the aid of a pump; these wells are known as *flowing artesian wells*. Flowing wells are not unlike the plumbing system in a house in that the water will flow out under its own pressure as long as the faucet is left open.

Geologic and topographic conditions combine to produce confined aquifers. In Figure 11.16, a classic example of a confined aquifer is illustrated. The aquifer in this case is exposed at the land surface at the left end of the cross section, where recharge occurs. To the right, the aquifer dips downward beneath a thick aquitard. The potentiometric surface shown above the aquifer indicates that just to the right of the recharge area, the aquifer develops confined conditions. The surface topography of the area determines where flowing artesian conditions will occur; as land surface slopes to the right more rapidly than the potentiometric surface, flowing well conditions are common. Figure 11.16 is actually a simplified version of the regional groundwater flow system in the Great Plains of the United States. The confined aquifer system, the Dakota Sandstone of Cretaceous age, crops out in the Black Hills region of western South Dakota. When wells were first drilled into the Dakota Sandstone during settlement of the eastern Dakotas, which is much lower in elevation than the Black Hills, heads well above land surface were encountered. Heads declined over the years as many flowing wells discharged water from the Dakota Sandstone aquifer.

EXAMPLE 11.2

The first well completed in the Dakota aquifer in South Dakota was drilled at Aberdeen by W. E. Swan in 1881. The well was 1100 feet in depth and had a closed-in pressure of 180 psi. (This is the pressure that developed at land surface when the well was closed and not allowed to flow.) How high above land surface was the potentiometric surface?

Solution

Fluid pressure is equal to the unit weight of the fluid times the height of the fluid column above the point of interest.

$$p = \gamma_{water} \times h$$

Since the unit weight of water in English units is usually reported as 62.4 lb/ft^3, pressure in psi can be converted to lb/ft^2.

$$180 \text{ lb/in.}^2 \times 144 \text{ in.}^2/\text{ft}^2 = 25,920 \text{ lb/ft}^2$$

The true unit weight of Dakota aquifer groundwater would be slightly more because it was somewhat mineralized. Head above land surface in feet can now be determined.

$$h = \frac{p}{\gamma} = \frac{25{,}920 \text{ lb/ft}^2}{62.4 \text{ lb/ft}^3} = 415.4 \text{ ft}$$

Thus, the potentiometric surface, or the level to which water would rise in a well (if the well were extended up into the air), was significantly above land surface in parts of eastern South Dakota.

A final aquifer type is the *perched aquifer*. Beds of material with low hydraulic conductivity in the unsaturated zone lead to the establishment of perched conditions (Figure 11.17). Meandering stream deposits (discussed later in this chapter) commonly contain perched aquifers. The downward flow of water through the unsaturated zone toward the regional water table is retarded by the clay unit. A localized water table and a perched aquifer then develop above the clay. Perched aquifers cannot usually sustain high well yields because of their limited areal extent.

The amount of time that groundwater is present in an aquifer as it moves from the recharge area to the discharge area is known as *residence time*. Residence time varies from days in unconfined aquifers, where local and intermediate flow systems are typical, to millions of years in deep confined aquifers in sedimentary basins (Figure 11.18). Confined aquifers are more likely to be part of intermediate and regional flow systems. Residence time is important for many reasons. For example, there is a much greater chance that surface contamination will affect shallow wells in unconfined aquifers than deeper confined aquifers because of the protection afforded by the aquitards above the confined aquifers and because of the short residence time of groundwater flow in unconfined aquifers.

Production of Water from Aquifers

Unlike piezometers, wells are designed to maximize the yield of aquifers. Larger diameters and greater lengths of screen are therefore utilized. If a well penetrating an unconfined aquifer is limited to that aquifer, and if vertical components of flow in the aquifer are not

◄ FIGURE 11.17
Development of a perched aquifer above a low-permeability unit that occurs within the unsaturated zone of a unit of higher permeability.

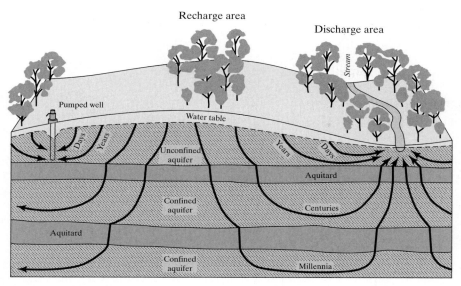

▲ FIGURE 11.18
Cross section illustrating the increase in residence time for groundwater flow in deeper confined aquifers in a multi-aquifer sequence including both unconfined and confined aquifers.

significant, the *static*, or unpumped, water level in a well may be a close approximation to the water table. Similarly, if a well in a confined aquifer is isolated from groundwater in overlying and underlying rock units, the water level in the well may be a good indication of the potentiometric surface in the aquifer. But if the well is pumped, changes in the hydraulic regime of the aquifer occur.

When an unconfined aquifer is pumped, the response to this stress is a decline of the water table in the vicinity of the well. These effects are usually assumed to be equal in all directions, so the area of influence of pumping is a cone-shaped region with greatest decline in the water table adjacent to the well (Figure 11.19). The affected region changes from a saturated to an unsaturated state during the drop in the water table. Because of its shape it is known as a *cone of depression*. As pumping continues, the cone expands radially outward from the well. The actual shape and dimensions of the cone depend on the hydraulic conductivity, thickness, and storage properties of the aquifer. If the pumping rate is not excessive, the cone of depression will eventually reach an equilibrium position. If pumping is excessive, the water table will continue to drop, causing increased pumping costs.

In a confined aquifer the effects of pumping are somewhat different. Here the cone of depression develops in the potentiometric surface without any dewatering of the aquifer (Figure 11.20). To explain the response of the aquifer to the lowering of the potentiometric surface we must invoke the *Principle of Effective Stress*, a concept that has many applications in geology. Referring to Figure 11.21, consider the distribution of stress in a confined aquifer. The total weight of rock, soil, and water above the aquifer exerted per unit area on the upper surface of the aquifer is designated as the *total stress* (σ_T). This stress is opposed in the aquifer by two stresses, the fluid pressure (p), which we have previously described, and a stress transmitted through the solid grains of the aquifer, the *effective stress* (σ_e). The equilibrium relationship defined by this group of stresses can be defined as the effective stress equation,

$$\sigma_T = \sigma_e + p \qquad (11.5)$$

▲ FIGURE 11.19
Development of a cone of depression in an unconfined aquifer by lowering of the water table. *Source:* From W. M. Alley, T. E. Reilly, and O. L. Franke, 1999, *Sustainability of Ground-Water Resources*, U.S. Geological Survey Circular 1186.

Despite its simplicity, this equation is of utmost importance in many areas of geology and engineering. When a confined aquifer is pumped and a cone of depression develops in the potentiometric surface, the fluid pressure is reduced. Since the total stress exerted on the aquifer is unchanged, the effective stress equation requires that the effective stress increase. When the grain-to-grain pressure in a material increases, the material tends to compact due to closer packing of the particles. This reduction in porosity is not possible unless water within the pores is expelled. Since the head has been lowered in the vicinity of the well, the pore water moves radially toward the well and the aquifer compacts vertically. Therefore, even though large amounts of water are removed from a confined aquifer, it remains saturated. An additional mechanism that accounts for water produced in the well is the expansion of water that occurs when the fluid pressure is decreased. As in unconfined aquifers, the cone of depression in the potentiometric surface in a confined aquifer expands with pumping until it approaches an equilibrium condition.

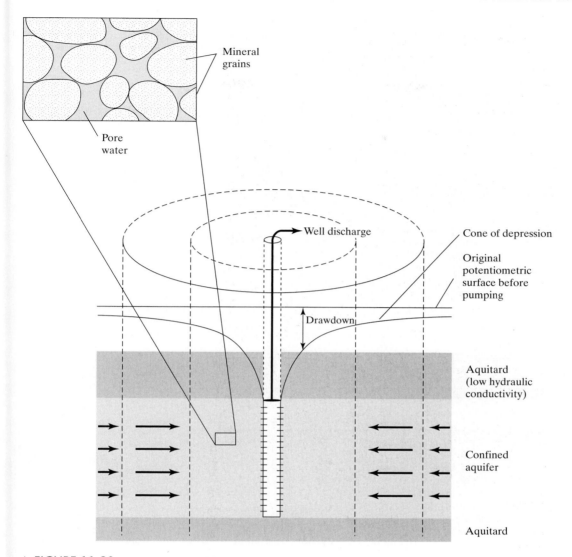

▲ FIGURE 11.20
The cone of depression produced in the potentiometric surface by pumping a confined aquifer.
Source: From W. M. Alley, T. E. Reilly, and O. L. Franke, 1999, *Sustainability of Ground-Water Resources*, U.S. Geological Survey Circular 1186.

The size of the cone of depression can be estimated using a form of Darcy's Law if some assumptions are made. For a confined aquifer of thickness B (Figure 11.22), let us assume that the piezometric surface is horizontal and that the aquifer extends to infinity in all directions. Let us further assume that the screen of the well installed in the aquifer penetrates the entire thickness of the aquifer. This assumption means that groundwater flow to the well will be horizontal and radial from all directions in the aquifer. Darcy's Law can now be written for flow passing through the surface area of a cylinder of radius *r* and depth B (Figure 11.23):

$$Q = KA \frac{dh}{dr} = K2\pi rB \frac{dh}{dr} \tag{11.6}$$

Land surface

Distribution of stress in a confined aquifer. Pumping lowers the fluid pressure and increases the effective stress.

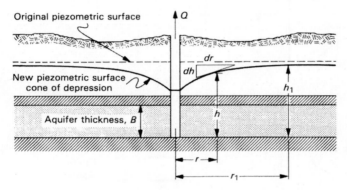

◀ FIGURE 11.22
Diagram showing parameters used in equation (11.6).

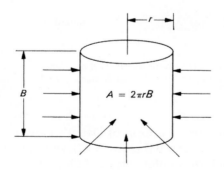

◀ FIGURE 11.23
Cylinder illustrating radial flow of groundwater to well.

This equation is similar to the form of Darcy's Law that we used previously, except the area used is the surface area of a cylinder and the hydraulic gradient is written in radial form. Equation (11.6) can now be written as an integral:

$$\int_{r}^{r_1} Q \frac{dr}{r} = 2\pi B \int_{h}^{h_1} dh \tag{11.7}$$

If we then integrate between arbitrary values r and r_1, which have heads of h and h_1, we obtain

$$Q \ln \frac{r_1}{r} = 2\pi KB(h_1 - h) \tag{11.8}$$

or

$$Q = \frac{2\pi KB(h_1 - h)}{\ln(r_1/r)} \tag{11.9}$$

A similar equation can be derived for an unconfined aquifer. The only difference is that the height of the cylinder is a variable because the saturated thickness of the aquifer decreases as the cone of depression is formed. This equation can be stated as

$$Q = \frac{\pi K(h_1^2 - h^2)}{\ln(r_1/r)} \tag{11.10}$$

EXAMPLE 11.3

A 3-m-thick confined aquifer is pumped at a rate of 1 L/s until the cone of depression in the potentiometric surface has reached equilibrium. The original potentiometric surface was at an elevation of 50 m. An observation well located 30 m from the pumping well has a drawdown of 1 m. If the hydraulic conductivity of the aquifer is 10^{-4} m/s and the diameter of the pumping well is 0.1 m, what is the drawdown in the potentiometric surface at the well?

Solution

Using the radius of the well, r_w, as the value of r, use equation (11.9):

$$Q = \frac{2\pi KB(h_1 - h_w)}{\ln(r_1/r_w)}$$

$$0.001 \text{ m}^3/\text{s} = \frac{2(3.14)\,(10^{-4}\text{ m/s})\,(3\text{ m})\,(49\text{ m} - h_w)}{\ln(30\text{ m}/0.05\text{ m})}$$

Solving for h_w gives

$$h_w = 45.7 \text{ m}$$

and the drawdown is equal to

$$50 \text{ m} - 45.7 \text{ m} = 4.3 \text{ m}$$

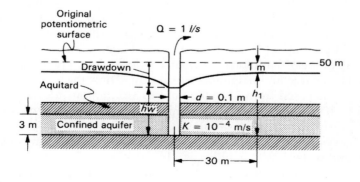

Sustainability of Groundwater Resources

As part of the hydrologic cycle, the groundwater system in an area is in a condition of long-term balance between inputs and outputs prior to extraction of groundwater by humans for drinking water supply, industrial use, irrigation, or other purposes. Recharge comes primarily from infiltration of rain and melted snow through the unsaturated zone to the water table and from there into shallow and deep aquifers. A smaller component of recharge originates from infiltration from lakes, streams, and wetlands. Discharge occurs in springs and streams, as we have seen, and also into lakes, wetlands, and the ocean. Over a long period of time, excluding human interference or climatic change, recharge equals discharge. The surface-water and groundwater systems are intimately connected in this balance and additions or subtractions of water from one will cause changes in the other.

Both natural changes and human activity can disrupt this balance between groundwater recharge and discharge. Droughts are one obvious example. With the sustained decreases in recharge that accompany droughts, water tables decline—and, because groundwater flow systems supply streams and lakes with water—stream flow decreases, wetlands dry up, and lakes shrink in size. Most of these changes are reversible, however, when the drought ends.

Groundwater pumpage adds another form of discharge to the balance between recharge and discharge. Water is pumped for many purposes and some of it is returned to the aquifer, but in most cases pumpage results in a net loss of water from the groundwater flow systems. Because the hydrologic system is now thrown out of its natural condition of balance, the responses include declining water levels and declining discharge to surface-water bodies. The consequences of these responses are not good—they include increased costs for pumping, loss of wildlife habitat as discharge is reduced, and ultimately a reduction or loss of the resource.

An excellent example of a situation in which current levels of groundwater extraction will most likely not be sustainable on a long-term basis is the High Plains aquifer, a shallow, mostly unconfined aquifer that occurs throughout much of the Great Plains region of the United States (Figure 11.24). The High Plains aquifer has been extensively pumped to support irrigated farming in this region since the agricultural disaster caused by the 1930s drought (Figure 11.25c). The development of irrigation has progressed to such an extent that annual groundwater withdrawals now exceed the annual recharge to the aquifer by 2 to 200 times. The result is a steady decline in water levels. These declines exceed 100 ft in some areas. In effect, the water is being "mined" like any other nonrenewable resource. On a short-term basis, the effect of water-table declines is to increase costs because of the greater pumping lift required. On a long-term basis, however, parts of the High Plains aquifer will be in serious trouble within the next 20 years. The point may be reached where irrigated farming will no longer be possible in some areas. This will have dire consequences for the economy of the affected areas.

Figure 11.24 shows the drop in water tables throughout the aquifer and the decreases in saturated thickness. The changes in the groundwater budget are also significant (Figure 11.25a, b). Prior to development, recharge and discharge were in balance at 24 million cubic feet per day (cfd). Since development began, the budget has changed dramatically. Not all of the 830 million cfd pumped is lost to plant use. Some of it is returned to the aquifer as recharge and irrigation return flow. Thus, the sum of natural and artificial recharge has increased from 24 million cfd to 510 million cfd. Natural discharge has been reduced to 10 million cfd. The remaining 330 million cfd, which is consumed by plants, is lost forever from aquifer storage.

Most aquifer systems under pumpage eventually reach a new balance between recharge and discharge and water levels stabilize at a lower level under pumping conditions. The High Plains aquifer is one of the best-known examples of long-term disequilibrium in a regional aquifer system because water table levels continue to decline.

▲ FIGURE 11.24

Changes in water-table levels (a) and saturated thickness (b) in the High Plains aquifer. Declines are cumulative from predevelopment to 1997. Changes are most significant in the southern part of the area, where the aquifer is thinner and less productive. *Source:* From W. M. Alley, T. E. Reilly, and O. L. Franke, 1999, *Sustainability of Ground-Water Resources,* U.S. Geological Survey Circular 1186.

System before development

Natural
recharge
24

Natural
discharge
24

(a)

System during development

Recharge
and irrigation
return flow
510

Pumpage
830

Decrease
in storage
330

Natural
discharge
10

(b)

(c)

▲ FIGURE 11.25
Water budgets before (a) and during (b) irrigation development. Values are in millions of cubic feet per day. *Source:* From W. M. Alley, T. E. Reilly, and O. L. Franke, 1999, *Sustainability of Ground-Water Resources*, U.S. Geological Survey Circular 1186. (c) Center pivot irrigation in the High Plains aquifer as seen from 30,000 ft. Four center pivot circles occupy one square mile. *Source:* Photo courtesy of the author.

Case in Point 11.1
Groundwater Recharge, Discharge, and Withdrawals in Long Island, New York

The effects of groundwater pumpage and other human activities on various components of the groundwater flow system are well illustrated by the Long Island, New York, aquifer system. The aquifer system is composed of permeable glacial deposits overlying bedrock that does not serve as an aquifer (Figure 11.26). The aquifers, which include both unconfined and confined or semiconfined aquifers where aquitards are present, are prolific groundwater producers. Because of the relatively low elevation of the island and the humid climate, the water table is close to the surface. Groundwater is derived from precipitation and flows from recharge areas at higher elevations in the interior of the island to the Atlantic Ocean, where it

▲ FIGURE 11.26
Model of Long Island, New York, groundwater flow system. Arrows with numbers refer to water-budget quantities in Table 11.1. *Source:* From W. M. Alley, T. E. Reilly, and O. L. Franke, 1999, *Sustainability of Ground-Water Resources*, U.S. Geological Survey Circular 1186.

mixes with seawater. Groundwater also discharges into streams that flow into the ocean. The linkage between groundwater levels and stream discharge is very obvious here (Figure 11.27)—when groundwater levels rise, stream discharge increases. The individual components of the hydrologic system are broken out in Table 11.1.

Human modification to the natural hydrologic system on Long Island was minimal for the first several centuries after European settlement when the island had a rural, agricultural character. Individual homes and farms had shallow wells and septic systems, so wastewater was returned to the aquifer. As long as the population density was low, these activities had little impact. In the first half of the 20th century, however, explosive population growth forced changes in the use and management of water resources. Contamination of the shallow aquifers by septic systems and agricultural and industrial chemicals led to the abandonment of individual domestic wells in favor of deeper municipal wells. Effects on groundwater levels were still minor, however, because of the predominance of domestic wastewater systems.

The major disruption of the natural system occurred in the 1950s, with the construction of a large-scale municipal sewage treatment system to replace the individual

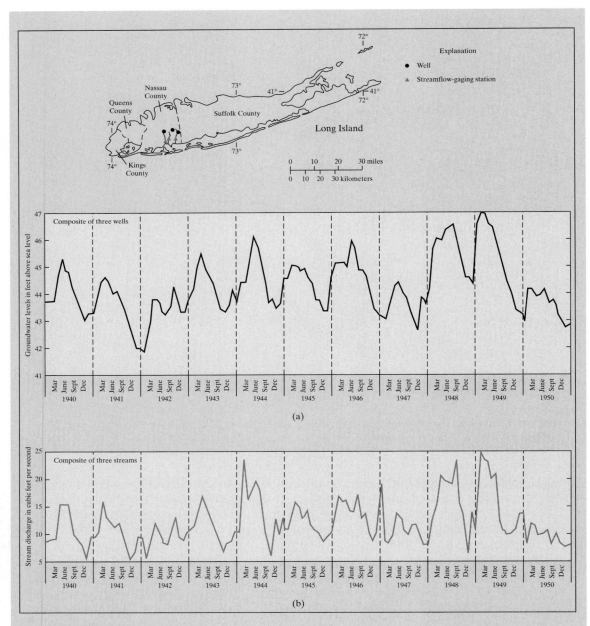

▲ FIGURE 11.27
Correlation of water levels in three wells and stream flow in three streams in Nassau County, Long Island, New York, between 1940 and 1950. Close correspondence indicates that the groundwater flow
system is closely coupled to surface water. *Source:* From W. M. Alley, T. E. Reilly, and O. L. Franke, 1999,
Sustainability of Ground-Water Resources, U.S. Geological Survey Circular 1186.

Table 11.1	Components of Water Budget for Total Hydrologic System (top) and for Groundwater Flow Systems (bottom) for Conditions Prior to Development

Overall Predevelopment Water-Budget Analysis

Inflow to Long Island Hydrologic System	Cubic Feet per Second
1. Precipitation	2,475

Outflow from Long Island Hydrologic System

2. Evapotranspiration of precipitation	1,175
3. Groundwater discharge to sea	725
4. Streamflow discharge to sea	525
5. Evapotranspiration of groundwater	25
6. Spring flow	25
Total outflow	2,475

Groundwater Predevelopment Water-Budget Analysis

Inflow to Long Island Groundwater System	Cubic Feet per Second
7. Groundwater recharge	1,275

Outflow from Long Island Groundwater System

8. Groundwater discharge to streams	500
9. Groundwater discharge to sea	725
10. Evapotranspiration of groundwater	25
11. Spring flow	25
Total outflow	1,275

Source: W. M. Alley, T. E. Reilly, and O. L. Franke, 1999, *Sustainability of Ground-Water Resources,* U.S. Geological Survey Circular 1186.

household systems. Wastewater that had been previously returned to the aquifer via septic systems was now treated and discharged to the ocean. Although the water quality in the aquifer was much improved by this new approach, it had a drastic effect on aquifer water levels. Water levels in Nassau County, which was extensively sewered, dropped significantly (Figure 11.28) to establish a new state of equilibrium because of the large amount of water that was pumped from the aquifer and discharged to the ocean.

Although this consequence was not extremely detrimental in itself, as in any system, a change to one part affects every other part of the system, and this change had unintended effects. As could have been predicted from Figure 11.27, the decline in groundwater levels led to a decline in groundwater discharge to streams. As a result, the headwaters of many of the streams dried up and no longer flowed during dry periods of the year. Another consequence was related to the loss of groundwater discharge to the ocean. The position of the interface between freshwater and saltwater represents a state of equilibrium between the force of the groundwater flowing toward the ocean and the seawater. When the groundwater levels declined, seawater advanced inland, a process known as *seawater intrusion*. There are many coastal areas in the world where sea-level intrusion is a major problem, because groundwater supply wells located near the coast become saltier and in some cases have to be abandoned.

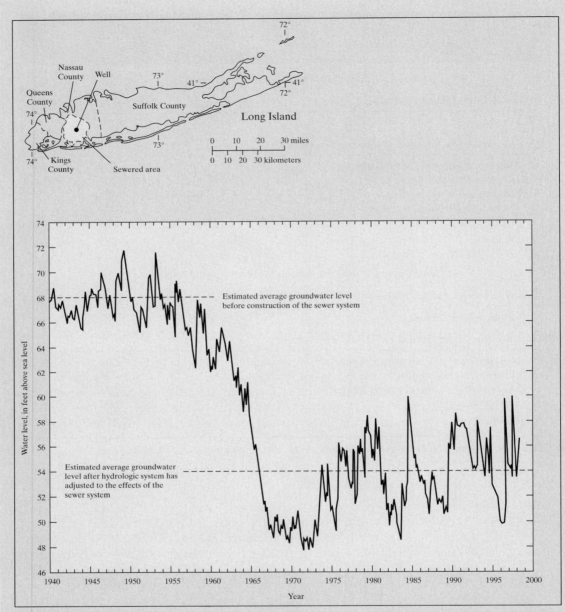

▲ FIGURE 11.28
Response of water levels in shallow wells to sewering of Nassau County, New York. After construction of the sewage treatment system, sewage effluent is discharged to the ocean, replacing the older method of individual domestic septic systems. *Source:* From W. M. Alley, T. E. Reilly, and O. L. Franke, 1999, *Sustainability of Ground-Water Resources,* U.S. Geological Survey Circular 1186.

Geologic Setting of Aquifers

Aquifers occur in vastly different geological settings, making exploration for groundwater challenging. For example, the United States has been divided into 15 groundwater regions, each of which contains aquifers with common geological conditions (Figure 11.29). Some of the most important aquifers occur in alluvial valleys, which are underlain by thick deposits of sand and gravel. The distribution of these aquifers in the United States is shown in Figure 11.30. This aquifer type is not included as a region in Figure 11.29 because many of the other groundwater regions contain alluvial valleys. The Mississippi Valley typifies this aquifer type (Figure 11.31). Thick deposits of coarse sand and gravel were deposited during the Pleistocene, when the valley served as a major glacial meltwater channel from glaciers that terminated in the upper Midwest. Sand and gravel aquifer units are interbedded with silt and clay aquitards deposited during interglacial periods such as the Holocene. In some parts of the Mississippi Valley, aquifers also occur in sedimentary bedrock underlying the alluvial valleys.

In regions where structural deformation has produced alternating mountain ranges and valleys by movement along faults, *alluvial basin aquifers* are present (Figure 11.32). The alluvium is deposited by streams eroding the uplifted mountain blocks. In arid regions, alluvial basin aquifers may provide an excellent supply of groundwater, although groundwater withdrawals during development of the aquifers commonly exceeds recharge and problems such as land subsidence (Chapter 10) result. Ephemeral streams, which are above the water table and therefore lose water to infiltration (losing streams), are most common in arid alluvial basins, although some do receive groundwater discharge (gaining streams).

The relatively low-relief glaciated terrain of the Glaciated Central region contains a variety of aquifer types (Figure 11.33). Deposits of continental glaciers (Chapter 16) contain statigraphically complex sequences of low permeability *till* deposited directly by the glacier itself, interbedded with permeable meltwater stream (*outwash*) or lacustrine sediment. These units constitute aquifers. Over the course of multiple glaciations, alluvium-filled valleys are buried by deposits of younger glaciations. Because there may be little or no surface expression of these valleys, they are known as *buried valleys*, and they may contain extremely productive aquifers. Glacial deposits in the Glaciated Central region range from 0 to hundreds of meters in thickness. Paleozoic rock formations underlying the glacial deposits also contain aquifers. Sandstone and limestone are the most productive rock types. Several deep sedimentary basins, where Paleozoic rock units accumulated in an area of crustal subsidence, underlie glacial deposits in the Glaciated Central region. Although deep sedimentary rock formations may be permeable, their formation waters are often too saline for drinking-water resources.

The problem of saltwater intrusion in coastal aquifers was mentioned in the context of aquifers on Long Island, New York. The aquifers throughout the Atlantic and Gulf Coastal Plain region (Figure 11.29) are characterized by a zone of interface between fresh groundwater and saline groundwater derived from the ocean (Figure 11.34). The sediments that contain these aquifers are poorly lithified marine sands, clays, and limestones ranging from Jurassic to present in age. Each aquifer has a diffuse interface between fresh and salty water that has migrated inland in the aquifer from the coast. The depth to the interface can be estimated by an approach called the *Ghyben-Herzberg relation*, which states that the depth below sea level of the interface at any point inland from the shore is equal to 40 times the elevation of the water table above sea level at that point. Thus, if the water table is 2 m above sea level, the saltwater-freshwater interface would lie 80 m below sea level. When the aquifers are developed by pumping, the lowering of the water table causes the upward and landward encroachment of the saline groundwater. In the example, if pumping from wells

▲ FIGURE 11.29

Groundwater regions of the United States. Region 12, Alluvial Valleys, is shown in Figure 11.30 because aquifers of this type occur in many of the other regions. *Source:* From R. C. Heath, 1984, *Ground-Water Regions of the United States,* U.S. Geological Survey Water Supply Paper 2242.

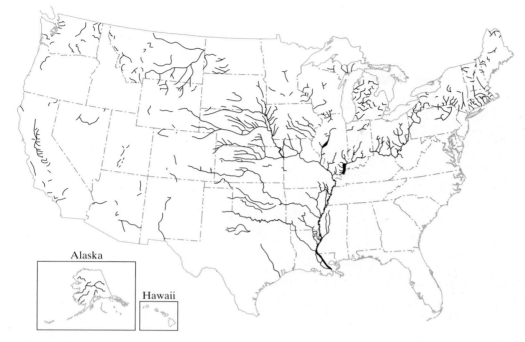

Alaska

Hawaii

▲ FIGURE 11.30
Distribution of alluvial valley aquifers in the United States. *Source:* From R. C. Heath, 1984, *Ground-Water Regions of the United States*, U.S. Geological Survey Water Supply Paper 2242.

in a coastal community lowered the water table by 1 m, the saltwater-freshwater interface would rise to an elevation of −40 m (40 m below sea level). Increasing salinity of drinking water can therefore be a serious problem for coastal cities that depend upon coastal-plain aquifers for their water supply.

Many rock types either have a low initial hydraulic conductivity because of their fine grain size or have a greatly reduced porosity because of cementing. These rocks may also function as aquifers, however, if they are fractured. Even shales can yield high quantities of groundwater if the network of joints, faults, and other fractures is dense, closely spaced, and interconnected. Coals also sometimes constitute aquifers through the development of such *fracture permeability*. Fractures are also important in carbonate rocks, because it is along these cracks that water circulation and, therefore, dissolution of the soluble calcium carbonate proceeds. One important aspect of groundwater production from fractured rocks is the problem of intersecting the fractures with a well screen. Well yields vary greatly in fractured rocks, depending on the number, size, and orientation of fractures in contact with the screen.

Igneous and metamorphic rocks, with several exceptions, are dense, interlocking aggregates of crystals with low initial porosity and permeability. Fracturing is the only mechanism that allows significant production from wells. The Piedmont Blue Ridge region is a good example of an area dependent upon fractured-rock aquifers in igneous and metamorphic rocks (Figure 11.35). The fractured bedrock lies beneath the *regolith*, which is a mantle of soil produced by in-place weathering of bedrock and downslope movement of weathered materials. The likelihood of drilling a productive well in this terrain is dependent upon the presence of fractures and fracture networks that are sufficiently interconnected

▲ FIGURE 11.31

Model of alluvial valley aquifer occurrence in the Mississippi Valley. *Source:* From R. C. Heath, 1984, *Ground-Water Regions of the United States,* U.S. Geological Survey Water Supply Paper 2242.

▲ FIGURE 11.32

Block diagram of alluvial basin aquifer, common in areas where thick alluvial deposits fill basins between fault-block mountains. *Source:* From R. C. Heath, 1984, *Ground-Water Regions of the United States,* U.S. Geological Survey Water Supply Paper 2242.

▲ FIGURE 11.33

Occurrence of aquifers in the Glaciated Central region. Deeper sedimentary rock units may contain water that is too salty for the drinking-water supply. *Source: From R. C. Heath, 1984, Ground-Water Regions of the United States, U.S. Geological Survey Water Supply Paper 2242.*

▲ FIGURE 11.34

Aquifer occurrence in the Atlantic and Gulf Coastal Plain region. The interface between fresh and saline water is located at a variable distance inland from the shore, dependent upon the elevation of the water table above sea level. Pumpage from these aquifers, which lowers the water table, will cause the interface to move farther inland, which is known as sea-water intrusion. *Source:* From R. C. Heath, 1984, *Ground-Water Regions of the United States,* U.S. Geological Survey Water Supply Paper 2242.

▲ FIGURE 11.35

Fractured-rock aquifers of the Piedmont Blue Ridge groundwater region. The rocks include igneous and metamorphic units with low primary porosity and permeability. *Source:* From R. C. Heath, 1984, *Ground-Water Regions of the United States*, U.S. Geological Survey Water Supply Paper 2242.

to supply groundwater to the well. The most productive wells are sited where major fractures intersect.

Although most igneous and metamorphic rocks have low permeabilities, an exception to this generalization are lava flows with vesicular texture; these rocks may be very porous and highly permeable. The Columbia Lava Plateau groundwater region (Figure 11.29) is characterized by thick sequences of nearly horizontal lava flows stacked upon each other (Figure 11.36). The main source of porosity and permeability is formed by vesicular zones caused by gas escape during cooling at the tops of the flows. Therefore, groundwater exploration is commonly focused on contacts between flows. A sequence of lava flows may contain aquifers at the flow contacts, separated by aquitards formed by the main body of the lava flows.

The occurrence of groundwater in carbonate-rock terrains differs from that in almost every other type of geological setting. In addition, these aquifers are similar to the Alluvial Valley region in that they occur in many of the other groundwater regions shown in Figure 11.29, including the Glaciated and Nonglaciated Central regions, the Southeast Coastal Plain region, and others. The unique hydrogeological properties of these regions owe their origin to the solution of limestone and dolomite by surface and groundwater. Following the exposure of carbonate rocks at or near the surface, the landscape undergoes a gradual transition in which the surface runoff and drainage of water is replaced by subsurface drainage. The unique topography that characterizes such areas is known as *karst*. A common form of karst topography associated with the surface exposure of nearly

▲ FIGURE 11.36
Aquifer occurrence in the Columbia Lava Plateau region. Highest permeability occurs at flow contacts because of the vesicular texture of the upper part of each flow produced during cooling. *Source:* From R. C. Heath, 1984, *Ground-Water Regions of the United States*, U.S. Geological Survey Water Supply Paper 2242.

◀ FIGURE 11.37
Portion of topographic
map of the Mitchell Plain
(portion of the U.S. Geo-
logical Survey Corydon
West 7.5 min Quadrangle),
southern Indiana. Each depression is a sinkhole on
the karst plain.

flat-lying beds of limestone is shown in Figure 11.37. The land surface consists of a mosaic of shallow, circular depressions called *sinkholes*, which make up a sinkhole plain. Except during short periods following heavy precipitation, there may be no surface runoff on a sinkhole plain. Variations of the topography shown in Figure 11.37 occur when limestone is overlain by beds of nonsoluble rock types.

Groundwater movement in karst aquifers has distinct aspects of recharge, flow, and discharge. In early stages of karst development, subsurface hydrology is similar to that of other aquifers. Groundwater movement is slow and follows minute cracks and joints within the rock. This condition is known as a *diffuse flow system* (Figure 11.38). Over time, flow paths grow unequally by solution so that flow becomes concentrated in subsurface conduits to form a *mixed flow system* (Figure 11.38), which retains some aspects of the diffuse system. Recharge becomes increasingly concentrated in point sources such as sinkholes and *swallets*, which are solution or collapse openings along stream channels through which surface streams lose flow to subsurface drainages. With further evolution, the flow system becomes a true *conduit flow system*, in which most recharge, flow, and discharge occur in well-defined subsurface conduits. Conduits may lie above or below the water table, if a water table can even be defined in such a flow system. Discharge occurs in springs of

▶ FIGURE 11.38

Cross sections showing diffuse, mixed, and conduit groundwater flow systems. Heavy black lines are cave passages. Flow systems evolve over time from diffuse to conduit. *Source:* From J. F. Quinlan and R. O. Ewers, 1986, Groundwater flow in limestone terranes: strategy, rationale, and procedure for reliable, efficient monitoring of groundwater quality in karst areas, in *Groundwater Flow in the Mammoth Cave Area, Kentucky, with Emphasis on Principles, Contaminant Dispersal, Instrumentation for Monitoring Water Quality, and Other Methods of Study*, National Water Well Association.

(a) Diffuse

(b) Mixed

(c) Conduit

various sizes and types. Large springs that discharge subsurface drainage to surface streams are called *rises*. Tracing the movement of groundwater in a conduit flow system is exceedingly complex. A common method involves injecting fluorescent dyes at recharge points and monitoring possible discharge points for the emergence of the dye.

An idealized cross section of a conduit flow system in the Mammoth Cave region of Kentucky is shown in Figure 11.39. Groundwater flows from recharge areas in the southeast to a discharge point in the northwest (Turnhole Spring), even though the surface topography is higher in the northwest! Notice that conduit flow is above the potentiometric surface toward the recharge end of the flow system and below the potentiometric surface throughout much of the downgradient area to the northwest. The strong control exerted by bedding on conduit location and orientation in these gently dipping rocks is apparent from the diagram.

Caves are simply large abandoned conduits formed by groundwater flow. Abandoned conduits are sometimes enlarged by roof collapse to form larger subsurface openings called *rooms*. Studies of major cave systems have shown that their evolution is quite complex and requires hundreds of thousands or even millions of years to achieve their current stage of development. Distinct levels of cave conduits can sometimes be related to episodes of landscape erosion, glacial cycles, or other geologic events.

▲ FIGURE 11.39

Cross section showing groundwater flow in the conduit flow system in the Mammoth Cave region. *Source:* From J. F. Quinlan and R. O. Ewers, 1986 Groundwater flow in limestone terranes: strategy, rationale, and procedure for reliable, efficient monitoring of groundwater quality in karst areas, in *Groundwater Flow in the Mammoth Cave Area, Kentucky, with Emphasis on Principles, Contaminant Dispersal, Instrumentation for Monitoring Water Quality, and Other Methods of Study*, National Water Well Association.

Case in Point 11.2
Groundwater Supply from a Karst Aquifer

The Edwards aquifer in Texas (Figure 11.40) is certainly one of the most important aquifers in the United States. In the San Antonio region (Figure 11.40), the aquifer supplies nearly all the municipal, domestic, and agricultural water needs. Within this area, San Antonio alone, the 10th-largest city in the United States, has a population of approximately 1 million. Because there are no other sources of water in the area, the U.S. Environmental Protection Agency (EPA) has designated the Edwards aquifer as a Sole Source Aquifer.

Three major hydrological divisions of the aquifer are recognized: the catchment area, the recharge area, and the artesian area (Figure 11.41). In the catchment area, surface streams carry water southward to the recharge area. The recharge zone coincides with a fault zone, the Balcones fault zone, and a steep topographic escarpment (Figure 11.42). Karst topography has developed in the recharge zone, and streams that collect water from the catchment area lose water to the groundwater flow system as they flow down the escarpment (Figure 11.43). Dams have been built on streams in the recharge area to enhance recharge to the aquifer. Some of these dams are designed to trap floodwater only, for recharge purposes, and do not contain a year-round reservoir. Natural discharge from the aquifer includes several large springs located near the base of the Balcones escarpment. The largest of these springs, Comal Springs, is located near the town of New Braunfels (Figures 11.40, 11.41). With an average discharge of 8000 L/s, Comal Springs is the largest spring in Texas.

In the recharge area, the aquifer is under unconfined conditions. South of the recharge area, in the Gulf coastal plain, the aquifer dips below younger formations and becomes confined to form the artesian area of the aquifer (Figure 11.41). Major withdrawals from

▲ FIGURE 11.40

Location of Edwards aquifer. *Source:* From R. K. Senger and C. W. Kreitler, 1984, *Hydrogeology of the Edwards Aquifer, Austin Area, Central Texas,* Bureau of Economic Geology, University of Texas at Austin, Report of Investigations 141.

▲ FIGURE 11.41

Map of Edwards aquifer showing catchment, recharge, and artesian areas. *Source:* From C. R. Burchett, P. L. Rethman, and C. W. Boning, 1986, *The Edwards Aquifer, Extremely Productive, But . . . .* U.S. Geological Survey.

Catchment area

During many storms, runoff from the catchment area is of such magnitude that much of it flows across the recharge area without percolating to the aquifer. Any type of engineered structure that holds or retards runoff on the plateau serves to prevent excessive runoff from flowing beyond the recharge area, and thus increases recharge to the aquifer.

Recharge area

Engineered structures that hold or retard runoff in the recharge area are the most beneficial for recharge enhancement.

Artesian area

Engineered structures built in the artesian area do not enhance recharge to the Edwards aquifer. The series of rocks above the Edwards Limestone do not readily transmit water, and serve as confining beds to the Edwards aquifer.

▲ FIGURE 11.42
Cross section of Edwards aquifer, showing components of flow system. *Source:* From C. R. Burchett, P. L. Rethman, and C. W. Boning, 1986, *The Edwards Aquifer, Extremely Productive, But....* U.S. Geological Survey.

◀ FIGURE 11.43
Sinkhole in recharge area of Edwards aquifer. *Source:* Photo courtesy of the author.

wells are made in this area for the city of San Antonio and smaller communities. Porosity and permeability in the confined area are very high due to the effects of the solution of the limestone. The southern boundary of the aquifer is a zone known as the "Bad Water Line" (Figure 11.41). This line represents a change in water quality rather than a

▶ FIGURE 11.44
Effects of pumping from Edwards aquifer on discharge at Comal Springs. *Source:* From C. R. Burchett, P. L. Rettman, and C. W. Boning, 1986 *The Edwards Aquifer, Extremely Productive, But* . . . . U.S. Geological Survey.

termination of the aquifer rocks. Groundwater south of the Bad Water Line is too mineralized for drinking purposes. One concern of water managers is that overpumping of the aquifer will cause mineralized water from the bad water zone to migrate up-dip into supply wells for the city of San Antonio. This would cause serious problems for the metropolitan area.

Although the Edwards aquifer is a prolific source of groundwater, droughts in the region can cause significant depletion of the water supply. One consequence of drought conditions is that flow to Comal Springs declines (Figure 11.44) as a result of decreased recharge and increased pumpage from the aquifer. Flow of the spring actually ceased for a short time in 1956. Maintenance of flow at Comal Springs is a high priority because the park containing the springs is a valued scenic and recreational area.

Groundwater Quality

Quality, or chemical composition of groundwater is as important as quantity in terms of groundwater supply. The most productive aquifer in the world will be useless if the water is highly mineralized or contains chemical compounds that make the water unsafe for drinking.

Groundwater almost always contains a higher dissolved mineral content than surface water. The minerals dissolved in groundwater, measured as *total dissolved solids* (TDS), are

Table 11.2 Classification of Groundwater Quality Based on TDS	
Water type	**Total dissolved solids (TDS) mg/L**
Fresh	0–1000
Brackish	1000–10,000
Saline	10,000–100,000
Brine	>100,000

derived from chemical reactions between the water and the soil and rock along its flow path from the point of infiltration to the point of sampling. All minerals and amorphous solids are soluble to some extent in water, despite great variations in the degree of solubility and the rates of dissolution. Minerals that originally precipitated from water, such as halite, gypsum, calcite, and dolomite, are most easily dissolved by circulating groundwater. The cementing agents of sedimentary rock, including calcium carbonate and silica, that originally precipitated from groundwater can return to solution as chemical conditions change through geologic time. A classification of water quality based on TDS is given in Table 11.2.

We have already considered one of the most important components of the groundwater chemical system, the weathering reactions that characterize the soil zone (Chapter 9). There, the CO_2-charged infiltrating water dissolves some minerals and reacts with others to produce clays. In the process, silica, bicarbonate, sulfate, and some cations may reach the water table during recharge. In the aquifer, additional chemical processes may modify the composition of the initial recharge.

Ultimately, the chemical compounds and concentrations in groundwater at any point in the flow system are determined by the type of solid materials it has encountered along the way. A water sample taken from an aquifer in limestone and dolomite will contain high concentrations of calcium, magnesium, and bicarbonate. The calcium and magnesium ions impart a condition called *hardness* to the water. Hard water, although not harmful to drink (and perhaps even beneficial to health), produces several undesirable effects, such as scale on plumbing fixtures and boilers and a lack of soap suds. Waters in contact with gypsum will be rich in sulfate. Sedimentary basins containing highly soluble rocks will yield groundwater with very high concentrations of sodium and chloride. Iron is another parameter of concern for reasons other than health. Iron concentrations above 0.3 mg/L will produce stains and solid precipitates on plumbing and clothing. The relative proportions of inorganic elements, ions, and compounds are listed in Table 11.3. In terms of concentration, various combinations of a small number of major constituents dominate groundwater quality.

Concentration Units

The most commonly encountered concentrations are *mass concentrations*. Laboratory analyses of groundwater usually report concentrations in mg/L, which means the mass of the solute in milligrams per liter of solution. For dilute groundwaters that fall within the fresh category in Table 11.2, milligrams per liter are numerically equal to *parts per million* (ppm) because a liter of water has a mass of 10^6 milligrams. As groundwater becomes progressively more mineralized, this approximate equality becomes less accurate. *Molar concentrations* are expressed as moles per liter or millimoles per liter, in which a mole is the gram molecular weight of a solute in a liter of solution. *Normal concentrations*, such as equivalents per liter (EPL), are derived by multiplying the molarity by the valence of the solute species. All three systems of concentration units are used in some situations.

Table 11.3   Classification of Dissolved Inorganic Constituents in Groundwater

Major constituents (>5 mg/L)

Bicarbonate	Magnesium	Sodium
Calcium	Silica	Sulfate
Chloride		

Minor constituents (0.01–10.0 mg/L)

Boron	Iron	Potassium
Carbonate	Nitrate	Strontium
Fluoride		

Trace constituents (<0.1 mg/L)

Aluminum	Iodide	Selenium
Antimony	Lanthanum	Silver
Arsenic	Lead	Thallium
Barium	Lithium	Thorium
Beryllium	Manganese	Tin
Bismuth	Molybdenum	Titanium
Bromide	Nickel	Tungsten
Cadmium	Niobium	Uranium
Cobalt	Phosphate	Vanadium
Copper	Platinum	Ytterbium
Gallium	Radium	Zinc
Germanium	Rubidium	Zirconium
Gold	Ruthenium	
Indium	Scandium	

Source: From R. A. Freeze and J. A. Cherry, *Groundwater*, © 1979 by Prentice Hall, Inc., Englewood Cliffs, N.J.

Conversion of units is often necessary. To convert mg/L to molar concentrations, the following formula is used:

$$\frac{\text{moles}}{\text{L}} = \frac{\text{mg/L (ppm)}}{1000 \times \text{formula weight}} \qquad (11.11)$$

It follows that

$$\frac{\text{mg}}{\text{L}}(\text{ppm}) = \frac{\text{moles}}{\text{L}} \times 1000 \times \text{formula weight} \qquad (11.12)$$

EXAMPLE 11.4

A water sample from an aquifer contains 60 mg/L calcium. What are the molar and normal concentrations?

Solution

Since calcium has an atomic weight of 40 g/mole, then

$$\frac{\text{moles}}{\text{L}} = \frac{60 \text{ mg/L}}{1000 \times 40} = 0.0015 \frac{\text{moles}}{\text{L}}$$

In solution, dissolved calcium is dominated by the calcium ion with a valence of +2. Therefore,

$$\frac{\text{equivalents}}{L} = 0.015\frac{\text{moles}}{L} \times 2 = 0.003 \text{ EPL}$$

Graphical Displays of Groundwater Quality

Groundwater concentrations can be plotted on several types of diagrams in order to create a visual image of water quality. This is particularly useful when comparing multiple water-quality analyses or looking for trends in an aquifer. The *Stiff graph* (Figure 11.45) is constructed by plotting certain concentrations on horizontal axes. The units used, equivalents per million, are numerically equal to equivalents per liter. Each geometric figure represents one water analysis. With practice, water-quality changes are apparent at a glance. The five Stiff diagrams in Figure 11.45, for example, represent samples taken along a groundwater flow path in which part of the aquifer has been contaminated. The figures at the top and bottom have the same shape and represent the natural water quality. The three figures between show the constituents affected by the contamination, as well as the changes in concentration.

Another type of graph is the *Piper diagram* (Figure 11.46), which contains two triangles and a diamond. Cations are plotted as a point on the left triangle and anions on the right. The values are calculated as percentages of cations or anions in equivalents per liter. For example, each apex of the cation triangle represents 100% calcium, magnesium, or sodium plus potassium. Waters that contain mixtures of the three plot somewhere in the center of the diagram. If a water plots near one of the apices, the water can then be classified as calcium, magnesium, or sodium plus potassium type. The same procedure is followed in plotting and classifying the anions. Each water analysis is then represented by one point on the cation triangle and one point on the anion triangle. These points can be projected

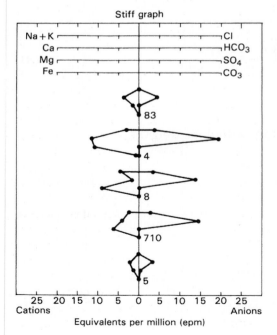

Stiff graph

Equivalents per million (epm)

◀ FIGURE 11.45
Stiff diagrams. The sequence from top to bottom shows changes in water quality caused by contamination from a waste-disposal site.

► FIGURE 11.46
Piper diagram. Each water analysis includes one point in the cation triangle and one point in the anion triangle. These points are projected into the diamond to form one point where lines intersect.

onto the diamond along lines parallel to the sloping lines shown (Figure 11.46). A point on the diamond is plotted where the two lines intersect. Many groundwater analyses can be plotted on a Piper diagram. Clusters or linear trends will sometimes provide information on chemical processes operating in the aquifer.

Chemical Evolution

Systematic changes in groundwater quality are sometimes observed in groundwater flow systems. The changes, which can be called *chemical evolution* of groundwater, are most prevalent in flow systems that penetrate deep into thick sedimentary rock sequences. A gradual increase in TDS is the most common trend along deep flow paths. Along with higher TDS, dominant ions change with distance in the direction of flow. In the recharge area, bicarbonate (HCO_3^-) is usually the dominant anion. As the groundwater moves deeper along the flow path, it is more likely to encounter evaporite-bearing rocks such as gypsum and halite. An idealized evolutionary sequence would proceed from bicarbonate to sulfate to chloride as dominant anions. Evolutionary trends in cations are not as well developed because of ion exchange and other chemical processes. Figure 11.47 shows the possible relationships between water quality and flow systems. The left side of the diagram has a uniformly sloping surface that yields bicarbonate-type waters in the upper left corner of the cross section, sulfate-type waters appear in the center, and chloride-type waters are beneath the discharge area. This schematic is based on the assumption that the appropriate rock types are present to supply the anions shown. On the right is a cross section with a hilly, or hummocky, land surface profile. This situation tends to produce an abundance of local flow systems (Figure 11.9), which involve recharge at individual topographic highs and discharge at adjacent topographic lows. The local flow systems dominate the shallow region of the cross section and commonly contain only bicarbonate-type groundwater. Flow is relatively rapid in the local flow systems and any salts that may have been present are flushed out of the system. The longer intermediate and regional flow systems differ in water quality from the local systems. Regional flow systems begin at the highest

▲ FIGURE 11.47
Chemical evolution of groundwater in sedimentary basins. The left side has a uniformly sloping water table and a predominance of regional flow. The right side has a hummocky water table and local, intermediate, and regional flow systems. *Source:* Modified from J. Tóth, *Proceedings, First Canadian/American Conference on Hydrogeology.*

recharge area (at the upper right-hand corner of the cross section) and follow deep paths to the lowest discharge point at the center of the diagram. Intermediate systems are recharged at intermediate topographic highs but descend lower than the local flow systems. Water quality is bicarbonate-type in the intermediate and regional systems, but evolves to sulfate- and chloride-types with distance along the flow path. Flow velocities are extremely slow in these deep systems and salts are not flushed out. TDS can reach the brine level (Table 11.2) in deep basins, and residence times on the order of millions of years are possible. Some of the water in these deep formations may even have originated as seawater trapped within the sediments during deposition.

Groundwater and Construction

In contrast to hydrogeologists who devote their careers to finding and developing groundwater resources, engineers involved in construction of dams, tunnels, buildings, and highways spend much of their time trying to control or dispose of groundwater.

Dams illustrate the potential problems of groundwater movement, or seepage, quite well. Dams can be designed to prevent internal seepage, as in the case of masonry dams, or, as in the case of earth dams, to allow controlled seepage through the embankment at a safe rate. In the former case, excessive seepage beneath or around the dam is the major concern. Because the impoundment of the reservoir behind the dam increases the head within the groundwater flow system, extremely high fluid pressures decrease the stability of the dam by generating uplift pressures on the base of the structure or by causing internal erosion, or piping, of material near the downstream toe of the dam, thus tending to undermine it. Most dam failures have resulted from inadequate control of groundwater seepage through and beneath the structure. The St. Francis Dam (Case in Point 6.1) and the Teton Dam (Case in Point 11.3) are examples.

Groundwater seepage beneath and through dams is studied by constructing flow nets of the type shown in Figure 11.48. From the flow net, seepage pressure and gradient can be calculated so that the dam design can be modified if necessary. Seepage control beneath dams is sometimes accomplished by injection of cement *grout* under high pressure into closely

► FIGURE 11.48
Flow nets showing groundwater flow systems beneath impermeable masonry dam (a) and through permeable earth dam (b). *Source:* From R. A. Freeze and J. A. Cherry, *Groundwater*, © 1979 by Prentice Hall, Inc., Englewood Cliffs, N.J.

(a)

(b)

► FIGURE 11.49
Dewatering by well points to lower the water table below the base of an excavation and prevent excessive inflow of water into the excavation.

spaced boreholes before the dam is constructed (Case in Point 8.1). After the grout penetrates cracks and joints in the rock, it sets up, providing a barrier to groundwater seepage. Various design measures are also incorporated in permeable earth dams for seepage control.

Groundwater inflow into tunnels and excavations often presents difficult engineering problems. In tunneling through rock, workers often suddenly encounter water in joints, fractures, or fault zones. Many lives are lost because of tunnel flooding when water under high pressure unexpectedly bursts into the tunnel. Seepage into excavations often leads to slumping and sliding of the walls of the cut. In Chapter 13 we will consider the importance of groundwater in landslide phenomena of all types. If seepage into an excavation is determined to be undesirably rapid, *dewatering* techniques can be applied. One method, as shown in Figure 11.49, is to install a system of shallow pumping wells, or *well points*, around the excavation to lower the water table until the project can be completed.

Case in Point 11.3
Failure of the Teton Dam

The failure of the Teton Dam in 1976 has proved to be one of the most costly dam failures in history. Constructed on the Teton River in southern Idaho, the dam was designed to provide water for irrigation and also to function as a flood control facility. The reservoir impounded by the dam was 17 miles long and had a capacity of 200,000 acre-feet. Construction of the dam was completed in 1975, and filling of the reservoir began in the fall of that year. On June 5, 1976, before reservoir filling was complete, the dam failed with very little warning, sending a flood wave down the Teton River. The first indications of problems were found on June 3, when two small seeps were detected 1000 and 1500 feet

▲ FIGURE 11.50
Increase in flow from the leak near the right abutment about 10:30 A.M. Dozers were lost in the pit about this time. *Source:* From U.S. Dept. of Interior, 1977, *Failure of Teton Dam*, U.S. Government Printing Office.

▲ FIGURE 11.51
Breach of the crest of the dam about 11:55 A.M., causing a great increase in the discharge rate. Complete failure came minutes after. *Source:* From U.S. Dept. of Interior, 1977, *Failure of Teton Dam*, U.S. Government Printing Office.

downstream from the dam. On June 4, another small seep was found 150 feet downstream of the right abutment, but as the day ended, no serious leaks were evident. The next morning, the first workers to arrive at the dam found small leaks at the toe of the right abutment and about halfway up the downstream face. As the morning progressed, the situation quickly became critical. About 10:30 A.M., the higher leak suddenly began to flow at a much greater rate following the collapse of a portion of the surface of the dam into a void below (Figure 11.50). Bulldozers were sent over the dam to push rock into the hole in an attempt to stop the leak. Instead, the flow grew stronger and at 11:30, the dozers slid into the hole and were washed downstream. Luckily, the operators were able to escape. The now-growing hole progressed quickly up the face of the dam toward the crest, and at 11:55, the crest was breached and total failure began (Figure 11.51). In

▲ FIGURE 11.52
Progressive failure of the dam by the enlarging area of seepage. Cross-ruled area and arrows show the location of seepage and piping. Sinkhole formation and collapse of the surface shown in (e). The black area represents material collapsed from above. Zone 1: silt core of dam; Zone 2: coarser alluvial cover. *Source:* From U.S. Dept. of Interior, 1977, *Failure of Teton Dam*, U.S. Government Printing Office.

▲ FIGURE 11.52
(*Continued*)

only 5 hours, the reservoir was completely emptied, releasing a discharge to the river below equal to the Mississippi River in flood.

The consequences of the dam failure included the loss of 11 lives and the displacement of 25,000 people from their homes due to the effects of the flooding. The monetary costs of the disaster were staggering. The $86 million cost of the dam, which was not rebuilt, was only the beginning. Damage claims paid by the U.S. Bureau of Reclamation were in the neighborhood of $400 million.

How could a dam designed and built with methods used many times in the past fail before it was filled for the first time? To answer this question, we must examine the geology of the site and the design of the dam.

The dam was constructed in an area of Late Cenozoic rhyolite welded ash-flow tuffs. These volcanic pyroclastic materials, which were thoroughly studied prior to design, were known to be highly fractured and jointed. Extensive grouting was carried out prior to construction, particularly in a cutoff trench excavated into the rhyolite tuff beneath the dam (Figure 11.52). The dam was designed as a zoned earth-fill embankment, with a core (Zone 1) of compacted silt derived from windblown deposits nearby. The selection of this nonplastic material may have contributed to the failure. Coarser alluvium (Zone 2) was emplaced over the core to protect it from surface erosion.

Although the exact cause of failure may never be known, the sequence of events illustrated in Figure 11.52 is a plausible explanation for the failure. The hypothesis includes the supposition that the Zone 1 material in the core of the dam was subject to a certain amount of cracking during settlement of the embankment. A more plastic core material would not have been susceptible to cracking. The result of the cracking would be to create preferential pathways for water movement. Figure 11.52a shows seepage through the fractured rock beneath the dam as well as through the Zone 1 soil in the cutoff trench. Because of the high hydraulic gradients caused by the reservoir behind the dam, Zone 1 soil at the contact with bedrock downstream from the cutoff trench began to be eroded by piping. The danger of a situation such as this is that removal of soil concentrates more flow in the zone of piping, leading to still more groundwater flow. This situation is illustrated in Figures 11.52c and 11.52d, where the zone of piping eventually breaks through Zone 2 soil, leading to the formation of sinkholes and collapse on both the upstream and downstream faces of the dam. After this point, failure is inevitable, as more and more of the core is washed away. Among the lessons learned from the Teton Dam failure is that even more thorough geologic investigations of dam sites are necessary prior to and during design and construction of these critical structures.

Summary and Conclusions

Groundwater is a vital resource; development and management of these resources are the keys to any major expansion of our water supplies in the future. Groundwater flow from recharge areas to discharge areas is governed by the parameters of Darcy's Law: aquifer cross-sectional area, hydraulic gradient, and hydraulic conductivity. Recharge involves the downward movement of water through the unsaturated zone to the water table. The water table is recognized as the level at which fluid pressure is equal to atmospheric pressure.

Geologic materials with high hydraulic conductivity are known as aquifers. Unconfined aquifers lie directly beneath the unsaturated zone and are bounded above by the water table. Confined aquifers, on the other hand, can occur at considerable depths. The conditions required for development of confined aquifers include confining aquitards both above and below. Water levels in wells penetrating confined aquifers rise to the potentiometric surface and can rise above the land surface in flowing artesian wells.

Pumping of wells creates a cone of depression in either the water table or in the potentiometric surface. Excessive pumping can lead to steadily dropping water levels and depletion of the resource.

Aquifers are common in surficial sediments, which are especially thick in tectonic valleys and coastal plains. Glacial deposits contain important unconfined aquifers in alluvial outwash sediments and buried-valley deposits. Aquifers in rocks often are located within fractured and jointed zones.

Groundwater quality depends upon the composition of rocks or soils with which the water comes in contact. Water analyses can be displayed on Stiff and Piper diagrams to aid in the classification of groundwater and in making comparisons with other analyses. Groundwater chemically evolves by increases in TDS and changes in major ion content. These evolutionary changes are most evident in intermediate or regional flow systems.

Construction activities often require groundwater control. Uncontrolled groundwater seepage threatens the stability of dams and leads to hazardous conditions in tunnels and excavations. Dewatering methods may be necessary for completion of the project.

Problems

1. For what reasons would you need to use Darcy's Law in the field rather than in the laboratory?

2. Flow occurs through a sand-filled pipe with a cross-sectional area of 60 cm^2. If K is 10^{-3} cm/s and the gradient is 0.01, what is the discharge from the pipe?

3. Water seeps vertically downward from the bottom of a lake through a uniform material to an aquifer below. The discharge is 0.005 ft^3/s through a 1 ft^2 area of the lake bottom. The lake is 50 ft deep and the elevations shown on the diagram are given in feet above sea level. What is the hydraulic conductivity of the material between the lake and the aquifer? Assume there is no horizontal flow in the aquifer. (*Hint*: The hydraulic head at the lake bottom is equal to the elevation of the lake.)

4. The bottoms of piezometers A and B are on the same flow line and 1000 m apart. The bottom of piezometer A is at 130 m and its water level is at 160 m. The bottom of piezometer B is at 100 m and the water level is 150 m. What is the hydraulic gradient?

5. Why isn't the water table defined as the upper limit of the saturated zone?

6. Is it correct to say that groundwater flows parallel to the water table? Why or why not?

7. What is an aquifer? Why aren't all saturated rocks and soils considered to be aquifers?

8. How do confined and unconfined aquifers differ in their response to pumping?

9. A well drilled into an artesian aquifer is 30 m deep. The fluid pressure in the aquifer is 392 kPa. Will the well be a flowing well?

10. The water table of a saturated, unconfined aquifer coincides with the land surface. The aquifer is 52 ft thick and the material has a unit wet weight of 135 lb/ft^3. What is the effective stress at the base of the aquifer?

11. What factors control the amount of groundwater recharge?

12. What rock units or sediments constitute the most productive aquifers in your area? How does groundwater impact the economy of the region?

13. How do groundwater flow systems in karst aquifers differ from those in other aquifers?

14. What would be the chemical characteristics of groundwater from a limestone aquifer?

15. Water analyses follow for samples from two wells in different areas. Calculate the molarity and normality of each constituent, and plot them on Stiff and Piper diagrams.

Ion (mg/L)	Well A	Well B
HCO$_3^-$	311	159
SO$_4^{2-}$	17.8	1902
NO$_3^-$	0.34	7
K$^+$	0.65	41
Na$^+$	2.26	574
Mg$^+$	22.4	171
Fe (total)	0.95	—

16. Why is groundwater sometimes a problem at construction sites? How can the problems be solved?

References and Suggestions for Further Reading

ALLEY, W. M., T. E. REILLY, and O. L. FRANKE. 1999. *Sustainability of Ground-Water Resources*. U.S. Geological Survey Circular 1186.

BURCHETT, C. R., P. L. RETTMAN, and C. W. BONING. 1986. *The Edwards Aquifer, Extremely Productive, But....* U.S. Geological Survey in cooperation with the Edwards Undergroundwater District.

FREEZE, R. A., and J. A. CHERRY. 1979. *Groundwater*. Englewood Cliffs, N.J.: Prentice Hall, Inc.

GUTENTAG, E. D., F. J. HEIMES, N. C. KROTHE, R. R. LUCKEY, and J. B. WEEKS. 1984. *Geohydrology of the High Plains Aquifer in Parts of Colorado, Kansas, Nebraska, New Mexico, Oklahoma, South Dakota, Texas, and Wyoming*. U.S. Geological Survey Professional Paper 1420-B.

HEATH, R. C. 1984. *Ground-Water Regions of the United States.* U.S. Geological Survey Water Supply Paper 2242.

HUBERT, M. K. 1940. The theory of groundwater motion. *Journal of Geology*, 48:785–944.

MASTERS, G. M. 1991. *Introduction to Environmental Engineering and Science.* Englewood Cliffs, N.J.: Prentice Hall, Inc.

QUNILAN, J. F., and R. O. EWERS. 1986. *Groundwater Flow in the Mammoth Cave Area, Kentucky, with Emphasis on Principles, Contaminant Dispersal, Instrumentation for Monitoring Water Quality, and Other Methods of Study,* Field Trip Guide Book. Dublin, Ohio: National Water Well Association.

SENGER, R. K., and C. W. KREITLER. 1984. *Hydrogeology of the Edwards Aquifer, Austin Area, Central Texas.* Bureau of Economic Geology, University of Texas at Austin, Report of Investigations No. 141.

STRAW, W. T., R. N. PASSERO, and A. E. KEHEW. 1993. Conceptual hydrogeologic glacial facies models: Implications for aquifers and agrichemicals in southwest Michigan, in *Environmental Impacts of Agricultural Activities: Hydrogeologic Investigations and Modeling*, Y. Eckstein and A. Zaporozec, eds. Water Environment Federation, Alexandria, Va. pp. 35–68.

TÓTH, J. A. 1963. A theoretical analysis of regional groundwater flow in small drainage basins. *Journal of Geophysical Research*, 68 (no. 16):4795–4811.

TÓTH, J. A. 1984. The role of regional gravity flow in the chemical and thermal evolution of groundwater, in *First Canadian/American Conference on Hydrogeology, Practical Applications of Groundwater Geochemistry*, B. Hitchon and E. I. Wallick, eds. Dublin, Ohio: National Water Well Association.

U.S. DEPARTMENT OF INTERIOR TETON DAM FAILURE REVIEW GROUP. 1977. *Failure of Teton Dam.* Superintendent of Documents, U.S. Government Printing Office, Stock No. 024-003-00112-1.

WHITE, W. B., and E. L. WHITE. 1989. *Karst Hydrology.* New York: Van Nostrand Reinhold.

CHAPTER 12

Subsurface Contamination and Remediation

Contamination of soil and water is a problem common to all human societies. Surface-water contamination probably has been recognized throughout the history of civilization. In industrialized countries, treatment of surface water for drinking became common in the late 19th century and health problems linked to impure drinking water are now rare. Less developed countries, however, are still faced with a lack of safe drinking water.

During the past several decades, a new problem has come to light: contamination of the subsurface. This problem is much more serious in developed countries because of their history of industrialization and the wide range of chemicals that have been introduced, either by accident or by design, to the underground environment. Much of the damage was caused by ignorance rather than intentionally. We simply did not comprehend the degree to which contaminants could migrate beneath the land surface or the difficulty we would encounter in tracing and removing them after discovery.

The response to subsurface contamination in the United States, as in many other countries, has been a massive effort to define the extent of contamination and to remediate the subsurface—starting with the most contaminated sites. This environmental campaign has been driven by governmental regulations dealing with waste handling and disposal, as well as with many other potentially contaminating activities. Billions of dollars have been spent on these efforts by the private sector and all levels of government. Although much remains to be done, many of the most hazardous sites have been partially or totally cleaned up.

Sources of Subsurface Contamination

The range of activities that cause subsurface contamination is much larger than most engineers and scientists would have guessed even a few years ago. A representative listing in Table 12.1 is divided into categories based on whether the source of the contamination originates on the land surface, within the unsaturated zone, or below the water table. Several of these activities are illustrated in Figure 12.1.

Subsurface Contamination from the Land Surface

Contamination derived from land-surface activities includes a variety of common practices and occurrences. Although contamination of rivers and other surface-water bodies is not the primary focus of this chapter, the linkages between groundwater and surface water described in Chapter 11 make it likely that contamination of one will lead to contamination of

Table 12.1   Sources of Groundwater Contamination

Groundwater Quality Problems That Originate on the Land Surface

1. Infiltration of contaminated surface water
2. Land disposal of solid and liquid waste materials
3. Stockpiles, tailings, and spoil
4. Dumps
5. Salt spreading on roads
6. Animal feedlots
7. Fertilizers and pesticides
8. Accidental spills
9. Particulate matter from airborne sources
10. Composting of leaves and other yard wastes

Groundwater Quality Problems That Originate Above the Water Table

1. Septic tanks, cesspools, and privies
2. Surface impoundments
3. Landfills
4. Waste disposal in excavations
5. Leakage from underground storage tanks
6. Leakage from underground pipelines
7. Artificial recharge
8. Sumps and dry wells
9. Graveyards

Groundwater Quality Problems That Originate Below the Water Table

1. Waste disposal in wet excavations
2. Agricultural drainage wells and canals
3. Well disposal of wastes
4. Underground storage
5. Secondary recovery
6. Mines
7. Exploratory wells and test holes
8. Abandoned wells
9. Water-supply wells
10. Groundwater development

Source: Modified from U.S. EPA, 1990, *Ground Water, Volume 1: Ground Water and Contamination.* EPA/625/6-90/016a.

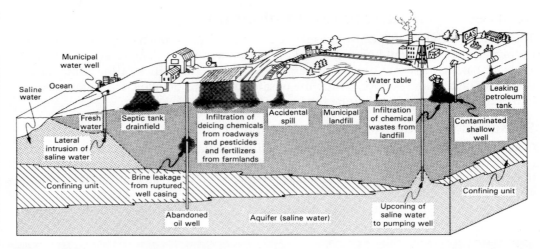

▲ FIGURE 12.1
Major sources and processes causing subsurface contamination. *Source:* From C. W. Fetter, *Contaminant Hydrogeology*, © 1993 by Macmillan Publishing Company. Reprinted with permission.

◀ FIGURE 12.2
Contamination of a well by induced infiltration from a stream.

the other. Infiltration of contaminated surface water may take place when shallow water-supply wells draw water from an alluvial aquifer adjacent to a stream (Figure 12.2). The cone of depression imposed by pumping the well or well field creates a gradient on the water table directed toward the well. In fact, wells are deliberately installed near streams and rivers in order to induce recharge from the stream and to provide high yields with low drawdowns. Contamination of the well field can occur if the stream is polluted. An example of this was documented at Lincoln, Nebraska, where shallow municipal water-supply wells are sited in an alluvial valley aquifer adjacent to the Platte River. Figure 12.3 shows the arrival in the Platte River of a large concentration peak of atrazine, a widely used preemergent herbicide for corn. Peaks of this type in streams are common after application of the herbicide in the spring. Wells near the river showed the same peaks in concentration, slightly delayed in arrival because of transport from the streambed through the aquifer to the well.

Land disposal or stockpiling of wastes or materials is a common practice. Liquid or solid wastes from sewage treatment plants, food processing companies, and other sources are applied to agricultural lands, golf courses, and other areas to serve as fertilizer and irrigation. The objective is to allow chemical and biological processes in the soil, along with plant uptake, to break down the waste products into harmless substances. Although the concept may be generally successful, if any of the wastes are water soluble and mobile, they are typically carried into the subsurface. A problem may arise if the fields are underlain by shallow aquifers.

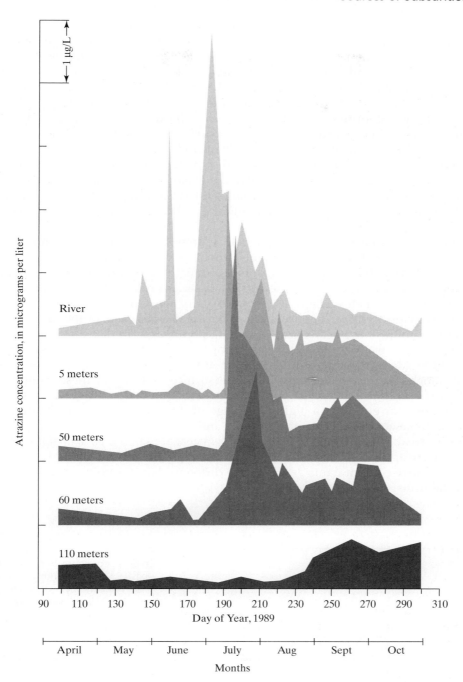

▲ FIGURE 12.3

Variations in atrazine concentrations with time in the Platte River near Lincoln, Nebraska and in wells located close to the river. *Source:* From D. Duncan, D. T. Peterson, T. R. Shepard, and J. D. Carr, 1991, Atrazine used as a tracer of induced recharge, *Ground Water Monitoring Review*, 11:144–150, reprinted by permission by the National Ground Water Association.

Stockpiles include substances such as road salt, which is highly soluble in water and which may enter the subsurface easily. Dumps could contain a wide variety of potential contaminants. One recent regulatory trend is to exclude leaves and yard wastes from land-fills in order to save landfill space. An alternative for these materials is large-scale surface composting. Because these leaves are collected from yards and streets, they may contain unwanted substances, such as animal wastes and petroleum contaminants.

Uncontrolled dumping is prohibited in most industrialized countries today, but old dumps may still constitute a threat of subsurface contamination. The spreading of salt on highways is a widespread practice in northern climates. In addition to causing deterioration of bridges, automobiles, and the roadway itself, the salt quickly leaches below the land surface. Figure 12.4 shows the steady rise in chloride, an ion derived from salt, in three city wells in a municipal well field in Kalamazoo, Michigan. The in-creases correlate with increases in the amount of salt applied to the city streets during the time period.

Fertilizers and pesticides yield several different types of contaminants. Nitrogen fertilizers are applied to stimulate plant growth, but often in greater quantities than plants can utilize at the time of application. Nitrate, the most common chemical form of these fertilizers, can be easily leached below the plant root zone by rainfall or irriga-tion. Once it moves below the root zone, it usually continues downward to the water table. Worldwide fertilizer applications increased dramatically in the last half of the 20th century to meet the needs of the growing population. In many areas, these rising application rates were accompanied by rising nitrate concentrations in groundwater (Figure 12.5). Most pesticides and herbicides are not as mobile as nitrate, but they are toxic at much lower concentrations. The subsurface mobility of these compounds is strongly related to the characteristics of the unsaturated and saturated zones. In some karst aquifers, for example, a high percentage of domestic wells in agricultural areas have detectable concentrations of pesticides and herbicides. Fertilizers and pesticides, along with road salt and land application of wastes, are known as *nonpoint sources*, be-cause they are applied to large surface areas. The regulatory and remedial strategies for nonpoint-source contaminants are therefore different than for point sources of wastes, such as landfills.

◀ FIGURE 12.4
Chloride concentrations in three wells (1, 2, and 3) in municipal well field in Kalamazoo, Michigan. *Source:* Data from Kalamazoo Dept. of Public Utilities.

► FIGURE 12.5
Fertilizer use (solid line) and median nitrate concentration in groundwater (shaded boxes) in the eastern San Joaquin Valley, California. *Source:* From U.S. Geological Survey Circular 1225.

Accidental spills, which occur at manufacturing plants and on roads and railroads, are disturbingly common. Containment and cleanup are relatively effective if initiated soon after the spill, but if the spill is discovered long afterward, the extent of migration of the pollutant can be great. One of the best documented spills was a pipeline rupture in Bemidji, Minnesota. The pipeline, carrying crude oil, ruptured in 1979. Before the rupture could be repaired, 10,700 barrels of oil were released on and just below the land surface. About a fourth of the oil remained in the subsurface after cleanup. As the oil gradually migrated through the unsaturated zone to the water table (Figure 12.6) it formed a pool at the top of the capillary fringe. Oil, being less dense than water, is classified as a "floater"

▲ FIGURE 12.6
Cross section of the Bemidji, Minnesota oil pipeline rupture. The oil pools at the top of the water table and more soluble compounds dissolve into groundwater and move downgradient. *Source:* From USGS Fact Sheet 084-98.

because it accumulates at the top of the water table rather than sinking into the saturated zone. Even though most of the compounds making up crude oil are not very soluble, a dissolved plume of the more soluble compounds formed and moved downgradient within the groundwater flow system. Distinct geochemical zones developed in the plume (Figure 12.6) as a result of biochemical reactions between the oil components, water, and aquifer solids. Most of these reactions involved microorganisms such as bacteria. The U.S. Geological Survey has used the Bemidji oil spill site as a natural laboratory to study the physical and chemical behavior of oil in the subsurface.

Particulate matter from airborne sources is thought to be minor compared with some pollutants from other sources of contamination, but it can be very serious in localized areas. Examples include the accumulation of heavy metals around smelters and exhaust fumes that are associated with auto emissions in urban areas.

Subsurface Contamination Sources in the Unsaturated Zone

Contamination originating in the unsaturated zone includes some of the most common types. Shallow burial of wastes, which was mostly unregulated before the 1970s, has been the most common method of disposal of most solid waste for many years. Experience has shown that the leaching of these wastes from the unsaturated zone to the saturated zone has been very common. In the United States, solid wastes are now separated for disposal into generally inert materials, municipal and household wastes, and hazardous wastes. Disposal in landfills of the latter type, in what are known as *secured landfills*, is subject to the most stringent regulations and requirements, including detailed site investigations, engineered leachate barriers, and regular monitoring of the unsaturated and saturated zones (Figure 12.7). *Sanitary landfills*, the type still used for municipal waste, can be designed in several ways (Figure 12.8). All designs include the construction of cells of compacted refuse with layers of soil added to minimize contaminant movement and to limit the infiltration of water (Figure 12.9). The current trend is toward greater isolation of these wastes by the use of impermeable liners and covers. Additional design components may include *leachate collection systems* in the secured landfills

▲ FIGURE 12.7
Design components of a secured landfill.

▲ FIGURE 12.8
Types of sanitary landfills include (a) the trench method, (b) the area method, and (c) the slope method.

(Figure 12.7), in which any leachate produced in the landfills is drained to a sump on one corner and then pumped to the surface for treatment. Some engineers and scientists have criticized these "dry tombs," however, because of the likelihood that even thick, plastic liners will tear or rupture at some time in the future. Thus, by sealing the cells and excluding water, we may be only delaying the eventual failure of the liners and subsequent contamination.

Old landfills continue to cause subsurface contamination in the United States, where there may be as many as 100,000 active and inactive sites. The Norman, Oklahoma landfill is a closed landfill that has produced a large leachate plume (Figure 12.10) that is moving toward the Canadian River under the influence of the groundwater gradient. The plume contains many inorganic and organic species. The distribution of nonvolatile dissolved organic carbon, which is a general indicator of the organic compounds in the landfill, is

◄ FIGURE 12.9
Photo of a waste cell being filled in a landfill of the area type in southwestern Michigan. *Source:* Photo courtesy of the author.

◄ FIGURE 12.10
Extent of the groundwater contaminant plume from the Norman, Oklahoma landfill. *Source:* From S. C. Christenson, and I. M. Cozzarelli, U.S. Geological Survey Fact Sheet FS-040-03.

shown along a cross-sectional view of the plume in Figure 12.11. Groundwater users with wells downgradient from landfills of this type must be protected from contaminated groundwater in the plume, either by cleanup of the plume, abandonment of the well in favor of alternative water sources, or both.

Surface impoundments (Case in Point 12.1) are used for storage and disposal of liquid wastes. Although they are now strictly regulated, in the past seepage from the base of the

▶ FIGURE 12.11
Cross section of the Norman, Oklahoma, landfill contaminant plume along line A-A' showing concentrations of nonvolatile organic carbon, a constituent leached from the organic waste in the landfill. Location of cross section shown in Figure 12.10. Concentration is expressed in millimoles per liter. *Source:* From S. C. Christenson, and I. M. Cozzarelli, U.S. Geological Survey Fact Sheet FS-040-03.

impoundments was extremely common and many examples of extensive subsurface contamination are known.

Septic systems, including leach fields and dry wells, are used for disposal of household liquid waste in unsewered rural and suburban locations. Waste solids are retained in a septic tank (Figure 12.12) and liquids are discharged to a leach field or dry well, which is simply a large buried container of some type with holes in the sides, to promote infiltration of the liquid into the soil. The original premise of these systems was that the movement of the waste liquid through the soil would renovate the water quality and serve as a treatment mechanism. Although some treatment does occur, depending on soil type and other conditions, contaminants are known to reach groundwater with a high frequency. Septic systems that are installed in unsuitable sites, where, for example, soils are impermeable or bedrock is close to the surface, tend to back up through the soils to the ground surface, creating an unsanitary situation. These systems are said to have "failed." In reality, the successful systems, with higher leaching capacity, are as much or more of a long-term environmental threat because of their impacts on groundwater.

The number of underground storage tanks (USTs) in the United States containing petroleum fuels or other hazardous liquids is currently estimated by the U.S. Environmental Protection Agency at about 680,000. This number is far fewer than were present prior to the EPA regulatory program that went into effect in 1988. The new standards required by this program included (1) tank replacement with a tank of improved performance standards; (2) closure with corrective action, if needed; or (3) an approved upgrade to the existing system. New tanks are composed of fiberglass-reinforced plastic or other corrosion-resistant designs. Numerous tank excavations have shown that corrosion and leakage were widespread among older tanks, most of which were constructed of steel. At the end of 2003, the total number of releases that had occurred from USTs either through leaks in the tank or connecting piping or by spills and overfills occurring during refueling stood at 439,385. Progress under the EPA regulatory program has been significant, with 303,120 cleanups of these releases completed by the end of 2003. Although this still leaves a backlog of more than 130,000 releases to be remediated, this number will decline, as the rate of detection of new releases is much lower than at the start of the program.

Other sources of pollution in the unsaturated zone are less common but may cause serious problems when they occur. Leakage from underground pipelines, such as the

▲ FIGURE 12.12

Design of a domestic septic system. Waste solids are captured in the septic tank and liquid effluent is discharged to the leach, or drain, field. Infiltration through the unsaturated zone is assumed to remove contaminants, although nitrates commonly reach the groundwater table in areas of coarse-grained soils. *Source:* From N. C. Brady and R. R. Weil, *The Nature and Properties of Soil*, 13th ed., © 2002 by Prentice Hall, Inc., Upper Saddle River, N.J.

Bemidji, Minnesota, pipeline rupture, is less common than leakage from underground tanks, but the amount of petroleum or other chemicals discharged from a pipeline leak is commonly very large.

Artificial recharge refers to practices such as the discharge of urban storm sewers into pits designed for disposal of the water and/or aquifer recharge. This water comes in contact with a variety of hazardous and nonhazardous compounds during runoff from roads, driveways, and parking lots. Even graveyards are a potential source of contamination. Of concern here are biological and chemical contaminants. For example, in the late 1800s and early 1900s, arsenic in very high concentrations was used as an embalming fluid.

Subsurface Contamination Sources in the Saturated Zone

Many practices involving the saturated zone can lead to direct contamination of groundwater without passage through the unsaturated zone. A common activity of the past, for example, was to use abandoned excavations associated with mines and gravel production for waste disposal. If the abandoned excavation was deeper than the water table, the emplaced waste materials became saturated as the water table rose to partially fill the excavation. Waste materials would therefore be in direct contact with the groundwater flow system.

Agricultural drainage wells and canals may involve such practices as draining contaminated shallow groundwater to deeper aquifers or inducing the upward flow of highly mineralized water as the water table is lowered by drainage. Canals excavated along coastal areas may lead to the inflow and infiltration of saltwater during tidal fluctuations.

Disposal of hazardous chemicals in wells was based on injection-well technology developed in the oil industry. There, it has been traditionally used as a means of disposing of brines produced with the oil, thereby overcoming the problems encountered with the disposal of brines in surface impoundments. The chemical industry adopted the method as a cheaper alternative than wastewater treatment or surface disposal. Disposal, or *injection, wells* are commonly hundreds to thousands of meters deep (Figure 12.13). Favorable conditions occur where a permeable zone containing nonpotable water is overlain by one or more aquitards to prevent upward migration of the waste into usable aquifers. Problems occur due to failure of the well casing or the sealing material around the casing, which may allow contaminants to move out of the target horizon. Other pathways for contaminants include fractures produced in the protecting aquitard caused by high injection pressures or by abandoned, unplugged, or poorly plugged wells that may have penetrated the aquitard. Abandoned test holes, which are commonly drilled in both petroleum and mineral exploration, can also facilitate migration of contaminants in the subsurface.

Mines, even without the addition of wastes, can be environmental problems. The introduction of oxygen to anoxic zones in the subsurface can cause undesirable reactions with minerals leading to the production of toxic leachates. Sulfide minerals, which occur both in coal-bearing rocks and in metallic ores, are commonly the culprits. Oxidation of sulfides will produce sulfuric acid, and the leachates formed in this manner are corrosive and highly mineralized. Migration out of the mines leads to degradation of both groundwater and surface water. One potential effect of a decrease in pH is the mobilization of toxic metals, which are more soluble in acid solutions than in neutral or basic waters.

Water-supply wells themselves can sometimes be a source of contamination. Problems are most commonly linked to poor well construction. If the *annulus* (the part of the well bore outside the casing) is not properly sealed, contaminants from surface or subsurface sources have a direct vertical pathway to the aquifer. Flooding is a good example of how contaminated surface water could be brought into contact with the top of a well. The main water-treatment plant for Des Moines, Iowa, was shut down during the summer floods of 1993 when floodwaters inundated the municipal wells. Groundwater development can lead to problems when pumping lowers the water table or potentiometric surface to such a degree that water from an undesirable source begins to migrate toward the well. The poor-quality water may be drawn through an aquitard from a different aquifer, from an offshore source by subsurface migration of seawater, or from other possible locations.

The number and diversity of activities and practices that lead to groundwater contamination is a serious concern for environmental regulators and water managers. Through additional research into these processes, we will be able to minimize their effects in the future.

Environmental Regulations

Our environmental experience over the past few decades has made it clear that industrial- and waste-disposal practices must be strictly regulated to prevent future contamination, and cleanups of past contamination sites must continue to reduce risks to humans and the

▲ FIGURE 12.13

Design features of a hazardous-waste disposal well. *Source:* From C. A. Wentz, *Hazardous Waste Management*, © 1989 by McGraw-Hill, New York.

environment. Several of the most important environmental laws dealing with subsurface contamination in the United States are summarized subsequently. Regulations based on these laws are administered by the U.S. Environmental Protection Agency (EPA). Other countries have similar agencies.

Clean Water Act (CWA)

Originally passed in 1972 as the Federal Water Pollution Control Act Amendments and amended in 1977, the bill that became known as the *Clean Water Act* (CWA) was designed to improve treatment of wastewater discharge to the environment. Although most wastewater is discharged to surface-water bodies, we have already seen how contaminated surface water can impact groundwater. Wastewater treatment is described in terms of three possible phases: *primary*, *secondary*, and *tertiary*. Primary treatment involves physical processes such as screening and settling. Floating and suspended solids are removed in several steps. Primary treatment results in the accumulation of *sludge*, which requires further treatment and disposal. Prior to the Clean Water Act, which requires secondary treatment of wastewater, primary treatment was the only type utilized in many communities.

Remaining in wastewater after primary treatment are many undesirable substances, including both organics and inorganics. The organic content of wastewater is characterized by the *Biochemical Oxygen Demand* (BOD), which is a measure of the amount of oxygen required by aerobic microorganisms to break down the organic compounds to less harmful substances such as carbon dioxide. The purpose of secondary treatment is to remove at least 85% of the BOD. At this level of treatment, normal municipal wastewater can be discharged to a surface-water body. Although there are several technologies used for secondary treatment, the most basic and inexpensive is the *waste-stabilization lagoon*, which simply is a surface impoundment in which the wastewater is contained until BOD removal is complete. Treatment is based on the exchange of carbon dioxide and oxygen between bacteria and algae (Figure 12.14). The bacteria (both aerobic and anaerobic species are present in some lagoons) break down the BOD to carbon dioxide, which promotes the growth of algae. The algae give off oxygen in return, to sustain the population of bacteria. Waste-stabilization lagoons are effective if they operate as designed. A problem arises, however, if the bottom of the impoundment is permeable enough to allow significant seepage downward toward the water table. Examples of groundwater contamination from this situation have been described in the literature.

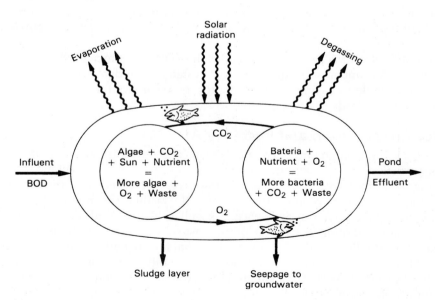

▲ FIGURE 12.14

Schematic of physical, chemical, and biological processes operating in a waste-stabilization lagoon.

A variety of chemical methods can be used in tertiary treatment. If necessary, very high quality water can be produced.

Safe Drinking Water Act (SDWA)

Public drinking-water-supply systems were regulated by the *Safe Drinking Water Act* of 1974 and its amendments passed in 1986. One reason that this law is important is that it established drinking-water standards for public supplies. These standards are based on the potential health risks of various waterborne substances. For cancer risk, for example, the standards are based on rodent exposure studies that are extrapolated to human beings. The methods assume a lifetime exposure to the water. The risk level used by the EPA is 10^{-6}. This means that the concentration of a particular carcinogen must be less than a level that would cause one additional cancer death in 1 million people. The methods used to arrive at these levels are extremely conservative and have generated controversy in the scientific community.

Three types of standards are reported: maximum contaminant-level goals (MCLGs), maximum contaminant levels (MCLs), and secondary maximum contaminant levels (SMCLs). The maximum contaminant-level goals are nonenforceable standards set at the most conservative levels with respect to adverse health effects. All carcinogens, for example, have MCLGs of zero. The maximum contaminant levels are the concentrations that must be met for water-supply systems, and the secondary maximum contaminant levels are nonenforceable standards set for aesthetic reasons. SMCLs include constituents such as iron and hardness, which may cause taste, staining, or scale problems but do not pose a known health risk.

Regulated contaminants are divided into four categories: organic chemicals, inorganic chemicals, microbiological contaminants, and radionuclides. The organic category, which contains by far the largest number of compounds, is further subdivided into *synthetic organic chemicals, trihalomethanes,* and *volatile organic chemicals.* Synthetic organics include pesticides and herbicides along with certain industrial chemicals. Trihalomethanes comprise a group of compounds that are formed during the disinfection of drinking water by the reaction between chlorine and organic compounds. It is somewhat disheartening to learn that carcinogenic compounds are being produced even in the treatment of drinking water. Volatile organics are compounds that readily vaporize at room temperature.

The number of chemicals regulated under the SDWA has increased drastically in the past few years. A number of municipal water systems have detected contamination that was not previously known simply by testing for these compounds. An abbreviated list of drinking-water standards is given in Table 12.2.

Resource Conservation and Recovery Act (RCRA)

The Resource Conservation and Recovery Act was enacted in 1976 to regulate the generation, transportation, storage, and disposal of hazardous wastes. The act created a system for tracking hazardous wastes from "cradle-to-grave," which is based on a manifest produced by the waste generator at the time it leaves the facility. This manifest contains information about the amount and type of waste, as well as identifies all companies involved in the transportation and ultimate disposal of the waste (Figure 12.15). With respect to disposal of hazardous waste, RCRA deals only with sites that are currently active. Amendments to RCRA in 1984 expanded its scope to include, among other components, the regulation of underground storage tanks (USTs). Under RCRA, the classification of a waste as hazardous is based on specific criteria. A test known as the Toxicity Characteristic Leaching Procedure (TCLP) includes a list of 39 organic and inorganic compounds. If a threshold value of any of these is exceeded, the waste is classified as hazardous and is therefore subject to RCRA.

Table 12.2   Selected Drinking-Water Standard

Chemical	MCLG (μg/L)	MCL (μg/L)	SMCL (μg/L)
Synthetic organic chemicals			
Alachlor	0	2	
Aldicarb	1	3	
Atrazine	3	3	
Benzene	0	5	
Benzo[*a*]pyrene	0	0.2	
Butylbenzyl phthalate	100	100	
Carbofuran	40	40	
Carbontetrachloride	0	5	
Chlorodane	0	2	
Chrysene	0	0.2	
Dibromochloropropane (DBCP)	0	0.2	
o-Dichlorobenzene	600	600	10
p-Dichlorobenzene	75	75	5
1,2-Dichloroethane	0	5	
1,1-Dichloroethylene	7	7	
cis-1,2-Dichloroethylene	70	70	
trans-1,2-Dichloroethylene	100	100	
1,2-Dichloropropane	0	5	
2,4-Dichlorophenoxyacetic acid (2,4-D)	70	70	
Di(ethylhexyl)phthalate	0	6	
Ethylbenzene	700	700	30
Ethylene dibromide (EDB)	0	0.05	
Hexachlorobenzene	0	1	
Lindane	0.2	0.2	
Methoxychlor	40	40	
Methylene chloride	0	5	
PCBs as decachlorobiphenol	0	0.5	
Styrene	100	100	10
2,3,7,8-TCDD (dioxin)	0	5×110^{-8}	
Tetrachloroethylene	0	5	
1,2,4-Trichlorobenzene	70	70	
1,1,2-Trichloroethane	3	5	
Trichloroethylene (TCE)	0	5	
1,1,1-Trichloroethane	200	200	
Toluene	1000	1000	40
Toxaphene	0	3	
2-(2,4,5-Trichlorophenoxy)-propionic acid (2,4,5-TP, or Silvex)	50	50	
Vinyl chloride	0	2	
Xylenes (total)	10,000	10,000	20
Inorganic chemicals			
Aluminum			50–200
Antimony	6	6	
Arsenic	50	50	
Asbestos (fibers per liter)	7×10^6	7×10^6	
Barium	2000	2000	
Beryllium	4	4	
Cadmium	5	5	
Chloride			250,000

(Continued)

Table 12.2 (*Continued*)

Chemical	MCLG (μg/L)	MCL (μg/L)	SMCL (μg/L)
Chromium	100	100	
Copper	1,300	1,300	
Cyanide	200	200	
Fluoride	4,000	4,000	2,000
Iron			300
Lead	0	15	
Manganese			50–200
Mercury	2	2	
Nickel	100	100	
Nitrate (as N)	10,000	10,000	
Nitrite (as N)	1,000	1,000	
pH			6.5–8.5 (pH units)
Selenium	50	50	
Silver			100
Sulfate			250,000
Total dissolved solids			500,000
Zinc			5,000
Microbiological parameters			
Giardia lamblia	0 organisms		
Legionella	0 organisms		
Heterotrophic bacteria	0 organisms		
Viruses	0 organisms		
Radionuclides			
Radium 226	0	20 pCi/L	
Radium 228	0	20 pCi/L	
Radon 222	0	300 pCi/L	
Uranium	0	20 μg/L (30 pCi/L)	

Source: U.S. EPA.

Among the many provisions of RCRA, the regulations of most relevance to subsurface contamination are those that apply to the final disposal of the waste. Each site undergoes an elaborate permit system that regulates the design of the facility as well as the groundwater and other monitoring systems. Design components similar to those shown in Figure 12.7 are required. The groundwater monitoring system consists of monitor wells located both upgradient and downgradient of the facility (Figure 12.16). Thus if any constituents contained in the waste are detected in groundwater downgradient of the site, they can be compared with the upgradient concentrations to determine if a significant increase has occurred. Statistical procedures are specified to quantify whether or not the increase is "significant."

Three phases of RCRA groundwater monitoring exist. *Detection monitoring* is the initial phase, which is carried out until a significant increase is documented. If this should happen, the owner/operator must begin a *compliance phase* of monitoring, which involves a more extensive hydrogeological investigation in order to determine the extent and severity of contamination. Finally, a *corrective action* may be required to remediate the contamination. When a disposal facility is closed, RCRA regulations do not end. The facility must submit a closure plan that includes continued monitoring for a period of time that usually lasts for 30 years.

Storage facility

Transporter

Treatment facility

UNIFORM HAZARDOUS WASTE MANIFEST

GENERATOR
TRANSPORTER
FACILITY
DESCRIPTION

SPECIAL HANDLING INSTRUCTIONS

RELIABLE

Disposal facility

Generator

EPA or state agency

▲ FIGURE 12.15
RCRA manifest used for "cradle-to-grave" tracking of hazardous wastes. *Source:* From G. M. Masters, *Introduction to Environmental Engineering and Science,* © 1991 by Prentice Hall, Inc., Upper Saddle River, N.J.

One of the most significant changes to the regulation of USTs under RCRA was the adoption in 1995 of new methods for assessing and comparing the risks of contaminated sites. This approach, called *Risk-Based Corrective Action (RBCA)* and pronounced "Rebecca," was necessary in order to deal with the large number of sites discovered under the UST program. RBCA involves a three-tier approach to assessing the risk of sites. A Tier-1 assessment, the most qualitative, evaluates the receptors of the site contaminants (homes, schools,

◀ FIGURE 12.16
Installation of a detection monitoring system for groundwater, containing upgradient and downgradient monitor wells.

streams, etc.), as well as the exposure pathways (drinking-water wells, vapor exposure, etc.). When these assessments are completed for a large number of sites, they can be compared and ranked so that resources can be allocated to sites with the highest risks. Tier-2 and Tier-3 assessments involve successively more complex and detailed investigations so that site-specific risks and exposures can be evaluated. Because of the increasing costs with higher tiers, Tier-3 assessments would only be conducted at the sites with highest risks.

Comprehensive Environmental Response, Compensation, and Liabilities Act (CERCLA)

CERCLA, better known as *Superfund*, was enacted in 1980 to handle the problem of abandoned hazardous-waste-disposal facilities along with emergency releases of hazardous materials. The term *Superfund* refers to a provision in the act to create a fund to initiate cleanup of high-priority sites until the liable private parties could be identified. The act was reauthorized in 1986 by the Superfund Amendments and Reauthorization Act (SARA), which expanded the funds available for cleanup. The trust fund itself is derived from environmental taxes of various types on corporations.

Known sites of contamination are included on a rank-ordered list called the *National Priority List* (NPL). The ranking is based on the characteristics and toxicity of the contaminants, the risk factors created for nearby residents, and other factors. Funds allocated for investigation and cleanup are then prioritized according to the NPL. Once a site is listed on the NPL, a complex series of steps is set in motion (Figure 12.17). The first phase involves the completion of a *Remedial Investigation/Feasibility Study* (RI/FS) to assess the contamination problem and evaluate remediation alternatives. These are extensive site investigations including installation and sampling of many monitor wells. Additional steps include the *Record of Decision*, which reports the remedial alternative selected, the *Remedial Design*, and the *Remedial Action*, which is the actual cleanup procedure. This process can take up to 5 years or more before the Remedial Action is even started, and decades for the completion of the cleanup if a large volume of aquifer is contaminated and groundwater must be pumped and treated.

As the technical procedure is put into action, the EPA is also pursuing a legal strategy to identify the *Potentially Responsible Parties* (PRPs) and force them to pay for the costs of cleanup. PRPs can include past or present owners of the site, as well as generators and

▶ FIGURE 12.17
Steps in the Superfund remedial process.

Site discovery

↓

NPL ranking and listing

↓

Remedial investigation/
feasibility study (RI/FS)

↓

Record of decision (ROD)

↓

Remedial design (RD)

↓

Remedial action (RA)

transporters of the waste. If the PRPs (sometimes there are many of them) refuse to participate in the cleanup, the EPA can legally collect more than 3 times the costs of the project. These amounts are not trivial as it is not unusual for remedial costs to reach tens of millions of dollars. The legal complexities of these operations have raised criticisms that the Superfund program spends an inordinate amount on legal costs as compared with the amount actually spent on cleanup.

Subsurface Fate and Transport of Contaminants

The movement of contaminants through the subsurface environment, which is known as *mass transport*, takes place by many complex and somewhat poorly understood processes. Because of the complexities of these phenomena, our discussion will serve only as a brief introduction to this important topic.

The most basic distinction to make with regard to mass transport mechanisms is the difference between aqueous and nonaqueous phases. *Aqueous-phase* transport occurs when the contaminant is dissolved in water, whereas *nonaqueous-phase* transport refers to the movement of a contaminant in a phase that is *immiscible* with water. Nonaqueous-phase liquids, or *NAPLs* (rhymes with "apples"), include petroleum products and other organic fluids that are spilled or leaked into the subsurface. The presence of NAPLs introduces additional complexity in the description and prediction of transport of subsurface contaminants. Most of the difficulties come into play with mass transport through the unsaturated zone. Transport of NAPLs through the saturated zone is less common because it can only occur when the NAPL is denser than water. In this case, it is referred to as a *DNAPL*, a dense nonaqueous-phase liquid. Petroleum products, among the most common nonaqueous-phase contaminants, are less dense than water and are therefore prevented from moving through the saturated zone as a distinct phase. These petroleum-derived fluids are known as *LNAPLs*, or light nonaqueous-phase liquids.

Aqueous-Phase Mass Transport

From a physical standpoint, there is little difference between the transport of an aqueous solution containing dissolved contaminants and the movement of pure water in the subsurface. Exceptions to this generalization include instances where there are large concentrations of highly soluble compounds. When evaporite minerals dissolve in water, for example, the aqueous solution can become significantly more dense than pure water because of the high solubility of these minerals. Density would then become an important factor in transport, in addition to the normal hydraulic gradients that we discussed in Chapter 11. Most contaminants, however, are much less soluble, and we can assume that the solutes do not have a significant effect on the movement of the aqueous fluid.

All inorganic and organic substances will dissolve in water to some extent. The range in solubility, however, is extremely large. The solution of minerals was discussed in several previous chapters. Minerals in the unsaturated zone will dissolve in water until a state of chemical equilibrium is reached, at which time solution of the mineral will cease. Inorganic contaminants are no different. Leachate plumes in groundwater beneath landfills in which building materials such as drywall (which contains gypsum) were disposed of commonly contain high concentrations of calcium and sulfate. The solubility of heavy metals and other elements is highly dependent upon the geochemical conditions of the leaching solution, including the pH and the *oxidation–reduction potential* (*redox potential*). Iron, for example, is much more soluble when the pH and redox potential are low. In the unsaturated zone, the pH is commonly close to neutral and the redox potential is high due to the presence of oxygen from the atmosphere. Therefore, iron concentrations are usually low in well-aerated unsaturated-zone solutions.

The subsurface behavior of organic compounds depends upon several physical properties. As with inorganics, the solubility of organic compounds varies over a wide range (Table 12.3). Solubility is an indication of the tendency of an organic compound to form an immiscible phase. Differences in solubility are due in part to the nature of chemical bonds in the compound. Compounds with a predominance of nonpolar, covalent bonds are less soluble in water, which is a strongly polar substance. These compounds are more likely to form a nonaqueous phase.

Table 12.3   Properties of Common Groundwater Subsurface Contaminants

Compound	Density, g/mL	Property[a] Solubility, mg/L	Henry's[b] Constant, atm	$Log_{10}K_{ow}$
Trichloroethene	1.4	1,100	550	2.29
Tetrachloroethene	1.63	200	1,100	2.88
Chloroform	1.49	8,200	170	1.95
Benzene	0.876	1,808	240	2.01
Toluene	0.876	535	308	2.69
1,2-Dichlorobenzene	1.305	145	90	3.38
Phenol	1.07	93,000[c]	0.04	1.49
1,1-Dichloroethene	1.013	250	1,400	0.73
1,1,1-Trichloroethane	1.435	480	860	2.49
Vinyl chloride	gas	1,100	35,500	0.60
Methyl ethyl ketone	0.805	260,000	1.5	0.26
Acetone	0.79	1,000,000	1.0	−0.24
Ethylene dibromide	2.18	3,400	26	1.80

Source: E. Bouwer, J. Mercer, M. Kavanaugh, and F. Digano, 1988, Coping with groundwater contamination, *Journal of the Water Pollution Control Federation*, 60:1415–1423.
[a] Values for temperature of 20° to 25° C.
[b] Values should be considered representative of literature data; wide ranges in values are common.
[c] Depends on pH.

Another diagnostic property is called the *octanol-water partition coefficient*. This quantity, K_{ow}, is expressed as

$$K_{ow} = \frac{C_{oc}}{C_w} \tag{12.1}$$

where C_{oc} is the equilibrium concentration of the compound in question dissolved in octanol and C_w is the equilibrium concentration in water. The higher the value of K_{ow}, which is usually expressed as a log value, the more likely it is to form a nonaqueous phase. Compounds with high log K_{ow} values are known as *hydrophobic* compounds.

The *Henry's constant* is useful in predicting the tendency of an organic compound dissolved in water to volatilize into the gas phase. This is particularly important in the unsaturated zone, where organic vapors can collect and migrate separately from their transport in the aqueous phase. Table 12.3 indicates, for example, that benzene is much more volatile than phenol. The differences in volatility between organic compounds have an important bearing upon the remedial method used.

Unsaturated Zone Transport

Because the majority of contaminants originate above the water table, we must first discuss the movement of water (along with dissolved contaminants) through the unsaturated zone. In order to do this, we must expand upon the comments made about the unsaturated zone in Chapter 11. In that discussion, it was mentioned that the fluid pressure in unsaturated zone water is less than atmospheric. These negative pressures, which hold the water against gravity by suction or tension, are known as the *matric potential*. The distribution of matric potential is shown in Figure 12.18, in which matric potential is plotted against

▶ FIGURE 12.18
Soil water retention curve and classification of soil moisture zones. *Source:* From S. A. Abdul, 1988, Migration of petroleum products through sandy hydrogeologic systems, *Ground Water Monitoring Review*, 8, (no. 4):73–81.

◀ FIGURE 12.19
Schematic diagram of a tensiometer.

the percent saturation. At the water table, where the fluid pressure is equal to atmospheric (matric potential = 0), the pores are saturated with water. Matric potential decreases above the water table, but within the capillary fringe, the pores remain saturated. Above the capillary fringe, the amount of water in pores decreases as the amount of air increases. This is known as the *funicular zone*. Water dominates pore volumes within this region (Figure 12.18). At some point the amount of water within the pores achieves a constant residual value as the matric potential decreases. In this region, called the *pendular zone*, pore water is limited to pore necks and may form only a very thin continuous layer over the particle surfaces. The bulk of the pore space is occupied by air.

The matric potential of the soil is measured by a device called a *tensiometer* (Figure 12.19). The tensiometer consists of a tube with a porous cup on the end that is inserted into the soil. The tube is totally filled with water and connected to a vacuum gauge at the top. Water is drawn out through the porous cup into the soil, creating a vacuum in the tube, which is measured by the gauge. Like flow in the saturated zone, flow in the unsaturated zone is controlled by the gradient of hydraulic head, which is the sum of the elevation head and the pressure head. The pressure head in the unsaturated zone is the matric potential, which is negative rather than positive, as it is below the water table. The direction of flow in the unsaturated zone can be determined by the installation of two adjacent tensiometers at different depths.

EXAMPLE 12.1

Tensiometer A, which is inserted to a depth of 50 cm below the land surface, has a measured matric potential of −100 cm. Tensiometer B, which is inserted at a depth of 70 cm, has a matric potential of −50 cm. Is water moving upward or downward in the unsaturated zone?

Solution

Total head or potential is the sum of elevation head and matric potential. If we use the ground surface as our datum and use depth for the elevation head, elevation head, z, is negative. The total head, $\theta = \psi + z$, for tensiometer A is $\theta = -100 + (-50) = -150$ cm. For tensiometer B, it is

$$\psi_A = -100 \text{ cm} \qquad \psi_B = -50 \text{ cm}$$

50 cm

70 cm

$\theta = -50 + (-70) = -120$ cm. Since the head at tensiometer B (-120 cm) is larger than the head at tensiometer A (-150 cm), the flow is in the direction of B to A, or upward.

When a substantial amount of rain falls to the surface, the infiltrating water flows into the pore spaces and increases the percent saturation. As a result the matric potential is increased (becomes less negative), and water moves downward through the soil. If the soil is homogeneous, the moisture will move downward as a slug, with a well defined *wetting front* (Figure 12.20). Any contaminants in solution would therefore move toward the water table at about the same rate. It is more likely, however, that the soil is not homogenous, but exhibits *preferential flow*. Preferential flow refers to the presence of more permeable pathways in the soil through which water and contaminants can move more rapidly than the wetting front. Preferential flow can follow *macropores*, which are large openings (such as root casts, animal

▶ FIGURE 12.20
Contrast between unsaturated flow through (a) uniform soils with isolated macropores and (b) fingering at a hydraulic conductivity boundary.

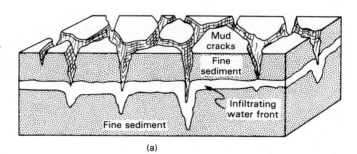

Mud cracks

Fine sediment

Infiltrating water front

Fine sediment

(a)

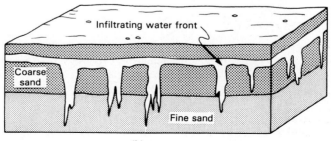

Infiltrating water front

Coarse sand

Fine sand

(b)

burrows, or shrinkage cracks), or it can occur through a process known as *fingering* (Figure 12.20), in which pathways of more subtle variation in permeability are exploited. Fingering is commonly observed to occur at the contact between a fine-grained soil and a coarse-grained soil beneath. The significance of preferential flow to mass transport is that it provides a very rapid journey for contaminants through the unsaturated zone to the water table.

Saturated Zone Transport

Once contaminants reach the water table, they travel in the direction of groundwater flow. If the contaminants originated as a point source, they move within the saturated zone as a coherent mass of degraded groundwater known as a *contaminant*, or *leachate*, *plume* (Figure 12.21). The direction of movement may be horizontal, vertical, or inclined, depending on the gradient of the flow system. If the contaminant plume is the approximate density of the groundwater, the major control on the direction of movement will be the hydraulic gradient; if the plume is significantly denser, however, it will have a downward component of motion separate from the hydraulic gradient (Figure 12.22).

The contaminant plume originating from discharge ponds from a sewage treatment plant at Otis Air Force Base on Cape Cod, Massachusetts, is a good example of a low-density plume in a shallow sand and gravel aquifer (Figure 12.23). Groundwater flow is nearly horizontal in the aquifer and the contaminants form a narrow plume elongated downgradient with very little vertical movement in the aquifer. Although the plume contains many different organic and inorganic constituents, the dimensions of the plume are well defined by boron concentrations (Figure 12.23).

There are two physical mechanisms that control the movement of dissolved contaminants in a contaminant plume. *Advection* is the transport of contaminants by the bulk movement of the flowing water, in response to the hydraulic gradient. The rate of movement of contaminants can be estimated by determining the *average linear velocity* of the groundwater. This velocity is described as *average* because of the nature of flow in porous media. Variation in size of the microscopic flow channels followed by individual particles of water leads to a difference in velocity, depending on which particular path a particle of water follows. Since it is not possible to determine the infinite variety of possible flow paths, an average velocity is used. When an average linear velocity is reported, it must be kept in mind that some groundwater, and therefore dissolved contaminants, will be traveling along certain flow paths at higher velocities. The average linear velocity can be calculated by the formula

$$\bar{v} = \frac{K}{n}\frac{dh}{dl} \tag{12.2}$$

▲ FIGURE 12.21
Control of leachate-plume movement by the hydraulic gradient. A well at *A* would not be contaminated because of the vertical component of groundwater motion in the flow system.

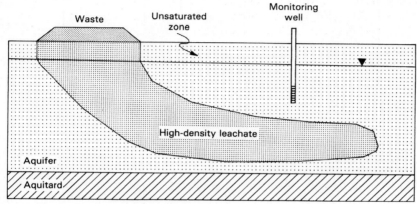

▲ FIGURE 12.22

Modification of contaminant flow by increased density of leachate

where

$\overline{v}$ is average linear velocity
K is hydraulic conductivity
n is the porosity of the aquifer
dh/dl is the hydraulic gradient

Hydrodynamic dispersion, the second physical transport mechanism, results in the spreading of the zone of contamination along the flow path. On a microscopic scale, several effects tend to move contaminants away from the center of mass of the plume (Figure 12.24). These effects include (a) higher flow rates in the center of a pore than along the margins, (b) more rapid flow through larger pores, and (c) lateral expansion of the contaminant because water molecules must flow around aquifer grains. The overall effect of hydrodynamic dispersion is to spread the contaminant over a larger volume of aquifer, both in the direction of flow and laterally to the direction of flow. When the aquifer is heterogenous, that is, when it is composed of beds or laminae of different hydraulic conductivity, even more spreading of the plume occurs. Figure 12.25 illustrates an aquifer that has lenses of high hydraulic conductivity. The contaminant moves more rapidly through these lenses with hydrodynamic dispersion superimposed on this preferential flow. The plume is spread in the direction of flow by this

▲ FIGURE 12.23

Cross sections oriented in direction of groundwater flow (A–A′) and transverse to groundwater flow (B–B′) and (C–C′) in the contaminant plume at Otis Air Force Base, Massachusetts. Treated wastewater is discharged to sand beds, from which it infiltrates to the water table and moves as a coherent plume in the aquifer. Contour lines represent equal concentrations of boron in micrograms per liter. *Source:* From D. R. LeBlanc, 1984, *Sewage Plume in a Sand and Gravel Aquifer, Cape Cod, Massachusetts,* U.S. Geological Survey Water-Supply Paper 2218.

process. Despite the spreading effect of hydrodynamic dispersion, the influence is much less than the mixing caused by turbulent flow in a river. It is correct to think of leachate plumes moving in compact masses through aquifers, with relatively minor spreading caused by hydrodynamic dispersion and aquifer heterogeneity. The plume shown in Figure 12.23 is a good example.

Because dispersion spreads a fixed mass of contaminant over a larger volume of aquifer, a reduction in contaminant concentration occurs. If the particular contaminant is very toxic at low concentrations, the dispersive tendency will have an undesirable result because more area is affected even though the concentration is decreased.

The ultimate path of contaminants in groundwater can be predicted only by knowing the nature of the flow system. Water-table and potentiometric-surface contour maps will

▶ FIGURE 12.24

Mechanisms of mechanical dispersion operating at the scale of individual pores and grains. *Source:* From R. W. Gillham and J. A. Cherry, 1982, Contaminant movement in saturated geologic deposits, in *Recent Trends in Geology*, T. N. Narasimham, ed, Geologic Society of America Special Paper 189:31–62.

(a)

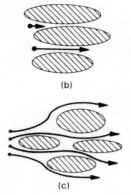

(b)

(c)

▶ FIGURE 12.25

More rapid flow of contaminants in thin lenses of high hydraulic conductivity.

give an indication of the horizontal component of movement. The importance of vertical components is illustrated in Figure 12.21, where a contaminant source is added to a flow system. Well A will not be contaminated because it is too shallow, even though it is in the direction of flow of groundwater from the waste-disposal site. Contamination of a well can occur even when the natural gradient suggests that groundwater flow will carry contaminants away from the well. A pumping well located upgradient of a disposal site, like the one shown in Figure 12.26, can become contaminated because the cone of depression it creates reverses the natural gradient and induces movement of the contaminated water toward the well.

▶ FIGURE 12.26

Contamination of an upgradient pumping well due to the gradient reversal caused by pumping.

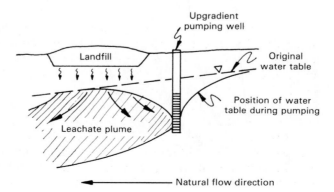

Attenuation Mechanisms

As dissolved contaminants move through the unsaturated and saturated zones, a number of physical, chemical, and biological processes may alter the concentration of the solute. Chemical processes, including acid–base and oxidation–reduction reactions, can change the chemical species of the contaminant and in doing so, alter its mobility.

Some important processes that decrease the concentration of contaminants in the subsurface, or *attenuation mechanisms*, are listed in Table 12.4. One of these, *adsorption*, is particularly important. Adsorption, the attraction of an ion to the surface of a solid particle in the unsaturated or saturated zones, is really a form of *ion exchange*. In order to maintain an overall neutral charge, the particles of clay and other substances that attract ions must release ions of the same charge. The most common reaction involves the exchange of cations. Contaminant cations are drawn to negatively charged particle surfaces in exchange for other cations released from the particle surface to the groundwater solution. Clays have the strongest ability to effect this exchange reaction, although other materials, including iron oxide coatings or particles, may also participate in adsorption reactions. Heavy-metal cations can be maintained at low dissolved concentrations by adsorption, although the pH of the solution is very important. Figure 12.27 shows that adsorption of each metal cation is controlled by a relatively narrow range of pH values. As the pH rises above the critical value, the environment changes from a condition where the metal is very mobile to a state where a high percentage of the metal is adsorbed to metal oxides.

Table 12.4 Attenuation Mechanisms
Hydrodynamic dispersion
Adsorption–desorption reactions
Acid–base reactions
Solution–precipitation reactions
Oxidation–reduction reactions
Microbial cell synthesis
Filtering, die-off, and adsorption of microorganisms
Radioactive decay
Biodegradation

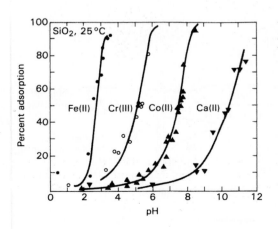

◄ FIGURE 12.27
Adsorption of metals to metal-oxide surfaces as a function of pH. *Source:* From R. O. James and T. W. Healy, 1972, *Journal of Colloid and Interface Science*, 40.

Adsorption of nonpolar organics is controlled by solid organic matter in the aquifer. Even though only minor amounts of organics may be present, nonpolar dissolved organic molecules have a strong affinity for these substances. The log K_{ow} value for the compound is a good indication of its mobility in groundwater. The higher the log K_{ow}, the less mobile the compound will be due to adsorption on solid organics.

Biodegradation, the breakdown of organic compounds by microorganisms in the contaminant zone, is also important in the attenuation of organic contaminants. Here again, the chemical environment is very important in the process. For example, *aromatic* compounds such as benzene, toluene, and xylene, which are common components of gasoline, dissolve in water but can be actively broken down to carbon dioxide under oxidizing conditions. These conditions are most common in the unsaturated zone. Biodegradation of the aromatics is much slower in the saturated zone. Other contaminants, like the chlorinated solvents trichloroethene (TCE) and tetrachloroethene (also known as perchloroethene, or PCE), biodegrade more rapidly under reducing conditions. Therefore, these compounds are resistant to biodegradation in the unsaturated zone but do break down more readily below the water table. Study of a plume of chlorinated solvents near St. Joseph, Michigan, yielded evidence for biodegradation of trichloroethene to its daughter products dichloroethene and vinyl chloride (Figure 12.28). In this instance, biodegradation is not a favorable outcome, because vinyl chloride is more toxic and more resistant to biodegradation than trichloroethene.

Contaminants may also be removed from solution by chemical precipitation, radioactive decay, and the die-off of microorganisms. Hydrodynamic dispersion is also an attenuation mechanism because spreading of the contaminant reduces its concentration.

▲ FIGURE 12.28
Distribution of chlorinated solvent compounds trichloroethene (TCE), dichloroethene (DCE), and vinyl chloride (VC) in a plume at St. Joseph, Michigan. DCE and VC are produced by biodegradation of TCE. *Source:* From P. L. McCarty and J. T. Wilson, 1992, Natural anaerobic treatment of a TCE plume, St. Joseph, Michigan, NPL site, in *Bioremediation of Hazardous Wastes*, U.S. EPA, EPA/600/R-92/126, pp. 47–50.

Case in Point 12.1
Industrial Waste Disposal of a Shallow Aquifer

Liquid chemical wastes were discharged to unlined surface impoundments at many locations in the past. The contaminants may move largely by aqueous-phase transport through the unsaturated zone to the saturated zone, where they form a contaminant plume. Perlmutter and Lieber (1970) presented an early example of this type of contamination on Long Island, New York. During and after World War II, an aircraft factory discharged aluminum-plating wastes into the disposal ponds shown in Figure 12.29. Because the area is underlain by a highly permeable unconfined aquifer with a high water table, the wastes rapidly infiltrated downward to the water table. From this point, the waste formed a plume of contaminated groundwater about 1300 m long, 300 m wide, and 21 m thick. The gradual spreading resulting from hydrodynamic dispersion is evident on the map. Calculations indicated that the contaminated groundwater moves rapidly at an approximate rate of 165 m per year. Some of the contaminated water discharges into Massapequa Creek in the southern part of the area.

Figure 12.30 shows a subsurface cross section of the plume. It is apparent that contamination is confined to the upper glacial aquifer and that movement is predominantly horizontal. The contours on the cross section show the distribution of hexavalent chromium in 1962. Hexavalent chromium is a toxic metal with an MCL of only 0.05 mg/L. At the levels present in the aquifer, use of the water for drinking would pose a very serious health threat. In addition to chromium, the plume contained other toxic contaminants.

◄ FIGURE 12.29
Map of part of Long Island, New York, showing size and extent of plume of contaminated groundwater originating from disposal pits at a metal-plating facility. *Source:* From N. M. Perlmutter and M. Lieber, 1970, *Disposal of Plating Wastes and Sewage Contamination in Ground Water and Surface Water, South Farmingdale–Massapequa Area, Nassau County, New York*, U.S. Geological Survey Water Supply Paper 1879-G.

▲ FIGURE 12.30
Cross section A–A' (location shown on Figure 12-29) through the center of the contaminant plume. Contours show the concentrations of hexavalent chromium in milligrams per liter. *Source:* From N. M. Perlmutter and M. Lieber, 1970, *Disposal of Plating Wastes and Sewage Contamination in Ground Water and Surface Water, South Farmingdale–Massapequa Area, Nassau County, New York*, U.S. Geological Survey Water Supply Paper 1879-G.

Nonaqueous-Phase Mass Transport

Unsaturated Zone Transport

The movement of a NAPL through the unsaturated zone is a complex phenomenon because it involves three or more phases: the NAPL, water, and air. If part of the NAPL evaporates into the soil atmosphere, a fourth phase is present. Additionally, the NAPL, or certain compounds contained within it, can dissolve into the pore water contained in the unsaturated zone. The discovery that spills and leaks have contaminated the subsurface on a routine basis has stimulated research into the properties and processes of flow through the unsaturated zone.

The flow of a NAPL through an unsaturated porous medium follows the same basic principles as does the flow of water. Below the level of saturation of the pores by the fluid, a capillary suction is exerted upon the NAPL similar to that exerted upon water, except that the adhesion of most nonpolar organics to soil surfaces is less than the adhesion of water. NAPLs are also subject to a condition similar to water in the pendular zone (Figure 12.18). Here water is said to be at *residual saturation* because the percent saturation does not decrease with decreases in the matric potential (greater suction). As a NAPL moves through the subsurface, it also must exceed the level of residual saturation prior to further movement. This level of NAPL will remain in the soil after drainage of the mobile fraction. The residual saturation depends upon the soil type (Table 12.5, where oil residual capacity is a measure of residual saturation). Fine-grained soils have a much higher residual saturation than coarse-grained soils.

The NAPL in a two- or three-phase unsaturated system is referred to as the *nonwetting fluid*, because water usually covers all solid surfaces within the medium. Water is known as the *wetting fluid*. The movement of a NAPL through a water-wet porous medium differs from movement through an initially dry medium. As a petroleum product, the nonwetting fluid, is released into the unsaturated zone, it begins to move downward under the influence of gravitational forces. These are resisted by capillary forces during the development of residual saturation. The NAPL may move downward as a distinct interface, although preferential flow is common. In order to progress through the medium, the NAPL must displace the existing pore fluids. This is relatively easy in the pendular zone, where air-filled pores are the norm. The rate of movement is largely a function of the viscosity of the NAPL, which can vary over a large range, and the grain size of the soil. Fine-grained soils

Table 12.5 Oil Retention Capacities for Kerosene in Unsaturated Soils

Soil Type	Oil Retention Capacity (R)	
	L/m³	g/yd³
Stone, coarse sand	5	1
Gravel, coarse sand	8	2
Coarse sand, medium sand	15	3
Medium sand, fine sand	25	5
Fine sand, silt	40	8

Source: CONCAWE, 1979, Protection of Groundwater from Oil Pollution, NTIS PB82-174608.

tend to have a higher residual saturation of water and it is therefore more difficult to drive a NAPL through them. When the NAPL encounters the funicular zone, its rate of movement slows, because pressure must be built up to overcome the resistance of the higher saturation percentage of water.

Several methods are available for prediction of the depth of penetration of a NAPL into the unsaturated zone. For example, the equation

$$D = \frac{1000V_{hc}}{ARC} \tag{12.3}$$

gives the approximate depth of penetration, D, in terms of V_{hc}; the volume of discharged NAPL in barrels (1 barrel = 42 gal); the area of infiltration, A; the oil-retention capacity or residual saturation, R (Table 12.5); and a correction factor, C, based on the viscosity of the NAPL (0.5 for gasoline to 2.0 for light fuel oil).

As the downward-moving NAPL finally reaches the top of the capillary fringe, a change in behavior occurs. If the source of the spill is large enough to provide a relatively continuous downward movement of product, and the product is an LNAPL (which would include gasoline, heating oil, kerosene, jet fuel, and aviation gas), the petroleum product begins to form a pool, or *pancake*, at the top of the capillary fringe (Figure 12.31). The layer is known as

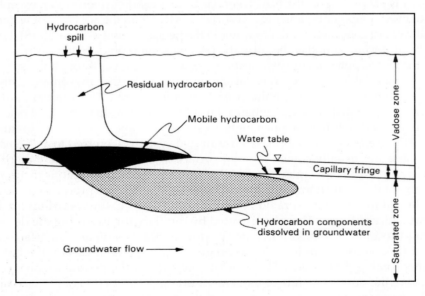

▲ FIGURE 12.31
Subsurface migration of LNAPL contaminants.

free product because it is not held under capillary pressure and is free to migrate along the top of the capillary fringe in the direction of slope of the water table. The top of the layer of free product forms an *oil table*, because product below is under positive pressure. With sufficient source, the pancake layer can actually depress the capillary fringe so that free product rests directly upon the water table. The layer of free product can remain in its subsurface position for a long period of time. Even after the source is terminated, a column of NAPL at a level of residual saturation will remain in the unsaturated zone above the free product.

Several other transport processes occur once an LNAPL has reached the water table to form a free product layer. One serious problem is that the more soluble components of the hydrocarbon will dissolve and form a groundwater contaminant plume that moves down-gradient (Figure 12.31), typically for a much longer distance than the extent of the free product. This is exactly what happened at the Bemidji oil pipeline rupture site (Figure 12.6). In gasoline, the dissolved contaminants include benzene, toluene, ethylbenzene, and xy-lene. Together these are referred to as *BTEX*. Additional processes include volatilization of hydrocarbon vapors from the free product into the unsaturated zone above. These gases can migrate within the unsaturated zone into basements and pose explosion hazards. The discovery of free product in the unsaturated zone and/or BTEX plumes in groundwater has become all too common since the advent of the UST program initiated under RCRA. More likely than not, the corner gas station that has been in business for several decades has had or will have a petroleum product release of some type.

Saturated Zone Transport

Transport of NAPLs in the unsaturated zone is limited to DNAPLs, those products denser than water (Table 12.3). The most common compounds involved in DNAPL spills and leaks are the chlorinated hydrocarbons, synthetic compounds such as 1,1,1-trichloroethane, carbon tetrachloride, and tetrachloroethene. These compounds are very commonly used as solvents and degreasers in industry, among other uses. The dry cleaning industry is a very good example of an industry that uses compounds of this type.

Migration of DNAPLs through the unsaturated zone is similar to the movement of LNAPLs. Once the residual saturation is exceeded, the product moves vertically downward to the water table. At that point, however, its behavior differs from that of LNAPLs. Although a layer of DNAPL may initially form at the top of the saturated zone, under sufficient pressure, the product will penetrate the water table and sink because of its higher density to the first layer of low-permeability material in the aquifer (Figure 12.32). The DNAPL will form a layer of free product on top of the low-permeability layer, and it can then migrate in the direction of slope of the layer. This direction is not necessarily in the direction of groundwater flow and, in fact, can be in the opposite direction if the low-permeability layer happens to slope

▶ FIGURE 12.32
Migration of DNAPL contaminants in the subsurface. *Source:* Diagram courtesy of Stan Feenstra.

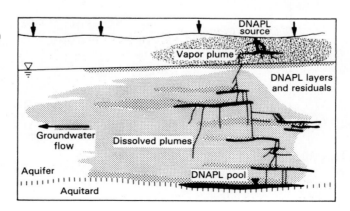

that way. Above the free product pool, the column through which the DNAPL moved retains product at residual saturation. Like LNAPLs, DNAPLs may be somewhat soluble and will therefore form contaminant plumes (Figure 12.32). Because the hydrocarbon is dissolved at low levels, these plumes move under the hydraulic gradient and may not display density control. Locating DNAPL free product layers is extremely difficult, particularly if the surface source is unknown. Since the free product layer does not follow the hydraulic gradient, a detailed knowledge of the subsurface geology must be developed by test drilling and other methods to track the movement of the free product layer.

Case in Point 12.2
The Lawrence Livermore National Laboratory (LLNL) Superfund Site

The degree of complexity of Superfund projects can reach immense proportions, not only in terms of the hydrogeology of the site but also in terms of the site's history and contamination. The Lawrence Livermore National Laboratory (LLNL) typifies the staggering task required in investigation and remediation of these sites (Thorpe et al., 1990).

The LLNL is a weapons research laboratory located southeast of Oakland, California (Figure 12.33). The site encompasses about 800 acres, including the facility itself and a

▲ FIGURE 12.33
Location of Lawrence Livermore National Laboratory.

buffer zone. The history of the site begins in World War II, when it was acquired by the U.S. Navy for use as an air station. Extensive repair and overhaul operations were conducted until the end of the war. Numerous organic solvents were used in these operations and, as was common at that time, handling and disposal were haphazard. Large amounts of solvents were undoubtably released at this time. In 1951, the property was transferred to the Atomic Energy Commission (AEC) for use as a thermonuclear weapons design facility. Later, the site was passed on to successors of the AEC, including the Energy Research and Development Agency and the Department of Energy. Weapons research continued under the management of the University of California, along with nonmilitary research. Hazardous materials of many types were handled at the site during this era.

Contamination was first detected in the early 1980s. An almost continuous series of hydrogeologic studies led to LLNL's inclusion on the NPL under the Superfund program in 1987. Volatile organic compounds (VOCs) were identified as the most serious contaminants, although a gasoline leak of 17,000 gallons over a 20-year period was documented for one portion of the site. Out of 130 possible sites for VOC releases, 14 were considered to

▲ FIGURE 12.34
Contour map showing concentrations of total VOCs in groundwater at LLNL site, November 1991.
Source: From F. Hoffman, 1993, Ground-water remediation using "smart pump and treat," *Ground Water*, 31:98–106.

be the most significant and were designated for further study. Hundreds of borings and monitor wells were installed. The purpose of the borings was to define the hydrogeological setting of the site and to determine the subsurface distribution of contaminants.

The LLNL is built upon several coalescing alluvial fans. The uppermost subsurface sediments beneath the site are composed of a complex series of alluvial units, consisting of alternating beds of fine and coarse materials. The coarser sediments comprise channel fills, formed as stream channels repeatedly migrated across the growing alluvial fan. The subsurface stratigraphy creates a very complex pattern of migration for contaminants. Overall, contamination appears to be limited to several large, diffuse, and overlapping plumes. Total VOC concentrations are shown in Figure 12.34, and the ranges in concentration for the major contaminants are shown in Table 12.6.

A sense of the vertical complexity of the site is given by Figure 12.35, which shows the generalized distribution of lenses of high hydraulic conductivity along one cross section.

▲ FIGURE 12.35
Generalized cross section showing concentration of VOCs in high hydraulic conductivity lenses in alluvial sediments beneath LLNL. *Source:* From R. K. Thorpe, W. F. Isherwood, M. D. Dresden, and C. P. Webster-Scholten, 1990, *CERCLA Remedial Investigations Report for the LLNL Livermore Site*, Lawrence Livermore National Laboratory.

Table 12.6   Summary of California State Action Levels and Maximum VOC Concentrations in Groundwater from LLNL Monitor Wells, June 15, 1988–June 15, 1989

VOC	EPA MCL (ppb)	DHS MCL (ppb)	DHS Action Level (ppb)	Maximum LLNL Concentration (ppb)
PCE	None	5	None	1800
TCE	5	5	None	5200
cis-1,2-DCE	None	None	6	—
trans-1,2-DCE	None	None	10	22
1,1-DCE	7	6	None	640
1,1,1-TCA	200	200	None	20
1,2-DCA	None	0.5	None	200
1,1-DCA	None	None	5	59
Carbon tetrachloride	5	0.5	None	140
Chloroform	100	None	None	650
Freon-113	None	None	1200	950

Source: From R. K. Thorpe, W. F. Isherwood, M. D. Dresden, and C. P. Webster-Scholten, 1990, *CERCLA Remedial Investigations Report for the LLNL Livermore Site*, Lawrence Livermore National Laboratory.

High concentrations of VOCs occur within these zones. Figure 12.35 suggests the difficulty of characterizing the three-dimensional distribution of contaminants.

Remediation of the site will be based on a pump and treat approach (Hoffman, 1993) that will include source removal, reinjection of treated water, and dynamic management of the remedial system in order to minimize the time required for cleanup.

Subsurface Contamination of Karst Aquifers

In the previous discussion of contaminant transport in the unsaturated and saturated zones, a basic assumption was made about the subsurface materials: The volume of material through which contaminants moved was considered to function as a granular porous medium, in which the flow of water and contaminants occurred through pores between closely packed solid particles. Preferential flow was mentioned as a condition in which larger interconnected pores are present. In some geologic media, however, most pore fluids move along linear discontinuities such as fractures, joints, or faults, and the intergranular flow is minor.

A well-developed karst aquifer with conduit flow represents the end member of this type of flow. Surface drainage moves rapidly through the unsaturated zone to a complex maze of subsurface conduits that may be partially or completely filled with water. Conduit flow is rapid, particularly during times of high precipitation. The consequences of this type of hydrological regime for contaminant movement are that surface contaminants move into the conduit flow system with very little opportunity for attenuation (Figure 12.36). In agricultural areas, contamination by pesticides and livestock waste is common. In unsewered suburban areas, septic-tank drainage moves rapidly downward through solution cavities. Cave explorers have documented direct seeps of septic-tank effluent into cave systems. Perhaps the worst circumstance, however, is the presence of a city directly over a karst aquifer system. Major sources of contamination include

▲ FIGURE 12.36
Schematic cross section showing contamination of a karst aquifer. *Source:* From N. C. Crawford, 1986, *Karst Hydrologic Problems Associated with Urban Development*, National Water Well Association.

storm water runoff that is discharged into the cave system and buried tanks that leak petroleum fuels and industrial chemicals (Figure 12.37). An example of some of the problems encountered by urban contamination of a karst aquifer is presented in Case in Point 12.3.

Case in Point 12.3
Subsurface Migration of Toxic and Explosive Fumes in Bowling Green, Kentucky

Difficulties with subsurface contamination in Bowling Green, Kentucky, typify the problems that can arise with urban development over a karst aquifer (Crawford, 1986). The city partially overlies the Lost River groundwater basin, a karst aquifer system with well-developed conduit flow (Figure 12.37). Figure 12.36 schematically illustrates the movement of contaminants through the system. The focus of this Case in Point is the problem of NAPL contaminants that float on the surface of underground streams in cave and conduit passages. Numerous sources for NAPL contaminants derive from the more than 500 buried tanks in the city.

As the contaminants float on the cave stream surfaces, toxic and explosive fumes evaporate into the cave atmosphere. In itself, this condition poses a nontrivial explosion threat. Storm drainage wells in parking lots and along streets provide an open connection to the cave system. A discarded cigarette or other source could trigger an underground explosion that could progress along the cave system. Even more disturbing, perhaps, is the upward movement of fumes through solution cavities in the limestone to basements, where they present a serious health hazard. Although the presence of hazardous fumes in houses in Bowling Green had occurred several times in the past, the problem became so

▲ FIGURE 12.37
Map of Lost River groundwater basin, which includes part of Bowling Green, Kentucky. *Source:* From N. C. Crawford, 1986, *Karst Hydrologic Problems Associated with Urbans Development*, National Water Well Association.

and (b) increasing tendency to sorb onto solid organic materials in an aquifer: (1) benzene, (2) phenol, (3) vinyl chloride.

8. A tensiometer installed to a depth of 60 cm reads −200 cm. An adjacent tensiometer is 150 cm deep and reads −100 cm. Which way is the soil moisture moving?

9. Discuss ways in which solutions can move downward in the unsaturated zone more rapidly than the average rate of movement.

10. A nonreactive contaminant is released from a landfill into a sandy aquifer with $K = 10^{-4}$ m/s, $n = 0.35$, and a hydraulic gradient of 0.001. Neglecting dispersion, how many years will it take for the contaminant to reach a well located 2000 m downgradient along a flow line extending from the point of release on the water table?

11. A monitor well 50 m from a waste lagoon first shows the presence of a nonreactive contaminant from the lagoon 10 months after waste is released into the lagoon. If the lagoon intersects the water table, the porosity of the aquifer is 0.3, and the hydraulic gradient is $10^{-3.5}$, what is the hydraulic conductivity of the aquifer? Assume dispersion is too small to be significant.

12. Discuss ways in which aqueous-phase contaminants can be attenuated during transport.

13. Explain how the movement of NAPLs differs from aqueous-phase transport both above and below the water table.

14. What special problems of contaminant movement are noted in karst terrains?

15. Discuss the advantages and disadvantages of pump and treat remediation systems.

References and Suggestions for Further Reading

ABDUL, S. A. 1988. Migration of petroleum products through sandy hydrogeologic systems. *Ground Water Monitoring Review*, 8 (4):73–81.

BOUWER, E., J. MERCER, M. KAVANAUGH, and F. DIGIANO. 1988. Coping with groundwater contamination. *Journal of the Water Pollution Control Federation*, 60:1415–1423.

CHRISTENSON, S. C., and I. M. COZZARELLI. 2003. *The Norma Landfill Environmental Research Site: What Happens to the Waste in Landfills?* U.S. Geological Survey Fact Sheet FS-040-03.

CONCAWE. 1979. *Protection of Groundwater from Oil Pollution.* The Hague, Netherlands, NTIS PB82-174608.

CRAWFORD, N. C. 1986. *Karst Hydrologic Problems Associated with Urban Development: Ground Water Contamination, Hazardous Fumes, Sinkhole Flooding, and Sinkhole Collapse in the Bowling Green Area, Kentucky*, Field Trip Guidebook. Dublin, Ohio: National Water Well Association.

DEVINNY, J. S., L. G. EVERETT, J. C. S. LU, and R. L. STOLLAR. 1990. *Subsurface Migration of Hazardous Wastes.* New York: Van Nostrand Reinhold.

DUNCAN, D., D. T. PETERSON, T. R. SHEPARD, and J. D. CARR. 1991. Atrazine used as a tracer of induced recharge. *Ground Water Monitoring Review*, 11:144–150.

FEENSTRA, S., and J. A. CHERRY. 1988. Subsurface contamination by dense non-aqueous phase liquid (DNAPL) chemicals, in *Proceedings, International Groundwater Symposium on Hydrogeology of Cold and Temperate Climates and Hydrogeology of Mineralized Zones*, C. L. Lin, ed. International Association of Hydrogeologists, Canadian National Chapter, pp. 62–69.

FETTER, C. W. 1993. *Contaminant Hydrogeology.* New York: Macmillan.

GILLHAM, R. W., and J. A. CHERRY. 1982. Contaminant movement in saturated geologic deposits, in *Recent Trends in Hydrogeology*, T. N. Narasimham, ed. Geological Society of America Special Paper 189:31–62.

HOFFMAN, F. 1993. Ground-water remediation using "smart pump and treat." *Ground Water*, 31:98–106.

LEBLANC, D. R. 1984. *Sewage Plume in a Sand and Gravel Aquifer, Cape Cod, Massachusetts.* U.S. Geological Survey Water-Supply Paper 2218.

MASTERS, G. M. 1991. *Introduction to Environmental Engineering and Science.* Englewood Cliffs, N.J.: Prentice Hall, Inc.

MCCARTY, P. L., and J. T. WILSON. 1992. Natural anaerobic treatment of a TCE plume, St. Joseph, Michigan, NPL site, in *Bioremediation of Hazardous Wastes*. U.S. EPA, EPA/600/R-92/126, pp. 47–50.

PERLMUTTER, N. M., and M. LIEBER. 1970. *Disposal of Plating Wastes and Sewage Contamination in Ground Water and Surface Water, South Farmingdale–Massapequa Area, Nassau County, New York.* U.S. Geological Survey Water Supply Paper 1879-G.

TESTA, S. M., and D. L. WINEGARDNER. 1991. *Restoration of Petroleum-Contaminated Aquifers.* Chelsea, Mich.: Lewis.

THORPE, R. K., W. F. ISHERWOOD, M. D. DRESEN, and C. P. WEBSTER-SCHOLTEN. 1990. *CERCLA Remedial Investigations Report for the LLNL Livermore Site.* Livermore, Calif.: Lawrence Livermore National Laboratory.

U.S. EPA. 1987. *Seminar on Transport and Fate of Contaminants in the Subsurface.* CERI-87-45.

U.S. EPA. 1989. *Seminar on Site Characterization for Subsurface Remediations.* CERI-89-224.

U.S. EPA. 1990. *Ground Water, Volume 1: Ground Water and Contamination.* EPA/625/6-90/016a.

U.S. GEOLOGICAL SURVEY. 1998. *Ground Water Contamination by Crude Oil Near Bemidji, Minnesota.* U.S. Geological Survey Fact Sheet 084-98.

U.S. GEOLOGICAL SURVEY. 1999. *Quality of Our Nation's Waters—Nutrients and Pesticides.* U.S. Geological Survey Circular 1225.

WENTZ, C. A. 1989. *Hazardous Waste Management.* New York: McGraw-Hill.

CHAPTER 13

Mass Movement and Slope Stability

Mass movement is the collective name for a variety of processes involving the downslope motion of soil and rock materials under the influence of gravity. Damages resulting from such movements have been estimated to exceed $3 billion per year in the United States. An even greater cost is the loss of life that commonly accompanies major slope movements. Prevention of damage from slope movements requires recognition and avoidance of potentially unstable slopes. Where construction takes place on slopes, a detailed geologic investigation coupled with a thorough engineering analysis of the stability of rocks and soils that underlie the slope must be the initial phase of the project. Table 13.1 is a compilation of some of the largest and most damaging mass movements between 1960 and 1998.

Types of Slope Movements

Mass movements are classified according to the type of movement and the type of slope material involved (Figure 13.1). Slope materials are divided into bedrock, soil composed of predominantly coarse particles (debris), and soil composed of predominantly fine clasts (earth). Six types of movement are utilized in the classification. Each slope movement, therefore, is given a two-part name that relates the type of movement and the type of material. Many slope movements cannot be assigned to a single process and thus must be included in the "complex" category, an indication that more than one type of movement has occurred.

Missing from the classification is any indication of the velocity of movement. Within each category, rates can range from imperceptibly slow to freight-train velocities (Figure 13.2). The velocity classes defined in Figure 13.2 are significant in that they can be shown to relate to the

Table 13.1 Major Catastrophic Mass Movements of the Late 20th Century

Year	Country (State/ Province)	Name and Types	Triggering Process	Vol of Material m³ Except Where Noted	Impact	Comments
1962	Peru (Ancash)	Nevados Huascaran debris avalanche	???	13×10^6	4–5000 killed; much of village of Ranrahirca destroyed	Major debris avalanche from Nevados Huascaran; average velocity 170 km/hr
1963	Italy (Friuli-Venezia-Griulia)	Vaiont Reservoir rockslide	???	250×10^6	2000 killed; city of Longarone badly damaged; total damages: US$200 million (1963 $)	High-velocity rockslide into Vaiont Reservoir caused 100-m waves to overtop Vaiont Dam
1964	United States (Alaska)	1964 Alaska landslides	Prince William Sound Earthquake M = 9.4	???	Estimated US$280 million (1964 $) damages	Major landslide damage in cities of Anchorage, Valdez, Whittier, Seward
1965	China (Yunnan)	Rockslide	???	450×10^6	Four villages; 444 dead	Occurred at "high speed"
1966	Brazil (Rio de Janeiro)	Rio de Janeiro slides, avalanches, debris/mud flows	Heavy rain	???	1000 dead from landslides and floods	Many landslides in Rio de Janeiro and environs
1967	Brazil (Serra das Araras)	Serra das Araras slides, avalanches, debris/mud flows	Heavy rain	???	1700 dead from landslides and floods	Many landslides in mountains SW of Rio de Janeiro
1970	Peru (Ancash)	Nevados Huascaran debris avalanche	Earthquake M = 7.7	$30–50 \times 10^6$	18,000 dead; town of Yungay destroyed; Ranrahirca partially destroyed	Debris avalanche from same peak as in 1962; attained average velocity of 280 km/hr
1974	Peru (Huancavelica)	Mayunmarca rockslide–debris avalanche	Rainfall? River erosion?	1.6×10^9	Mayunmarca village destroyed, 450 killed; failure of 150-m-high landslide dam caused major downstream flooding	Debris avalanche with average velocity of 140 km/hr dammed Mantaro River
1980	United States (Washington)	Mount St. Helens rockslide–debris avalanche	Eruption of Mount St. Helens	2.8×10^9	World's largest historic landslide; only 5–10 killed, but major destruction of homes, highways, etc.; major debris flow; deaths low because of evacuation	Evacuation saved lives; began as rockslide; deteriorated into 23-km-long debris avalanche with average velocity of 125 km/hr; surface remobilized into 95-km-long debris flow

(Continued)

Table 13.1 (*Continued*)

Year	Country (State/Province)	Name and Types	Triggering Process	Vol of Material m² Except Where Noted	Impact	Comments
1983	United States (Utah)	Thistle debris slide	Snowmelt and heavy rain	21×10^6	Destroyed major railroad and highways; dammed Spanish Fork flooding town of Thistle; no deaths	Total losses: US$600 million (1983 $)—50% direct losses, 50% indirect losses
1983	China (Gansu)	Saleshan landslide	???	35×10^6	237 dead; buried four villages; filled two reservoirs	Loess landslide
1985	Colombia (Tolima)	Nevado del Ruiz debris flows	Eruption of Nevado del Ruiz	???	Four towns and villages destroyed; flow in valley of Lagunillas River killed more than 20,000 in city of Armero.	Death toll unnecessarily large because hazard warnings not passed to residents
1986	Papua, New Guinea (East New Britain)	Bairaman rockslide–debris avalanche	Bairaman earthquake M = 7.1	200×10^6	Village of Bairaman destroyed by debris flow from breached landslide dam; evacuation prevented casualties; huge effect on local landscape	Debris avalanche formed 210-m-high dam that impounded 50-million m^3 lake; dam failed, causing 100-m-deep debris flow–flood downstream
1987	Ecuador (Napo)	Reventador landslides	Reventador earthquakes M = 6.1 and 6.9	$75–110 \times 10^6$	1000 killed; many km of trans-Ecuadorian oil pipeline and highway destroyed; total losses: US$1 billion (1987 $)	Land sliding mainly in saturated, residual soils on steep slopes; thousands of thin slides remobilized into debris flows in tributary and main drainages.
1994	Colombia (Cauca)	Paez landslides	Paez earthquake, M = 6.4	250 km²	Several villages partially destroyed by landslides; 271 dead; 1700 missing; 158 injured; 12,000 displaced	Thousands of thin, residual-soil slides on steep slopes turned into damaging debris flows in tributary and main drainages
1998	Honduras, Guatemala, Nicaragua, El Salvador	Hurricane Mitch flooding/landslides/debris flows	Hurricane Mitch		Approximately 10,000 people killed in the flooding and landslides, which occurred throughout the region. Casitas volcano in Nicaragua experienced large debris flows. Impossible to differentiate deaths from landslides from deaths due to flooding.	180-mile per hour winds affected Honduras primarily. Torrential rains occurred, at the rate of 4 in. per hr. Large landslides in Tegucigalpa and elsewhere.

Source: U.S. Geological Survey, National Landslide Information Center, *landslides.usgs.gov.*

Type of movement			Type of material		
			Bedrock	Engineering soils	
				Predominantly coarse	Predominantly fine
Falls			Rock fall	Debris fall	Earth fall
Topples			Rock topple	Debris topple	Earth topple
Slides	Rotational	Few units	Rock slump	Debris slump	Earth slump
	Translational		Rock block slide	Debris block slide	Earth block slide
		Many units	Rock slide	Debris slide	Earth slide
Lateral spreads			Rock spread	Debris spread	Earth spread
Flows			Rock flow (deep creep)	Debris flow	Earth flow (soil creep)
Complex			Combination of two or more principal types of movement		

▲ FIGURE 13.1

Classification of slope movements. *Source:* From D. J. Varnes, 1978, Slope movement types and processes, in *Landslides: Analysis and Control*, R. L. Schuster and R. J. Krizek, eds., TRB Special Report 176, Transportation Research Board, National Research Council, Washington, D.C.

▶ FIGURE 13.2

Velocity scale for slope movements. *Source:* From D. M. Cruden and D. J. Varnes, 1996, Landslide types and processes, in *Landslides: Investigation and Mitigation*, A. K. Turner and R. L. Schuster, eds., TRB Special Report 247, Transportation Research Board, National Research Council, Washington, D.C.

Velocity class	Description	Velocity (mm/sec)	Typical velocity
7	Extremely rapid		
		5×10^{3}	5 m/sec
6	Very rapid		
		5×10^{1}	3 m/min
5	Rapid		
		5×10^{-1}	1.8 m/hr
4	Moderate		
		5×10^{-3}	13 m/month
3	Slow		
		5×10^{-5}	1.6 m/year
2	Very slow		
		5×10^{-7}	16 mm/year
1	Extremely slow		

Table 13.2   Relative Level of Destructiveness for Slope Movements of Each Velocity Class

Landslide Velocity Class	Probable Destructive Significance
7	Catastrophe of major violence; buildings destroyed by impact of displaced material; many deaths; escape unlikely
6	Some lives lost; velocity too great to permit all persons to escape
5	Escape evacuation possible; structures, possessions, and equipment destroyed
4	Some temporary and insensitive structures can be temporarily maintained
3	Remedial construction can be undertaken during movement; insensitive structures can be maintained with frequent maintenance work if total movement is not large during a particular acceleration phase
2	Some permanent structures undamaged by movement
1	Imperceptible without instruments; construction possible with precautions

Source: D. M. Cruden and D. J. Varnes, 1996, Landslide types and processes, in *Landslides: Investigation and Mitigation*, A. K. Turner and R. L. Schuster, eds., TRB Special Report 247, Transportation Research Board, National Research Council, Washington, D.C.

amount and type of damage from slope movements (Table 13.2). The boundary between velocity classes 7 and 6, 5 m/s, or about the velocity a person can run, is important in that class 7 events usually involve total loss of life, whereas survivors are common in class 6 movements.

Falls and Topples

When a mass of rock, debris, or soil separates from a steeply sloping surface and rapidly moves downslope by free fall, bounding, or rolling, the movement is termed a *fall*. These phenomena range from massive bodies of rock on mountain peaks set in motion by earthquakes (Figure 13.3) to small blocks of soil that fall down a river bank when lateral erosion by the stream undercuts the bank to the point where the overhanging section collapses. Hazards are greatest in narrow mountain valleys with steep rock walls. Residential construction in areas of rock fall hazard has accompanied population growth in the Rocky Mountains, primarily associated with recreational development. The resort town of Vail, Colorado, provides an excellent example. Development had encroached into hazardous areas prior to adoption of hazard mapping and zoning. Figure 13.4a shows the slope geology of a rock fall area near a ski resort. Nearly vertical cliffs form at the outcrop of resistant formations near the top of the 300-m-high slope. Rock falls are generated by weathering processes, including freeze and thaw. Rock slabs that break off from the upper limestone cliff gain momentum by rolling down the upper shale acceleration zone and are launched into the air over the vertical sandstone ledge below. Rocks from both source areas accelerate on the steep slope below the sandstone ledge and come to rest in the runout zone, where houses are located (Figure 13.4b). After a falling mass of rock becomes detached from the slope, its behavior upon impact depends upon the amount of slope. At slope angles between 76° and 45°, bouncing predominates; at angles below 45°, rolling is more common.

A *topple* is a rotational movement that occurs as a block of material pivots forward about a fixed point near the base of the block (Figure 13.5). Topples develop in rock or cohesive-soil slopes divided into blocks by vertical fractures or joints oriented parallel to the slope face. Horizontal discontinuities, such as bedding, may affect the process if

▲ FIGURE 13.3

Damage to town of Friuli, Italy, from a rock fall caused by an earthquake in 1976. *Source:* Photo courtesy of Mario Panizza.

differential erosion allows a more resistant column of overlying rock to be undermined (Figure 13.6). The thin slabs of material are pushed outward by lateral forces exerted by adjacent material or by water freezing and expanding in the cracks. If the base of the slab is fixed, the lateral forces cause an overturning moment on the slab, and a topple may occur. The gradual pivoting motion prior to final failure may take a long period of time.

Slides

One of the most common types of movement is *sliding*—that is, shearing displacement between two masses of material along a surface or within a thin zone of failure. The basic difference between slides and flows is that slides initially move as a unit with little or no deformation within the sliding mass, whereas *flows* are thoroughly deformed internally during movement.

Slide types are further divided by the nature of the failure surface (Figure 13.7). Failure surfaces are often curved or circular. These slides, or *slumps*, are very common in soils or rocks of low shear strength (Figure 13.7). The rotational movement of the slump mass or block may result in a significant tilt of the upper surface of the moving block, or *head*, backward toward the exposed upper part of the failure plane, which is called the *scarp* (Figure 13.7). Trees, telephone poles, and other objects on top of the sliding blocks are also tilted by the rotational movement (Figure 13.8). Slumps can attain truly immense proportions. Recent mapping of the seafloor around the island of Hawaii, for example, has shown that huge slumps extend from the coasts out onto the seafloor for more than 100 km. Major fault scarps on land, some with more than 700-m displacement, are actually the head scarps of the gigantic submarine slumps (Figure 13.9).

(b)

▲ FIGURE 13.4
Diagram (a) and photo (b) of slope in Vail, Colorado, that produces rock falls. *Source:* From B. K. Stover, 1988, Booth Creek Rockfall Hazard Area, in *Field Trip Guidebook*, Geological Society of America. *Source:* Photo courtesy of the author.

If failure surfaces are planar rather than curved, the resulting movement is a *translational slide* (Figure 13.7). *Rock block slides*, consisting of the motion of a mass of rock that remains in a small number of blocks, can be distinguished from *rockslides* (Figure 13.10), which are composed of multiple rock fragments of various sizes. The actual configuration of the slide mass usually is determined by joints and planes of foliation or bedding. The most hazardous situation occurs when predominant planes of weakness dip in the same direction as the slope. A disastrous slide that at least began with translational movement localized

▶ FIGURE 13.5

Topples occurring in jointed bedrock. *Source:* Modified from D. J. Varnes, 1978, Slope movement types and process-es, in *Landslides: Analysis and Control*, R. L. Schuster and R. J. Krizek, eds., TRB Special Report 176, Transportation Research Board, National Research Council, Washington, D.C.

▶ FIGURE 13.6

Column of rock beginning to pivot down-slope. A topple will eventually occur. Petri-fied Forest National Park, Arizona. *Source:* Photo courtesy of the author.

along planes of weakness dipping toward the center of a valley occurred in the Italian Alps in 1963. During the slide, 270 million cubic meters of rock and soil traveled at a ve-locity as high as 30 m/s (velocity class 7) into a reservoir impounded by Vaiont Dam, the world's highest thin-arch dam (267 m) (Figure 13.11). Slow movements of the southern wall of the gorge (creep) were detected during filling of the reservoir, indicating that in-creases in pore pressure in the dipping rocks were destabilizing the slope. A smaller land-slide occurred during the initial filling of the reservoir in 1960. Three years later, while the site engineers were still trying to control the creep by gradually filling and drawing down the reservoir, the catastrophic slide broke loose from the hillside. The slide filled the reser-voir for a distance of 2 km and generated a huge wave of water that overtopped the dam by 245 m and flooded the valley below. Nearly 3000 people were killed by a wall of water as much as 70 m high. The destruction would have been even greater had the dam not re-mained intact during the tremendous stress placed upon it. A major factor in the Vaiont slide was the presence of planes of weakness in the valley wall rocks dipping toward the valley floor (Figure 13.11b). These planes included bedding planes in the sedimentary rock units at the site, as well as fractures produced by unloading of the rocks in the slopes—fractures similar to those caused by exfoliation.

▲ **FIGURE 13.7**
Types of slides. (a) Rotational rockslide (rock slump). (b) Rotational earth slide (earth slump).
(c) Translational rockslide (upper part is rock block slide). (d) Debris slide. (e) Translational earth slide.
Source: Modified from D. M. Cruden and D. J. Varnes, 1996, Landslide types and processes, in *Landslides: Investigation and Mitigation*, A. K. Turner and R. L. Schuster, eds., TRB Special Report 247, Transportation Research Board, National Research Council, Washington, D.C.

Debris slides (Figure 13.7) are common on slopes where bedrock is overlain by *colluvium*, a mixture of residual and transported soil slowly moving downslope under the influence of gravity (Figure 13.12). The failure plane in these situations is either the interface between less-weathered bedrock and the debris above or a plane within the debris. The common association of excess pore-water pressure and slope movement is recognized in this type of setting. The effect of increased pore-water pressure is to produce a buoyant force acting upward on the slide mass, thus decreasing the natural stability of the slope. In addition, heavy rainfall can increase the weight of the potential slide mass by saturating the upper part of the debris and creating a perched water table within the slope debris. The role of water will be examined in more detail later in this discussion.

A type of translational slide that caused major damage in Anchorage, Alaska, during the 1964 earthquake is illustrated in Figure 13.13. Translational movement was initiated above a plane in the weak Bootlegger Cove clay (Figure 13.7e). Here the triggering mechanism was

▲ FIGURE 13.8
Telephone poles tilted by the rotational movement of a slump located in northern Idaho. *Source:* Photo courtesy of the author.

▶ FIGURE 13.9
Island of Hawaii showing huge submarine slumps (stippled pattern). Contours are water depth in kilometers. Faults on land (shown by fine lines with ball on downdropped side) are scarps of slumps. *Source:* From J. G. Moore and R. K. Mark, 1992, Morphology of the island of Hawaii, *GSA Today*, 2:257–262.

▲ FIGURE 13.10
The scar of a rockslide near Lake Tahoe that occurred in granitic igneous rocks. *Source:* Photo courtesy of the author.

the violent ground shaking by a major earthquake, although the subsurface materials created a potentially unstable situation.

Lateral Spreads

The slow-to-rapid lateral extensional movements of rock or soil masses are known as *lateral spreads*. Liquefaction and flowage of a weak soil layer within a slope is the cause of most lateral spreads in the debris and earth categories of the slope-movement classification. The stronger material above the failure is rafted along without intense deformation, although it may be broken into blocks that can subside or rotate as the spread progresses (Figure 13.14). The most susceptible materials are sensitive clays (Chapter 10), like the Bootlegger Cove clay in Anchorage. These clays exhibit the tendency to instantaneously lose shear strength upon rearrangement of the clay particles (remolding).

The danger of quick clays (clays with the highest sensitivity values) has been proven by many spreading failures in such places as the St. Lawrence River valley and the fjords of Scandinavia, in addition to the Anchorage region. A hypothesis for the mechanism of failure of the clays in these areas is based on their common geologic histories. When glaciers advanced to coastal regions, the crust was depressed under the great weight of the continental ice sheets. Offshore marine clays deposited during these times were later raised

▲ FIGURE 13.11
(a) Map of Vaiont Reservoir area showing extent of the landslide of October 1963. Black dots show the locations of villages in the flooded area. (b) Cross section of the valley showing the structure and stratigraphy of the slide area. *Source:* Modified from G. A. Kiersch, 1964, Vaiont Reservoir disaster, *Civil Engineering*, 34:32–39.

above sea level by the slow, but continuous, rebound of the crust after the glaciers retreated. Because of their origin, the clays were originally deposited with a flocculated structure (Figure 13.15). After the clay deposits became exposed to nonmarine conditions, the salty pore water was gradually displaced by fresher water derived from rainfall and snowmelt. This freshwater then began to leach away the salt ions that provided the bonds between particles or to cause geochemical reactions that weakened the particle bonds. By this process a very unstable situation is created. In its flocculated state, the clay has significant shear strength. However, after leaching or geochemical changes, the clay fabric can be instantaneously changed from a flocculated state to a dispersed state. The dispersed clay has only a fraction of the shear strength it previously possessed and will behave as a liquid, with the ability to flow on extremely low slope angles. Thus lateral spreads are generated.

Earthquakes often provide the shock that induces the sudden liquefaction of quick clays. Other possible causes are construction blasting and the vibrations caused by the movement of heavy equipment. Once a lateral spread is initiated, buildings, and everything else on the land surface, are rafted along above the flowing clay with the more coherent surficial material. A good example of this type of movement is the lateral spread that occurred in Nicolet, Québec, in 1955 (Figure 13.16). The costs of this event were three lives lost and several millions of dollars in property damage.

◄ FIGURE 13.12
Shallow debris slide near Monterey, California, resulting from the severe winter storms of 1982–1983. *Source:* G. F. Wieczorek; photo courtesy of U.S. Geological Survey.

The lateral spreads just described involve the rapid flowage of earth materials. They are assigned to a separate category in the classification because of their predominantly lateral motion and because of the thick blocks of surficial material that are passively transported along above the flowing clay like boxes on a conveyor belt. There is really a complete transition from earth slides like the Anchorage slope failure, where remolding of sensitive clay occurred in very thin zones within more competent clay; to lateral spreads, where the zone of flowage is thicker; to certain types of rapid earth flows, where the entire mass of sediment is flowing. It is this latter type of movement to which we now turn our attention.

Flows

Flows are complex physical phenomena. Flows of various types of rock and soil may exhibit viscous or plastic behavior, as well as variations and combinations of both. Velocities of flows span the entire velocity class scale and densities are also highly variable, reaching 80% by weight. High-density flows are able to transport boulders that are many meters in diameter, creating the potential for great destructiveness. The overriding criterion in distinguishing flows from slides is that flow must involve continuous internal deformation of the moving material.

▲ FIGURE 13.13

A complex translational slide in the Turnagain Heights area of Anchorage, Alaska, resulting from the 1964 earthquake. Notice that some of the houses remained nearly intact during the predominantly lateral movement. A block model of the slide is shown in Figure 13.7e. *Source:* W. R. Hansen; photo courtesy of U.S. Geological Survey.

Firm clay

Soft clay with water-bearing silt and sand layers

Firm clayey gravel

▲ FIGURE 13.14

A lateral spread involves the rapid outward movement of blocks of firm material because of the liquefaction and flowage in weaker material below. *Source:* From D. J. Varnes, 1978, Slope movement types and processes, in *Landslides: Analysis and Control*, R. L. Schuster and R. J. Krizek, eds., TRB Special Report 176, Transportation Research Board, National Research Council, Washington, D.C.

◀ FIGURE 13.15
Scanning electron microscope image of flocculated fabric of Bootlegger Cove clay, which liquefied in the Anchorage earthquake. *Source:* From R. G. Updike et al., 1988, *Geologic and Geotechnical Conditions Adjacent to the Turnagain Heights Landslide, Anchorage, Alaska,* U.S. Geological Survey Bulletin 1817.

▲ FIGURE 13.16
Aerial view of the Nicolet, Québec, lateral spread of 1955. Buildings and trees were rafted along in an upright position. *Source:* Henry Laliberté; photo courtesy of Governement du Québec, Ministére de l'Environnement.

When a flow travels at velocities at the slow end of the velocity scale, the process is called *creep*. Creep may occur in rock or surficial debris. In rock, creeplike flow can be a very slow, steady process, persisting over long periods of time. Alternatively, slow creep may accelerate to the point where a dramatic failure results, as happened in the Vaiont Reservoir slide. Instruments can be installed in vertical bore holes to measure downslope movement; it can also be observed by the displacement of trees or fences. Creep had been monitored in the slopes above the Vaiont Reservoir preceding the main slide. The rapid acceleration of the movement alerted officials to the impending disaster. Unfortunately, desperate attempts to draw down the reservoir were too late to avert the slide.

A common type of creep is often observed within weathered bedrock and soil on steep slopes (Figure 13.17). This process displays a pronounced seasonal variation characterized by expansion of soil materials perpendicular to the slope (Figure 13.18). The expansion, primarily due to freezing or swelling caused by increases in water content, is only the first phase of the movement. When either thawing or shrinkage takes place, the movement is vertically downward under the controlling influence of gravity. Thus, during each cycle of expansion and contraction the soil undergoes a small downslope component of movement.

The more rapid examples of flow phenomena are variously termed *debris, earth,* or *mud* flows. These processes involve the rapid to very rapid flow of materials down steep slopes (Figure 13.19). The similarity in process among the flow of debris, lava, and glacial

▲ FIGURE 13.17
Creep in weathered bedrock, Maryland. The initially vertical beds at the top of the exposure have been tilted by slow downslope movement. *Source:* G. W. Stose; photo courtesy of U.S. Geological Survey.

▶ FIGURE 13.18
Creep is often a seasonal process in which freezing or swelling causes movement of particles perpendicular to the slope. Upon thawing or shrinkage, vertical movement results in a net downslope component of motion.

◄ FIGURE 13.19
An earth flow that occurred in 1953 in Okanogan County, Washington. The flow lasted 7 minutes and destroyed a house. The source of the flow is the hill in the background. *Source:* F. O. Jones; photo courtesy of U.S. Geological Survey.

ice is reflected in the production of similar landforms by all three flow phenomena. Like glacial ice and lava, debris and earth flows follow channels, spread out to form lobate *toes* when not confined to channels (Figure 13.19), and develop conspicuous *lateral ridges* along the sides of the channels (Figure 13.20).

◄ FIGURE 13.20
Morphology and features of an earth flow. *Source:* From D. K. Keefer and A. M. Johnson, 1983, *Earth Flows: Morphology, Mobilization, and Movement,* U.S. Geological Survey Professional Paper 1264.

▶ FIGURE 13.21
Cross section of channelized debris flow showing the development of a
nondeforming plug at the center of the flow.

Many observations of debris flows have confirmed that the flows have a high water content, and thus are highly fluid, yet can transport boulders and other large objects many times the size of anything that even the fastest rivers can move. While in transit, debris flows treat boulders and other surficial objects with remarkable care. For example, large boulders that apparently have been transported intact within a debris flow have broken apart along fractures by weathering processes within a relatively short time period after deposition by the flow. These observations require careful analysis to determine the mechanical nature of debris-flow processes.

The visual similarity of rivers and rapid debris flows seems to suggest a similar flow mechanism. Thinking in these terms, we could consider debris flows simply to be rivers of high-density, highly viscous fluid in which boulders are supported by *turbulent eddies* (random velocity fluctuations) in the flow. Laboratory tests have shown, however, that dense debris flows are *laminar* rather than turbulent, and laminar flow lacks high-velocity turbulent eddies (Chapter 14). As an alternative, plastic behavior could be called upon to explain the concentration of boulders near the tops of the flow masses. Plastic flow could prevent the sinking of boulders into the center of the flow by the strength of the supporting flow material. An ideal plastic, however, would accelerate infinitely once the yield strength has been exceeded. A reasonable composite model combines elements of plastic flow and laminar viscous flow. This type of behavior is the type exhibited by a Bingham substance (Chapter 7). It seems to best fit the observations of debris flows and debris-flow deposits with respect to their ability to transport boulders in a laminar flow regime.

A further implication of the flow of a Bingham material is the existence of a rigid "plug" near the center of the channel and at the top of the flow (Figure 13.21). This plug is rafted along as a nondeforming unit in the debris flow because the shear stress within the plug is less than the strength of the material.

Case in Point 13.1
La Conchita California Debris Flows

La Conchita, California, became a major media event in January 2005 when exceptionally heavy rains caused a steep slope behind the town to fail, mobilizing a devastating debris flow that swept through the town, burying about 15 houses and killing at least 10 residents (Figure 13.22). Tragic as it was, this event was not a surprise to geologists who had studied the area. In fact, a smaller flow of the same type occurred in 1995, similarly triggered by heavy rains, and destroyed 9 homes, although no lives were lost (Figure 13.23). A retaining wall was built after that event in an attempt to mitigate future damages. The 2005 event was actually a remobilization of the 1995 debris mass.

The geologic setting of La Conchita includes a set of conditions that make the location of the town extremely unfortunate (Figure 13.24). The city lies upon a narrow marine terrace between the Pacific Ocean and a steep mountain front. The California coast is tectonically active in many areas and the La Conchita area is crossed by an active fault and is being uplifted relative to sea level. The terrace serving as the site of the city is a Holocene

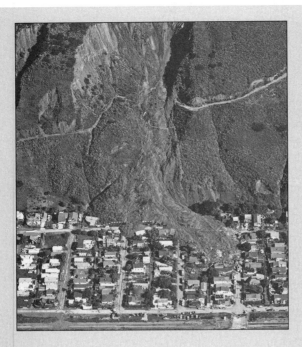

◀ **FIGURE 13.22**
Debris flow of January 10, 2005 in La Conchita, California. Heavy rains caused remobilization of a flow that occurred 10 years earlier (Figure 13.23). Ten lives were lost in this sudden and rapid event. *Source:* From AP Wide World Photos.

◀ **FIGURE 13.23**
Landslide of March 3, 1995, which destroyed 9 houses in La Conchita, California. *Source:* Compliments of U.S. Geological Survey.

feature formed by the uplift. The mountain front was steepened by wave erosion prior to the most recent uplift. Geologic mapping of the area has shown for decades that the site is very dangerous. An older, much larger landslide along the slope lies directly behind La Conchita. It is this area that has failed in the two recent events, and will most likely fail again when heavy rainfall again drenches the area.

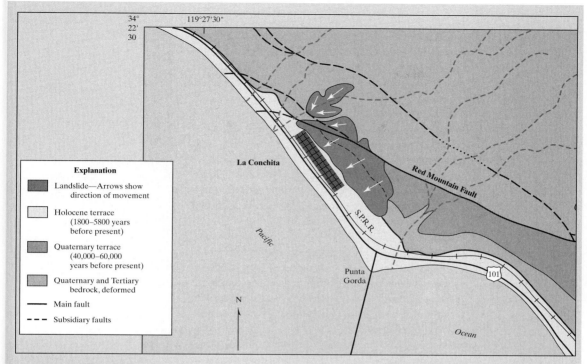

▲ FIGURE 13.24
Geological map of La Conchita. *Source:* From U.S. Geological Survey.

Now that the hazard potential is known without any doubt, decision makers must agree on a course of action. The safest solution would be to relocate the entire town to a less hazardous site. Many residents would strongly object to this solution and compensation for residents for the loss of their property would be extremely expensive given the cost of coastal California real estate. One alternative would be an engineering approach to attempt to protect the town by retaining walls, cutting benches or terraces into the slope, slope drainage, or other means. This would also be extremely costly. The least viable response would be to do nothing at all and wait for the next debris flow to impact the town.

Complex Slope Movements

Slope failures in nature that involve a single type of movement are probably the exception rather than the rule. A very common occurrence is the initiation of a slope movement as a slide, followed by conversion to a flow during the course of the event. A slump-earth flow, for example, is illustrated in Figure 13.25. Topographic features resulting from the slump include a *main scarp*, separating the *crown* from the *head*, as well as one or more *minor scarps*. Beyond the main body in the downslope direction the lobate *toe* indicates that the dominant type of movement in this section is flowage. *Transverse ridges* in bands parallel to the toe cross the earth flow. The La Conchita debris flows may have been this type of a movement.

Some of the greatest natural disasters associated with slope failures have resulted from complex slope movements that began as rock falls or rockslides and then transformed into

Nomenclature

Main scarp—A steep surface on the undisturbed ground around the periphery of the slide, caused by the movement of slide material away from undisturbed ground. The projection of the scarp surface under the displaced material becomes the surface of rupture.

Minor scarp—A steep surface on the displaced material produced by differential movements within the sliding mass.

Head—The upper parts of the slide material along the contact between the displaced material and the main scarp.

Top—The highest point of contact between the displaced material and the main scarp.

Toe of surface of rupture—The intersection (sometimes buried) between the lower part of the surface of rupture and the original ground surface.

Toe—The margin of displaced material most distant from the main scarp.

Tip—The point on the toe most distant from the top of the slide.

Foot—That portion of the displaced material that lies downslope from the toe of the surface of rupture.

Main body—That part of the displaced material that overlies the surface of rupture between the main scarp and toe of the surface rupture.

Flank—The side of the landslide.

Crown—The material that is still in place, practically undisplaced and adjacent to the to the highest parts of the main scarp.

Original ground surface—The slope that existed before the movement which is being considered took place. If this is the surface of an older landslide, that fact should be stated.

Left and right—Compass directions are preferable in describing a slide, but if right and left are used they refer to the slide as viewed from the crown.

Surface of separation—The surface separating displaced material from stable material but not known to have been a surface on which failure occurred.

Displaced material—The material that has moved away from its original position on the slope. It may be in a deformed or undeformed state.

Zone of depletion—The area within which the displaced material lies below the original ground surface.

Zone of accumulation—The area within which the displaced material lies above the original ground surface.

▲ FIGURE 13.25
Diagram of a slump-earth flow. *Source:* From D. J. Varnes, 1978, Slope movement types and processes, in *Landslides: Analysis and Control*, R. L. Schuster and R. J. Krizek, eds., TRB Special Report 176, Transportation Research Board, National Research Council.

rapid debris flows, or avalanches, along their downslope paths. Particularly deadly examples of these events originated from the glacier-mantled Peruvian peak Nevados Huascaran in 1962 and 1970 (Figure 13.26). The first avalanche traveled 14 km in 5 minutes and buried the village of Ranrahirca, killing 3500 residents in the process. An even larger avalanche swept down the mountain in 1970, overtopping a ridge too high for the 1962 avalanche, and devastated the town of Yungay. More than 18,000 lives were lost on this tragic

▶ FIGURE 13.26
Aerial view of Nevados Huascaran area, Peru. Rock avalanches in 1962 and 1970 buried the villages of Yungay and Ranrahirca, killing thousands of people. *Source:* Photo courtesy of U.S. Geological Survey.

day. Unfortunately, rock avalanches are also common in the United States and Canada in the Rocky and Cordilleran Mountain ranges. The Madison Canyon "slide" of 1959 (Figure 13.27), which was triggered by an earthquake, created a lake as it flowed across a valley and blocked the Madison River. A lake quickly developed in the valley behind the debris dam, raising fears that the rapid rise of water level would soon cause a catastrophic failure of the natural dam. Quick work by the U.S. Army Corps of Engineers, which excavated an emergency spillway through the debris, prevented the dam from failing and causing massive floods downstream.

The mechanics of movement of rock avalanches have been debated by geologists in recent years. One theory, which was widely accepted in the 1960s and 1970s, attributed the mass movements to a sliding mechanism (Shreve, 1968). According to this hypothesis, the mass moves as a relatively intact block that slides above a thin layer of compressed air. The compressed air would sufficiently lower the coefficient of friction to allow the great runout distances observed in these movements (Figure 13.28b). A sliding mechanism was invoked to explain the observation that the stratigraphic distribution of rock types in the mass of material at the base of the slope was identical with the distribution of rock types in

◀ FIGURE 13.27
The Madison Canyon "slide." Actually a rock avalanche, this devastating slope movement blocked the Madison River to form a lake and buried a campground at the base of the valley on the side opposite the avalanche. *Source:* Photo courtesy of the author.

◀ FIGURE 13.28
Possible flow mechanisms of rock avalanches: (a) A block of rock on a steep mountain slope is suddenly launched into motion by an earthquake or other triggering factor. (b) Travel of the debris by sliding on a layer of compressed air. Initial stratigraphic order within the moving mass is preserved. (c) Movement of the avalanche by grain flow. After suspension of blocks in the air, motion and stratigraphic order are maintained by inertial collisions between trailing blocks and those fragments ahead of them. Acoustic fluidization is similar to (c) except that rocks are closer together.

the source area of the avalanche high upon the mountain. In other words, the mass appeared to have moved without extensive internal disruption. This stratigraphic order within the moving mass could be explained if the mass moved as an intact block (Figure 13.28b). Stratigraphic order would not be preserved in a viscous flow because the material near the top of the flow moves faster than material near the base. Thorough mixing of rock types would therefore be expected in viscous flow.

An alternative hypothesis was later put forth (even though it was first suggested in the 1880s by Swiss geologist Albert Heim) to explain the problem of stratigraphic order. In this type of flow, called *grain flow,* solid grains dispersed throughout the avalanche achieve the ability to flow without being carried along by a flowing viscous fluid (Hsu, 1975). Although both water and compressed air may be involved in the flow, their presence is not necessary for the flow to occur. This is best illustrated by the identification of giant rock avalanches on the Moon and Mars, bodies without an earthlike atmosphere or water. As grains collide in a grain flow, they transfer momentum to grains ahead of them (Figure 13.28c) and therefore do not pass their counterparts in the direction of flow. Thus, stratigraphic order of rock types will be preserved in a grain flow. This model seems to fit evidence observed in rock avalanches better than the compressed-air hypothesis because of the topographic form of the deposits. Rock-avalanche deposits have the physical appearance of flow, including lobate toes and transverse flow ridges.

Still another hypothesis suggests a process called *acoustic fluidization* for the mobility of rock avalanches (Melosh, 1987). The process is similar to grain flow, except that the particles are in closer contact and are fluidized by the propagation of elastic (acoustic) waves rather than by collisions. The elastic waves, which are similar to seismic waves, are generated as the mass of debris initially moves over an irregular surface. As the waves propagate through the flowing mass, they transmit pressure in the same way as particle collisions.

Controversy over the mechanics of movement of rock avalanches is likely to continue because none of the hypotheses just described has yet been proven.

Case in Point 13.2
The Frank Slide, Southwestern Alberta

One of the best known North American examples of a rock avalanche occurred in 1903 in the Canadian Rockies near the mining town of Frank, Alberta. The avalanche started suddenly when a 30 million m^3 mass of rock broke loose from Turtle Mountain and flowed across the valley below (Figure 13.29). In all, an area of 3 km^2 was buried in rock debris to a depth of 14 m. The entire event took only 100 seconds. Part of the town of Frank was

▶ FIGURE 13.29
Photo of Frank "slide." Whitish area at the base of the slope is the area of deposition of rubble from the rock avalanche. *Source:* Photo courtesy of C. B. Beaty.

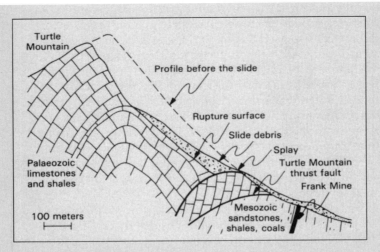

◀ FIGURE 13.30
Cross section of Turtle Mountain, Alberta, showing conditions before and after the rock avalanche. *Source:* From D. M. Cruden and C. B. Beaty, 1987, The Frank Slide, southwestern Alberta, *Geological Society of America Centennial Field Guide—Rocky Mountain Section.*

destroyed and 70 people were killed. The entrance to a coal mine near the base of the slope was buried, but miners trapped below the debris were able to dig their way out.

It is likely that the structure of Turtle Mountain played an important role in the launching of the avalanche. The crest of the mountain forms the core of a large anticline (Figure 13.30). Thus, on the eastern side, beds of limestone were steeply dipping in the same direction as the slope of the mountain. Fractures near the region of maximum curvature of the anticline could also have been important in facilitating weathering processes that may have weakened the rock mass. The triggering event is not known, but freeze and thaw may have been involved. The avalanches occurred on a very cold spring morning following several weeks of warm, wet weather. These conditions would have caused freezing of meltwater from the heavy snowpack on the mountain.

All the mechanisms discussed in the preceding section have been used to explain the mechanics of movement of the Frank slide. The appearance of the debris (Figure 13.29) is strongly suggestive of some type of flow. The lobate form of the rubble has been described as resembling a sack of flour spilled onto the ground. Further research into the cause and mechanics of the Frank slide and other rock avalanches may yet yield an accurate understanding of these awe-inspiring natural processes. As mountainous areas become more and more developed, the ability to predict the locations and the effects of rock avalanches will achieve great importance.

Causes of Slope Movements

Determining the cause of a slope movement may be more difficult than it appears. Rarely is there a single cause; in most cases, although there may be an obvious triggering mechanism, the interaction of many variables controls the initiation of a slope movement.

A basic approach to evaluating slope stability in terms of simple sliding mechanisms is to identify the forces tending to cause movement, or *driving forces*, and those that tend to resist failure, or *resisting forces*. The ratio of the resisting forces to the driving forces is a quantitative indication of stability known as the *safety factor*. Thus,

$$\text{Safety factor (S.F.)} = \frac{\Sigma \text{ resisting forces}}{\Sigma \text{ driving forces}} \tag{13.1}$$

The greater the safety factor, the less likely the slope is to fail. The minimum value of the safety factor is 1 because below this value the driving forces are greater than the resisting forces, and the slide will be in progress. When the safety factor is exactly 1, the slope is in a state of incipient failure and any increase in driving forces or decrease in resisting forces will serve to trigger the movement.

Geologic Setting of Slopes

Before considering factors that effect changes in the safety factor of a slope, it is worthwhile to identify some of the conditions that render a slope susceptible to mass movement. The slope angle, for example, plays an important role in establishing the initial ratio of resisting forces to driving forces. This can be illustrated by analyzing the forces acting on a block of rock resting upon a slope (Figure 13.31a). The weight of rock mass, W, acting vertically downward can be resolved into a force acting parallel to the slope in the downslope direction, $W \sin \theta$, and a force acting normal to the slope, $W \cos \theta$. Opposing the block's tendency to slide down the slope is the frictional force F, which is dependent upon the normal force and the coefficient of friction (μ) between the rock and the slope, so that $F = \mu(W \cos \theta)$. As the slope angle increases (Figure 13.31b), the force acting downslope increases and the normal force decreases. When the force acting downslope, or driving force, equals the frictional resisting force, motion of the block is incipient.

The lithology of the materials composing the slope is probably the most important component of the slope-stability problem. Geologic materials of low strength or materials that tend to weather to materials of low strength present the greatest hazard. Included in this group are clays, shales, certain volcanic tuffs, and rocks containing soft platy minerals such as mica. Clays, and other rocks and soils that contain clays, are particularly notorious for their association with slope failures. Clays have an initially low strength and tend to become even weaker with increases in water content. The ability of clays to absorb water and swell can lead to a significant loss in strength by the materials on a slope.

A special type of slope failure occurs in clays that are classified as *overconsolidated*. Overconsolidated clays have been subjected to large overburden loads during their geologic history. Under these loads, clays become denser by compaction and expulsion of pore water. When overburden pressures are eased, either by erosion of overlying materials

μ — Coefficient of friction
$F = \mu (W \cos \theta)$

(a)

(b)

▲ FIGURE 13.31
Forces acting on a block of rock tending to slide downslope. In (a) the weight of the block is resolved into a downslope force ($W \sin \theta$) and a normal force ($W \cos \theta$). Downslope movement is opposed by the frictional force, F, which is equal to the coefficient of friction times the normal force. The same block is more likely to slide on a steeper slope (b) because the downslope force is greater and the normal force is less than in (a).

◄ FIGURE 13.32
In overconsolidated clays, the residual strength—the resistance to failure along existing fissures in the clay—is less than the peak strength. The peak strength is the resistance to failure obtained by tests on nonfissured, massive samples of clay.

or by excavation, the compacted clays tend to expand and, in the process, develop a network of fractures, or *fissures*. The clay is then classified as overconsolidated because it currently supports a lower weight of overburden than it has at some time in its geologic past. The effect of these fissures is to lower the strength of the clay along the discontinuities. When small samples are taken for lab strength tests, they are frequently trimmed from intact blocks of clay between the fissures. The strength values obtained from these tests, described as the *peak strength*, give a misleading estimate of the strength of the clay mass, because mass movements on the slope are initiated by shear failure along the fissures. The strength mobilized along these discontinuities, the *residual strength*, is significantly lower than the strength measured from intact samples (Figure 13.32). There are numerous examples of excavated slopes such as road and railway cuts that have failed because the slope angle was designed using the peak instead of the residual strength.

In rock slopes a situation similar to that with overconsolidated clays exists, because it is the discontinuities in the rock that control the stability of the rock mass. Discontinuities along which strength is reduced are common in almost every type of rock. In sedimentary rocks, bedding planes constitute significant planes of weakness within the rock mass. In terms of slope stability, the orientation of the planes is critical. Bedding that dips in the same direction as the slope and at about the same amount is the most dangerous situation (Case in Point 6.1 and 13.1). Roadcuts and other excavations can decrease the stability of rock slopes by increasing the stress or decreasing strength along bedding planes in several different ways.

Planes of weakness other than bedding that are present in all rock types include joints and faults. Careful preliminary studies of slopes are necessary to determine the spacing, width, and orientation of these fractures.

Influence of Water

Water is just as important as type and attitude of slope materials as a major controlling factor in the occurrence of mass movements. Its importance can be seen in the fact that most rapid mass movements occur during and after periods of heavy rainfall.

Water can interact with slope materials in several different ways. An obvious result of the addition of water to a slope is an increase in the weight of a potential sliding block. With reference to Figure 13.31, the effect of the addition of water is to increase both the downslope force, $W \sin \theta$, and the normal force, $W \cos \theta$. Although these changes will not drastically alter the safety factor of the slope, the water has a totally different effect on the normal force

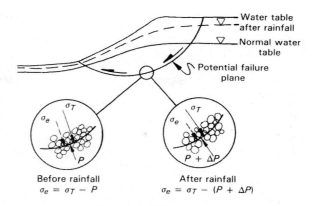

► FIGURE 13.33
Effect of pore water on slope movements. Rainfall causes a rise in the water table and increases pore pressure $(P + \Delta P)$ throughout the subsurface. The stresses on a potential failure plane show the magnitude of effective stress (σ_e). Strength (S) is proportional to effective stress. After rainfall (right), effective stress decreases because of the increase in pore pressure. The strength of the soil along the potential failure plane is therefore lower.

Before rainfall
$$\sigma_e = \sigma_T - P$$

After rainfall
$$\sigma_e = \sigma_T - (P + \Delta P)$$

$$S = C + \sigma_e \tan \theta$$

than was suggested previously. To understand this relationship, we must recall the effective stress equation presented in Chapter 11. In Figure 13.33, the stresses are shown on a potential failure plane within the slope material. Because the fluid pressure acts upward against the soil and rock load above the failure plane, the effective stress equation can be rearranged as

$$\sigma_e = \sigma_T - P \tag{13.2}$$

to show the amount of load borne by the soil particles. The name *effective stress* is appropriate because it determines the frictional resistance to sliding that could be mobilized by the soil particles along a potential failure plane.

Figure 13.33 illustrates the application of the effective stress equation to sliding-type slope movements. The fluid pressure is determined by the thickness of the saturated zone above the failure plane. So when the fluid pressure *increases*, as it would when the water table rises in response to heavy rainfall, the effective stress equation indicates that the effective stress will *decrease*. The rise in water table has therefore decreased the *effective* normal stress (σ_e) resisting sliding movement along the failure plane.

When effective stress is inserted into the Mohr-Coulomb equation for shear strength, the equation becomes

$$S = C + \sigma_e \tan \phi \tag{13.3}$$

It is obvious from equations (13.2) and (13.3) that the shear strength of a material will decrease when the fluid pressure increases.

Another way to think of the effect of fluid pressure on sliding is to compare the process with the buoyant force exerted on an object submerged in water. Just as the submerged weight of any object is less than its weight in air, the effective weight of soil particles submerged by a water table rise will be less than their unsaturated weight. Thus, the resisting forces on the failure plane are decreased and the safety factor is reduced because the driving forces are unchanged or even increased by the weight of the water added to the soil.

Debris slides and flows in the steep hillslopes of California provide an excellent example of the relationship between rainfall, pore pressure, and slope failures. Regionally, extensive slope failures have occurred during winters with exceptionally high rainfall, including the winters of 1968–69, 1982–83, and 2004–05. The most serious failure-generating events consist of periods of very intense rain occurring within a wet period. For example a 32-hour period of intense rainfall in 1982 triggered 18,000 debris flows in the San Francisco Bay region. Figure 13.34 shows cumulative rainfall records for gauges in the San

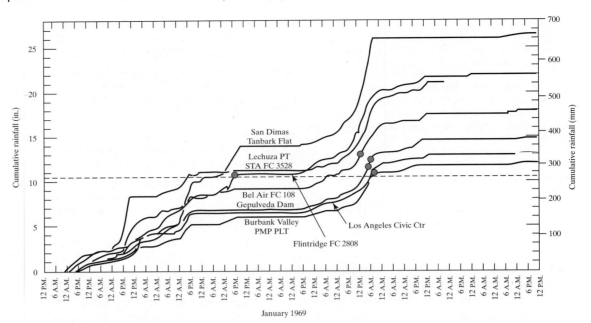

▲ FIGURE 13.34

Cumulative rainfall records from rain gauges in the Santa Monica and San Gabriel Mountains in southern California. Dots represent times of debris flows. These occurred during periods of intense rainfall when the cumulative curves steepened. *Source:* Modified from R. H. Campbell, 1975, *Soil Slips, Debris Flows, and Rainstorms in the Santa Monica Mountains and Vicinity, Southern California*, U.S. Geological Survey Professional Paper 851.

Gabriel and Santa Monica Mountains in southern California in 1969, along with the times of debris flows. All of the debris flows took place where the cumulative rainfall curves steepen, indicating periods of very intense precipitation.

A problem encountered with reservoirs provides an interesting application of the relationship of pore pressure and mass movements (Figure 13.35). The water table in the slopes adjacent to a reservoir rises and falls to adjust to the water level in the impoundment. When

◀ FIGURE 13.35

Reservoir bank failure due to rapid drawdown of reservoir level. When the reservoir is drawn down (b), lateral support provided by the water is lost. Pore pressure in the slope remains high because the water table adjusts to its new level more slowly than the reservoir. These conditions combine to cause slope failure.

the water table rises, the decrease in resisting forces is offset by the lateral support exerted by water in the reservoir on the reservoir banks (Figure 13.35a). If the water level in the reservoir is rapidly lowered, perhaps in anticipation of a flood that must be contained or partially contained in the reservoir, the lateral support provided by the water is removed. The water table in the adjacent slopes is much slower to respond, and the high pore pressures now tend to cause slumping of bank material into the reservoir (Figure 13.35b).

Processes That Reduce the Safety Factor

A natural or constructed slope exists under a certain ratio of resisting and driving forces (the safety factor) at any particular time. Many processes, however, acting over long or short periods of time, may decrease the slope's safety factor. Recognition and identification of these processes are necessary for the prediction of slope instability. In Table 13.3 processes are grouped into those that increase the shear stress on a failure plane and those that reduce shear strength within the materials on a slope.

One of the most common ways to increase the shear stress on a potential failure plane is to remove material from the base of the slope. As shown in Figure 13.36, construction of the roadcut has removed material from an area where the failure plane is nearly horizontal or even curving upward. Material in this part of the potential slide mass acts against

---

Table 13.3 Processes Causing Changes in the Safety Factor

Processes that cause increased shear stress

1. Removal of lateral support
   a. Erosion by rivers
   b. Previous slope movements such as slumps that create new slopes
   c. Human modifications of slopes such as cuts, pits, canals, open-pit mines
2. Addition of weight to the slope (surcharge)
   a. Accumulation of rain and snow
   b. Increase in vegetation
   c. Construction of fill
   d. Stockpiling of ore, tailings (mine wastes), and other wastes
   e. Weight of buildings and other structures
   f. Weight of water from leaking pipelines, sewers, canals, and reservoirs
3. Earthquakes
4. Regional tilting
5. Removal of underlying support
   a. Undercutting by rivers and waves
   b. Construction of underground mines and tunnels
   c. Swelling of clays

Processes that reduce shear strength

1. Physical and chemical weathering processes
   a. Softening of fissured clays
   b. Physical disintegration of granular rocks such as granite or sandstone by frost action or thermal expansion
   c. Swelling of clays accompanied by loss of cohesion
   d. Drying of clays resulting in cracks that allow rapid infiltration of water
   e. Dissolution of cement
2. Increases in fluid pressure within soil
3. Miscellaneous
   a. Weakening due to progressive creep
   b. Actions of tree roots and burrowing animals

(a)

Original slope

Road cut

(b)

◀ FIGURE 13.36
Stability of the slope shown in (a) is decreased by removal of material from the base of the slope and by steepening the slope (b).

movement by resisting the tendency for rotational sliding. When the material is removed, the driving forces on the remainder of the slide mass are increased. In addition, the decrease in area of the failure plane after the excavation results in less shear strength developed along the plane. In this way, the resisting forces have been decreased. The combination of the increased shear stress and the decreased shear strength significantly lowers the safety factor of the slope.

The effect of *surcharge* (overloading) applied upslope from the line of action of the center of gravity of the sliding mass is to increase the rotational tendency of the mass. Surcharge loads introduced by humans, including mine wastes and construction fill, frequently initiate slope failures (Figure 13.37).

Processes that decrease the resistance to failure, with the exception of increases in fluid pressure, are usually long-term phenomena. These changes gradually modify the stability of the slope for many years prior to movement. The safety factor is then lowered to the point where a sudden triggering mechanism—an earthquake (Figure 13.38) or unusually wet period, for example—rapidly equalizes the driving forces and resisting forces, causing failure.

The large number of factors affecting both the driving and resisting forces supports the statement made earlier concerning the complexity of mass movements. Investigations of mass movements made after the fact as well as those aimed at predicting stability

◀ FIGURE 13.37
Slumping of surface mine spoils into a newly excavated mine pit was caused by piling too much spoil material at the edge of the pit. *Source:* Photo courtesy of the author.

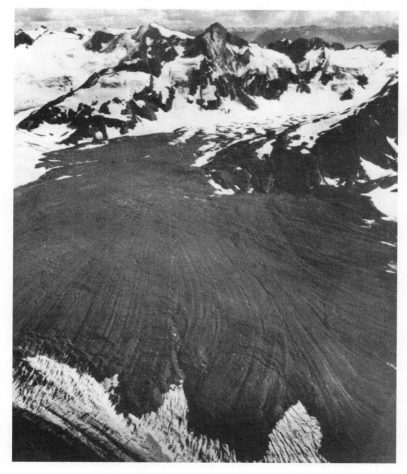

▲ FIGURE 13.38
A huge debris avalanche covering part of Sherman glacier, Alaska. The avalanche was triggered by the 1964 earthquake. *Source:* Austin Post; photo courtesy of U.S. Geological Survey.

relationships must not overlook any of the long- or short-term geologic and human influences on slopes.

Slope Stability Analysis and Design

The most important step preceding the construction of any engineering project that will alter or interact in any way with a slope is to determine the stability of the slope, both in its natural state as well as in its altered state after completion of the project. Embankments, earth dams, and other constructed slopes must likewise be designed with slope stability in mind. We must begin our discussion of stability analysis, therefore, with some comments about the existing state of stability of natural slopes in and near the area of excavation.

Recognition of Unstable Slopes

Areas of potential slope instability are most easily defined if evidence exists for previous slope movements. These areas can be delineated and evaluated by preliminary geologic studies. Several techniques can be used. First, all existing geologic reports of an area should be examined. Some geologic maps are specifically made for the purpose of identifying hazardous areas (Figure 13.39). Other maps show existing slope movements or the

Explanation of Map Units

Least

I Areas least susceptible to landsliding. Very few small landslides have formed in these areas. Formation of large landslides is possible but unlikely, except during earthquakes. Slopes generally less than 15%, but may include small areas of steep slopes that could have higher susceptibility. Includes some areas with 30% to more than 70% slopes that seem to be underlain by stable rock units. Additional slope stability problems; some of the areas may be more susceptible to landsliding if they are overlain by thick deposits of soil, slopewash, or ravine fill. Rockfalls may also occur on steep slopes. Also includes areas along creeks, rivers, sloughs, and lakes that may fail by landsliding during earthquakes. If area is adjacent to area with higher susceptibility, a landslide may encroach into the area, or the area may fail if a landslide undercuts it, such as the flat area adjacent to sea cliffs.

II Low susceptibility to landsliding. Several small landslides have formed in these areas and some of these have caused extensive damage to homes and roads. A few large landslides may occur. Slopes vary from 5–15% for unstable rock units to more than 70% for rock units that seem to be stable. The statements about additional slope stability problems mentioned in I above also apply in this category.

III Moderate susceptibility to landsliding. Many small landslides have formed in these areas and several of these have caused extensive damage to homes and roads. Some large landslides likely. Slopes generally greater than 30% but includes some slopes 15–30% in areas underlain by unstable rock units. See I for additional slope stability problems.

IV Moderately high susceptibility to landsliding. Slopes all greater than 30%. These areas are mostly in undeveloped parts of the County. Several large landslides likely. See I for additional slope stability problems.

V High susceptibility to landsliding. Slopes all greater than 30%. Many large and small landslides may form. These areas are mostly in undeveloped parts of the County. See I for additional slope stability problems.

VI Very high susceptibility to landsliding. Slopes all greater than 30%. Development of many large and small landslides is likely. Slopes all greater than 30%. The areas are mainly in undeveloped parts of the County. See I for additional slope stability problems.

L Highest susceptibility to landsliding. Consists of landslide and possible landslide deposits. No small landslide deposits are shown. Some of these areas may be relatively stable and suitable for development, whereas others are active and causing damage to roads, houses, and other cultural features.

Most

Definitions: Large landslide — more than 500 ft in maximum dimension
Small landslide — 50 to 500 ft in maximum dimension

(b)

▲ **FIGURE 13.39**
A portion of a landslide-susceptibility map for San Mateo County, California. *Source:* From E. E. Brabb, E. H. Pampeyan, and M. G. Bonilla, 1972, *Landslide Susceptibility in San Mateo County, California.* U.S. Geological Survey Miscellaneous Field Studies Map MF-360.

distribution of rock and surficial material types that are known to be associated with slope movements.

If more detailed studies of the project area are needed, these should be initiated by analysis of topographic maps and aerial photographs. Air photos are invaluable for the analysis and interpretation of landforms related to mass movements. Recognizable features include scarps, slump shoulders, and disrupted drainage patterns. Hummocky topography—a random pattern of hills and depressions containing ponds or wetlands—is a good indicator of recent slump or flow activity. Deposits of flows occur as lobate or tongue-shaped landforms in association with such other features as lateral levees and transverse flow ridges. Other relevant slope conditions, including fracture and joint patterns, the locations of springs and seeps, and the oversteepening of slopes by rivers and waves, can be noted on air photos.

Field studies complete the site investigation. Detailed geologic maps of the area can be made by examining rock and soil exposures and plotting data on topographic base maps. Subsurface information is frequently needed for evaluation and design. Some types of data can be obtained using surface geophysical techniques like seismic refraction and earth resistivity. In seismic refraction, seismic energy is generated near land surface using hammers or explosions. Seismic waves are refracted from harder layers of rock or soil at depth back to the surface, where they are recorded and analyzed to calculate the depth of the subsurface layer. Earth resistivity is a technique based on the variable resistance of subsurface materials to an electrical current passed through the ground. The most accurate method available for obtaining subsurface data (and also the most expensive) is test drilling. There are many methods and types of equipment that can be used for various subsurface conditions. Samples taken during drilling are useful not only for determining the distribution of soil and rock units but also for conducting lab tests to characterize the properties of the materials. For slope-stability analysis, the shear strength of the soil, including its lateral and vertical variations, is a necessary type of data. After test holes have been drilled, piezometers can be installed for monitoring groundwater conditions.

Stability Analysis

The methods used for analyzing stability are based on finding the safety factor for a particular slope movement. For soil slopes composed of uniform, cohesive clays, rotational slump is considered to be the dominant failure mechanism. Successive circular failure surfaces are evaluated in order to determine the potential failure surface with the lowest safety factor. If this failure surface, the *critical circle*, has a safety factor near 1.0, the slope is considered to be unstable and must be redesigned or modified. Several trial surfaces and their computed safety factors are shown in Figure 13.40.

One of the most basic methods of calculating the safety factor for circular failure surfaces is shown in Figure 13.41. The safety factor for the indicated failure surface is expressed as the ratio of the resisting moments to the driving moments about the center of rotation, point O. Each cross section of the slope, such as the one illustrated, can be analyzed separately by treating the failure surface as a line rather than as a plane. The resisting moment is the product of the soil shear strength, S, which resists downslope movement of the mass, times the length, L, of the failure circle, times the moment arm, R, about the center of rotation. The driving moment is expressed as the weight, W, of the slump mass acting through its center of gravity times its moment arm, X, about the center of rotation. The safety factor for this particular slump would then be

$$\text{S.F.} = \frac{\Sigma \text{ resisting moments}}{\Sigma \text{ driving moments}} = \frac{SLR}{WX} \tag{13.4}$$

strength value is equivalent to the cohesion measured in an undrained test (Chapter 10). It can only be used in the field for estimates of slope stability while undrained conditions are approximated in the slope.

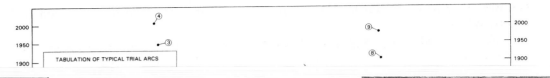

dolomite is directly underlain by less resistant shales of the Rochester Formation, which is part of the unit labeled as the Clinton Shale in Figure 14.10.

The cutting of the Niagara Gorge began during the last retreat of the Pleistocene glaciers about 12,000 years ago. At that time, meltwater from the ice sheet produced greater discharge through the Niagara River and the lower gorge was rapidly cut by headward retreat of Niagara Falls from the Niagara escarpment at a rate of approximately 1.6 m/yr. Meltwater flow decreased about 10,000 years ago and the rate of retreat of the falls was consequently lower. The upper gorge, which is about 3700 m long was cut at an approximate rate of 1 m/yr. Since the 17th century, the retreat of the falls has been measured by repeated surveys. Successive positions of the Horseshoe Falls are shown in Figure 14.11. The actual rate of recession may be very irregular. Several relict crest lines are shown in Figure 14.11 from the time that the position of the Horseshoe Falls was in the vicinity of the American Falls. A horizontal arch spanning the gorge was probably the most stable form of cross section. During the time that an arch existed, the rate of retreat was relatively low. Separation of the falls about 600 years ago was initiated by the formation of a notch (Figure 14.11), which greatly accelerated the rate of retreat. As a result, Horseshoe Falls separated from the American Falls and has since retreated at a much greater rate due to its higher proportion of the total discharge.

The specific processes that contribute to retreat of the falls include weakening due to wetting and drying of the Rochester shale, freeze and thaw, and penetration of water into joints in the Lockport Dolomite, causing elevated hydrostatic pressures at depth. The more rapid erosion of the Rochester shale produces an overhang of Lockport dolomite.

▲ **FIGURE 14.11**
Historic and relict crest lines of Niagara Falls. *Source:* From P. E. Calkin and P. J. Barnett, 1990, Glacial geology of the eastern Lake Erie Basin, in *Quaternary Environs of Lakes Erie and Ontario*, D. I. McKenzie, ed., Escart Press, Waterloo, Ont. After S. S. Philbrick.

▲ FIGURE 14.12
Failure mechanisms, American Falls. *Source:* From American Falls International Board, 1974, *Preservation and Enhancement of the American Falls at Niagara,* Final Report.

The actual failures occur as large rock falls when the overhang finally collapses. Two large rock falls at the American Falls are shown in Figure 14.12. On the left, the 1931 rock fall occurred by gradual undermining of the dolomite by erosion of the shale beneath. When the overhanging block of dolomite was unable to support itself, it collapsed. The 1954 rock fall (Figure 14.12, right), however, involved horizontal shearing within the shale. Hydrostatic pressure (hydraulic head), which increases with depth in joints and fractures as shown in the diagram, may have contributed to this failure.

Huge dolomite blocks from rock falls at the American Falls accumulate as talus at the base of the American Falls because the relatively low discharge is insufficient to rapidly erode and transport them. No such accumulations are found at the base of the powerful Horseshoe Falls, however, because rock fall debris is quickly pulverized and carried away by the larger current flows.

Stream Hydraulics

Flow Types

We are all familiar with the variety of flow conditions that exist in river channels. Although the rapid rush of water down a boulder-filled mountain stream channel appears very different from the sluggish, almost imperceptible, movement of a winding coastal-plain river, the flow conditions in all rivers are the balance between the gravitational driving forces and the resistance of the stream channel and the flowing water itself. Two basic flow conditions can be defined by considering the flow paths of individual water particles. At low velocities, the parallel flow paths of adjacent particles constitute *laminar* flow. Figure 14.13 illustrates laminar flow using *streamlines*, which are imaginary lines parallel to the velocity vectors of representative water particles. Resistance to flow under laminar conditions is generated by viscous interaction between particles within the flow.

As velocity increases, the parallel laminar flow paths can no longer be maintained. Instead, random velocity fluctuations in all directions arise within the flow. These perturbations of flow are called *turbulence* and the resulting flow is *turbulent* flow. Turbulence

Laminar
flow

Turbulent
flow

◀ FIGURE 14.13
Laminar and turbulent flow.

generates additional resistance by these secondary motions, initiating a component of viscosity called *eddy viscosity*.

A dimensionless number called the *Reynolds number (Re)* is used to distinguish laminar from turbulent flow. Its definition is

$$Re = \frac{\rho_w v R}{\mu} \tag{14.2}$$

where

ρ_w is the density of water
v is velocity
μ is dynamic viscosity
R is the hydraulic radius of the channel

Hydraulic radius, R, is a geometric property of the river channel defined as the cross-sectional area of the channel, A, divided by the wetted perimeter, P:

$$R = \frac{A}{P} \tag{14.3}$$

In many natural channels, cross-sectional area can be approximated by the product of width and depth (Figure 14.14) and the wetted perimeter as $P = W + 2D$. The boundary between laminar and turbulent flow occurs within the Reynolds number range of 500–750. In nature, laminar flow usually occurs in groundwater flow systems, and even the slowest of rivers fall into the range of turbulent flow.

A second dimensionless number used to characterize the flow of rivers is the *Froude number (Fr)*:

$$Fr = \frac{v}{\sqrt{gD}} \tag{14.4}$$

in which v is stream velocity, g is the acceleration of gravity, and D is depth of flow. Froude number values are used to classify flow into categories defined as subcritical ($Fr < 1$),

◀ FIGURE 14.14
Hydraulic geometry of a natural stream channel.

$P = W + 2D$
$A = W \times D$
$R = \frac{A}{P}$

critical ($Fr \sim 1$), and supercritical or rapid ($Fr > 1$). Since velocities of natural streams rarely exceed several meters per second, examination of equation (14.4) indicates that only shallow, high-velocity streams can achieve critical or supercritical flow. Gravelly mountain streams occasionally meet these criteria.

Froude numbers can also be used to predict the type of *bedform* that will develop on a stream bottom composed of loose, cohesionless sediment such as sand. Bedforms include wavelike accumulations of bed material that form in response to the frictional drag exerted by the flowing water on the particles lying on the streambed. The continued flow of water over the bedforms erodes, deposits, and transports sand, so that the entire train of sand waves migrates across the bed, usually in a downstream direction, while maintaining its wavelike form. Figure 14.15 illustrates that, with gradually increasing flow velocities and Froude numbers, bedforms progress through a sequence of *ripples, dunes*, plane-bed conditions, and *antidunes*. Ripples are small bedforms less than 60 cm in length from trough to trough, whereas dunes and antidunes are larger bedforms that usually range in size from 60 cm to several meters in length. Antidunes migrate in an upstream direction because particles are eroded from the downstream side of one antidune and deposited on the upstream side of the adjacent bedform.

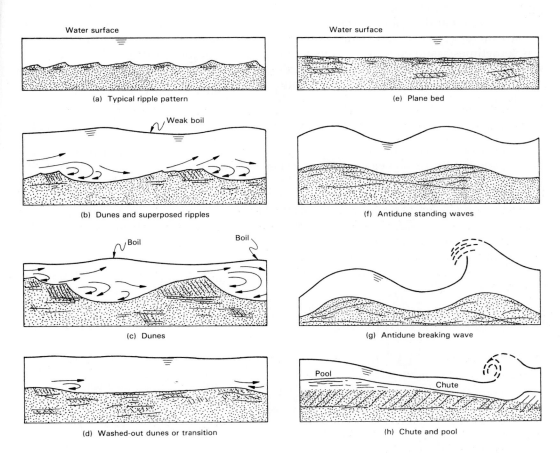

▲ FIGURE 14.15
Sequence of bedforms associated with increasing flow velocities and Froude numbers. Froude number is less than 1 for diagrams (a) through (c), approximately 1 for diagrams (d) and (e), and greater than 1 for diagrams (f) through (h). *Source:* From D. B. Simons and E. V. Richardson, 1966, *Resistance to Flow in Alluvial Channels*, U.S. Geological Survey Professional Paper 422-J.

Discharge and Velocity

The amount of water moving past a certain point in a channel in a given amount of time is a very important characteristic of flow. This quantity, known as *discharge*, is measured perpendicular to the direction of flow. The usual units of measurement for discharge are cubic meters per second. Flow is considered to be either *steady* or *unsteady*, depending on whether discharge is constant or variable with time at a certain point. *Uniform* flow exists when the water depth is constant along a stream course and the slope of the water surface is equal and parallel to the slope of the bed. Even though steady, uniform flow is never truly present in a natural river, this assumption is often made for short periods of time so that uniform flow equations can be used.

The velocity of a stream at any point in the channel is determined by the slope of the channel and the distance from the point in the flow to the sides or bed of the channel. The influence of the bed on velocity is shown in Figure 14.16. There is a very thin layer of water adjacent to the bed in which the water velocity is zero. The variation in velocity with vertical distance above the bed is shown in Figure 14.16a. The channel sides exert a similar influence on velocity (Figure 14.16b). These two effects are combined in Figure 14.16c, which shows contours of relative velocity (lines connecting points with an equal fraction of the maximum velocity). It is clear that the maximum velocity occurs at the water surface in the middle of the channel. The mean velocity of the stream (v) occurs somewhere below the water surface in the middle of the channel. It is sometimes assumed that the velocity at six-tenths of the depth as measured downward from the water surface in the middle of the stream is the mean velocity of the cross section. If this velocity can be measured or estimated, it is possible to determine the discharge of the river at a point using the formula

$$Q = A\bar{v}$$

(14.5)

where Q is the discharge in cubic meters per second, A is the cross-sectional area in square meters, and $\bar{v}$ is the mean velocity in meters per second.

◀ FIGURE 14.16
Velocity distribution in a river channel.

Longitudinal cross section
(a)

Plan view
(b)

Transverse cross section
(c)

Other commonly used flow formulas can be derived by equating the driving forces of streamflow (gravity) with the resisting forces (channel boundaries) and solving for the resulting velocity. For example, the Chezy formula,

$$\bar{v} = C\sqrt{RS} \qquad\qquad (14.6)$$

expresses the velocity (v) in terms of the hydraulic radius (R), the slope (S), and a constant (C). The Manning equation (for SI units),

$$\bar{v} = \frac{1}{n}R^{2/3}S^{1/2} \qquad\qquad (14.7)$$

is a similar approach, although it includes a parameter n, which describes the roughness of the streambed. A value for *Manning's n*, which typically ranges from 0.01 to 0.07, can be chosen from published descriptions of channels that consider variation in grain size of channel-bottom material, vegetation along the stream banks, and other factors. Both the Manning and Chezy formulas are used for natural streams even though they assume uniform flow.

EXAMPLE 14.1

During a flood, the water depth at a stream station was 2 m. The slope of the water–surface profile was 0.005 and the Manning's n value was 0.05. If the channel cross section can be approximated as a rectangle with a width of 20 m, what was the discharge?

Solution

From equation (14.3),

$$R = \frac{A}{P} = \frac{(20 \text{ m} \times 2 \text{ m})}{(20 \text{ m} + 4 \text{ m})} = 1.7 \text{ m}$$

Mean velocity can be obtained from the Manning equation:

$$\bar{v} = \frac{1}{n}R^{2/3}S^{1/2} = \frac{1}{(0.05)}(1.7)^{2/3}(0.005)^{1/2} = 2 \text{ m/s}$$

Using Equation (14.5), discharge is equal to

$$Q = A\bar{v} = (40 \text{ m}^2)(2 \text{ m/s}) = 80 \text{ m}^3/\text{s}$$

Stream Sediment

Part of the work that streams accomplish is the transportation of sediment produced by weathering and erosional processes. Over millions of years, entire mountain ranges are leveled by these processes. In a particular reach of a stream, sediment can simply be transported by the stream. Alternatively, the stream, by the forces it exerts on its channel boundaries, can *entrain* particles from the channel sides and bottom into the flow, or it can deposit sediment that is being carried. In this way rivers can deepen or fill their channels over time.

Entrainment

Erosion of a stream's sides or bottom can be accomplished in several ways. If the boundary material is composed of soluble rock such as limestone, solution by the stream is possible. In addition, the sediment carried by the stream can do erosional work. Particularly in the case of large particles that slide and roll across the stream bottom, *abrasion* can occur. Laboratory experiments indicate that streams with a high sediment load are more erosive under some conditions than sediment-deficient streams. When large particles are dropped to the stream bottom as a result of velocity fluctuations, the impact can be great enough to break off projecting fragments of the streambed material.

Although streams may actively erode their channels by the preceding mechanisms, the process of entrainment usually refers to the ability of the river to pick up loose particles of material from the bed by the direct hydraulic action of flowing water. As the stream flows in its channel, the moving water exerts shearing forces upon the stationary bed. If these forces are larger than the gravitational forces tending to hold a particular particle on the stream bottom, the particle will be picked up off the bed and entrained into the flow.

Flume experiments, in which river processes are simulated by the flow of water in long tanks, demonstrate that as the velocity increases, progressively larger particles are entrained from the bed. The Hjulstrom diagram, shown in Figure 14.17, relates water velocity and particle size. The dashed lines divide the areas into fields of erosion, transportation, and sedimentation. One important point shown by the diagram is that higher velocity is required to entrain a particle of a given size than is needed to transport the particle once it has entered the flow. Another aspect of the curves that initially may not seem logical is that the velocity needed for erosion increases as particle size decreases near the left side of the diagram. This effect can be explained by the properties of clay particles that fall in that size range. The cohesion between clay particles leads to the entrainment of clay-particle aggregates into the flow rather than individual clay particles.

The relationship between velocity and particle size is often difficult to work with because of the variation of velocity in the channel. For a particle of a certain size, it is the flow velocity just above the bed, or the *critical bed velocity*, that must be large enough to move the particle. The relationship between the critical bed velocity and the mean velocity of the stream is not always clear. For these reasons, the relationship between particle size and *critical shear stress* is often used instead of velocity. The critical shear stress, τ_c, or the shear stress just large enough to move a particle of a certain size, is expressed as

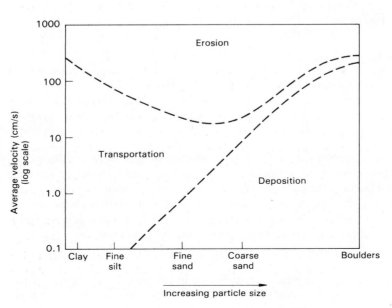

◀ FIGURE 14.17
The Hjulstom diagram, relating stream velocity to particle size for conditions of erosion, transportation, and deposition.

▶ FIGURE 14.18
An empirical relationship between shear stress (τ) and particle size (D), using data from natural streams and canals. *Source:* Modified from V. R. Baker and D. F. Ritter, 1975, Competence of rivers to transport coarse bedload material, *Geological Society of America Bulletin*, 86:975–978.

$$D = 65\tau^{0.54}$$

$$\tau_c = \rho_w gRS \tag{14.8}$$

where

ρ_w is the water density
g is gravity
R is hydraulic radius
S is the energy slope of the stream

The energy slope can be approximated by the mean slope of the streambed and the hydraulic radius can be approximated by the stream depth for wide, relatively shallow streams. This approach presents fewer problems in application because hydraulic radius and slope can be measured more accurately than critical bed velocity. An empirical relationship between critical shear stress and particle size is shown in Figure 14.18 for data from natural streams and canals.

Transportation

The total amount of sediment a stream carries is known as its *load*. Sediment load can be divided into three types based on the process by which it is transported.

   Dissolved load is carried totally in solution by a stream. Much of the dissolved load of a stream is derived from the base flow—that is, the groundwater component of the total streamflow. During its relatively long residence time in the groundwater flow system, the base-flow component of streamflow chemically reacts with the solid materials it encounters through the various types of weathering processes. From these reactions it derives the dissolved load that it carries to the surface stream in a groundwater discharge area. The surface runoff component of a stream is generally lower in dissolved solids content because of its short travel time to the stream channel.

   Of the nondissolved load of a river, the smaller particles are carried as *suspended load*. This form of transport implies that particles are prevented from settling to the bottom of the stream by the movement of the water. The rate at which particles tend to fall through a static liquid column is known as the *terminal velocity*. This is a constant value that represents a balance between the opposing forces of gravity and viscous resistance by the fluid.

◄ FIGURE 14.19

Transportation of stream sediment includes rolling and sliding of bed load, saltation of small bed-load particles, and suspended and dissolved load transport.

The equation relating these forces is known as *Stokes' Law*, from which the terminal velocity (v_t) can be stated (in SI units) as

$$v_t = \frac{1}{18}D^2\frac{(\rho_s - \rho_w)g}{\mu}$$ (14.9)

where

D is particle diameter
ρ_s and ρ_w are the densities of the particle and fluid
g is the acceleration of gravity
μ is the dynamic viscosity of the fluid

Stokes' Law indicates that the terminal velocity of a particle is proportional to the square of the particle diameter. For the particle to remain in suspension, an upward velocity greater than the terminal velocity must be imparted to it. Upward velocities are provided by turbulent eddies and other velocity fluctuations distributed throughout the stream. These turbulent effects are proportional to the mean velocity of the stream, so that the higher the stream velocity, the larger the particle size that can be carried in suspension.

The final type of load is the fraction of the total load that generally remains in contact with the bed during transportation. *Bed load* includes the particles that are too large to be transported by suspension. These particles move by rolling or sliding along the bed or by a process known as *saltation*, in which turbulent eddies can temporarily lift a particle off the bed. When the eddy dissipates, the particle can no longer be kept in suspension and is dropped to the bed (Figure 14.19). The particle thus moves downstream in a series of small jumps or bounces.

Two general terms characterize the load transported by a stream. *Competence* refers to the largest particle that a stream can carry. It is a function of the velocity and shear stress exerted by the stream. *Capacity* is the total amount of sediment that a stream can carry by all mechanisms. Thus, mountain streams have high competence but low capacity, and large rivers like the lower Mississippi may have high capacity but low competence.

Depositional Processes

When the velocity of a stream decreases, particles begin to drop from suspension or cease movement along the streambed. Deposition of this type is usually a very temporary situation. Sediment is merely stored in the channel for short periods of time until the velocity increases again. Periodic fluctuations in stream velocity are common.

More permanent deposition within and adjacent to a stream channel is also possible. When a river floods, for example, it overflows its channel and inundates the flat *floodplain* that adjoins the channel. Velocity drops significantly once water leaves the channel. Floodplains are characterized by parallel layers of fine-grained sediment deposited by successive floods.

Climatic conditions or the existence of an excessive sediment supply can promote a gradual accumulation of sediment in a river valley. The nature of the deposits corresponds to the particular type of channel pattern developed by the river. Channel patterns, which can be classified as *straight, meandering*, and *braided*, will be discussed in the next section.

A different type of depositional control is provided by decreases in channel gradient (slope). Gradient and velocity decrease where a stream valley extends from a mountain range out onto a plain and when a river enters a body of standing water like the ocean or a lake. Continuous sediment deposition occurs at these locations.

Meandering Streams

The factors that control the type of channel pattern include the gradient of the channel, the type of material the channel is incising, and the type and amount of sediment that the river carries. Straight channels are not very common, although meandering streams can have straight segments. Meandering streams, the most common type, tend to develop in gently sloping areas with low sediment supply and a high percentage of cohesive sediment. The channel pattern also correlates with the cross-sectional shape of the channel (Figure 14.20), with meandering streams characterized by narrow, deep channels.

The intensity of meandering is quantified by the use of the parameter called *sinuosity*, which is measured as the ratio of channel length to valley length. Sinuosity is apparently related to the type of material that composes the channel banks, the type of sediment load transported, and the gradient. The highly sinuous river shown in Figure 14.21 has a low gradient, is cut into sediment with a high percentage of silt and clay, and transports mostly

▶ FIGURE 14.20
Map and cross-sectional views of meandering and braided streams.

Meandering   Braided

▶ FIGURE 14.21
Vertical air photo of a meandering stream with high sinuosity in eastern North Dakota. Area shown by photo is approximately 3.5 km by 2 km. *Source:* Photo courtesy of the author.

◀ FIGURE 14.22
Meandering thalweg in a nearly straight channel.

fine-grained sediment as suspended load rather than bed load. Streams with the opposite characteristics have a much lower sinuosity.

The tendency of a stream to meander appears to be a basic property of fluid flow, although the factors that cause meandering are not totally understood. Even in a straight channel (Figure 14.22), the *thalweg*, the deepest part of the channel, establishes a meandering pattern. Straight excavated channels like canals or drainage ditches often erode their banks in order to develop a natural meandering pattern.

The flow of water in a meandering channel is shown in Figure 14.23. The position of the thalweg impinges on the outside banks of the meanders just downstream from the point of maximum curvature. Because the thalweg follows the deepest part of the channel, it also represents the point of highest velocity in the channel. As a result of this higher velocity, the meandering channel tends to be eroded on the outside bank of the meander to form a *cut bank*. A secondary pattern of circulation is also imposed upon the flow in the vicinity of meanders. In the cross section of the lower part of Figure 14.23, the secondary flow is shown to move across the channel toward the cut bank at the water surface and then back along the stream bottom. Because the water is also moving downstream, the resulting overall motion resembles a corkscrew and is termed *helical flow*. On the inner bank of the meander, where the depth and velocity are less, deposition of the coarser fraction of stream sediment occurs. The resulting sediment accumulation is called a *point bar*.

This pattern of erosion and deposition of a meandering stream causes lateral migration of the channel meanders across the floodplain. The characteristic set of landforms and deposits associated with meandering rivers (Figure 14.24) records the sequence of events

◀ FIGURE 14.23
In a meandering channel, erosion occurs on the outer sides, or cut banks, of meander bends. Point-bar deposition is present on the inner bends. A secondary flow pattern superimposed on the downstream motion gives the overall flow a helical circulation.

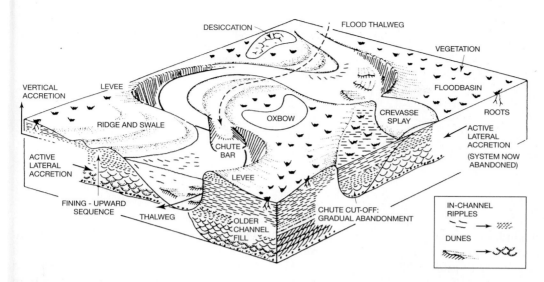

▲ FIGURE 14.24

Landforms and deposits in a meandering stream depositional environment. *Source:* From R. G. Walker and D. J. Cant, 1984, Sandy fluvial systems, in, R. G. Walker (ed), *Facies Models*: Geoscience Canada Reprint Ser. 1, Fig. 1, p. 72, reprinted by permission of Geological Association of Canada.

of recent geologic time. Many of the topographic features are related to the lateral migration of meanders. The continuous process of cut-bank erosion and point-bar deposition produces a series of parallel concentric ridges and swales (Figure 14.24). The ridges represent old point bars and therefore are composed of coarser sediment. The intervening swales are underlain by organic-rich, fine-grained sediment. The dominantly coarser sediment in point bars and active channels is classified as a *lateral accretion* deposit, and the sediment deposited by floods that spread out over the floodplain is known as a *vertical accretion* deposit. Vertical accretion deposits are fine-grained because of the velocity decrease that accompanies flow out of the channel during floods onto the broad floodplain. The coarsest available sediment is deposited adjacent to the channel margins to form *natural levees*, as shown in Figure 14.25. Grain size decreases with distance from the channel. The areas behind the natural levees are often called *backswamps* because of their organic-rich, fine-grained sediment, high water table, and swampy vegetation. One exception to the deposition of fine-grained sediment on floodplains occurs when floodwater breaks through the natural levee and forms a *crevasse splay* as the water rushes out of the channel onto the floodplain and drops its coarsest sediment. Notice that the position of the thalweg under low-flow conditions is not necessarily the flood thalweg. Channels cut by high-velocity flow across point bars are known as *chutes*.

Lateral migration of meanders cannot occur indefinitely because rivers tend to maintain a relatively constant value of sinuosity. The pattern of erosion and deposition in a

▶ FIGURE 14.25

Natural levee deposits on a floodplain. *Source:* From S. Judson, M. E. Kauffman, and L. D. Leet, *Physical Geology*, 7th ed., © 1987 by Prentice Hall, Inc., Upper Saddle River, N.J.

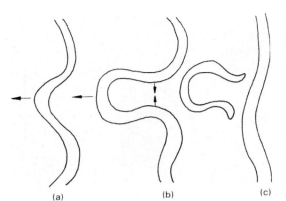

◄ FIGURE 14.26
Meander migration leading to abandonment and formation of an oxbow lake. Arrows show direction of cut-bank migration.

(a) (b) (c)

meander leads to growth of the meander followed by its abandonment. After abandonment, or meander *cutoff*, the old channel loop, now bypassed by the river, becomes an *oxbow lake* (Figures 14.26, 14.27, and 14.28). Figure 14.28(a) shows a recently abandoned meander and Figure 14.28(b) shows an oxbow lake. Eventually, the oxbow lake fills with highly organic silty and clayey sediments. These old channel fills, or *clay plugs*, present foundation problems in construction because of the high compressibility of the sediment.

The floodplain is so-named for a good reason, for stream discharge periodically exceeds the channel capacity and floodwater flows out of the channel and onto the adjacent surfaces. Typically, streams attain bankfull flow every 1 to 2 years on the average. The floodplain is the river system's natural storage area for the water and sediment available

▲ FIGURE 14.27
Topographic map of part of the White River floodplain in Arkansas, showing meander scrolls and an oxbow lake. The river will soon form another meander cutoff in the upper right part of the map.
Source: Augusta SW Quadrangle, Arkansas, U.S. Geological Survey, 7.5-minute series.

▶ FIGURE 14.28
(a) Recently abandoned meander along meandering stream. (b) Oxbow lake formed in an abandoned meander. (Red River, southern Manitoba.) *Source:* Photos by Greg Brooks. Reproduced with the permission of the Minister of Public Works and Government Services Canada, 2005, and courtesy of Natural Resources Canada, Geological Survey of Canada.

(a)

(b)

when the discharge exceeds bankfull flow. The complex distribution of sediments in the alluvial valley of a meandering stream presents many engineering problems because floodplains often are highly developed. Engineering problems arise both in the control of the periodic floods that inundate these areas and in the construction attempted on floodplains. High water tables frequently pose troublesome conditions. In addition, the rapid lateral and vertical variation in sediment type requires a great deal of site investigation and soil testing for foundation design. The engineering properties associated with specific fluvial landforms and deposits are listed in Table 14.1.

Braided Streams

Streams that flow in broad, shallow channels of low sinuosity and consist of multiple subchannels separated by islands or bars are called *braided streams* (Figures 14.20 and 14.29). The channel pattern within a braided stream—where subchannels continually divide and recombine around obstacles in the main channel—can be referred to as *anastomosing*. The anastomosing channel network of a braided stream forms in response to a different set of controlling variables from those associated with meandering streams. Conditions that favor the formation of braided streams include an abundant source of coarse-grained sediment, a high gradient, and cohesionless, nonresistant sediment in the channel banks. The cross-sectional profile that develops under these conditions is broad and shallow (Figure 14.29). In a broad, shallow stream, higher velocities are closer to the streambed, and transport of coarse particles is facilitated.

The formation of braids involves the deposition of bars of sediment within the channel. A bar is a large-scale bedform. Once initiated, bars grow and migrate downstream because

Table 14.1	Engineering Properties of Alluvial Deposits in the Lower Mississippi Valley					
Environment	Soil Texture (USC Classification)	Natural Water Content (Percent)	Liquid Limit	Plasticity Index	Cohesion (lb/ft^2)	Angle of Internal Friction (Degrees)
Natural levee	Clays (CL)	25–35	30–45	15–25	360–1200	0
	Silts (ML)	15–35	Nonplastic (NP)–35	NP–5	180–700	10–35
Point bar (ridges)	Silts (ML) and silty sands (SM)	25–45	30–45	10–25	0–850	25–35
Abandoned channel	Clays (CL and CH)	30–95	30–100	10–65	300–1200	0
Backswamp	Clays (CH)	25–70	40–115	25–100	400–2500	0
Swamp	Organic clay (OH)	110–265	135–200	100–165	Very low	
Marsh	Peat (Pt)	160–465	250–500	150–400	Very low	
Lacustrine	Clay (CH)	45–165	85–115	65–95	75–150	0
Beach	Sand (SP)	Saturated	NP	NP	0	30

Source: From C. R. Kolb and W. G. Shockley, 1957, Mississippi Valley geology, *Journal of Soil Mechanics and Foundations,* 83 (no. SM3): 1–14. Used by permission of the American Society of Civil Engineers.

▲ FIGURE 14.29
A braided stream in Alaska. *Source:* T. L. Péwé; photo courtesy of U.S. Geological Survey.

sediment is transported over the bar and then deposited on the downstream end, where the depth increases. As the bar grows, it occupies more of the channel cross-sectional area that was previously available for water. As a result, the stream must erode its banks laterally to enlarge the channel to maintain a sufficient capacity for the discharge (Figure 14.30). As the discharge drops from a period of very high flow when much of the bed load was in transportation, previously deposited bars are dissected by erosion and the sediment is redistributed through the system. The channel pattern in a braided stream constantly changes with fluctuations in discharge.

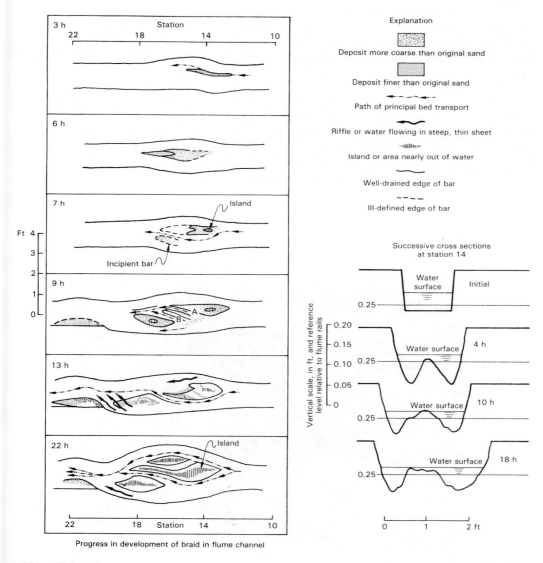

▲ FIGURE 14.30

Development and modification of braid bars in a flume channel. Note changes in cross section during flow period. *Source:* From L. B. Leopold and M. G. Wolman, 1957, *River Channel Patterns*, U.S. Geological Survey Professional Paper 282-B.

Alluvial Fans

If the gradient of a stream suddenly decreases, equations (14.6) and (14.7) indicate that the velocity will decrease if other channel parameters remain the same. A river responds to a decrease in velocity by depositing the coarser fraction of its sediment load. A good example of a gradient decrease is the point where a stream emerges from a steep mountain front onto the floor of the adjacent basin. The stream gradient and velocity are suddenly decreased and sediment is deposited to form a hemiconical landform called an *alluvial fan* (Figure 14.31). The velocity decrease is also partly caused by the increase in width of the channel on the fan surface. The width can increase to the point where flow is no longer channelized but instead moves as a thin layer across a broad surface. This type of flow is known as *sheet flow*. If the fan surface is permeable, water rapidly infiltrates into the fan, and overland flow ceases.

Alluvial-fan deposits are quite variable (Figure 14.31). A major factor is the water to sediment (w/s) ratio of the floods that flow over the fan surface. If the w/s ratio is high, streamflow dominates and the deposits consist of coarse-grained, relatively well-sorted bars, channel fills, and sheets of sand-to-boulder-size material. Extremely coarse, well-sorted gravels are known as *sieve deposits* because the material is so coarse and lacking in sand and finer-size material that water can flow through them rather than over them (Figure 14.32). As the w/s ratio decreases, the flows become debris flows, which deposit raised levees along the channel sides (Figure 14.31). Alluvial fans range from dominantly streamflow to dominantly debris-flow types with intermediate mixture of the two types common.

Alluvial fans are best developed where recent faulting has produced steep mountain fronts with adjacent basins (Figure 14.33). Arid climatic conditions, where flow in drainage networks occurs mainly during infrequent rainfall events of large magnitude, also facilitate

Key

Debris-flow lobes

Sieve deposits

Debris-flow levee deposits

Stream-flow channel deposits

Stream-flood & old channel deposits

▲ FIGURE 14.31
Schematic diagram of an alluvial fan showing different types of deposits on the fan surface. *Source:* Modified from D. R. Spearing, 1974, *Summary sheet of sedimentary deposits*, Mc-8; Sheet 1, Fig. 1-C. Geological Society of America.

▶ FIGURE 14.32
Sieve deposits in alluvial fan at mouth of Wadi Isla, South Sinai, Egypt. *Source:* Photo courtesy of the author.

▲ FIGURE 14.33
Alluvial fans in Death Valley National Monument, California. *Source:* H. E. Malde; photo courtesy of U.S. Geological Survey.

the deposition of alluvial fans. The hemiconical shape of some alluvial fans makes them easily recognizable on topographic maps.

Deltas

Where a river enters a lake or the sea, a *delta*—the coastal counterpart of an alluvial fan—is constructed. Deposition occurs in a delta because the river current slows and eventually dissipates as it moves progressively offshore. The gradual reduction in velocity is reflected in the distribution of sediment types in a delta, with deposition of coarse bed-load materials in the vicinity of the mouth of a river. As the river deposits its sediment load, the delta extends or *progrades* in a seaward direction.

Deltaic environments are highly variable. Marine deltas are influenced by the fluvial input of water and sediment, the wave energy imparted to the coastline, and tidal currents. The relative importance of one of these three factors determines the form and depositional

▲ **FIGURE 14.34**
Idealized sedimentation in a delta, including topset, foreset, and bottomset beds. *Source:* From S. Judson, M. E. Kauffman, and L. D. Leet, *Physical Geology*, 7th ed., © 1987 by Prentice Hall, Inc., Upper Saddle River, N.J.

processes of the delta, leading to a classification of deltas as *river-dominated, wave-dominated,* and *tide-dominated* deltas. A basic type of delta sedimentation recognized by G. K. Gilbert more than 100 years ago is shown in Figure 14.34. This type of delta is most common in lakes. As the delta is gradually constructed, the river extends across accumulating delta sediments to a slope called the *foreset slope.* Coarse sediments are deposited on the foreset slope in long, dipping *foreset beds* (Figures 14.34 and 14.35). Progradation (offshore migration) of the delta results from the continual accumulation of foreset beds. Beyond the foreset slope, slow-moving currents carry the finer fractions of the river's suspended load across the lake floor, where they are eventually deposited in nearly horizontal beds called *bottomset beds.* As the river extends its channel across the prograding delta to the foreset slope, it deposits horizontal beds called *topset beds.*

As a river approaches a delta, the single main channel divides repeatedly into a network of *distributary* channels, in which sediments are transported across the delta (Figure 14.36). Distributary channels are maintained by natural levees that project slightly above the delta surface. Sediment is conveyed through the distributary channels unless a break in the natural levee is present. These breaks, called *crevasses,* allow the diversion of water and sediment through the natural levee to a lateral position between distributary channels. The minidelta that is deposited as water flows through the crevasse is called a *crevasse splay.*

Sediment is deposited within a particular distributary network for a certain period of time. Eventually, that network becomes raised in elevation because of sediment accumulation.

◀ **FIGURE 14.35**
Exposure of dipping deltaic foreset overlain by nearly horizontal topset beds in a glaciomarine delta in Maine. The delta was formed near the coastline from meltwater and sediment from a retreating glacier. Postglacial isostatic rebound raised the delta above sea level. *Source:* Photo courtesy of the author.

▶ FIGURE 14.36
In a delta, the trunk stream branches into distributaries. When distributaries are breached, crevasse-splay sediments are deposited.

1	Sale Cypremort		> 4600 yr B.P.
2	Cocodrie	ca.	4600–3500 yr B.P.
3	Teche	ca.	3500–2800 yr B.P.
4	St. Bernard	ca.	2800–1000 yr B.P.
5	Lafourche	ca.	1000–300 yr B.P.
6	Plaquemine	ca.	750–500 yr B.P.
7	Balize		< 550 yr

▲ FIGURE 14.37
Evolution of the Mississippi delta by the growth and abandonment of delta lobes. *Source:* From J. P. Morgan, 1970, Deltas—a résumé, *Journal of Geologic Education*, 18, National Association of Geology Teachers.

The river will then abandon the active part of the delta in preference for an adjacent part of the delta with a steeper gradient. The Mississippi River delta is a good example of this process. Over the past 5000 years, active sedimentation has successively deposited seven major lobes (Figure 14.37).

Case in Point 14.2
Future Shift of the Mississippi River?

The history of the Mississippi River delta indicates that the position of the river at any particular time is a transient situation. As a river and its distributary system build up a delta lobe by sediment deposition on natural levees, crevasse splays, and in backswamps, the delta lobe gradually becomes higher in elevation than adjacent, inactive parts of the delta. The river finally reacts by rapidly changing its course to begin deposition of a new lobe of the delta.

Over the course of the past several centuries, both natural and human changes to the lower Mississippi River made it evident that the river began a shift of course to one of its distributaries, the Atchafalaya River (Chen, 1983). In doing so, the river would follow a steeper and more direct path to the Gulf of Mexico, a distance of 142 miles down the Atchafalaya versus 335 miles along the current channel of the Mississippi. If such a diversion occurred, the Atchafalaya (Figure 14.38) would become the main channel of the Mississippi and the current channel would become a distributary with a greatly reduced flow. The consequences of such a shift would be disastrous to the Louisiana cities of New Orleans and Baton Rouge because of the loss of shipping trade and also freshwater. The greatly reduced flow of the main channel of the Mississippi would not provide the cities with a sufficient supply of freshwater, and, in fact, saltwater from the Gulf of Mexico would move up the river channel. Saltwater encroachment would occur because the Mississippi channel bottom is below sea level for more than 100 km upriver from Baton Rouge. Dramatic changes would also occur in the Atchafalaya Basin. The town of Morgan City, already a flood-prone area, would be flooded out of existence.

Between 1850 and 1950, the percentage of flow of the Mississippi that entered the Atchafalaya distributary increased from less than 10% to 30%. To prevent a channel shift, which was predicted to occur as early as 1990, the U.S. Army Corps of Engineers constructed the Old River Control Structure (ORCS) in the 1950s, including the Low Sill Structure and the Overbank Structure, at the head of the newly excavated Outflow Channel (Figure 14.39) near the head of the Atchafalaya to maintain 70% of the flow down the main channel. In 1973, a large flood partially undermined the Low Sill Structure, which nearly failed. If it had collapsed, the river probably would have shifted catastrophically and could not have been restored to its prior condition. As a result, the Low Sill Structure was repaired and a new inflow channel containing the $219 million Auxiliary Control Structure was added to preserve the ability of the ORCS to block the channel shift into the Atchafalaya.

◀ FIGURE 14.38
Map of Louisiana showing the Mississippi River and the Atchafalaya River, the valley to which the Mississippi may shift its course.

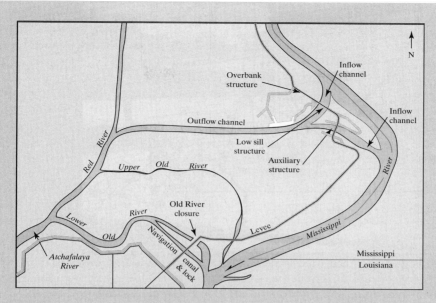

▲ FIGURE 14.39
Map of the Old River area showing the location of components of the Old River Control Structure, which together act to prevent the Mississippi from shifting to a new main channel down the Atchafalaya River.

Thus the stage is set for a continuing struggle between the Mississippi River and the U.S. Army Corps of Engineers. As the present course of the Mississippi becomes more and more unstable with respect to a course to a different part of the delta, more and more extensive (and expensive) control measures will be required. Only time will reveal the outcome of this situation.

Equilibrium in River Systems

The Graded Stream

River systems provide the primary means by which continents elevated to great heights by tectonic processes are gradually worn down to low-relief plains projecting only slightly above sea level. Throughout this process, rivers transport sediment from the topographically high areas of continents to the ocean. Along the way, stream channels progressively grow in size to accommodate the larger discharge supplied by the addition of tributary streams carrying the flow from drainage basins of lower stream order. Many studies of river systems have indicated that sediment grain size decreases in a downstream direction. The sediment transported by a stream ranges from large boulders in mountainous areas to silt and clay at the river's mouth. The slope, or gradient, of a stream also decreases progressively in the downstream direction. The change in gradient of a stream from headwaters to mouth is known as the *longitudinal profile*. As shown in Figure 14.40, longitudinal profiles have a characteristic concave shape. *Base level* is the elevation that the downstream segment of the curve asymptotically approaches. Sea level is the ultimate base level for all streams, although temporary base levels may be established by lakes or reservoirs at intermediate positions along a stream's course.

◀ FIGURE 14.40
Idealized longitudinal profile of a stream. The gradient decreases as the stream approaches base level.

It is generally accepted that the concave longitudinal profiles of most streams represent a state of equilibrium in the river system. Under this assumption the longitudinal profile of a river adjusts to provide the velocity necessary to transport the sediment supplied to the channel. An equilibrium, or *graded*, stream tends to maintain its profile for a period of time during which there are no drastic changes in the prevailing climatic or tectonic conditions affecting the basin. Although the idea of equilibrium in stream systems has merit, it is now known that the interactions between the numerous variables controlling the discharge, sediment load, and profile of rivers are complex. For example, the velocity of a river actually increases in the downstream direction despite the decrease in slope evident in the longitudinal profile.

Causes and Effects of Disequilibrium

The concept of equilibrium must be considered in terms of the period of time over which it is proposed. The graded stream represents equilibrium during some intermediate-length period of time (tens to hundreds of years), during which the slope and channel characteristics are adjusted to the discharge and sediment load supplied to the channel segment. There are many factors that tend to disturb the fragile balance that may exist.

Tectonic activity can be a major cause of disequilibrium in stream systems. An example of tectonic initiation of disequilibrium is shown in Figure 14.41. A fault has downdropped part of a stream valley in relation to the upstream reach. The response of the stream to this disturbance is to reestablish an equilibrium condition. By erosion and downcutting upstream from the fault, and by deposition downstream from the fault, the stream can eventually develop a new equilibrium profile. Gradient is not the only variable that can be adjusted, however. Channel characteristics such as width and depth may also be altered. Other tectonic influences on river equilibrium involve changes in base level. If a

◀ FIGURE 14.41
Alteration of a stream profile after fault displacement. The stream reestablishes a smooth, concave profile by erosion above the fault and deposition downstream.

coastal region is uplifted so that the elevation of sea level relative to that particular coastline drops, streams must steepen their gradients in the vicinity of the coast in order to adjust to the changing conditions.

Climatic change is of vital importance to stream equilibrium. The climate in a drainage basin, along with the geology, controls the discharge and sediment load supplied to a river system. Increases or decreases in precipitation within a drainage basin, aside from short-term variations, will cause changes in vegetation and will also cause the streams to adjust their gradient and channel characteristics. The extreme manifestation of climatic change—an advance of a glacier into the drainage basin—can lead to drainage system disruptions that persist for thousands of years after the glacier has retreated. The Missouri River and many of its tributaries, for example, flowed northward to Hudson Bay before the Pleistocene Epoch. The southward advance of glaciers blocked and diverted the course of the Missouri into the Mississippi River drainage basin.

Stream Terraces

One of the best types of evidence available for inferring past changes in river system equilibrium is the existence of *stream terraces*, which are remnants of former floodplains that stand above modern floodplains in stream valleys. Terraces attest to a former state of equilibrium that was abandoned during a period of erosional instability. Figure 14.42 illustrates the general morphology of terraces. Terraces may be *paired* or *unpaired*, depending upon whether corresponding surfaces at the same elevation exist on opposite sides of the valley. Morphologically, the terrace consists of a flat surface, the *tread*, bordered by a slope called the *scarp*. Multiple terraces are common in many valleys (Figure 14.43).

In terms of genetic processes, terraces consist of two types. During the formation of an *erosional terrace* (Figure 14.44), stream downcutting is dominant. Possible explanations for downcutting could include tectonic or climatic factors. The origin of *depositional terraces* includes a time interval in which the stream fills its valley with sediments. This period of *aggradation* can be caused by decreases in discharge, increases in sediment load, or changes in base level. A frequent cause of aggradation in northern latitudes was the abundant sediment load supplied to the stream system by Pleistocene glaciers. Following the period of aggradation, a stream must then erode its channel to form a depositional terrace. The sudden decrease in sediment load accompanying the retreat of a glacier or other climatic factors can lead to the incision of a previously deposited valley fill.

Human Interaction with Equilibrium

The changes in equilibrium we have considered thus far have not included those that are primarily induced by human activity. Among these, the most important examples are changes in river systems brought about by land-use changes in the drainage basin.

▶ FIGURE 14.42
Morphology of river terraces.

▲ FIGURE 14.43
Multiple river terraces in the Madison River valley, Montana. *Source:* Photo courtesy of the author.

◀ FIGURE 14.44
Differences between erosional and depositional terraces. Depositional terraces are underlain by thick deposits of alluvium.

The increased sediment yield caused by the conversion of natural areas to agricultural production and urban zones was discussed in Chapter 9 and illustrated in Figure 9.21. The consequences of increased sediment yield in a drainage basin can be severe. Sediment is temporarily stored in river channels, causing a decrease in channel capacity, which in turn results in more frequent and damaging flooding. The effect of the artificial drainage network in a city (storm sewers) is to supply more water to river channels in a shorter time, as compared with rivers in nonurbanized areas. The magnitude of

▶ FIGURE 14.45
Construction or removal of reservoirs causes streams to adjust their gradients in order to approach a new condition of equilibrium. *Source:* From S. Judson, M. E. Kauffman, and L. D. Leet, *Physical Geology*, 7th ed., © 1987 by Prentice Hall, Inc., Upper Saddle River, N.J.

flooding from a given rainstorm is also increased in an urbanized area. The river's response to greater sediment load, as well as greater discharge, is to enlarge its channel to accommodate the increased flow. Serious bank erosion may result, with floodplain landowners sustaining property losses.

Another class of human adjustments to equilibrium includes such engineering works as dams and levees that seek to control the river for such purposes as flood protection, water supply, power supply, and recreation. Reservoirs interrupt the longitudinal profile of streams by establishing temporary base levels. Figure 14.45 shows some of the effects. The sediment load of the stream is dropped as it flows into the reservoir. The decrease in gradient causes aggradation upstream from the impoundment as well, as the stream attempts to reconstruct an equilibrium profile utilizing the reservoir as base level. The outflow from the reservoir, having deposited its sediment load, is likely to erode as it flows down the remainder of the river channel and thus dissipate energy available for sediment transport. Dams along the Missouri and other rivers have required extensive downstream bank-stabilization projects to limit loss of floodplain area by bank erosion.

Flood-control projects often involve alterations to river channels. These may include levees and other structures to contain the flow within the channel so that flooding will be minimized. Alternatively, the channel may be enlarged, or its gradient steepened, in order to convey floodwater at a greater discharge or higher velocity through a particular reach. Unfortunately, these changes often are answered by undesirable and costly adjustments by the river system. Levees may decrease flooding along one reach of the channel, but because natural floodplain storage of water is decreased, they may aggravate the downstream flood problem. The increase in channel size or gradient for attempted control of the river is called *channelization*. This disturbance of the river's equilibrium is propagated both upstream and downstream from the channelized reach. An interesting example documented by Daniels (1960) is the Willow River in Iowa. After artificial straightening of the channel for flood-control purposes, the gradient was increased to the point at which the river began to incise its channel. Eventually, the destabilized channel was twice as wide and deep as its original size. Changes of this type are transmitted upstream in a river system. Tributary downcutting following channelization can lead to the destruction of farmland throughout the basin.

Case in Point 14.3
Arroyo Cutting in the American Southwest

The most significant geomorphic change in the semi-arid Colorado Plateau region of the American Southwest in the past century has been the rapid and extensive entrenchment of streams in alluvial reaches to form deep valleys, or *arroyos*. The process was well documented because it caused great hardship among the ranchers and farmers that had just recently settled the area in the mid-19th century. A prime example of arroyo cutting is the history of Kanab Arroyo (Webb et al., 1992), a tributary of the Colorado River located in southern Utah and northern Arizona (Figure 14.46).

Kanab Creek, and its tributary Johnson Wash, alternate between reaches that flow in bedrock canyons and reaches that flow across thick alluvial deposits (Figure 14.46). The small town of Kanab, Utah, which was settled in the 1860s, lies along an alluvial reach of Kanab Creek. At the time of settlement, the stream flowed in shallow channels that the settlers could literally step across. The topography along the stream consisted of broad meadows with a rich cover of natural grasses. The water table was only about 1 m below land surface and watering holes for stock could be created easily by digging a shallow hole. In places, drainage ditches had to be dug in order to drain the land to cut grasses for hay. The conditions were ideal for grazing, and cattle and sheep were quickly introduced to the area.

Beginning in 1882, a series of unusually large floods caused drastic changes in Kanab Creek. By 1885, the channel, which was several meters deep only a few years before, had been downcut to a depth of 18 m and a width of 21 m. A series of dams was built along the valley to supply domestic and irrigation needs for water, but these all failed in subsequent floods. By 1900, the arroyo had reached its current dimensions, 25 m deep and 80–120 m wide (Figure 14.47). A similar sequence of events has occurred on other streams in the Southwest. Erosion was so severe in places that towns were abandoned. Bedrock reaches of the streams, however, were only slightly altered, in contrast to the extreme changes in the alluvial reaches.

The question that now must be answered is, what was the cause of this remarkable disequilibrium in Kanab Creek and other streams? A hypothesis that has been put forth several times is that overgrazing and other agricultural practices caused a decrease in vegetation and an increase in soil erosion. This explanation is attractive because the entrenchment began soon after settlement and agriculture were introduced. Other evidence, however, leads to a different conclusion. Analysis of climatic data from stations in the Southwest with long-time records suggests that precipitation in the 1880s and subsequent decades was unusually high. This may have been related to an anomalous El Niño event, a condition in the Pacific Ocean that alters weather patterns over broad areas. The causes for arroyo cutting could therefore have been climatic changes on a global or hemispheric scale. Additional supporting evidence comes from studies of the alluvial sediments in the

▲ FIGURE 14.46
Map of the Kanab Creek drainage basin. *Source:* From R. H. Webb, S. S. Smith, and V. A. S. McCord, 1992, Arroyo cutting at Kanab Creek, Kanab, Utah, in *Paleoflood Hydrology of the Southern Colorado Plateau: Field Trip Guidebook*, J. Martinez-Goytre and W. M. Phillips, eds., Arizona Laboratory for Paleohydrological and Hydroclimatological Analysis, University of Arizona, Tucson, Ariz.

walls of the arroyos. These studies, using radiocarbon dating of organic matter, have documented cycles of arroyo cutting and filling events dating back several thousand years, long before the impacts of grazing occurred.

The final verdict on the cause of arroyo cutting is still not known. The case for natural climatic cycles is becoming stronger, but it is also possible that the effects were accentuated or localized by agriculture, dam building, and other human interferences with the complex fluvial system.

▲ FIGURE 14.47
Appearance of Kanab Arroyo in 1992. The arroyo is about 25 m deep and 80–120 m wide. *Source:* Photo courtesy of the author.

Flooding

Flooding is the most serious hazard posed by river processes (Figure 14.48). In 1972, 238 people were killed by flooding in Rapid City, South Dakota. During the same year, flooding in the eastern United States caused by precipitation from Hurricane Agnes claimed 113 lives and resulted in more than $3 billion in property damage. More recently, the residents of towns along the Red River of the North, which forms the border between North Dakota and Minnesota, suffered through the worst flood on record in 1997. Particularly hard hit by this flood were the towns of Grand Forks, North Dakota, and East Grand Forks, Minnesota, which adjoin each other on opposite sides of the river. At the time of the crest of the flood, the national media were focused on these towns, as protective dikes were over-topped, flooding most of the area of the cities. About 55,000 people were forced to evacuate their homes and 11,000 homes and businesses were damaged in Grand Forks. Damage from the flood was estimated at $2 billion (1999 dollars), but remarkably no lives were lost.

◀ FIGURE 14.48
Flooding periodically damages structures built on unprotected flood-plains. *Source:* Photo courtesy of the author.

▲ FIGURE 14.49

The U.S. portion of the Red River drainage basin. Shaded area is the portion of drainage basin that contributes water to the river upstream from Grand Forks. *Source:* From J. A. LeFever, J. P. Bluemle, and R. P. Waldkirch 1999, *Flooding in the Grand Forks–East Grand Forks North Dakota and Minnesota Area*, North Dakota Geological Survey Education Series 25.

The U.S. portion of the north-flowing Red River drainage basin is shown in Figure 14.49. The Canadian problems with this river will be discussed later. Citizens of the region who live in the floodplain are used to periodic flooding—in fact, a series of permanent dikes lines the channel through all major towns and some of the rural areas. When a flood comes along bigger than anything ever recorded, however, there is very little that can be done about it. The causes of floods are complex and highly variable from river to river. The Red River cuts through a very flat ancient lake plain underlain by very clayey soils. Therefore, there is a high proportion of runoff relative to infiltration. Cities are vulnerable because there is no high ground on which to build. The north-flowing orientation of the river is also significant. Spring melting occurs earlier in southern headwaters of the basin. As a rising flood wave moves north, ice jams often exacerbate the flooding. Finally, seasonal factors are also very important in producing a flood. Rainfall was abnormally high in the fall preceding the winter of 1978–1979 in the Red River Basin, and winter snowfall was 2 or 3 times the normal amount, with the last of eight major snowstorms occurring in April, just before a rapid temperature rise and thaw.

The Red River of the North flood of 1997 illustrates the dilemma of flood hazards: How can we predict how large floods will be, how often they will occur, and what kind of protective measures we need? The lessons learned from many floods suggest that we can protect ourselves against moderate events, but when extreme events occur (which will invariably happen over time), there is little that we can do other than get out of the way.

Flooding typifies many hazardous geologic processes in that the obstacles to mitigating the threat to human life and property are largely psychological rather than of a hydrologic or engineering nature. The public perception of floods is that they are "acts of God" that are impossible to predict or explain. People are reluctant to relocate their homes and businesses out of flood-prone areas because of events that they believe may never happen in their lifetimes. In reality, the magnitude and frequency of floods can usually be predicted with some degree of certainty, at least on a statistical basis. In addition, flood-prone areas can be accurately identified and mapped. In fact, floodplain maps exist for most developed areas in the United States.

Flood Magnitude

Flood magnitude can be measured as the elevation to which a river rises during a flood, but more commonly it is reported as the maximum discharge of the stream during the event. Changes in river height, or *stage*, are measured by continuous-recording gauges installed along streams throughout most developed countries (Figure 14.50). At each gauging station, measurements are made of cross-sectional area and stream velocity through the range of possible stages. Through the use of equation (14.5), a *rating curve* can be constructed for each station relating river stage to discharge. With the rating curve and the stage measurement made during a flood, a stream hydrograph can be made to show the relationship between discharge and time (Figure 14.51). A stage height discharge curve for the Red River at Grand Forks, North Dakota, is shown in Figure 14.52a and a stage height hydrograph is shown in Figure 14.52b. The peak discharge for the 1997 flood was measured at 136,900 cfs. This discharge actually occurred prior to the maximum stage height, at a stage height 52.21 feet above datum, because the river was still confined by the dikes at the time of peak discharge and therefore flowing faster. When the flood overtopped the dikes, the velocity and therefore discharge decreased so that at its maximum stage height of 54.4 ft, its discharge was 114,000 cfs. A view of downtown Grand Forks is shown in Figure 14.53.

◀ FIGURE 14.50
Streamflow gauging station. *Source:* B. W. Thomsen; photo courtesy of U.S. Geological Survey.

▶ FIGURE 14.51
(a) Hydrograph of river stage versus time during a flood. (b) A rating curve for a gauging station. (c) A hydrograph showing discharge versus time constructed from (a) and (b).

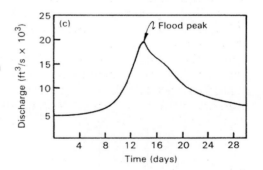

Variables that control the magnitude of a flood can be divided into two types: transient factors that depend on the meteorological conditions, and permanent factors that deal with the hydrology and geology of the drainage basin. There are a large number of meteorological conditions that must be considered. If the flood is caused by a rainfall event, the factors include the size, direction, and rate of movement of the storm; the intensity, duration, and location of the rain; and the moisture conditions prevailing in the soil before the storm. If the flood is caused by a snowmelt event, transient factors include amount of snow, rate of melting, and the soil-moisture conditions. Permanent controls within a basin include type and distribution of vegetation, topography, type of soil and rock at the surface, and type of drainage network present in the basin. From the number of factors involved, it is easy to understand why it is so difficult to predict the magnitude of a particular flood so that safety precautions can be taken. The most basic method of

◀ FIGURE 14.52
(a) Rating curve for the Red River at Grand Forks, North Dakota. Notice that flood stage is reached at an elevation of 28 ft above the datum and the 1997 flood reached a crest height of 54.4 ft. The discharge associated with this crest height is actually slightly lower than the peak discharge, which occurred prior to the crest when the river was still confined by its levees. When the flood overtopped the levees, the velocity and discharge declined.
(b) Stage height hydrograph for the 1997 flood. *Source:* From J. A. LeFever, J. P. Bluemle, and R. P. Waldkirch, 1999, *Flooding in the Grand Forks–East Grand Forks North Dakota and Minnesota Area*, North Dakota Geological Survey Education Series 25.

predicting the peak discharge from a particular rainfall event is by use of equations such as the *Rational Equation*, which can be stated as

$$Q = CIA \qquad (14.10)$$

where Q is peak discharge in cubic feet per second, C is a runoff coefficient whose values are given in Table 14.2, I is the rainfall intensity in inches per hour, and A is basin area in acres.

► FIGURE 14.53
Aerial view of downtown Grand Forks, North Dakota, looking eastward toward East Grand Forks, Minnesota, during the 1997 flood. Bridges at center of view show the normal size of the river channel.
Source: Photo courtesy of North Dakota Geological Survey.

Table 14.2 Values of C (Runoff Coefficient) for the Rational Equation	
C	**Land Cover or Land Use**
0.1–0.3	Forest
0.1–0.5	Grasslands
0.3–0.6	Row crops, plowed ground
0.5–0.8	Bare ground, smooth
0.5–0.9	Suburban lands
0.7–0.95	Urban lands
0.9–0.99	Water body, full1
0.0	Water body, empty

Source: From C. C. Mathewson, *Engineering Geology,* © 1981 by Charles E. Merrill Publishing Co., Columbus, Ohio.

Flood Frequency

Perhaps as important as predicting the magnitude of a flood generated by a particular precipitation event is the problem of predicting the frequency of large floods. Because it is impossible to predict when a flood of a certain magnitude will occur, the only alternative is to attempt to determine the probability of a particular flood within a given year. A general relationship that seems to hold true is that the magnitude of natural processes is inversely proportional to the probability of their occurrence. Thus, the larger the flood under consideration, the less likely it is to happen in a particular amount of time.

Another way of defining frequency is in terms of *recurrence interval* (R.I.), which is the average period of time between floods of a certain magnitude. It is related to probability through the following relationship:

$$\frac{1}{p} = \text{R.I.} \tag{14.11}$$

Thus, if a flood has a probability (p) of 0.01 of occurring in a single year, its recurrence interval is 100 years. It cannot be overemphasized that the recurrence interval refers to a statistical average over a long period of time. A town that experiences a 100-year flood has the same risk of a similar flood in every following year.

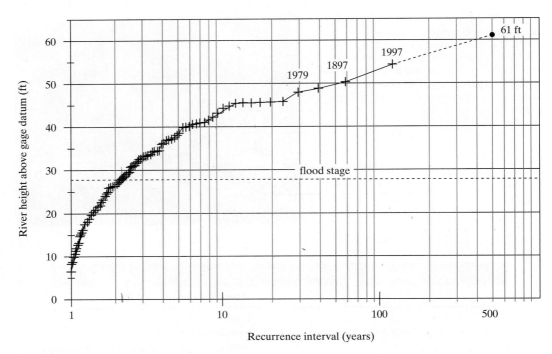

▲ FIGURE 14.54
Flood frequency diagram for the Red River of the North at Grand Forks, North Dakota. The dashed line is the extrapolated portion of the curve to indicate the largest flood that could theoretically occur. *Source:* From J. A. LeFever, J. P. Bluemle, and R. P. Waldkirch, 1999, *Flooding in the Grand Forks–East Grand Forks North Dakota and Minnesota Area*, North Dakota Geological Survey Education Series 25.

Flood-frequency curves are established from gauging-station records to show the relationship between magnitude and frequency. A basic way of constructing these diagrams is to plot magnitude expressed as discharge or stage height versus recurrence interval for the maximum discharge for each year of the historical record. An example for the Red River of the North at Grand Forks is shown in Figure 14.54. Recurrence interval (R.I.) is established by ranking each yearly discharge from highest to lowest and using the relationship

$$\text{R.I.} = \frac{N + 1}{M} \tag{14.12}$$

where N is the number of years of record and M is the magnitude rank. This method can be successful only after many years of record are available. For example, the gage at Grand Forks has 117 years of record. Therefore, the 1979 flood, which was the largest historical flood, has a recurrence interval of 118 years. Flood-frequency curves can be extrapolated to predict larger floods than have been historically recorded. The Grand Forks record has been extrapolated in Figure 14.54 to a recurrence interval of 500 years, thought to represent the flood that would occur if the most favorable conditions for flooding for all possible factors happened simultaneously. This hypothetical flood, which would have a stage height of 61 feet, is considered the worst flood that could possibly happen at Grand Forks. Various statistical methods are available for fitting points on a flood-frequency diagram and extrapolating the curve to higher-magnitude values. The greater the extrapolation, however, the more uncertain are the predicted flood magnitudes and recurrence intervals.

EXAMPLE 14.2

The following table shows the five largest annual flood discharges for Seneca Creek at Dawson, Maryland, for the period 1928–1958. Calculate the recurrence interval for the 1953 flood.

Year	Discharge (m^3/s)	Rank Order
1956	427.7	1
1933	263.3	2
1953	207.6	3
1928	107.6	4
1958	103.1	5

Solution

The record contains 31 years of annual flood peaks. For the 1953 flood,

$$\text{R.I.} = \frac{N+1}{M} = \frac{31+1}{3} = 10.7 \text{ yr}$$

Effect of Land-Use Changes

The flood-frequency analysis presented involves the assumption that flood magnitudes are random samples from a population of possible events and that the underlying factors that cause floods do not change. As Case in Point 14.3 illustrated, however, we know that yearly mean temperature and precipitation amounts can change through time, making floods more or less frequent and severe for a certain period of time. Land-use changes that have been mentioned also influence flood magnitude and frequency. Most important among possible land-use changes is urbanization. The cumulative effects of urbanization, including less infiltration and more rapid runoff, can alter a flood hydrograph in the manner shown in Figure 14.55. The result is flooding of greater magnitude and shorter lag time. Such floods are termed *flashy*. The flood-frequency analysis made from natural conditions in the stream basin before urbanization may not accurately predict the magnitude-frequency relationships after urbanization.

Flood Control

The unpredictability of floods and the great damages that they can cause present a major dilemma to society. How can we best protect our population against a nebulous and poorly understood threat? In answer to this question there are basically two alternatives that can be considered. First, river systems can be controlled and regulated by means of channel improvements, levees, and dams. This is known as the *structural approach*. Second, land use on flood plains can be regulated by zoning ordinances so that flood-prone areas are not occupied by buildings susceptible to great damage during flooding. This alternative is known as the *nonstructural approach*.

   Until recently, the United States has chosen to combat floods through the structural approach. Costa and Baker (1981) report that the federal government had spent more than $14 billion on flood control between 1936 and 1981. The advantage of the structural approach is that flood-control projects can offer protection against floods of a certain magnitude (the design flow) if they are properly designed and constructed. In addition, they are politically attractive to legislators because they create jobs in a flood-prone area and are perceived by the public as a positive step toward solving the problem. Unfortunately,

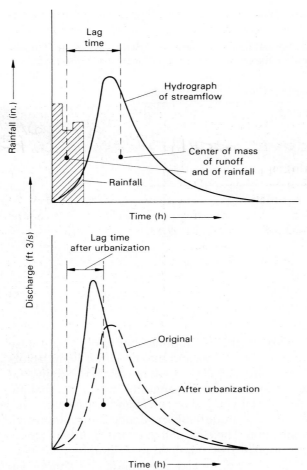

◀ FIGURE 14.55
Changes in a flood hydrograph after urbanization of a drainage basin. *Source:* After L. B. Leopold, 1968, *Hydrology for Urban Land Planning*, U.S. Geological Survey Circular 554.

there are a number of disadvantages to flood-control structures. Most important, a flood greater than the magnitude for which the structure is designed can cause greater damage than would have occurred without the structure. If a dam fails, for example, the resulting flood can assume catastrophic proportions. Most dam failures have been caused by inadequate site investigations or design relative to the geologic conditions at the site. The St. Francis and Teton Dam failures are prime examples.

We have already discussed the potential effect of altering river channels by levees, channelization, or other projects. These projects change the natural balance of the stream and cause it to adjust other variables in the system. A final comment about structural measures is that they may actually encourage development of flood-prone areas. If people believe that there is no danger from floods, they will readily develop floodplains. This only increases the potential for loss and damage in the event of a rare flood of large magnitude.

The nonstructural approach is an attempt to integrate geologic processes into land use and development. The method of flood protection in this approach is to delineate flood-prone areas and restrict development in those areas to uses that are compatible with periodic flooding. In urban areas, these uses include parks, golf courses, and other natural or open areas. In addition to flood protection, these zones enhance the aesthetic value of a river in a crowded urban setting. The costs for land acquisition for these uses may even be less in the long run than the costs of repeated flood damage. The limitations to this approach are that it requires the relocation of homes and businesses. This is considered by some to be an infringement of their right to live wherever they please. In highly developed areas, it may be possible to apply this approach only after a major flood has occurred.

Case in Point 14.4
Flooding and Land Use

Two examples that illustrate differing approaches to flood control are provided by the cities of Winnipeg, Manitoba, and Rapid City, South Dakota.

The city of Winnipeg, with a population of 650,000 compared to about 60,000 in Grand Forks and East Grand Forks, lies at the confluence of the Assiniboine River and the Red River of the North (Figure 14.56), both of which flood regularly. Winnipeg lies about 225 km downstream from Grand Forks, North Dakota. The topographic setting of the city is similar to that of Grand Forks, lying within the same glacial lake basin. Large areas of the city were inundated by past floods because of the shallow valleys of the rivers and the low relief of the areas adjacent to the rivers. As in Grand Forks, most major floods in Winnipeg generally occur during the spring snowmelt. Snow depths and the rate of melting are critical factors in determining whether or not a flood will be forthcoming. After the devastating flood of 1950, the city government decided to construct a floodway to divert floodwaters from the Red River of the North around the city. An inlet control structure in the channel of the Red River consists of a gate that can be raised to block the flow of the river. Water then rises behind the structure and flows into the floodway. The artificial channel, constructed in 1966, has an average depth of 9 m, a maximum width of 18 m, and can carry a maximum discharge of 1700 m^3/s (Figure 14.57). The floodway, along with other structural and nonstructural flood-protection measures, had been successful in preventing flood damage in the city in the years before 1997. About 2 weeks after the crest of the 1997 flood passed Grand Forks, it reached the Winnipeg floodway. Apprehension was widespread because the flood actually exceeded the design capacity of the floodway. With temporary diking, however, the floodway survived its greatest test and there was

▶ FIGURE 14.56
Map of the Winnipeg, Manitoba, area showing the location of the Red River floodway.

▲ FIGURE 14.57
The Red River floodway. *Source:* Photo courtesy of the author.

little damage in the city. Had it failed, the damage would have been many times what was experienced in Grand Forks. After the 1997 flood, the Province of Manitoba began planning for an expansion of the floodway to accommodate even larger floods.

Rapid City, South Dakota, lies along Rapid Creek at the base of the Black Hills. On the night of June 9, 1972, heavy thunderstorms over the Black Hills dumped as much as 38 cm of rain in less than 6 h. The result was a 100-year flood that killed 238 people and cost $128 million in property damage (Figure 14.58). In response to this event, the city adopted a nonstructural plan for flood control (Rahn, 1984). The floodplain today consists of parks, natural areas, golf courses, and other recreational-area developments. A few large commercial buildings were allowed to remain. The main objective was to remove private

◀ FIGURE 14.58
Map of part of Rapid City, South Dakota, showing extent of flooding (dark area) June 9–10, 1972. *Source:* From O. J. Larimer, 1973, *Flood of June 9–10, 1972, at Rapid City, South Dakota*, U.S. Geological Survey Hydrologic Investigations Atlas 511.

dwellings from the flood-prone area. As the mayor of Rapid City remarked, "Anyone who sleeps on a floodplain is crazy." Unfortunately, outside of the city limits, areas along Rapid Creek that were devastated by the 1972 flood were rebuilt with the same type of structures in the same hazardous locations. New construction also has occurred in the floodplain. The "act of God" mentality is difficult to combat.

Summary and Conclusions

Individual rivers are components of an integrated stream network in a drainage basin. A stream network develops a particular spatial stream pattern that is controlled by the geology, soils, topography, climate, and vegetation in a drainage basin. The concept of stream order has led to quantitative analysis of the variables characterizing stream networks. A hydrologic budget can be established for a drainage basin as a whole.

The flow of water in rivers can be studied using hydraulic parameters such as the Reynolds and Froude numbers. The Chezy and Manning equations relate the velocity of the stream to the gradient, channel shape, and boundary roughness. The velocity and shear stress, in turn, control the entrainment and transportation of stream sediment. Sediment is transported as bed load, suspended load, and dissolved load until the velocity drops and deposition removes the suspended and bed load. Deposition and other stream characteristics depend on whether the river is meandering or braided. These stream types result in different sediment distributions and landforms. Alluvial fans and deltas develop at points of major change in gradient in stream systems, where much of the stream load is deposited in a valley or in an ocean or lake.

All the geologic and climatic variables that control the discharge and sediment load in a stream can be considered as a system that can achieve a state of equilibrium when considered over a period of years. If natural or artificial changes in equilibrium are induced, the system attempts to reestablish an equilibrium condition. In this process many of the variables may be affected. Stream terraces constitute evidence that natural changes in equilibrium have occurred. Urbanization, dams, and other human alterations to the system may produce far-reaching changes to the channel and to the drainage basin.

Flooding is a natural, periodic process in river systems. Flood magnitude depends on a large number of geologic and climatic factors. Flood-frequency graphs show the inverse relationship between flood magnitude and recurrence interval. The problem of flood control can be approached in two ways. A certain degree of protection can be achieved by the construction of dams, levees, and other projects. This method is called the structural approach. A better alternative in most cases is the nonstructural approach, whereby land use on flood plains is restricted to those uses that are most compatible with periodic inundation.

Problems

1. Why do hydrologists use the drainage basin in the study of river systems?

2. How is drainage density related to the geology of a drainage basin?

3. How does stream capture modify river systems?

4. If Niagara Falls maintains an average rate of headward retreat of 1 m/yr, and the distance to the outlet of Lake Erie is 29 km, when will the falls reach Lake Erie? Based on Figure 14.10b, what is the eventual fate of Lake Erie?

5. How do the Reynolds and Froude numbers characterize flow in rivers?

6. If the depth of a stream is 2 m, what must the velocity be to attain critical flow?

7. What factors determine the velocity of flow in a stream?

8. A particular channel has a semicircular cross section with a radius of 5 m. If the slope is 10^{-3} and Manning's roughness coefficient is 0.04, what is the stream discharge?

9. Describe the two parameters that are used to predict entrainment of particles as flow conditions change.

10. A channel cross section has a hydraulic radius of 1.6 m and a slope of 0.002. Calculate the critical shear stress exerted by the stream and the largest particle size that could be transported, based on

Table 14.3   Annual Flood of the Red River at Grand Forks–East Grand Forks—Highest Level Reached During Each Year

Date	Level	Date	Level	Date	Level
1882	48.00	1921	20.90	1960	28.88
1883	42.20	1922	28.72	1961	9.75
1884	31.10	1923	26.60	1962	34.45
1885	23.10	1924	8.20	1963	21.23
1886	20.60	1925	19.00	1964	22.71
1887	16.30	1926	18.10	1965	44.92
1888	29.50	1927	21.70	1966	45.55
1889	12.00	1928	21.80	1967	37.50
1890	10.60	1929	28.30	1968	20.03
1891	17.70	1930	18.90	1969	45.69
1892	33.40	1931	6.48	1970	34.42
1893	45.50	1932	22.07	1971	27.86
1894	26.90	1933	15.18	1972	38.73
1895	9.90	1934	10.02	1973	27.32
1896	32.00	1935	13.07	1974	40.25
1897	50.20	1936	25.00	1975	43.30
1898	15.00	1937	11.57	1976	34.58
1899	20.90	1938	15.49	1977	8.71
1900	13.20	1939	20.13	1978	45.73
1901	26.30	1940	21.88	1979	48.81
1902	26.00	1941	27.86	1980	31.01
1903	28.00	1942	24.10	1981	14.68
1904	40.65	1943	38.16	1982	37.18
1905	26.11	1944	19.79	1983	29.17
1906	36.00	1945	32.00	1984	37.06
1907	39.95	1946	33.23	1985	25.90
1908	32.80	1947	40.71	1986	37.00
1909	18.80	1948	41.68	1987	33.19
1910	30.70	1949	29.11	1988	21.16
1911	10.70	1950	45.61	1989	44.37
1912	12.73	1951	33.52	1990	17.56
1913	26.70	1952	33.60	1991	17.63
1914	17.50	1953	24.63	1992	23.30
1915	30.80	1954	18.63	1993	36.39
1916	41.00	1955	26.17	1994	34.30
1917	32.50	1956	32.43	1995	39.80
1918	11.30	1957	24.67	1996	45.90
1919	23.20	1958	16.03	1997	54.35
1920	41.00	1959	16.10	1998	39.84

Source: From J. A. LeFevre, J. P. Bluemle, and R. P. Waldkirch, 1999, *Flooding in the Grand Forks–East Grand Forks North Dakota and Minnesota Area*, North Dakota Geological Survey Education Series 25.

the relationship shown in Figure 14.18. (Figure 14.18 uses units of kg_{force}/m^2; $1\ kg_f = 9.8\ N$.)

11. Describe the ways in which sediment is transported in streams.

12. Compare and contrast braided and meandering stream systems.

13. What are the similarities and differences between deltas and alluvial fans?

14. How can a stream be considered in a state of equilibrium when discharge, velocity, and other parameters are constantly changing?

15. How do human activities affect equilibrium?

16. Two nearby drainage basins have areas of 5 acres. One basin is urbanized and the other basin is forested. Calculate the stream discharge from each basin for a 1 in./h rainfall.

17. Table 14.3 gives the annual flood record (stage height of highest event in each year) for Grand Forks, North Dakota for a 117-yr period. Rank order the 10 largest floods on record. When did the 10th largest flood occur and what is its recurrence interval?

18. Based upon diagrams included in the chapter, what is the approximate recurrence interval of floods that just reach the flood stage at Grand Forks? What is the discharge of these floods?

19. In the aftermath of the 1997 flood, local and federal agencies began to make plans for flood prevention in Grand Forks. What measures would you suggest to alleviate this problem? Discuss the pros and cons for each potential action.

References and Suggestions for Further Reading

AMERICAN FALLS INTERNATIONAL BOARD. 1974. *Preservation and Enhancement of the American Falls at Niagara*. Final Report to Niagara International Joint Commission (Appendix C—Geology and Rock Mechanics, and Appendix D—Hydraulics.)

BAKER, V. R., and D. F. RITTER. 1975. Competence of rivers to transport coarse bedload material. *Geological Society of America Bulletin*, 86:975–978.

BLOOM, A. L. 1978. *Geomorphology*. Upper Saddle River, N.J.: Prentice Hall, Inc.

BRUSH, L. M., Jr. 1961. *Drainage Basins, Channels, and Flow Characteristics of Selected Streams in Central Pennsylvania*. U.S. Geological Survey Professional Paper 282-F.

CALKIN, P. E., and P. J. BARNETT. 1990. Glacial geology of the eastern Lake Erie basin, in *Quaternary Environs of Lakes Erie and Ontario*, D. I. McKenzie, ed. Waterloo, Ont.: Escart Press.

CALKIN, P. E., and T. A. WILKINSON. 1982. Glacial and engineering geology aspects of the Niagara Falls and Gorge, in *Geology of the Northern Appalachian Basin, Western New York*, E. J. Buehler and P. E. Calkin, eds. Field Trips Guidebook for the New York State Geological Association, 54th Annual Meeting, Amherst, N.Y., pp. 245–247.

CHEN, A. 1983. Dammed if they do and dammed if they don't. *Science News*, 123:204–206.

COSTA, J. E., and V. R. BAKER. 1981. *Surficial Geology: Building with the Earth*. New York: John Wiley.

DALRYMPLE, T. 1960. *Flood-frequency Analysis*. U.S. Geological Survey Water Supply Paper 1543-A.

DANIELS, R. B. 1960. Entrenchment of the Willow drainage ditch, Harrison County, Iowa. *American Journal of Science*, 225:161–176.

HANSEN, D. E., and J. KUME. 1979. *Geology and Ground Water Resources of Grand Forks County, North Dakota. Part I: Geology*. North Dakota Geological Survey Bulletin 53, and North Dakota State Water Commission County Ground Water Studies 13.

HARRISON, S. S., and J. P. BLUEMLE. 1980. *Flooding in the Grand Forks–East Grand Forks Area*. North Dakota Geological Survey, Educational Series 12.

HOWARD, A. D. 1967. Drainage analysis in geologic interpretation: A summation. *American Association of Petroleum Geologists Bulletin*, 51:2246–2259.

JUDSON, S., M. E. KAUFFMAN, and L. D. LEET. *Physical Geology*, 7th ed. Upper Saddle River, N.J.: Prentice Hall, Inc.

KOLB, C. R., and W. G. SHOCKLEY. 1957. Mississippi Valley geology: Its engineering significance. *Journal of Soil Mechanics and Foundations*, American Society of Civil Engineers, 83 (SM3):1–14.

LARIMER, O. J. 1973. *Flood of June 9–10, 1972, at Rapid City, South Dakota*. U.S. Geological Survey Hydrologic Investigations Atlas 511.

LEFEVER, J. A., J. P. BLUEMLE, and R. P. WALDKIRCH, 1999. *Flooding in the Grand Forks–East Grand Forks North Dakota and Minnesota Area*. North Dakota Geological Survey Education Series 25.

LEOPOLD, L. B. 1968. *Hydrology for Urban Land Planning: A Guidebook on the Hydrologic Effects of Urban Land Use*. U.S. Geological Survey Circular 554.

LEOPOLD, L. B., and M. G. WOLMAN. 1957. *River Channel Patterns: Braided, Meandering, and Straight*. U.S. Geological Survey Professional Paper 282-B.

MATTHEWS, R. K. 1984. *Dynamic Stratigraphy*, 2nd ed. Upper Saddle River, N.J.: Prentice Hall, Inc.

MATHEWSON, C. C. 1981. *Engineering Geology*. Columbus, Ohio: Charles E. Merrill.

McKNIGHT, T. L. 1984. *Physical Geography: A Landscape Appreciation*. Englewood Cliffs, N.J.: Prentice Hall, Inc.

MORGAN, J. P. 1970. Deltas: A résumé. *Journal of Geologic Education*, 18:107–117.

RAHN, P. H. 1984. Flood-plain management program in Rapid City, South Dakota. *Geological Society of America Bulletin*, 95:838–843.

RENDER, F. W. 1970. Geohydrology of the metropolitan Winnipeg area as related to groundwater supply and construction. *Canadian Geotechnical Journal*, 7:244–274.

RITTER, D. F. 1986. *Process Geomorphology*, 2nd ed. Dubuque, Iowa: Wm. C. Brown.

SIMONS, D. B., and E. V. RICHARDSON. 1966. *Resistance to Flow in Alluvial Channels*. U.S. Geological Survey Professional Paper 422-J.

SPEARING, D. R. 1974, *Summary sheet of sedimentary deposits*, Mc-8; Sheet 1, Fig. 1-C. Geological Society of America.

STRAHLER, A. N. 1957. Quantitative analysis of watershed geomorphology. *Transactions, American Geophysical Union*, 38:913–920.

WALKER, R. G., and D. J. CANT. 1984. Sandy fluvial systems, in *Facies Models*, R. G. Walker, ed. Geoscience Canada Reprint Ser. 1, Geological Association of Canada.

WEBB, R. H., S. S. SMITH, and V. A. S. McCORD. 1992. Arroyo cutting at Kanab Creek, Kanab, Utah, in *Paleoflood Hydrology of the Southern Colorado Plateau: Field Trip Guidebook*, J. Martinez-Goytre and W. M. Phillips, eds. Tucson, Ariz.: Arizona Laboratory for Paleohydrological and Hydroclimatological Analysis, University of Arizona, pp. 17–39.

15 CHAPTER

Oceans and Coasts

The importance of the oceans can be appreciated when we realize that more than 70% of the earth's surface is covered by water. The significance of the oceans, however, goes well beyond their areal extent. Life on the earth began in the oceans. Today, microscopic organisms in the sea form the base of the oceanic food chain, which supports the population of fish and other higher organisms in the oceans. This huge reservoir of food is desperately needed by the exponentially growing human population and we must ensure that fish populations are managed properly so they are sustainable in the future.

As mentioned in Chapter 1, oceanic circulation is being increasingly recognized as a factor in climatic stability. Rapid and severe changes in the earth's climate—from greenhouse to icehouse flips and vice versa—are linked to fluctuations in deep circulation of the oceans. Evidence suggests that this circulation exists in two stable states that are separated by rapid global climate change. Under some scenarios, our current global warming trend could cause a partial shutdown of deep circulation and switch the earth's climate into a much colder state.

In recent decades we have begun to tap the vast storehouse of mineral and energy resources in the oceans. Throughout geologic time the remains of tiny marine organisms have accumulated to form the earth's supplies of petroleum. Exploration for petroleum is increasingly focused on deep offshore targets. The problems involved in the construction of offshore drilling platforms and other structures make an understanding of the oceans a matter of practical importance.

Coastlines, where the oceans and continents meet, are among the most dynamic areas of the earth. The combination of active geologic processes and dense development of coastal areas leads to a wide spectrum of engineering and environmental problems. The great devastation caused by the Indian Ocean tsunami of 2004 is one sign that we must develop a better understanding of coastal processes and implement warning systems to prevent future catastrophes of this type. Environmental degradation of oceans and coasts is a more subtle, but growing threat. The recent occurrence of large oil spills from tanker accidents has highlighted the need to understand the geologic, chemical, and biological interactions in coastal waters.

The Ocean Basins

Oceanic Circulation

Large-scale motions of seawater are partly driven by prevailing winds over the sea surface. Surface currents—the Gulf Stream, for example—travel long distances and transport huge quantities of water. Individual currents are actually parts of huge loops called *gyres* (Figure 15.1). These currents are important mechanisms for transferring heat, in the form of warm water, from the equatorial regions toward the poles.

The oceans also undergo vertical circulation. When surface currents diverge, or flow away from a continent, deep water moves upward to replace the surface water. This movement, known as *upwelling* (Figure 15.2), is of great economic importance because deep water is rich in nutrients. Organisms, including fish, proliferate in areas of upwelling. The opposite of upwelling is sinking. Vertical movement is also driven by density differences. Cold, saline polar water sinks and moves very slowly toward the equator. North Atlantic deep water is forced to flow over the even denser Antarctic bottom water (Figure 15.2). The global oceanic circulation of deep waters, known as *thermohaline* circulation, is closely linked to climate. In the Atlantic Ocean, warm surface currents flow north into the North Atlantic. Warm air masses, carrying moisture that evaporates from the warm surface currents, keep the northern midlatitudes, such as Europe, much warmer than they would be without these currents. As moisture is removed, however, the waters become more saline and denser, a process that is accentuated by cooling of the currents as they move northward. Eventually, the surface waters become dense enough to sink to the bottom to form

▲ FIGURE 15.1

Current movement in the oceans. *Source:* From S. Judson, M. E. Kauffman, and L. D. Leet, *Physical Geology* 7th ed., © 1987 by Prentice Hall, Inc., Upper Saddle River, N.J.

▶ FIGURE 15.2
(a) Upwelling caused by surface flow away from a continent. (b) Vertical distribution of selected water masses.

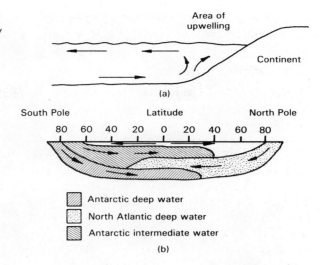

North Atlantic bottom water (Figure 15.2), which then flows southward toward the equator along the ocean floor. It is this large-scale, conveyor-like circulation that distributes heat from the equatorial regions and moderate climates in the midlatitudes. At times in the past, perhaps during periods of accelerated melting of midlatitude glaciers, cold, fresh waters spread over the North Atlantic Ocean, preventing the formation of North Atlantic bottom waters. Thus, with the deep circulation conveyor belt shut down or weakened, the climate in the northern and midlatitudes became much colder. A fluctuation between these two states—a strong conveyor belt versus a weak one—is thought to be related to major climate changes during Quaternary time. Under the warm climatic conditions of the Holocene, the conveyor belt has been strong and steady. If, however, the climate continues upon its current warming trend, melting sea ice in the North Atlantic and glacial runoff from high-latitude glaciers could conceivably weaken the conveyor belt and throw the earth's climate into a colder state. This is only one of many possible scenarios associated with global warming, but one that would have major implications for the human population if it occurs.

Topography

The modern study of the oceans began with the voyage of the research vessel H.M.S. *Challenger* in 1872. Prior to this voyage, the ocean basins were assumed to be broad, featureless plains. Instead, the *Challenger* expeditions brought back evidence of huge submarine mountain ranges, deep trenches, and other topographic features. Detailed study of the ocean basins accelerated during and after World War II. This work has provided much of the evidence for plate tectonics.

The topography of the ocean basins is dominated by midoceanic ridges (Figure 15.3). These ridges rival the size and extent of continental mountain ranges. At the crest of the midoceanic ridges lie the faulted rift valleys, where new oceanic crust is formed in the seafloor spreading process.

The ocean bottom slopes downward away from the rugged midoceanic ridges into a region of more subdued relief containing *abyssal hills*. The relatively high elevation of the midoceanic ridges is the result of volcanism and thermal expansion of the lithosphere; as newly formed crustal material moves away from the ridge crest it cools and subsides. Isolated volcanic mountains, or *seamounts*, rise above the floor of the oceans in many places. Some seamounts have flat tops rather than the typical conical shape of a volcano. Drilling

▲ FIGURE 15.3
Topographic features of the ocean basins. Notice the position and size of the midoceanic ridges.
Source: Reproduced from the *World Ocean Floor* panorama by Bruce C. Heezen and Marie Tharp, © 1977 by Marie Tharp.

of these flat-topped seamounts, termed *guyots*, has produced evidence to explain their emergence in the geologic past. The flat-topped form is attributed to wave erosion near sea level. The present depth of guyots is an indication of the great amount of subsidence that has taken place.

The greatest depths of the ocean basins are located at the *trenches*, long narrow troughs associated with subduction zones. Oceanic crust is pushed or dragged downward toward the mantle beneath these linear troughs.

The surface of the ocean bottom changes considerably in the vicinity of the continents. Rapid rates of deposition of continental sediment bury the landscape, producing *abyssal plains* of low relief.

Oceanic Sedimentation

Sediments deposited on the ocean floors beyond the influence of the continents, called *pelagic sediments*, consist of inorganic and organic types. The inorganic pelagic sediments are composed mostly of clay-size (<0.004 mm) particles of quartz, feldspar, clay, and other common rock-forming minerals. These particles are originally derived from the continents but can be transported great distances by ocean currents or winds because of their extremely small size. Large areas of the ocean floors are covered by red or brown muds composed of the inorganic pelagic sediments. Sedimentation rates are very low, often less than 1 cm per 1000 years.

Organic sediments are also deposited on the ocean floor. These include the shells of microscopic organisms living near the surface of the ocean, as well as fecal pellets and other organic remains. The shells secreted by organisms are composed of calcium carbonate or silica. The shells of organisms that secrete calcium carbonate, such as the one-celled foraminifera, are the most abundant in the oceans, although these shells are never found in sediments deposited below a water depth of approximately 4 km. The reason for this is that deeper waters tend to be undersaturated with respect to calcium carbonate, and therefore steadily dissolve

▲ FIGURE 15.4
Carbonate compensation depth in the oceans. *Source:* From F. Press and R. Siever, *Earth,* 3rd ed., © 1982 by W. H. Freeman and Co., San Francisco.

shells that fall to the bottom. The depth at which seawater becomes undersaturated is called the *carbonate compensation depth* (Figure 15.4). Carbonate-rich *foraminiferal oozes* are common where the ocean bottom lies above this critical level. *Radiolarian* or *diatom oozes* are deposited where organisms that secrete siliceous shells are abundant. These sediments are not affected by the carbonate compensation depth.

Continental Margins

Topography

The margins of the continents have several topographic divisions (Figure 15.5). Extending seaward from the coastline are gently sloping platforms called *continental shelves*. Continental shelves vary in width from less than 2 km to more than 1000 km. Although arbitrary depth limits have been used to define the extent of the shelves, the best method is to define the limit of the continental shelf as the point where the slope suddenly increases. A water-depth range of 200 to 650 m will include a large percentage of the shelves.

Continental shelves are correctly considered to be parts of the continents submerged by the present elevation of sea level. A variety of sediment types covers the shelves, although the sediments were reworked and redistributed by waves and currents during the postglacial sea-level rise. Valleys called *submarine canyons* cross the continental shelves. Most of these valleys are the offshore continuation of river valleys on land. Many submarine canyons were cut across the continental shelves during Pleistocene time when rivers flowed across the exposed shelves.

The economic value of the continental shelves is very great because of the fishing industry that they sustain and the petroleum reserves that they overlie. Construction of offshore

▶ FIGURE 15.5
Topographic divisions of the continental margins.

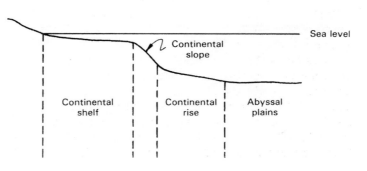

drilling platforms on the continental shelves is an engineering problem that requires detailed knowledge of the distribution and engineering properties of shelf sediments. The platforms must be designed to withstand the huge waves that are generated as storms pass over the continental shelves. Offshore platforms in the Gulf of Mexico were damaged by hurricanes in 2004 and 2005.

The continental shelves terminate at the upper margins of the *continental slopes*. The most steeply sloping segments of the continental slopes average 70 m per kilometer in comparison to the mean value of about 2 m per kilometer for the slope angle of the shelves. Submarine canyons dissect the continental slopes, although these valleys could not have been eroded by rivers because the sea level has never dropped low enough to expose them. At the base of the continental slopes, the slope angle again decreases to a more gentle decline along the surface of the *continental rises*. The continental rises gradually merge seaward with the flat abyssal plains of the ocean floors.

Turbidity Currents

The continental slopes and rises are blanketed by thick accumulations of continental sediment that was transported to the edge of the continental shelves by rivers and wave action during the recent geologic past. On the continental slopes these unconsolidated sediments are somewhat unstable at the greater slope angle found on that part of the continental margin. Slumps are common, sometimes triggered by earthquakes.

When slope movements occur on the continental slope, a very important type of current can be generated (Figure 15.6). The current consists of a dense cloud of suspended sediment that accelerates and flows rapidly downslope because it is more dense than the surrounding seawater. The name given to this type of flow is *turbidity current*. When turbidity currents flow out onto the continental rises, the velocity decreases because of the decrease in slope. Particles begin to drop out of suspension in order of decreasing size as the current slows, so that graded bedding develops in the layer of sediment deposited by the turbidity current.

Dramatic evidence for the existence of turbidity currents was provided when data were analyzed from the 1929 earthquake of the Grand Banks of Newfoundland. The data included the times of breakage of submarine transatlantic telephone cables following the earthquake. Some of the cables were broken hours after the event, so the earthquake was ruled out as a direct cause. Cables were broken sequentially with increasing distance from a point on the continental slope. The only reasonable explanation for this pattern of cable breakage is that the earthquake generated a powerful turbidity current that flowed down the continental slope at a velocity of 40 to 55 km/h, systematically breaking each cable it crossed along its path.

Turbidity currents are important because they transport huge quantities of continental sediment onto the adjacent abyssal plains to form *deep-sea fans* (Figure 15.7). The submarine

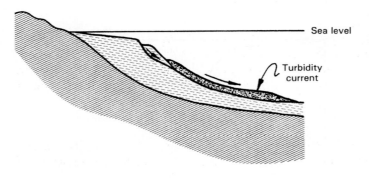

Sea level

Turbidity current

◀ FIGURE 15.6
Generation of a turbidity current by slumping in the sediments of the continental slope.

▲ FIGURE 15.7
Model of a deep-sea fan deposited at the end of a submarine canyon that cuts both the continental shelf and continental slope. Turbidity currents can be generated both within the canyons and from slumps on the continental slope. *Source:* From H. G. Reading and M. Richards, 1994, Turbidite systems in deep-water basin margins classified by grain size and feeder system, *American Association of Petroleum Geologists Bulletin*, 78:792–822.

canyons incised into the continental slopes are most likely eroded by these dense, sediment-laden flows. Ancient sequences of marine sedimentary rocks exposed on land have now been interpreted to be the result of turbidity-current deposition (Chapter 5). These sequences of turbidites consist of alternating layers of sandstone or siltstone and shale (Figure 5.24). The coarse-grained beds display graded bedding and are interpreted to be deposits of turbidity currents. The interbedded shales represent normal pelagic sedimentation between turbidity currents. The wide distribution of these sedimentary sequences attests to the importance of turbidity currents in marine sedimentation along continental margins.

Shorelines

Waves

Among the processes that shape our coastlines, those involving waves are probably the most important. The impingement of waves upon a shoreline governs coastal erosion, the formation of beaches, and the distribution of sediment in the shoreline environment. Coastal engineering and management therefore must be based on an understanding of wave processes.

Generation and Movement

The basic parameters used to describe wave motion are illustrated in Figure 15.8. *Wavelength* is the distance between adjacent crests or troughs. The amount of time between passage of

◀ FIGURE 15.8
Definition of wavelength and wave height.

successive crests relative to a fixed point is the *wave period*. The *wave velocity, C*, then, is simply the wavelength, *L*, divided by the period, *T*:

$$C = \frac{L}{T} \tag{15.1}$$

Although waves move through the water at a particular velocity, the water through which the waves propagate does not travel forward in the direction of wave movement. Instead, particles of water beneath a passing wave describe circular orbits with decreasing diameters as depth increases below the surface (Figure 15.9). This pattern of motion can be observed in the behavior of a cork bobbing in the water as waves pass beneath. When we speak of wave velocity, then, we are referring to the velocity of the moving waveform rather than the water. Additional parameters needed to characterize wave movement are *wave height*—the vertical distance from trough to crest—and water depth. The relationship between water depth and wavelength is an important aspect of wave motion.

Waves are produced by the transfer of energy from wind to the water surface. The amount of energy transferred, which is reflected in the size and period of the waves generated, is dependent on three parameters illustrated in Figure 15.10. The storm duration

◀ FIGURE 15.9
The circular motion of water particles in waves in deep water. *Source:* After U.S. Hydrographic Office Publication 604, 1951.

◀ FIGURE 15.10
Factors that determine the size and period of wind waves. *Source:* From P. D. Komar, *Beach Processes and Sedimentation*, © 1976 by Prentice Hall, Inc., Upper Saddle River, N.J.

and wind speed are functions of the size and intensity of the storm. The *fetch*, the distance over which the wind blows, is a function of the storm and the body of water. Fetch is a limiting factor for the generation of waves in small lakes and ponds. The generation of waves by storms at sea produces a complex group, or spectrum, of waves, with large variation in period. Under the influence of the generating storm, these *sea waves* are highly irregular. Individual wave crests are random summations of smaller waves of different periods.

The amount of energy transferred to the water from the wind is a function of the wave period and the square of the wave height. Long-period waves, those with periods of 20 s or more, are generated only when wind speed, storm duration, and fetch are all at their maximum values. Wave heights of 30 m or more have been observed in the open ocean. As waves move out of the area of generation, wave groups of similar periods sort themselves out and the wave motion becomes more regular. These waves, which travel great distances from their area of generation, are termed *swell*. Although the equations that describe wave motion are fairly complex, simplifications can be made under certain conditions. For example, in deep water—in which the depth is greater than one-half the wavelength—both wavelength and velocity can be expressed as functions of period. Thus

$$L = \frac{gT^2}{2\pi} \tag{15.2}$$

where g is the acceleration of gravity. By substituting equation (15.2) into (15.1), wave velocity can be expressed as a function of period:

$$C = \frac{gT}{2\pi} \tag{15.3}$$

After conversion to swell, wave groups of similar period travel forward until they encounter a coastline. Along this path the waves lose energy by several processes, including lateral spreading of the waves from the area affected by the storm to a large area of the open ocean. The potential coastal erosion or damage to structures depends upon the characteristics of the waves originally generated by the storm and the changes to the wave groups that occur along their path to the coast. Charts are available for prediction of wave heights and periods from known storm characteristics.

Breaking Waves

As waves approach shore, they undergo a gradual transition from sinusoidal profiles with low, rounded crests to waves with sharp peaks and higher wave heights (Figure 15.11). This transitional process begins when the water depth decreases to about one-half the wavelength. As the waveform begins to interact with the bottom, the circular orbits of water particles (Figure 15.9) become compressed to elliptical shapes. The wavelength and the velocity both decrease, while the period remains the same. As the wavelength decreases,

▶ FIGURE 15.11
Changes in waves as they approach shore. *Source:* From P. D. Komar, *Beach Processes and Sedimentation,* © 1976 by Prentice Hall, Inc., Upper Saddle River, N.J.

▲ FIGURE 15.12
Waves breaking on shore. *Source:* R. Dolan; photo courtesy of U.S. Geological Survey.

the wave crest must build in order to maintain a constant mass of water within a single wavelength. The ratio of the wave height to wavelength, called *wave steepness*, gradually increases until a critical value of stability is exceeded. This occurs when the steepness reaches a value of about $\frac{1}{7}$. At this point, the wave breaks (Figure 15.12) because it has become too steep to maintain its shape. In addition, at the point of breakage the horizontal orbital velocity at the crest exceeds the velocity of the waveform, and the crest collapses ahead of the wave.

Most breakers can be classified as either *spilling* breakers or *plunging* breakers (Figure 15.13). Spilling breakers, in which the crest collapses into bubbles and foam that rush down the shoreward face of the wave, occur over gently sloping bottoms. The plunging breaker creates a spectacular visual effect as the crest curls outward and downward in front of the wave, forming a tunnel for a few seconds prior to total collapse of the wave.

Spilling breakers

Foam

Nearly horizontal beach

◀ FIGURE 15.13
Form of spilling and plunging breakers. *Source:* From P. D. Komar, *Beach Processes and Sedimentation*, © 1976 by Prentice Hall, Inc., Upper Saddle River, N.J.

Plunging breakers

Steep beach

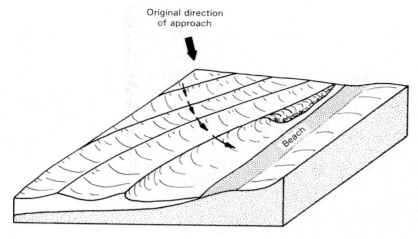

Original direction
of approach

Beach

▲ FIGURE 15.14
Refraction of wave crests approaching shore at an angle to become parallel with the beach.

Wave Refraction

The decrease in velocity experienced by waves as they enter shallow water results in a change in direction of the wave called *wave refraction*. Along a straight shoreline in which the offshore bottom contours parallel the shore, wave refraction is evident when waves approach the coast at an angle (Figure 15.14). The portion of a wave crest that reaches shallow water first begins to decrease in velocity compared with the portion of the same wave that has not yet reached shallow water. Consequently, the seaward part of the wave travels faster and the wave crest appears to bend, so that it becomes nearly parallel to the beach by the time it breaks.

Wave refraction has an important influence on erosional and depositional processes affecting irregular coastlines (Figure 15.15). In a coastline composed of headlands, or points jutting out into the ocean between adjacent bays, the bottom contours often parallel the shoreline. In other words, waves approaching the shore first begin to interact with the bottom directly offshore from the headland because water is shallower there relative to the bay on either side. As shown in Figure 15.15, wave refraction tends to focus the wave rays (lines drawn perpendicular to wave crests) toward the headland. The spacing of the wave rays, which are equally spaced and parallel in open water, is an indication of the amount of wave energy expended against a particular part of the coastline. Close spacing in the vicinity of headlands indicates that wave energy is concentrated in those areas. The rate of shore erosion, therefore, is much greater on headlands than in adjacent bays. The low-energy environment of bays, in fact, often results in net deposition of sediment there. This explains why headlands are commonly rocky and lack good beaches, whereas nearby bays contain broad expanses of sandy beaches. If the shoreline materials are nonresistant to erosion, these processes tend to straighten out an initially irregular coast by erosion of headlands and deposition in bays.

Tides

In coastal areas, tides are an important factor in many processes. This is particularly true where the conditions and forces that influence tides combine to produce a large difference in elevation, or *tidal range*, between high and low tides. The Bay of Fundy between New Brunswick and Nova Scotia, for example, has a maximum tidal range of 16 m, the highest in the world (Figures 15.16).

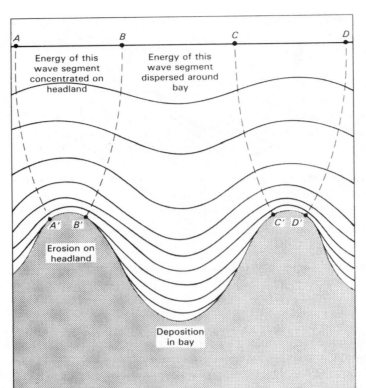

◀ FIGURE 15.15
Concentration of wave energy on headlands of an irregular shoreline. The energy of wave segments *AB* and *CD* is distributed over a short distance at the headland, whereas the energy of segment *BC* is dispersed over a long shoreline distance in the bay. *Source:* From S. Judson, M. E. Kauffman, and L. D. Leet, *Physical Geology*, 7th ed., © 1987 by Prentice Hall, Inc., Upper Saddle River, N.J.

(a)

(b)

▲ FIGURE 15.16
Hopewell Rocks, Bay of Fundy at low (a) and high (b) tide. The tidal range here is about 14 m. *Source:* Photo courtesy of Tourism New Brunswick, Canada.

Although many factors influence tides, the dominant control is the gravitational attraction between the earth and the moon. In terms of Newton's law of gravitation, the net attractive force can be expressed as

$$F = \frac{M_m M_e}{R^2} G \qquad (15.4)$$

The terms M_m and M_e represent the masses of the moon and the earth, respectively, and R is the distance between the center of the earth and the moon. G is the universal gravitation constant. Although equation (15.4) represents the net attractive force between the earth and the moon, the attractive force between the moon and a unit of mass on the earth's surface is slightly different from the net force because the distance, R, between the earth's surface and the moon is not the same as the distance between the centers of the two bodies. It is the difference between the total force of attraction and the local attractive force that is responsible for tides.

If the earth were covered with a uniform layer of water over its surface, tidal forces would produce the effect shown in Figure 15.17. Two tidal bulges would be created. The bulge facing the moon would be generated because the water on the moon's side of the earth is attracted more than the earth as a whole. The bulge on the opposite side of the earth is produced in a similar way: The moon's attraction for water on the opposite side of the earth is less than its attraction for the main body of the earth. As the earth rotates, the tidal bulges travel over the surface. A fixed point on the earth would therefore experience two high and two low tides in a little more than 24 h because the moon revolves around the earth in the same direction as the earth's rotation. It therefore takes an additional 50.47 min for the fixed point on the earth to "catch up" to its original position relative to the moon because the moon also has moved along its path around the earth during the day.

A number of factors complicate this simplified explanation of tide-generating forces. One of these factors is the influence of the sun. Despite its huge mass, the sun's great distance from the earth lessens its effect upon the tides. The influence it does have, however, must be considered in terms of the positions of the three bodies. When the earth, the moon, and the sun are aligned (Figure 15.18), the size of the tidal bulge on the earth is maximized. These periodic high tides are called *spring tides*, although they have nothing to do with the seasons. When the sun and the moon are aligned at right angles with respect to the earth (Figure 15.18), the lower tides produced are called *neap tides*.

These factors still do not explain the huge tidal ranges of the Bay of Fundy and other areas. For the reasons that cause these tides, we must consider the distribution of continents and ocean basins and the topography of bays and estuaries along the coastline. Because the earth is not covered by a uniform layer of water, tidal movements must occur in the confined

▶ FIGURE 15.17
Tidal bulge produced on opposite sides of the earth by the gravitational attraction of the moon.

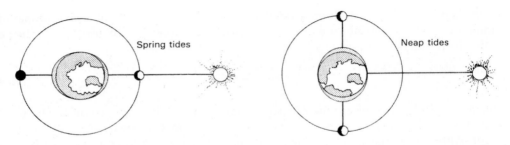

▲ FIGURE 15.18
Origin of spring and neap tides. *Source:* From E. A. Hay and A. L. McAlester, *Physical Geology: Principles and Perspectives*, 2nd ed., © 1984 by Prentice Hall, Inc., Upper Saddle River, N.J.

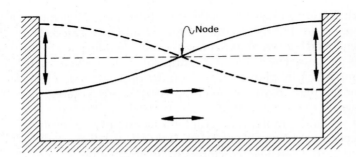

◀ FIGURE 15.19
Standing wave in an enclosed basin. Certain bays and estuaries experience very high tides due to resonant amplification when the natural period of the basin and the period of the tidal wave are similar. *Source:* From P. D. Komar, *Beach Processes and Sedimentation*, © 1976 by Prentice Hall, Inc., Upper Saddle River, N.J.

basins that contain the world's oceans. An ocean basin can be likened to a rectangular container of water that is suddenly disturbed by being lifted at one end (Figure 15.19), forming a wave of water that moves toward the opposite side of the tank. When it reaches the side, it is reflected back toward its original side. The process continues and a standing wave called a *seiche* is generated as the wave moves back and forth across the tank. The period of the wave is determined by the length and depth of the tank. In other words, every container, including ocean basins, bays, and estuaries, has a characteristic natural period. As the earth rotates, tides move across ocean basins as long-period waves. If the period of a *tidal wave* (not to be confused with the misnamed "tidal waves" that cause great destruction along coasts) is similar to the natural period of the basin in which it moves, resonance occurs and the wave is amplified. A process of this type takes place in bays such as the Bay of Fundy (Figure 15.20); the natural period of the bay is close to the tidal period, so a resonant standing wave moves back and forth across the bay. Other unique phenomena can be observed in bays with high tidal ranges. These include reversing waterfalls in rivers that drain into the bay such as one near St. John, New Brunswick (Figure 15.20). At low tide, the river drains into the bay with rapids forming over a steep section of the bedrock channel. At high tide, the huge tidal wave flowing into the bay reverses flow in the river and rapids form with water flowing upstream. As the tide rises, distinct wave fronts called *tidal bores* can be observed moving upstream in rivers and streams draining into the bay. In an irregular coast with many smaller bays and islands like the Bay of Fundy, tides coming from different directions can converge to form a *tidal whirlpool*. One of these forms as tides flow around Campobello Island from two directions (Figure 15.20) and converge on the back side. As the tidal currents meet each other in the narrow channel between Campobello and Deer Islands, a whirlpool known as Old Sow is generated, and the spinning vortex is strong enough to draw boats into the center. According to local lore, boats have been lost and people drowned in Old Sow.

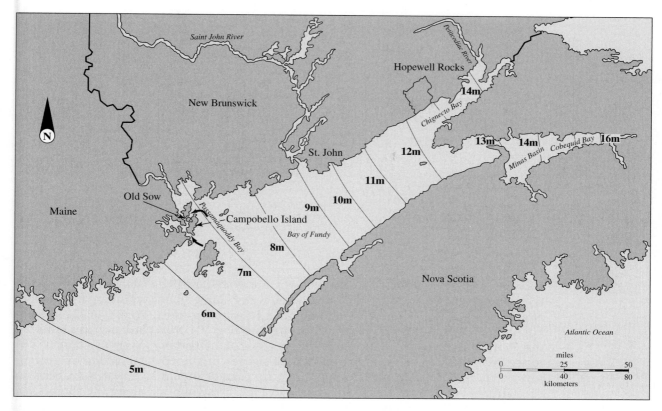

▲ FIGURE 15.20
Tidal ranges in the Bay of Fundy (lines crossing the bay show the maximum tidal range at that location). Locations of Hopewell Rocks (Figure 15.16), St. John reversing waterfall, and Old Sow are shown. Arrows indicate directions of tidal waves that converge on the northwest side of Campobello Island.

Longshore Currents

Although wave refraction tends to change the direction of wave movement so that waves approach the shore more directly, an oblique approach angle is often observed (Figure 15.21). This movement results in a net flow of water, or *longshore current*, parallel to the shore. Longshore currents are apparent to swimmers in the surf zone who may

▶ FIGURE 15.21
Longshore current associated with waves approaching coast at an angle.

◄ FIGURE 15.22
A longshore current cell bounded by rip currents.

suddenly realize that they have drifted along the shore for some distance from the point on the beach at which they swam out into the water. Rather than continuing indefinitely along the shore, longshore currents are often broken into cells (Figure 15.22) that are bounded by rapidly moving currents flowing directly away from the beach in a seaward direction. These *rip currents* are dangerous because of their ability to carry swimmers out to sea. Many swimmers make the mistake of swimming directly against the rip current toward shore. Instead, a person caught in a rip current should swim laterally, because rip currents are narrow zones that can be evaded by swimming along the shore for a short distance. Bottom topography can control the position of rip currents; for example, they often occur near the heads of submarine canyons. The direction and characteristics of incoming waves, however, can also influence the development of longshore cells. Rip currents have been observed on shores with a very regular bottom topography.

One of the most important effects of longshore currents is the transportation of sand along the shore. This *longshore drift* is the movement of sand in the shore zone by longshore currents. Transport actually occurs in a zig-zag fashion. Sediment is carried up the beach at the oblique angle of the approaching wave by the wave *swash*, but it is carried back to the water at right angles to the beach in the *backwash*. Thus there is a component of motion along the shore. The interaction of coastal structures with longshore drift is one of the most significant problems in coastal engineering.

Coastal Morphology

Coastlines are characterized by continuous change. The changes that occur present a great challenge to coastal engineers and managers because coastal processes can be rapid and powerful. In addition to the drastic changes that can be caused by a single damaging storm, long-term processes are also at work. These include worldwide changes in sea level or regional changes in the elevation of the land relative to sea level caused by tectonic forces. A study of the geomorphology of a coast can frequently indicate the types of processes that are currently influencing the coast.

Erosional Features

Coastlines rapidly eroding under the relentless attack of wave erosion provide examples of steep *wave-cut cliffs* (Figure 15.23). The cliffs maintain steep slope angles as they retreat because they are undermined at their bases by wave erosion that removes support for the overlying material. A rock fall, slump, or other type of mass movement can occur when the safety factor of the slope decreases to a value of 1.0 (Figure 15.24). The debris that accumulates at

▲ FIGURE 15.23
Wave-cut cliffs along the Oregon coast. A stack remains as an erosional remnant of the cliff. *Source:* J. S. Diller; photo courtesy of U.S. Geological Survey.

▶ FIGURE 15.24
A mass movement caused by coastal erosion. *Source:* J. T. McGill; photo courtesy of U.S. Geological Survey.

the base of the slope is further eroded and transported away from shore by wave action. The flat surface at the base of wave-cut cliffs beveled by wave erosion is called a *wave-cut platform* (Figure 15.25). As the coastline retreats, resistant rock units may project out of the water as *stacks or arches* (Figures 15.16 and 15.23). Eventually, these remnants are also worn down and removed by wave erosion.

Depositional Features

In addition to erosional processes, shorelines undergo deposition from waves, wind, and longshore currents. The landforms that result from these processes are dominant along some coastlines. Some of these features are illustrated in Figure 15.26. A detailed discussion of the beach, probably the most important depositional landform, will be reserved for the following section.

▲ FIGURE 15.25
Wave-cut platform in Pleistocene carbonate reef rock, Puerto Rico. *Source:* C. A. Kaye; photo courtesy of U.S. Geological Survey.

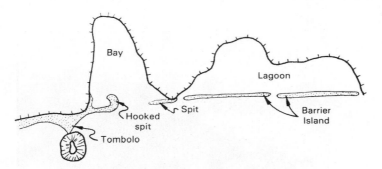

◀ FIGURE 15.26
Depositional landforms along coasts.

The transportation of sand by longshore currents is responsible for the deposition of a *spit* (Figure 15.27), a thin strip of beach that extends from a point or headland across the mouth of a bay or other coastal indentation. Frequently, the ends of spits curve inward toward the bay, forming a *hooked spit*. The presence of an offshore island close to the shore creates a low-energy zone shoreward of the island that is somewhat protected from wave processes. A spit commonly forms in such a protected location, and when the spit connects the beach and the offshore island, it is given the name *tombolo* (Figure 15.26).

Most depositional coastlines are characterized by long, narrow islands that lie just offshore and trend parallel to the coast. Such *barrier islands* are common off the East and Gulf coasts of the United States (Figures 15.26 and 15.28). Periodic breaks in the barrier islands, or *tidal inlets*, allow tidal flow between the ocean and the narrow *lagoons* that lie between the barrier islands and the mainland. The origin of barrier islands has been a matter of considerable geologic debate. One of the most likely recent hypotheses relates the formation of barrier islands to the recent rise in sea level accompanying the melting of Pleistocene glaciers. Linear ridges may have been built along coastal plains by wave or wind processes and isolated as barrier islands when the low-lying areas shoreward of the ridges were flooded to form the lagoons. Once formed, barrier islands can migrate in response to changes in sea level and other factors.

▲ FIGURE 15.27
Spit built across a bay in Nova Scotia. *Source:* Photo courtesy of the author.

▶ FIGURE 15.28
Barrier islands off the North Carolina
mainland.

Beaches

The *beach* is the actual point of interaction between the sea and land. As such, its form adjusts constantly to the energy expended upon it by tides, wind, and waves. Because the term beach describes only the part of the shoreline that is exposed above the mean low tide position, we must expand our zone of interest to consider all the significant processes in the nearshore environment. The zone that includes the beach as well as the shallow offshore area is known as the *littoral zone*. Its subdivisions and landforms are shown in Figure 15.29. The *offshore* begins just at the breaker zone and extends seaward. The *inshore* describes the zone between the breaker zone and the lowest point on the beach exposed at low tide. The *foreshore* includes the area from the low tide point to the top of the *berm crest*, and the term *backshore* is applied to the remainder of the beach shoreward to a major topographic break such as a dune or sea cliff. The littoral profile includes one or more flat *berms* on the shoreward extent of the beach, the sloping *beach face* that is under constant wave action, and one or more *longshore bars* (Figure 15.30) and *troughs* in the vicinity of the breaker zone.

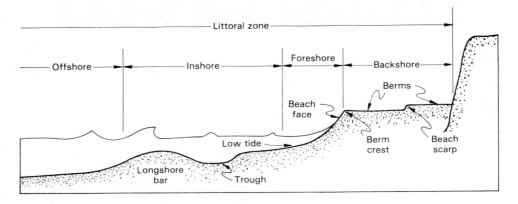

▲ FIGURE 15.29
Topographic divisions of the beach profile. *Source:* From P. D. Komar, *Beach Processes and Sedimentation,* © 1976 by Prentice Hall, Inc., Upper Saddle River, N.J.

◄ FIGURE 15.30
Longshore bars (arrows) running parallel to the Lake Michigan shoreline. *Source:* Photo courtesy of the author.

▲ FIGURE 15.31
The summer and winter beach profiles—the response of a beach to different energy conditions. *Source:* From P. D. Komar, *Beach Processes and Sedimentation*, © 1976 by Prentice Hall, Inc., Upper Saddle River, N.J.

The profile illustrated in Figure 15.29 is a very generalized representation. The beach, or littoral, profile is constantly changing in response to changing wave and tide conditions. The beach must therefore be thought of as a dynamic system in which changes in wave inputs initiate responses that tend to establish a new condition of equilibrium under those particular wave conditions. The adjustments that take place in order to reach equilibrium are changes in the beach profile. The size of the particles making up the beach is also an important factor in this process.

An example of this system at work is the change in profile that occurs from the predominantly swell-wave condition that exists in summer to the influence of storm waves that are more common in winter. Figure 15.31 shows the profiles that develop under each of these conditions. The broad berm that characterizes the summer profile is a depositional landform that is gradually built up by the onshore sand transport by swell waves. A berm is a very desirable attribute for a beach, not only because it provides wide expanses for sunbathers but because it protects the sea cliff or dunes from wave erosion. The powerful storm waves of winter attack and erode the berm that was constructed during the summer. The sand that was incorporated into the berm is transported offshore and deposited as part of one or more longshore bars. Longshore bars are long, narrow, submerged ridges of sand that develop in the vicinity of large breakers (Figure 15.30). Without the protection of the berm, storm waves are free to attack the sea cliff. Many beachfront property owners have learned about this dynamic system as they watched destructive storm waves pounding against the foundations of their houses. In the summer the process is again reversed: Gentle swell waves transport sand from the longshore bars up onto the beach to form a new berm.

Another manifestation of equilibrium between form and process is the angle of slope of the beach face. This angle is related to the size of the incoming waves and the particle size and angularity of the beach material. Even though strong waves transport sand from a beach in the offshore direction, they also have the ability to transport large particles toward the beach. Thus an exposed beach that is frequently subjected to strong waves is commonly composed of coarse particles. The movement of a wave onto the beach face is illustrated in Figure 15.32. Swash transports particles upslope until deposition when the velocity decreases. Before the return flow down the beach face in the backwash can occur, some of the water infiltrates into the beach face. More water infiltrates into the beach as the particle size increases. Because of this loss of water, the backwash tends to be weaker than the swash. In order for an equilibrium profile to develop, the same amount of sand must be transported down the beach slope as up. Backwash must be assisted by an additional

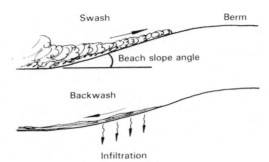

Swash   Berm

Beach slope angle

Backwash

Infiltration

◀ FIGURE 15.32
The stronger swash transports sediment up the beach face. Backwash is weakened by infiltration of water into the beach. In order for an equilibrium profile to develop, equal amounts of sediment must be transported up and down the beach face. The coarser the particles, the steeper the beach slope angle must be.

◀ FIGURE 15.33
A beach in Iceland composed of large boulders because of the high wave energy. *Source:* Photo courtesy of the author.

force, namely, the downslope component of gravity. The larger the beach particles are, the steeper the slope must be, because of infiltration losses. Beaches composed of large particles, such as the beach illustrated in Figure 15.33, must therefore develop steeper slopes in order to establish an equilibrium state.

Case in Point 15.1
Where Not to Build a Swimming Pool

Location of buildings in areas of active geologic processes usually leads to rapid destruction. The elaborate pool in Figure 15.34a extends too far out onto a beach along the Virginia coastline. The movement of waves onto the beach adjacent to the structure is obvious in the photograph. Construction of anything on a beach indicates a total lack of understanding of coastal processes. The tranquil waves of summertime become much more powerful and destructive in the winter. A storm of moderate intensity in 1979 demonstrated the foolishness of constructing such a project (Figure 15.34b).

► FIGURE 15.34
(a) A structure that extends out onto a beach is vulnerable to storm damage. (b) The same house after a storm of moderate intensity. *Source:* C. Alston; photos courtesy of Virginia Sea Grant at VIMS.

(a)

(b)

Types of Coasts

Classification of coasts is useful because the particular types of landforms present on a coast gives an indication of geologic processes that are currently active or have been significant during the development of the coast.

Depositional coasts (Figure 15.26) are characterized by deltas, barrier islands, and other depositional features. Gently sloping coastal plains underlain by nonresistant materials favor the development of depositional coasts. The East and Gulf coasts of the United States are good examples of depositional coasts.

Glacial coasts are produced by glacial erosion of valleys that meet the coastline. Glacial valleys tend to have very steep walls and a U-shaped cross-sectional profile. Scouring of these valleys took place at a time when sea level was much lower than at present, so the valleys extend well below the present sea level. The postglacial rise in sea level drowned these glacial valleys, producing the long, narrow bays with steep valley sides called *fjords*. The coast of Norway is famous for the picturesque scenery of its fjords (Figure 15.35).

Some coasts have been raised or uplifted relative to sea level. These *emergent coasts* can be produced in several ways. Tectonic uplift associated with faulting is an important mechanism. Parts of the California coast have a series of raised wave-cut platforms (Figure 15.36), indicating that the coast has been subjected to a succession of tectonic episodes. Each platform represents a temporary period of stability. The La Conchita debris flows (Case in Point 13.1) originate on a wave-cut cliff and flow onto the most recent raised wave-cut platform.

Glaciated regions may also have emergent coastlines. In this instance, the mechanism producing emergence is *isostatic rebound*. The huge weight of a glacier several kilometers

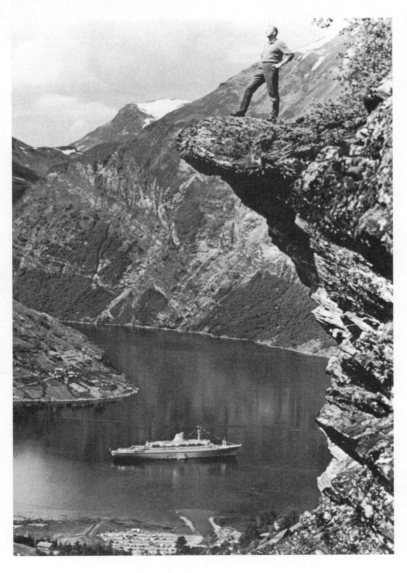

◀ FIGURE 15.35
The steep valley walls of the Norwegian fjords were produced by glacial erosion. *Source:* Photo courtesy of Norwegian Tourist Board.

▶ FIGURE 15.36
Wave-cut terraces on the west side of San Clemente island, California. *Source:* John S. Shelton.

thick causes the rigid crust to sink slightly deeper into the plastic mantle of the earth. Isostatic rebound is the readjustment of the crust after the glacier's retreat. The fact that glaciated regions are still rising relative to sea level 10,000 years after the last ice retreat demonstrates that isostatic rebound is a very slow process. A series of beaches or other shoreline features above sea level is an indication of an emergent coastline produced by postglacial rebound.

A final type of coastline is the *biogenic coast*, characterized by the presence of carbonate reef structures in the shallow offshore zone along the coast. Carbonate reefs are composed of the shells of living organisms called corals that grow in colonies, where individual organisms are connected to form a three-dimensional reef framework. Coral reefs are classified according to their position relative to the mainland. Reefs that actually connect with the shoreline are called *fringing reefs*. If the reef is separated from the coast by a lagoon, the name *barrier reef* is used. A circular reef without the presence of land is known as an *atoll*. More than 100 years ago, Charles Darwin provided a perceptive hypothesis for the origin of atolls that has been confirmed by deep-sea drilling only in the decades since World War II. Noting that most islands in the tropics of the Pacific Ocean were of volcanic origin, Darwin proposed that fringing reefs originally developed around these volcanoes (Figure 15.37). Gradual subsidence of the ocean crust caused the volcanoes to sink until only a small island remained surrounded by barrier reefs. Upward growth of the reef structure is necessary to counteract the subsidence of the crust, because corals can live only in warm, shallow water. With continued subsidence, the eroded remnants of the volcano disappeared altogether, leaving only a circular atoll to mark its original position.

Sea-Level Changes

Far from being a static surface, the elevation of sea level has undergone major fluctuations in geologic time. One of the major causes for *eustatic* sea-level changes, those that are synchronous throughout the world, is a change in the mass of the polar ice caps. When polar ice caps expand and advance into the midlatitudes, as they did repeatedly during the Pleistocene Epoch, sea level drops accordingly. During the last glaciation, sea level dropped approximately 100 m. While sea level was lowered, a greater area of the continental shelves was exposed than at present. Rivers eroded valleys across these broad coastal plains

◀ FIGURE 15.37
Evolution of atolls. (a) Fringing reefs form around an active volcano. (b) Reefs grow vertically upward to form barrier reefs as volcano subsides and is eroded. (c) Further subsidence leaves only reefs above sea level. (d) Map view of atoll.

underlain by unconsolidated sediment. Later, as glaciers retreated and released huge volumes of water to the oceans, the rising sea level inundated these newly formed valleys, creating estuaries. An irregular coastline in which estuaries extend up previous river valleys is sometimes referred to as a *drowned* coastline. Chesapeake Bay along the eastern coast of the United States is an example of an estuary that was formed by the drowning of a river valley during the postglacial sea-level rise. Rising sea levels related to our current episode of global warming are raising fears that low-lying coastal areas throughout the world may be more frequently inundated during storms, causing major disruptions to the growing coastal population. The measures being taken by Venice, Italy, to combat this problem are described later.

Coastal Management

Coastal Hazards

Throughout human history, a large percentage of the population of coastal nations has chosen to reside along the coasts. Unfortunately, various coastal processes have produced events of extreme magnitude that have inflicted great disasters upon coastal population centers. Today, the steady migration of the U.S. population to the coasts of the "Sunbelt" makes an understanding of coastal hazards even more important.

Tropical cyclones are potentially destructive storm systems in which winds move around centers of low pressure in a circular fashion. In the Northern Hemisphere, wind circulation is counterclockwise around low-pressure areas. Regional names for intense tropical cyclones

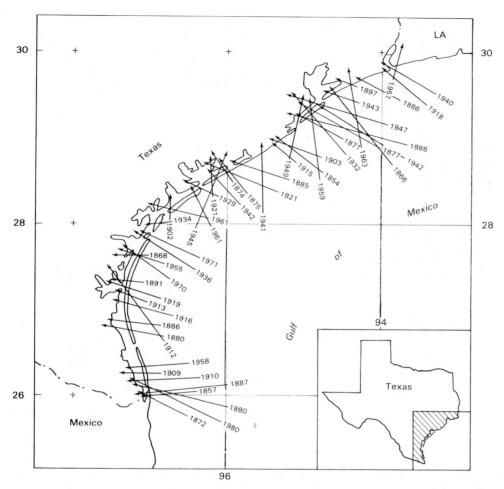

▲ FIGURE 15.38
Map of hurricane landfalls along the Gulf Coast of Texas between 1850 and 1980. *Source:* From J. P. Giardino, P. E. Isett, and E. B. Fish, 1984, Impacts of Hurricane Allen on the Morphology of Padre Island, Texas, *Environmental Geology and Water Sciences*, 6:39–43. Used by permission of Springer-Verlag, Inc., New York.

include hurricane, typhoon, and monsoon. These storms are generated over the ocean in the tropics and then migrate thousands of kilometers before they dissipate. In North and Central America, the Gulf Coast region is the prime land target of hurricanes. The map of part of this coastal area (Figure 15.38) indicates that the 20th-century distribution of storms was quite uniform. The storm season of 2004 was an unprecedented year for Florida, as four hurricanes passed over the state. Certain areas were impacted by hurricane-force winds 2 and even 3 times (Figure 15.39). Tropical cyclones are classified by the Saffir-Simpson hurricane scale (Table 15.1).

The arrival of a hurricane at a particular coastline is preceded by a 1- to 1.5-m rise in water level called a *forerunner*. The forerunner is caused by swell waves produced by the storm in its generating area, as much as several thousand kilometers distant from the coast at the time of arrival of the abnormally high forerunner swells.

Most damage from a tropical cyclone is produced by the *storm surge*, a rapid rise in water level generated by the winds of the storm as it approaches shore. The storm-surge ranges for various categories of storms are shown in Table 15.1. Because of the counterclockwise

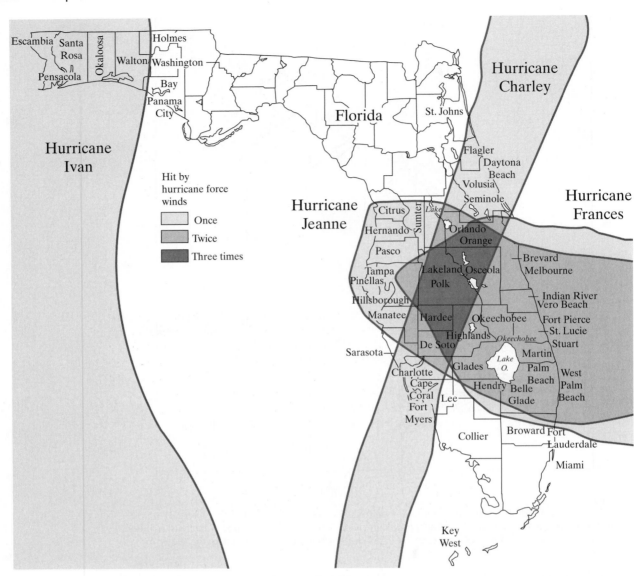

▲ FIGURE 15.39
Distribution of hurricane-force winds experienced in Florida in the 2004 hurricane season. *Source:* National Weather Service.

Table 15.1	Saffir-Simpson Hurricane Scale		
Category	Pressure (mb)	Winds (mph)	Surge (ft)
Depression	—	<39	—
Tropical storm	—	39–73	—
Hurricane 1	>980	74–95	4–5
Hurricane 2	965–980	96–110	6–8
Hurricane 3	945–965	110–130	9–12
Hurricane 4	920–945	131–155	13–18
Hurricane 5	<920	>155	>18

► FIGURE 15.40
A hurricane approaching shore in the Northern Hemisphere will produce maximum storm surge to the right side of the storm center as viewed in the direction of movement of the storm.

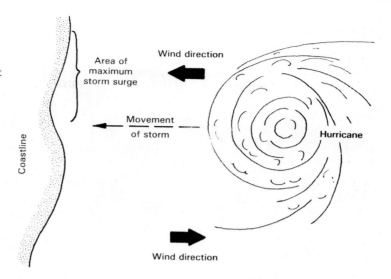

circulation of winds, the storm surge reaches its maximum to the right (north, in the northern hemisphere) of the storm center, where the winds are blowing in the direction of the storm's movement (Figure 15.40). The actual rise in water level in a storm surge is also a function of the tides at the time the storm strikes the shore (Figure 15.41). Thus, the magnitude of storm damage on the shore depends on the time of day and month at which the maximum storm surge arrives. A rising trend in sea level will gradually lead to higher storm surges under the same storm and tidal conditions. Effects of the storm surge include flooding of low-lying coastal areas and intense wave erosion of beaches, dunes, sea cliffs,

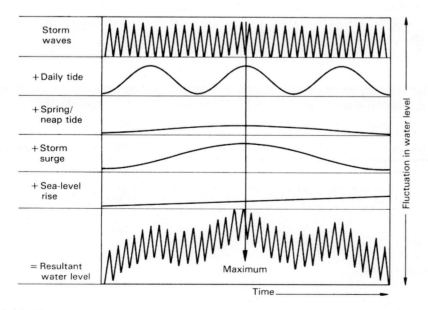

▲ FIGURE 15.41
Dependence of water-level rise in storm surge on tides and other factors. *Source:* From R. Dolan and H. Lins, 1985, *The Outer Banks of North Carolina*, U.S. Geological Survey Professional Paper 1177-B.

◀ FIGURE 15.42
Severe beach erosion along the Virginia coastline by a storm. *Source:* Scott Hardaway; photo courtesy of Virginia Sea Grant at VIMS.

and any structures that happen to be in the way (Figure 15.42). Additional effects of hurricanes include wind damage to structures and coastal flooding from the intense rainfall that occurs as the storm moves onshore. Low-lying, highly populated coastal areas are most susceptible to loss of life from storm surges. A tragic example is the coast of Bangladesh, in which a large population lives on islands that rise only a few meters above sea level in the Ganges-Brahmaputra delta plain. The disproportionate loss of life in Banglahesh, compared to other major storm surges, is shown in Table 15.2. The Galveston, Texas, storm surge of 1900 is the worst disaster of this type to occur in the United States.

Barrier islands are particularly susceptible to damage from large storms. In the absence of human modification, barrier islands are commonly narrow, low-lying ridges. During large storms, water can flow over the island to the lagoon, thus dissipating the energy of the storm waves. The erosion that accompanies these *washovers* is often intense enough to cut new inlets through the islands. Severe damage is inflicted on roads, bridges, buildings, and utilities in washover zones. An example of storm effects on barrier islands is shown in Figure 15.43.

Table 15.2   Major Historical Storm Surges

Date	Shelf Region	Estimated Maximum Surge Height	Estimate of Lives Lost
November 1218	Zuider Zee (Dutch North Sea)	Unknown	100,000
October 1737	India and Bangladesh	12 m	300,000
1864	Bangladesh	Unknown	100,000
October 1876	Bangladesh	15 m	100,000
1897	Bangladesh	Unknown	175,000
September 1900	Galveston, Texas (Gulf of Mexico)	4.5 m	6000
Jan./Feb. 1953	Southern North Sea	3.0 m	2000
March, 1962	Atlantic Coast U.S.A.	2.0 m	32
November, 1970	Bangladesh	9.0 m	500,000

▶ FIGURE 15.43
Washover damage near Cape Hatteras, North Carolina. *Source:* R. Dolan; photo courtesy of U.S. Geological Survey.

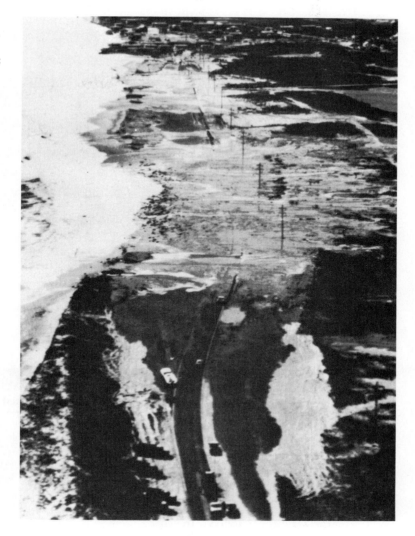

Perhaps the most dreaded coastal hazard is the tsunami. As the December 26, 2004 tsunami in the Indian Ocean demonstrated, coastal areas throughout the world are vulnerable to these events. These waves, sometimes incorrectly referred to as tidal waves, have nothing to do with astronomical tides. The actual cause of the waves is the rapid displacement of water by submarine earthquakes, volcanic eruptions, or submarine landslides. Because of their common association with earthquakes, they are called seismic sea waves or, from the Japanese term, *tsunamis*. Tsunamis move through the deep water of the ocean basins at velocities of several hundred meters per second. At these rates they can cross major oceans in a matter of hours. In deep water, tsunamis are barely detectable because their amplitudes reach only several meters. They are also characterized by wavelengths ranging from 150 to 250 km and periods of 10 to 60 min. When tsunamis encounter shallow water, however, they undergo the same transformations as wind waves. The wave height increases to the point where the run-up is able to inundate nearshore areas and cause great damage (Figure 15.44). The 10 most damaging tsunamis in history are listed in

▲ **FIGURE 15.44**
An Acehnese woman and her child look around the rubble of their house in the tsunami-hit city of Banda Aceh, Jan 8, 2005.

Table 15.3	Ten Most Destructive Tsunamis in History	
Year	**Location**	**Estimated Deaths**
2004	Indian Ocean	~300,000
1782	S. China Sea	40,000
1883	S. Java Sea	36,500
1707	Tokaido-Nankaido, Japan	30,000
1896	Sanriku, Japan	26,350
1868	N. Chile	25,674
1792	SW Kyushu Island, Japan	15,030
1771	Ryukyu Trench	13,486
1976	Moro Gulf, Phillipines	8000
1703	Tokaido-Kashima, Japan	5233

Table 15.3. In the Pacific Ocean, a large destructive tsunami has an average return period of about 10 years. Smaller tsunamis happen several times per year in the Pacific.

The great loss of life from the 2004 tsunami was unexpected. As indicated in Table 15.3, most tsunamis occur in the Pacific Ocean. The triggering event for the tsunami is officially known as the Sumatra-Andaman Islands earthquake, whose epicenter and aftershocks are shown in Figure 15.45. The magnitude of the event was 9.0, making it the fourth-largest earthquake since magnitude measurement began in 1899. The progression of the tsunami across the Indian Ocean is shown in Figure 15.45. Great loss of life occurred in Indonesia, Malaysia, Thailand, India, Sri Lanka, and on numerous small islands. Unlike the Pacific, which has a tsunami warning system, the countries bordering the Indian Ocean were totally unprepared for a tsunami. If such a warning system had been in place, many lives could have been saved, especially in the western Indian Ocean.

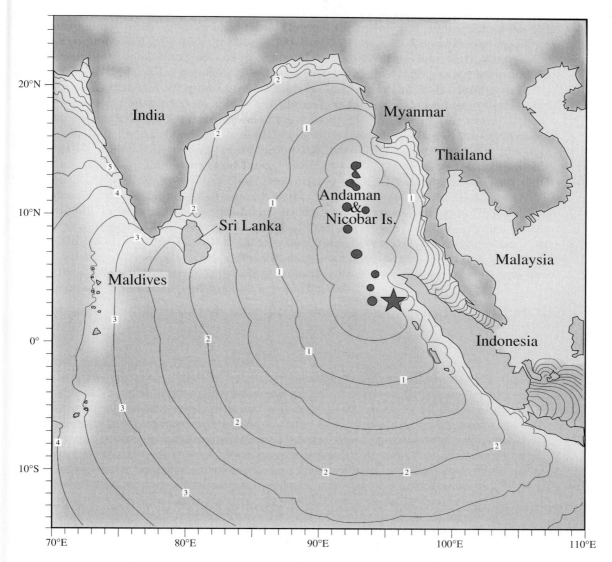

▲ FIGURE 15.45
Epicenter of the Sumatra-Andaman Islands earthquake of December 26, 2004 and the progression of the waves across the ocean. Time is in hours.

Case in Point 15.2
"The Storm(s) of the Century"

Early in the morning of September 22, 1989, Hurricane Hugo, a category 4 storm, made landfall near Charleston, South Carolina. The devastation it produced represented the most costly natural disaster in the history of the United States to that point. Three years later, a storm of similar magnitude, Hurricane Andrew, struck the coast of Florida (Figure 15.46) and caused even greater damage. Andrew and Hugo still rank as the first and second most

▲ FIGURE 15.46
Radar image of Hurricane Andrew approaching the coast of Florida. *Source:* Image courtesy of Harold Wanless.

costly hurricanes in U.S. history. To put Andrew's magnitude of losses into perspective, the four hurricanes that hit Florida in 2004 combined may equal or slightly exceed Andrew's damage.

Despite a similar degree of destruction, the effects of Hugo and Andrew were quite different. For an explanation for the differences in these storms, we must examine the storms themselves, and the geology of the coastlines. The winds of both storms exceeded 131 mph; gusts may have been significantly higher. Hugo's maximum storm surge was about 20 ft, while Andrew's surge was much less—in the range of 8 to 10 ft. Herein lie the major differences of the storms. Hugo caused major damage to beaches by erosion and redeposition of sediment. Structures that happened to be within the zone of impact of the storm surge were damaged or destroyed. Andrew, on the other hand, caused most of its damage to structures inland as well as on the beach by winds. The geologic effects of Andrew were therefore much less than Hugo's.

There are several differences in the storm tracks of the two hurricanes that controlled their effects. Andrew crossed the Bahama platform prior to hitting the Florida coast. Some of the wave energy was dissipated over this shallow carbonate shelf. In addition, the Florida coast itself is very flat at the point of the hurricane's landfall, a condition less conducive to the

▲ FIGURE 15.47
Storm track of Hurricane Hugo, September 1989. Line type indicates storm severity.
Dots = depression; dashes = tropical storm; solid line = hurricane; ++ = extratropical cyclone. *Source:*
From R. H. Willoughby and P. G. Nystrom, Jr., 1990, Some effects of Hurricane Hugo on shoreline landforms,
South Carolina coast, Columbia, September 21–22, 1989, *South Carolina Geology*, 33:39–58.

formation of large waves. Hugo's greater damage can be attributed to the longer fetch between Puerto Rico and South Carolina and the steeper South Carolina coastline.

Hugo's storm track is shown in Figure 15.47. Winds were distributed around the eye,
which came ashore at Charleston, as shown in Figure 15.40. As a result, the maximum
storm surge was north of the eye (Figure 15.48). Several barrier islands along the coast
were heavily impacted. Hugo's most significant effect was beach erosion. Beach profiles
were flattened and lowered by erosion of sand (Figure 15.49). The erosion extended landward to natural dunes, which were truncated and steepened. Any structures that happened to be seaward of the natural dunes were weakened by removal of sand from
around their pile foundations. Structures that were located landward of forested ridges
and higher in elevation sustained much less damage. Buildings near the shore were commonly protected by riprap consisting of large granite blocks. In some places, the riprap
afforded some protection, but in others, 4- to 6-ft blocks were moved shoreward, and the
buildings were destroyed. Damage was also sustained to roads and other structures
(Figure 15.50). Sand eroded from the beaches was deposited both offshore in offshore
bars and landward to form washover fans and storm ridges. Normal waves and currents
began to redistribute the sand as soon as the storm passed. A new inlet was cut in a narrow section of one barrier island as the storm surge, temporarily impounded landward of

◀ FIGURE 15.48
Approximate storm-surge contours (ft) for Hurricane Hugo. *Source:* From R. H. Willoughby and P. G. Nystrom, Jr., 1990, Some effects of Hurricane Hugo on shoreline landforms, South Carolina, Columbia, September 21–22, 1989, *South Carolina Geology*, 33:39–58.

▲ FIGURE 15.49
Flat storm beach produced by erosion of sand and damage to buildings at Surfside Beach, near Charleston, S.C. by Hurricane Hugo. *Source:* From R. H. Willoughby and P. G. Nystrom, Jr., 1990, Some effects of Hurricane Hugo on shoreline landforms, South Carolina, Columbia, September 21–22, 1989, *South Carolina Geology*, 33:39–58.

the island, escaped back to the sea. The inlet was subsequently filled by the U.S. Army Corps of Engineers.

The damage by Hugo was predictable. Until development is tailored to the natural configuration of the shoreline and prevented on beaches, hurricanes will continue to take a terrible toll of life and property.

▲ FIGURE 15.50
Destruction of road and damage to building foundations by Hurricane Hugo at northern Folly Island, S.C. *Source:* From R. W. Willoughby and P. G. Nystrom, Jr., 1990, Some effects of Hurricane Hugo on shoreline landforms, South Carolina, Columbia, September 21–22, 1989, *South Carolina Geology*, 33:39–58.

Coastal Engineering Structures

Coastal engineering works date back to biblical times. Recently, attempts have been made to control and manage coastal processes on a larger scale. Some of these projects have failed because of a lack of understanding of coastal processes. Coastal systems tend at all times to establish a condition of equilibrium between form and process. Whenever natural processes are interfered with, adjustments throughout the system are a result. These adjustments often take the form of undesirable erosion and deposition of sediment.

Coastal engineering structures are designed to enhance or maintain navigation, to prevent erosion, or to promote deposition. *Jetties* are walls built perpendicular to the shoreline in pairs at the mouths of inlets (Figure 15.51). Their purpose is to prevent blockage of the inlet by longshore drift. In doing so, they may prevent the movement of sand by longshore currents past the jetty. This will cause deposition of sand along the up-drift side of the jetty and erosion of sand from the down-drift side. Inlets serve as the location for the inflow and outflow of tidal currents. Therefore, the size and channel geometry imposed by jetties must be carefully designed to provide the proper velocity of tidal currents. If the channel opening is too large, deposition of sand and shoaling of the inlet channel occurs, and dredging is required to maintain the necessary channel depth. If the opening is too small, the high current velocity causes scouring of the channel and potential undermining and failure of the jetties.

▶ FIGURE 15.51
Types of coastal engineering structures.

▲ FIGURE 15.52
Patterns of erosion and deposition produced by the Santa Barbara breakwater. *Source:* After J. W. Johnson, 1975, The littoral drift problem at shoreline harbors, *Journal of Waterways and Harbors Division*, 83:1–37, American Society of Civil Engineers.

When adequate natural harbors are lacking, *breakwaters* can be constructed (Figure 15.51). Breakwaters provide protection from waves, but they also interact with the longshore drift in the same manner as jetties. A classic example of this interaction is the Santa Barbara, California, breakwater. As illustrated in Figure 15.52, enlargement of the beach on the up-drift side of the breakwater occurred until longshore currents were able to transport sand along the breakwater and deposit it directly into the harbor. Dredging was then required to maintain the harbor. At the same time that the deposition was occurring in the harbor, erosion of the beach was in progress on the down-drift side of the harbor. Eventually, shoreline dwellings were destroyed by wave erosion because the sand source for the beaches had been eliminated.

Detached breakwaters have also been constructed (Figure 15.53) in an attempt to provide protection from waves while at the same time allowing the passage of longshore drift. These structures usually fail to achieve their purpose because the protected area behind the breakwater becomes a low-energy zone for the longshore drift, and sand deposition ensues at that site.

Seawalls are structures built parallel to the shore to prevent erosion and slumping of sea cliffs or bluffs. Construction materials range from concrete to timber to riprap (Figure 15.54). Seawalls provide short-term protection against moderate events, but erosion continues on both sides of the protected zone.

Groins (Figure 15.55) are walls built perpendicular to the beach for the purpose of trapping sand from longshore drift. Although a beach can be considerably widened by the

◀ FIGURE 15.53
Construction of detached breakwaters usually results in sand deposition between the structure and the shoreline.

▶ FIGURE 15.54
Tiered seawall designed to prevent erosion along the shore of Lake Michigan. *Source:* Photo courtesy of the author.

▶ FIGURE 15.55
Aerial view of groin field along the Maryland coast. Arrows show individual groins in the foreground. *Source:* R. Dolan; photo courtesy of U.S. Geological Survey.

construction of a groin, down-drift erosion is also induced. This may necessitate a series of groins along a particular segment of the coastline.

One of the most ambitious coastal engineering projects in recent years is the Moses Project, a scheme to prevent the increasing amount of tidal flooding of Venice, Italy. The origin of this problem lies in the siting of the city on marshes in the Venice lagoon (Figure 15.56). The combination of subsidence of the city on these compressible sediments along with rising sea level has caused increasing impact of high tides upon the city. Settlement has amounted to 23 cm over the last 100 years and 50 high tides occurred in the period 1993–2002 compared to just 5 between 1923 and 1932. An engineering solution to this problem was

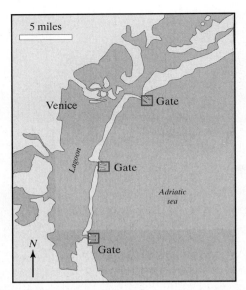

◄ FIGURE 15.56
Conceptual design of Project Moses, intended to close the Venice lagoon to high tides that cause damage to the city.

conceived after a particularly severe tide in 1966. A decision was made that the buildings, monuments, and canals, perhaps Italy's greatest tourist attraction, must be saved from further damage. After more than three decades of debate and planning, construction began in 2003 on a series of gates at the mouths of tidal inlets through the barrier island enclosing the lagoon (Figure 15.56). The $7 billion project will include 78 water-filled steel gates that lie on the sea bottom under normal tidal conditions. Individually, the gates are from 5 m high to 28 m high. When high tides are predicted, compressed air forces the water out of the gates and they rise to the surface to form a barrier to the lagoon. Strong opposition to the project has arisen from environmentalists and the shipping industry, which is concerned that ships will be prevented from entering the harbor. Like the Old River Control Structure on the Mississippi, the long-term success of this challenge to geological processes is questionable.

Implications for Coastal Development

With an understanding of the natural processes occurring along a coastline, development that minimizes the potential for damage from hazardous processes can be implemented. Individual processes and the area that they may affect must be identified. When this is accomplished, zoning regulations can be utilized to prevent development of particularly hazardous areas. In some cases, development of marginal areas may be allowed if specific protective design criteria are met. In assessment of risks, however, the possibility of extreme events must not be neglected. The magnitude and frequency of these occurrences is a major uncertainty in any risk analysis.

In evaluation of coastline processes, historical records are of invaluable assistance. Maps and charts show changes in shoreline configuration during the period of record. Air photos are particularly useful in this respect.

The lessons of past coastal management projects must not be lost when considering future plans. In the following Case in Point some of the problems of large-scale coastal management schemes are illustrated.

Case in Point 15.3
Human Interference with Coastal Processes

The coast of North Carolina is a depositional coast with a nearly continual system of barrier islands (Figure 15.28). An illustrative comparison of altered and unaltered sections of these islands has been provided by Dolan, Godfrey, and Odum (1973).

A typical profile of a barrier island in its natural state is shown in Figure 15.57a. This profile, consisting of a broad berm and low, irregular dunes, is adjusted to the impact of large storms. Wave energy is efficiently dissipated by frequent overwash over the low natural dunes. Salt-tolerant vegetation species well adapted to periodic overwash inhabit the dunes and overwash terraces behind the dunes.

Although the barrier islands are well adjusted to extreme events that characterize their environment, frequent overwash is not compatible with human occupation of the islands. To rectify this situation, nearly 1000 km of sand fence was constructed behind the beach

▲ FIGURE 15.57
Cross-sectional profiles of barrier islands in their (a) natural state and (b) after construction of barrier dunes. *Source:* From R. Dolan and H. Lins, 1985, *The Outer Banks of North Carolina*, U.S. Geological Survey Professional Paper 1177-B.

berm in the 1930s. The purpose of the fences was to trap sand, thereby creating a system of high, artificial barrier dunes to protect the islands from overwash. Non-salt-tolerant vegetation was introduced behind the dunes and fertilized in order to stabilize the dunes.

These measures accomplished the short-term objective of making the islands safe for development. Unfortunately, the disturbance of the natural equilibrium state initiated serious long-term changes in the beach profile. Figure 15.57b shows the altered profile 30 years after construction of the artificial dunes. The major change has been the destruction

◀ FIGURE 16.13
Map of glacial Lake Agassiz at its maximum extent. The dark areas are modern remnants of the lake. *Source:* From J. P. Bluemle, 1991, *The Face of North Dakota,* North Dakota Geological Survey Educational Series 21.

and other deformational structures produced during the melting of the glacier after deposition of the sediment. Isolated bodies, lenses, or beds of ice-contact stratified drift may be buried or interbedded with till in a glacial sequence. Location of these deposits is relevant to excavation, foundation design, and the successful siting of water wells.

Once the meltwater from the glacier reaches the snout, it may either be ponded in *proglacial lakes* or flow along or away from the ice in meltwater streams. The sediment deposited in these environments is known as *glaciolacustrine* sediment and *outwash*, respectively. When continental ice sheets retreated, large volumes of water were trapped in proglacial lakes. The largest of these in North America was Glacial Lake Agassiz (Figure 16.13). Glaciolacustrine sediment is mostly fine-grained, bedded silt and clay. Low strength and high compressibility are characteristic of these sediments. Glaciolacustrine sediment is sometimes interpreted to be *varved*. Varves are thin, alternating laminations of clay and silt or sand (Figure 16.14). The dark-colored clay beds are thought to be deposited during the winter, when the lakes are ice covered; and the coarser, light-colored silt or sand beds are deposited

◀ FIGURE 16.14
Varved glacial lake deposits (light- and dark-colored banded unit), Sanpoil Valley, Washington. The rock at center (arrow) was dropped from an iceberg floating on the lake. *Source:* Photo courtesy of the author.

▶ FIGURE 16.15
Cross-bedded, coarse-grained outwash exposed in a gravel pit, north-central North Dakota. *Source:* Photo courtesy of the author.

during the summer, when glacial meltwater carries more sediment into the lake. Each varve pair, or couplet, therefore implies one year's sedimentation in the lake.

Glacial outwash (Figure 16.15) is alluvial sediment deposited by braided meltwater streams. Outwash is coarse-grained and may be quite thick in the valleys that carried meltwater. Outwash is a good source of aggregate material for road construction, concrete, and other purposes. Outwash deposits are also highly permeable and may contain excellent aquifers.

Eolian sediment and *glaciomarine* sediment are less common but may be extremely important in some areas. Eolian sediment is derived from outwash and other types of glacial drift and then transported and deposited by wind. It is very well sorted and varies in grain size from sand to silt. Deposits of eolian silt are called *loess*. Loess has the unusual property (for nonlithified sediment) of forming stable vertical cuts. The stability is provided by bonds between clay particles composed of calcium carbonate, clay, or moisture. If these soils become saturated or subjected to excessive disturbance, the previously stable slopes may fail or compaction may result (Chapter 10). Glacial-marine sediment is deposited where glaciers terminate at the ocean. Large blocks of ice calve off from the glacier and float away. As they melt, debris sinks to the seafloor and is deposited as glacial-marine sediment. This material is clay-rich and stratified, but it also contains large particles that were incorporated into the ice.

Types of Glaciers

The major types of glaciers are shown in Table 16.2. *Ice sheets* and *ice caps* are large expanses of ice that have a dome-shaped profile (Figure 16.16) with a much greater thickness near the center than at the edge. Ice builds up to such great thicknesses near the center of the dome that it completely buries the topography, including whole mountain ranges. Expansion and contraction of the dome is therefore not controlled by topography, but instead by the mass

Table 16.2   Types of Glaciers
Ice sheets and ice caps
Ice fields
Valley glaciers and outlet glaciers
Cirque glaciers
Ice shelves

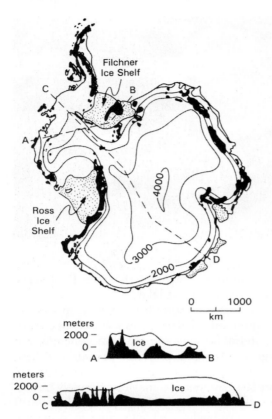

◄ FIGURE 16.16
Map and profile of the Antarctic ice sheet.

balance of the ice sheet. The difference between an ice sheet and an ice cap is simply based on size. Ice sheets must be greater than 50,000 km² in size. Only two ice sheets, the Antarctic and Greenland Ice Sheets, remain on the earth at the present time. During the Pleistocene Epoch, other ice sheets included the Laurentide, which was centered on Hudson Bay and covered most of Canada and the northern United States east of the Rocky Mountains.

Ice fields are also very large, wide glaciers, but lack the dome-shaped profile of ice sheets and ice caps. Ice fields may have a gently undulating surface profile, and mountain peaks may project above the glacier (Figure 16.17).

▲ FIGURE 16.17
View from top of Wapta ice field, Alberta. *Source:* Photo courtesy of the author.

▶ FIGURE 16.18
Network of valley glaciers near Mount McKinley (Denali), Alaska. *Source:* Photo courtesy of Glenn Oliver.

Valley glaciers and *outlet glaciers* are long, narrow glaciers whose shape is controlled by the valley or trough in which they flow (Figure 16.18). Valley glaciers, also known as mountain glaciers and alpine glaciers, flow downslope in narrow valleys from high mountain peaks. Outlet glaciers are similar in shape, but they begin at the edges of ice fields, ice sheets, or ice caps. The valley that contains an outlet glacier may extend back beneath the ice cap or ice sheet as a subglacial trough, producing a narrow zone of more rapid flow in the ice sheet called an *ice stream.*

Common only in Antarctica, *ice shelves* are thin slabs of ice that extend beyond the coastline in bays and float over the denser saltwater, while maintaining a connection with glaciers on shore (Figure 16.16).

Valley Glaciers

Along with vanished ice sheets from past glaciations, modern valley glaciers are much more limited in size and extent than they were during the Pleistocene Epoch. Today, extensive valley glaciers develop only in high latitudes in areas of substantial precipitation. These glaciers form networks analogous to river systems, with tributary glaciers joining larger glaciers that occupy major valleys.

A region that has contained valley glaciers can be recognized by characteristic landforms of both erosional and depositional origin. The most distinctive erosional landform is the shape of the glaciated valley itself. Glacial erosion modifies the cross-sectional profile of a valley from the V-shape common to valleys produced by stream erosion to a U-shape, characteristic of glaciated valleys (Figure 16.19). The U-shaped profile of glaciated valleys results in valley sides that are steeper than their counterparts in V-shaped stream valleys. Glaciated valleys are said to be oversteepened because the steep valley sides are prone to rock fall, rock slide, and other forms of mass wasting. These conditions may be hazardous for highway construction or other development in glaciated valleys.

In a glaciated mountainous area, valley glaciers extend, or once extended, downslope in all directions from the central region of highest elevation. In addition to U-shaped valleys, other erosional landforms are characteristic of these areas. The bowl-shaped depression at the base of a mountain peak where glaciers form and begin to flow downvalley is called a *cirque* (Figure 16.20). If a glacier is confined by temperature or lack of precipitation to the cirque, it is known as a *cirque glacier* (Table 16.2). Due to erosion by the cirque glacier and the action of physical weathering processes such as freeze and thaw, cirques are gradually

◀ FIGURE 16.19
The U-shape of this valley in Yosemite National Park indicates that it was once occupied by a valley glacier. *Source:* Photo courtesy of Dexter Perkins III.

enlarged by headward erosion and deepened. Small lakes, or *tarns*, are common in the scoured floors of cirques that formerly contained glacial ice. The narrow ridge that separates two glaciated valleys often is quite steep and sharp. It is called an *arête*, a French word meaning "sharp edge." When a peak contains cirques on several sides, headward erosion in the cirques begins to modify the topography of the peak itself. A gap produced in a ridge by the headward erosion of glaciers in two cirques from opposite sides of the ridge is a *col*. Where more than two glaciers erode headward at the base of a mountain peak, an isolated, sharply pointed pinnacle known as a *horn* is formed (Figure 16.20). The Matterhorn in the Alps of Switzerland is a horn.

The rock and sediment eroded from the cirque and valley walls by plucking and abrasion is transported downvalley toward the glacier terminus and is deposited as drift in several types of landforms. A ridge of drift deposited by the glacier at its point of maximum advance is called a *terminal moraine* (Figure 16.21). Terminal moraines are usually composed of till and ice-contact stratified drift and may be deposited by subglacial or ablation processes. Frequently, meltwater emerges from the terminus in tunnels or from the glacier surface and deposits outwash in front of the glacier. These deposits are formed by braided fluvial systems because of the extremely high sediment load, and they may occupy the entire width of the valley bottom. In this case, the outwash deposit is known as a *valley train*. Valley-train sediments are composed of coarse sand and gravel and decrease rapidly in grain size with distance from the glacier (Figure 14.29). As a valley glacier retreats, the terminal moraine may impound a proglacial lake (Figure 16.22). In addition, during a period of net retreat, the glacier terminus may stabilize or even readvance short distances. The result may be a series of concentric *recessional* moraines in a glaciated valley.

A narrow ridge of drift is also deposited along the lateral edge of valley glaciers and of outlet glaciers against the valley wall. The elevation of these *lateral moraines* is an indication of the maximum elevation of a particular glacier in the valley (Figure 16.23). In a series of active, coalescing valley glaciers, the lateral moraines from glaciers in tributary valleys can be traced as longitudinal debris bands in the glacier in the main valley. These *medial moraines* (Figure 16.24) may or may not be preserved as ridges extending down the center of the valley after glacial retreat.

Ice Sheets

The distribution of deposits of Pleistocene glaciers in the Northern Hemisphere (Figure 16.1) indicates that huge ice sheets, also known as continental glaciers, were not limited to high

(a)

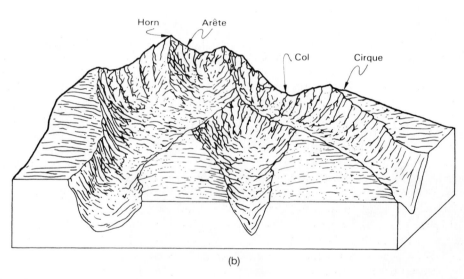

(b)

▲ FIGURE 16.20

(a) Cirques and arêtes in the Sierra Nevada Mountains, California. *Source:* F. E. Matthes; photo courtesy of U.S. Geological Survey. (b) Diagram of valley-glacier landforms. *Source:* From W. M. Davis, 1911, The Colorado Front Range, reproduced by permission from the *Annals of the Association of American Geographers*, 1: 57.

latitudes. These immense glaciers flowed southward across the Canadian border into the United States (Figure 16.25). The largest ice sheet, the Laurentide, penetrated southward to the approximate positions of the Ohio and Missouri Rivers. It was as much as 4 km thick at its center in the Hudson Bay region.

The processes of erosion and deposition associated with continental ice sheets are similar to those of valley glaciers, although they vary tremendously throughout the extent of the glaciated area. Landforms and deposits present in any particular area are the result of the characteristics of the topography, bedrock material, and the glacier-bed thermal regime. For example, broad areas of the Canadian Shield, underlain by hard Precambrian igneous and metamorphic rocks, were subjected to relatively shallow erosion. Other

▲ FIGURE 16.21
A terminal moraine (arrow) near Ennis, Montana, deposited by a valley glacier emerging from the U-shaped valley in the background. Faint channel markings can be seen on the gently sloping outwash plain in front of the moraine. *Source:* Photo courtesy of the author.

◀ FIGURE 16.22
A small proglacial lake in southern Iceland impounded between a retreating outlet glacier and its terminal moraine. *Source:* Photo courtesy of the author.

regions, such as the basins of the Great Lakes, were deeply scoured into the bedrock, and deposits of glacial drift tens of meters to more than 100 meters thick blanket some parts of the plains of the United States and Canada.

Deposits of both lodgement till and ablation till are common. Thick accumulations of ablation till are common in terminal moraines and in other areas of decelerating flow, where large amounts of sediment were sheared upward into the ice and concentrated at the glacier surface. The topography of ablation-till deposits is called *hummocky* (Figure 16.26) because it consists of alternating mounds (hummocks) and depressions, filled with lakes

▲ FIGURE 16.23
Lateral moraine (arrow) at edge of Athabasca glacier, Alberta. *Source:* Photo courtesy of the author.

▲ FIGURE 16.24
Medial moraines (M) in Ruth Glacier, Alaska. *Source:* Photo courtesy of Glenn Oliver.

or swamps. Local relief can be as much as tens of meters in areas of thick ablation till. These deposits represent thick covers of sediment emplaced on top of stagnant glacial ice that slowly melted from beneath.

Glacial meltwater deposits are concentrated in *outwash plains,* where braided fluvial systems transport sediment away from ice margins. Broad plains of outwash sediment

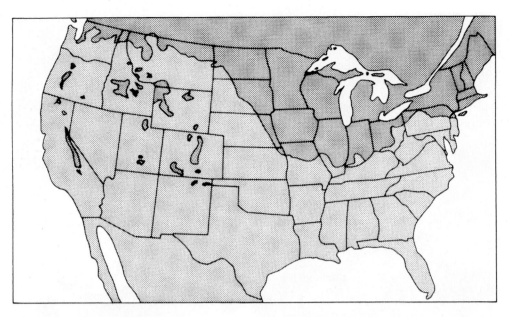

▲ FIGURE 16.25
Extent of Pleistocene glaciation in the United States (darker area). *Source:* From R. F. Flint, 1945, *Glacial Map of North America*, Geological Society of America Special Paper 60.

◄ FIGURE 16.26
High-relief hummocky to-pography in southern Saskatchewan resulting from deposition of thick ablation till. *Source:* Photo courtesy of the author.

often contain abundant resources of shallow groundwater. Unfortunately, these uncon-fined aquifers are highly susceptible to contamination.

Several types of continental glacial landforms deserve special mention. *Drumlins* are streamlined hills produced by wet-based, sliding glaciers (Figure 16.27). These hills, which occur in groups or fields (Figure 16.28), are the subject of more research and debate than perhaps any other glacial landform. Characteristics include an asymmetrical "stream-lined" shape elongated in the direction of glacier flow and a composition of till sometimes deposited over a core of bedrock. The controversy surrounding the origin of drumlins may

► FIGURE 16.27

A drumlin near Morley, Alberta. Glacial flow from right to left. *Source: Photo by Lionel Jackson. Reproduced with the permission of the Minister of Public Works and Government Services Canada, 2005 and courtesy of Natural Resources Canada, Geological Survey of Canada.*

► FIGURE 16.28

Map of drumlins in northwestern Michigan. Ice moved toward the south-south-east. *Source: From F. Leverett and F. B. Taylor, 1915, The Pleistocene of Indiana and Michigan and the History of the Great Lakes, U.S. Geological Survey Monograph 53.*

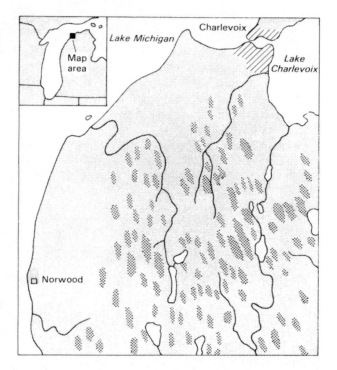

have arisen because there probably are multiple origins for these landforms. Drumlins may be formed by erosion, deposition, or a combination of both.

The origin of several glacial landforms is related to deposition of meltwater flowing within or beneath the ice. An *esker* (Figure 16.29) is a sinuous ridge composed of poorly sorted sand and gravel. Eskers are formed by meltwater streams flowing in tunnels within or beneath a glacier. Because the stream channel is confined by ice walls and roof, the deposits remain as a raised mound after the ice melts. The sinuous, meandering shape of eskers is similar to subaerial stream channels. *Kames* are isolated, commonly conical hills of sand and gravel (Figure 16.30). One way in which kames are formed involves the flow of meltwater into a cavity in the ice similar to a sinkhole in karst topography. Gradually, the hole is filled by the stream deposits.

◀ FIGURE 16.29
The Dahlen esker (arrows) located in eastern North Dakota. *Source:* Photo courtesy of J. R. Reid.

▲ FIGURE 16.30
A kame located in Erie County, New York. *Source:* Photo courtesy of the author.

Case in Point 16.1
Subsurface Complexity in Glacial Terrain

Complex glacial stratigraphy beneath deceptively simple glacial landforms can cause frustrating and expensive foundation problems. Boston's historic Beacon Hill was traditionally interpreted as a drumlin. Under this assumption, soils composed mainly of till with favorable conditions for foundations and excavations were predicted. The site investigations for several large modern buildings, however, revealed a much more complicated subsurface stratigraphy (Kaye, 1976). One of these structures was a garage constructed near the south end of the cross section shown in Figure 16.31. Till was expected to be present for the entire

▲ FIGURE 16.31
Cross section of Beacon Hill showing the complex structure and stratigraphy. Ice advanced from the north and transported frozen slabs of bed sediment, which were emplaced by ice thrusting at the glacier terminus near Beacon Hill. *Source:* From C. A. Kaye, 1976, Beacon hill end moraine, Boston: A new explanation of an important urban feature, in *Urban Geomorphology*, D. A. Coates, ed., Geological Society of America Special Paper 174, pp. 1–20.

13-m depth of the foundation excavation. Instead, subsurface folds containing gravel were encountered. Groundwater inflow from these gravels into the excavation required the installation of an expensive dewatering system and caused months of construction delay. The construction of a large office building on the north side of the hill was also affected by inadequate subsurface information. The building was designed to rest upon piles driven into dense till beneath a surficial bed of clay. When test piles were driven, the surficial clay was found to extend to a much greater depth than anticipated because of folding and faulting of the section. The piles were driven 15 m beyond the predicted depth of the till unit with very little resistance to penetration.

Data gathered from these and other projects indicated that Beacon Hill is not a drumlin composed mostly of till. The hill is stratigraphically and structurally complex. Beneath a surficial layer of till, north-dipping slabs of sand, gravel, clay, and till are stacked against each other like books standing tilted to one side on a bookshelf. The depositional units are separated by thrust faults, and beds are intensely folded in some parts of the hill.

A reasonable explanation for the origin of Beacon Hill is that it consists of a series of individual thrust sheets emplaced by a glacier frozen to its bed near its terminus. Under this hypothesis, slices of frozen bed sediment were sheared into the ice and transported as intact slabs to the ice margin. There, the blocks were thrust upward against other slabs of subglacial bed material. Therefore, the landform is a type of terminal moraine rather than a drumlin. Construction experience in this area exemplifies the necessity for detailed subsurface site investigations in glaciated terrain.

Engineering in Glaciated Regions

Glaciated regions present several problems relevant to construction and other types of land use. In general, these include problems associated with differences in engineering properties among various types of deposits, lateral and vertical variability in subsurface stratigraphy, and variations in depth to bedrock. Of major importance are differences in engineering properties among till, glacial-lacustrine, glacial-marine, and outwash deposits. Glacial-lacustrine deposits are clayey, compressible materials with low shear strength. The role of these materials in the collapse of the Transcona, Manitoba, grain elevator was described in

Chapter 10. The low bearing capacity of glacial-lake sediments may require piles or other deep foundation types for heavy structures. Piles are usually driven until they encounter a dense till or bedrock. Other engineering problems associated with glacial-lacustrine deposits include settlement, shallow water tables, shrink-swell behavior, and instability of natural or constructed slopes. Glacial-marine deposits are similar in texture to glacial-lacustrine sediments. The special slope-stability problems associated with these soils were discussed in Chapters 10 and 13. The Nicolet, Quebec slope failure (Figure 13.16) is an excellent example of the construction hazards that must be considered.

The geotechnical properties and behavior of till are dependent upon its mode of deposition. Lodgement tills become dense and overconsolidated under the load of the overlying glacier. As a result, they tend to have favorable strength and compressibility in comparison with ablation tills. On the negative side, lodgement tills may be so dense that excavation may require ripping or blasting. An important characteristic of lodgement, and to some extent ablation tills, is that they are commonly observed to be jointed or fissured (Figure 16.32). In the case of lodgement tills, fissures may have been caused by unloading and rebound during and after the retreat of the glacier. Shrinkage associated with drying after deposition is another possible cause of fissure formation. Fissures have several important effects upon the engineering properties of the soil mass. First, fissures weaken the material. This can be shown by comparing the results of strength tests on samples of two sizes from the same construction site (Figure 16.33). The strength values measured from the larger samples are lower because of the greater number of fissures included in the sample. Fissures also influence the permeability of glacial deposits. Laboratory permeability tests conducted on small intact samples consistently yield lower permeability values than field tests because of the absence of fissures in the small lab samples. These discontinuities also provide preferred paths of movement for water through the deposit. A consequence of this condition is that tills may not be so favorable for waste-disposal sites as they may appear based on lab permeability tests on small samples. The presence of isolated large boulders can sometimes prove troublesome in construction projects involving till. Difficulties associated with excavation, test drilling, and pile driving have been experienced. Boulders may be concentrated in *boulder pavements* between till units in sequences containing multiple tills.

◄ **FIGURE 16.32**
Exposure of a joint in till, southwestern North Dakota. The surfaces of the joint are covered by manganese oxide. *Source:* Photo courtesy of the author.

▶ FIGURE 16.33
Effect of fissures on the strength of till. Samples of two different sizes from identical materials were tested. Points plot below the line of equality because the smaller samples are stronger than the larger samples, which have more fissures. *Source:* From W. F. Anderson, Foundation engineering in glaciated terrain, in *Glacial Geology: An introduction for Scientists and Engineers*, N. Eyles, ed., © 1983 by Pergamon Press, Ltd., Oxford.

Outwash deposits generally are composed of cohesionless materials of variable sorting and high permeability. They frequently constitute excellent unconfined aquifers and favorable soils for irrigation if they occur at the surface. Outwash deposits are of great value as construction materials unless they contain excessive amounts of poor-quality material, such as shale. Therefore, states and provinces expend considerable effort in exploration for and evaluation of these deposits.

Sand and gravel deposits have moderate to high strength and low compressibility. Therefore, they provide high bearing capacities for foundations. Problems arise with high water inflow in excavations below the water table. Dewatering techniques (Chapter 11) may then be required. Glaciofluvial materials make very poor waste-disposal settings because of their high permeability and their potential utilization as sources of groundwater supplies.

Case in Point 16.2
The Big Dig: Urban Engineering in Glaciated Terrain on a Massive Scale

The Central Artery/Tunnel (CA/T) project, also known as the Big Dig, is one of the largest and most complex highway construction projects ever undertaken in the United States. The main purpose of this project was to replace the Central Artery, an elevated six-lane highway that passed through the heart of Boston as part of Interstate 93. When it opened in 1959, the Central Artery carried about 75,000 vehicles per day. By the 1990s, traffic had increased to about 200,000 vehicles per day, making it one of the most congested highways in the country.

The challenges in replacing this highway were daunting—traffic along the existing route had to be maintained during construction, which lasted from 1991 to 2005. The design selected consisted of construction of an underground eight- to ten-lane expressway directly below the existing elevated Central Artery (Figure 16.34) In addition, the project includes a new section of the Massachusetts Turnpike (I-90), which formerly ended at the I-93/I-90 interchange. The new section of I-90 crosses beneath the Fort Point Channel, rises back to the surface, and then passes through the Ted Williams Tunnel under the Boston

◀ FIGURE 16.34
Map of the Central Artery/Tunnel project in downtown Boston. *Source:* Massachusetts Turnpike Authority.

Inner Harbor to Logan Airport. Construction on the project began in 1991 and was 95% complete at the end of 2004. The overall cost was originally estimated at $4 billion, but rose to more than $14 billion as construction neared completion, making it the most ambitious and costly urban highway project in the nation's history.

The unprecedented geotechnical challenges of the Big Dig included dealing with the highly variable, complex glacial geology of the city. Bedrock consists of variably weathered metamorphic rocks. The oldest glacial deposit in the area is a till about 70,000 to 80,000 years old. The till is quite dense and stiff. The ice then retreated from the area and readvanced to its Late Glacial Maximum position about 20,000 years before present. At its greatest extent, ice was several hundred meters thick on the site of Boston. The weight of the glacier isostatically depressed the crust, allowing the ocean to transgress across the city site as the ice retreated. Marine clays were deposited during the time when the land was below sea level. These are generally weak materials that cannot carry heavy foundation loads. Although sea level rose as the great ice sheets melted, the isostatically depressed land rose faster as the weight of the ice was removed. Eventually, the area emerged above sea level. Freshwater deposits, sometimes with high organic content, overlie the marine clays. The uppermost stratum is a layer of fill ranging in age from the past few decades to before the Revolutionary War. A representative cross section is shown in Figure 16.35.

Boston, like other cities, has a maze of underground utilities, consisting of different ages and different materials. All of the utilities buried beneath the Central Artery, some 46 km of gas, electric, telephone, sewer, water, and other lines, had to be relocated prior to construction of the tunnel. Other procedures that were necessary in a project of this type included archeological surveys, which uncovered several significant sites. The fundamental

▲ FIGURE 16.35
Geological conditions along a section of the Central Artery Tunnel (heavy dark line). *Source:* From T. L. Neff, K. K. See-Tho, S. S. C. Liao, and L. Bedingdield, 1989, Integration of design and construction for Boston's Central Artery/Third Harbor Tunnel Project, in *Proceedings of the International Congress on Progress and Innovation in Tunnelling*, vol. 2, K. Y. Lo, ed., pp. 719–726, Toronto Canada, 1989.

challenge of the Central Artery project was the necessity to maintain the existing elevated highway, while simultaneously digging a tunnel beneath it. The unique *underpinning* process that was designed to solve this problem is shown conceptually in Figure 16.36. The foundation of the existing highway was based on footings supported by piles, all of which were within the footprint of the tunnel. The first step in construction was to install concrete *slurry walls*, which would form the basis of the support system as well as the permanent wall of the tunnel. They also form a permeability barrier and prevent water inflow into the tunnel and excavations. Slurry walls are built by pumping concrete into a trench about a meter wide, as much as 36 m deep in this case, and then lowering steel soldier piles into the slurry to provide support. Cross beams are added between the walls (Figure 16.36b) and the dead load is carefully transferred to the beams by a jacking system. Then the old piles can be cut and the excavation of the tunnel can proceed. The northbound and southbound tunnels were built separately and when finished, the old elevated artery was removed (Figure 16.36d).

Problems were encountered at specific locations on the project. For example, in one section of the artery, the tunnels had to pass beneath two separate overlying tunnels (Figure 16.37). The Red Line Tunnel is a subway tunnel that was built in 1915. Settlement of this tunnel could not be allowed as this would damage the subway line. Excavation down to the level of the tunnel, as used in most sections of the artery, would not work in this location. Another complicating factor at this location was the presence of large buildings, all of different ages and different foundation types that could potentially experience

▲ FIGURE 16.36
(a) Existing foundations of the elevated artery. (b) Slurry walls are installed, along with transverse grade beams and surface decking. The load is transferred from the old to the new foundation system. (c) Tunnels for the new artery are excavated and constructed. (d) The old artery is removed after the underground highway is in service. *Source:* From T. L. Neff, K. K. See-Tho, S. S. C. Liao, and L. Bedingdield, 1989, Integration of design and construction for Boston's Central Artery/Third Harbor Tunnel Project, in *Proceedings of the International Congress on Progress and Innovation in Tunnelling*, vol. 2, K. Y. Lo, ed., pp. 719–726, Toronto Canada, 1989.

damaging settlement as the tunnel was constructed below. An ingenious underpinning system was constructed in this area (Figure 16.38), where the geology consists of several tills overlying bedrock. Two vertical access shafts, similar to huge elevator shafts, were excavated to a depth of 6 m below the bottom of the subway tunnel using the slurry wall construction described earlier. Two horizontal access tunnels, called grouting galleries, were drilled from the vertical access shafts. From the grouting galleries, hundreds of small shafts about 5 cm in diameter were drilled to bedrock 15 m below and grout was pumped in under pressure to harden the soils. These shafts were arranged in a fanlike pattern to stabilize a large area. Then three horizontal drifts (Figure 16.38) were drilled one on top of another and filled with concrete. Finally, roof girders spanning the access galleries and directly beneath the subway tunnel were constructed by drilling horizontal drifts and installing the roof girders. The last step was filling the grouting galleries with concrete. The end result (Figure 16.39) was two massive concrete walls 30 m long and 15 m high supporting the roof girders that in turn support the subway tunnel. To address the concern of the adjacent building owners, extensive monitoring systems were installed that detect even minute horizontal and vertical movements.

▲ FIGURE 16.37
Conceptual cross section along the northbound Central Artery tunnel showing the location of the Red Line Subway Tunnel and the Transitway Tunnel above. *Source:* From B. B. Goyal, P. K. Das, C. K. Shah, C. W. Wood, and B. R. Brenner, Underpinning of Red Line South Station, in *Engineering Geology of the Big Dig Project*, D. Bobrow and C. Daugherty, leaders, in *Guidebook for Geological Field Trips in New England*, D. P. West and R. H. Bailey, eds., 2001 Annual Meeting of the Geological Society of America.

▶ FIGURE 16.38
Underpinning system for the Central Artery tunnel under the Red Line Subway Tunnel. *Source:* From B. B. Goyal, P. K. Das, C. K. Shah, C. W. Wood, and B. R. Brenner, Underpinning of Red Line South Station, in *Engineering Geology of the Big Dig Project*, D. Bobrow and C. Daugherty, leaders, in *Guidebook for Geological Field Trips in New England*, D. P. West and R. H. Bailey, eds., 2001 Annual Meeting of the Geological Society of America.

The tunnels running under Boston Harbor created design challenges that were met by similarly innovative techniques. The Big Dig is a truly monumental highway construction project. A number of "firsts" achieved by the project include the most extensive geotechnical investigation and testing program in North America and the largest use of slurry wall construction in one location in North America.

▲ FIGURE 16.39
Cross section showing completed underpinning system for the Red Line Subway system and the tunnel below for the Central Artery. *Source:* From E. R. Drooff, D. J. Dobbels, and J. R. Wheeler, 1999, Role and performance of chemical grout in the underpinning of an active subway station, in *Proceedings of the Rapid Excavation and Tunnelling Conference*, D. E. Hilton and K. Samuelson, eds., Society for Mining Metallurgy and Exploration www.smenet.org.

Permafrost and Related Conditions

The development of petroleum resources in arctic regions and the transport of oil southward through pipelines have required new concepts of construction in cold regions. The likelihood of continued resource development in these regions requires a better understanding of this unique environment.

Construction problems in arctic regions result from the presence of permanently frozen ground, or *permafrost*. The area of permafrost on the earth is extensive. Approximately 20% of the earth's land surface is underlain by permafrost, including 50% of Canada and the former U.S.S.R. and 85% of Alaska.

Extent and Subsurface Conditions

Permafrost areas can be classified as either continuous or discontinuous (Figure 16.40). Discontinuous permafrost can be identified by adjacent tracts of frozen and unfrozen ground. Continuous permafrost extends to great depths at high latitudes and gradually thins to the south (Figure 16.41).

▲ FIGURE 16.40

Extent of permafrost in the Northern Hemisphere. The light gray area represents continuous permafrost; dark gray indicates discontinuous permafrost. *Source:* From O. J. Ferrians, R. Kachadoorian, and G. W. Greene, 1969, *Permafrost and Related Engineering Problems in Alaska*, U.S. Geological Survey Professional Paper 678.

Immediately below the land surface is the *suprapermafrost zone*, which is several meters thick and may contain pockets of material called *taliks* that never freeze, as well as ground that thaws during the summer and refreezes in the winter (Figure 16.42). This upper part of the suprapermafrost layer is called the *active zone*. At the base of the suprapermafrost layer is the *permafrost table*. This is the upper boundary of the zone that remains frozen year-round, although taliks also may be present below the permafrost table.

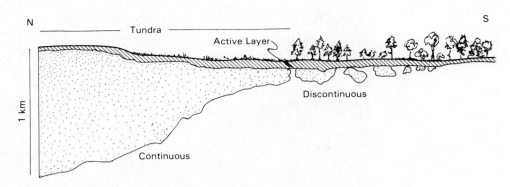

▲ FIGURE 16.41
Distribution of permafrost along a northwest transect in northwest Canada. Numbers above the diagram represent degrees of latitude. *Source:* From N. Eyles and M. A. Paul, Landforms and sediments resulting from former periglacial climates, in *Glacial Geology: An Introduction for Engineers and Scientists*, N. Eyles, ed., © 1983 by Pergamon Press, Ltd., Oxford.

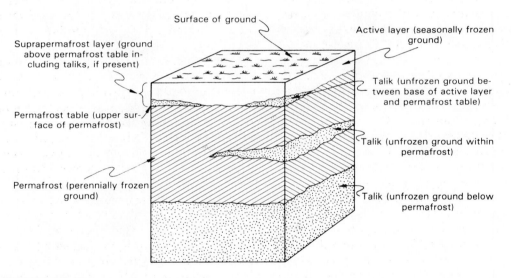

▲ FIGURE 16.42
Subdivision of permafrost below land surface. *Source:* From O. J. Ferrians R. Kachadoorian, and G. W. Greene, 1969, *Permafrost and Related Engineering Problems in Alaska*, U.S. Geological Survey Professional Paper 678.

Freeze and Thaw

Processes that produce distinctive landforms in climates subject to ground freezing are called *periglacial* processes. These processes, which are dominated by freeze and thaw activity, occur in any cold climate whether or not a glacier is nearby. Therefore, areas of high elevation in the midlatitudes experience periglacial activity. During the Pleistocene, periglacial climates extended southward as far and even beyond the advance of the ice sheets. Periglacial, like glacial, phenomena can reach equatorial regions if the elevation is

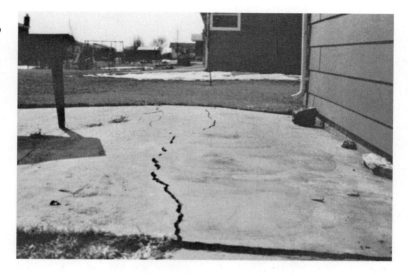

► FIGURE 16.43
Cracking of a cement slab by frost heave, Grand Forks, North Dakota. Notice the bulging of the slab in the vicinity of the crack. *Source:* Photo courtesy of the author.

sufficiently high. Although the effects of periglacial phenomena during the Pleistocene are sometimes relevant to engineering projects in the midlatitudes, we will focus the bulk of our discussion on the regions of current permafrost. In addition to the direct effects of freeze and thaw, special types of mass movement occur under periglacial conditions. They are discussed in the following section.

The effects of ground freezing are extremely variable. Lithology of the soil is an important factor. Under appropriate conditions, isolated *ice lenses* form in the soil and begin to grow by drawing moisture by capillary movement toward the expanding ice mass. Soils of predominantly silt-size particle composition are most susceptible to ice-lens formation. Clay-rich soils retard the movement of soil water toward the ice lens, and coarse-grained soils contain pores that are too large for capillary movement. *Frost heave*, or uplift of the land surface, is most damaging in silty, frost-susceptible soils (Figure 16.43). Frost heave is not limited to permafrost regions, but is a problem in any area where temperatures dip below freezing for prolonged periods in the winter.

As the ground freezes in the cold season, vertical cracks may form by thermal contraction of the soil. These cracks, which extend below the bottom of the active zone, commonly form a polygonal pattern on the land surface. In the summer, meltwater from the active zone percolates downward into the crack and freezes during the following winter. Over a period of years, large ice wedges can develop by repeated frost cracking in the same location (Figure 16.44). The polygons at the surface are known as *ice-wedge polygons* (Figure 16.45). They are a good indication of permafrost conditions. Other characteristic permafrost landforms include *pingos* and *patterned ground*. Pingos are large circular or conical mounds, tens of meters high, that form by the vertical growth and heaving of an ice core (Figure 16.46). Patterned ground develops by the sorting of surficial particles into coarse and fine fractions (Figure 16.47). The resulting shapes include polygons, circles, and stripes.

Mass Movement

A characteristic form of mass movement in permafrost terrain is known as *gelifluction*. During the summer, meltwater in the active zone is prevented from draining downward by the impermeable frozen soil below. Therefore, the soil within the active zone becomes saturated and flows as lobate shallow masses on very gentle slopes. The velocities

◄ FIGURE 16.44
Ice wedge in soft sediment, Yukon Territory. *Source:* T. L. Péwé; photo courtesy of U.S. Geological Survey.

◄ FIGURE 16.45
Aerial view of ice-wedge polygons near Barrow, Alaska. *Source:* T. L. Péwé; photo courtesy of U.S. Geological Survey.

of the gelifluction lobes are variable, from imperceptibly slow to tens of centimeters per day. Internally, gelifluction masses contain poorly sorted debris with a wide range of particle sizes.

Alpine regions contain moving masses of debris called *rock glaciers* (Figure 16.48), which originate at high elevations in cirques or on slopes leading to mountain peaks. They consist of angular blocks of rubble that are produced by frost action on rock outcrops and accumulate on slopes by rockfall, snow avalanches, or other processes. Rock glaciers resemble true glaciers in form and movement. In addition to rock debris, they also contain ice, either as a core beneath the debris or as an interstitial matrix between rock fragments.

▲ FIGURE 16.46

Pingoes near Tuktoyaktuk, Northwest Territories. *Source:* Photo by Greg Brooks; reproduced with the permission of the Minister of Public Works and Government Services Canada, 2005 and courtesy of Natural Resources Canada, Geological Survey of Canada.

► FIGURE 16.47
Patterned ground near
Keflavik, Iceland, formed
by sorting of surficial material. Coarse particles out-
line the polygons. *Source:*
Photo courtesy of the author.

Engineering Problems in Permafrost Regions

Engineering problems in permafrost terrains result from unwanted thawing of frozen ground and frost heaving during seasonal freezing of the active zone. Problems are most severe in low-lying areas underlain by fine-grained soils. Upon thawing, these soils become waterlogged and lose bearing capacity because the frozen soil beneath the active zone prevents drainage. A lesson learned from many mistakes is that permafrost exists in a fragile state of equilibrium. Vegetation insulates the frozen ground beneath, and if it is removed, thawing and depression of the permafrost table will result. Roads and railroads are constructed by building an insulating pad of non–frost-susceptible soil on the ground after vegetation has been removed. Failure to design roads correctly leads to expensive failures (Figure 16.49). Buildings are often constructed on insulated piles driven below the permafrost table. The bottom of the floor is elevated about a meter above land surface to allow for the circulation of air and dissipation of heat.

◀ FIGURE 16.48
An active rock glacier moving downslope, Copper River region, Alaska. *Source:* F. H. Moffit; photo courtesy of U.S. Geological Survey.

◀ FIGURE 16.49
Disruption of railroad resulting from permafrost thaw after removal of vegetation, near Valdez, Alaska. *Source:* O. J. Ferrians, Jr.; photo courtesy of U.S. Geological Survey.

Case in Point 16.3
The Trans-Alaska Pipeline

Discovery of a huge oil field near Prudhoe Bay on the north coast of Alaska in 1968 posed a unique challenge to the United States. Here was a golden opportunity to reduce the increasing dependence on foreign oil, a factor that became even more apparent during and after the energy crisis of 1973. However, Prudhoe Bay is accessible to oil tankers for only a short period each year. The alternative was to build a 1300-km pipeline across the state of Alaska (Figure 16.50). Williams (1986) has summarized the history of this and other pipeline projects.

The trans-Alaska pipeline route was mostly wilderness and crossed two mountain ranges, several large rivers, and vast areas of permafrost. Construction problems in an uninhabited area with an inhospitable climate, not to mention the special construction techniques that would be required for the permafrost, were quickly appreciated. The environmental movement sensed an ecological disaster in the destruction of wilderness and the disruption of wildlife such as caribou that freely migrate across the route. When it was finally built at a cost of $7 billion, more than 8 times the original estimate, the pipeline had become the largest private engineering project in history.

The major dilemma in the design of the pipeline was the necessity to conduct oil at a temperature of 65°C through a 48-in.-diameter steel pipe without thawing the permafrost. The engineers realized that destruction of the permafrost would cause land subsidence and the possibility of rupture of the pipeline. The potential of rupture from ground shaking due to earthquakes was also taken into consideration. After extensive test drilling along the route to assess the soil and permafrost conditions, the decision was made to construct about one-half of the line (610 km) aboveground to prevent contact between the pipe and the frozen soil. Of the remainder, 640 km were buried in areas where soil conditions

▶ FIGURE 16.50
Route of the trans-Alaska pipeline.

◀ FIGURE 16.51
Design of aboveground pipeline support. *Source:* From P. J. Williams, 1986, *Pipeline & Permafrost: Science in a Cold Climate*, Ottawa: Carleton University Press.

were favorable and frost heave or subsidence was not anticipated to be a problem. In sensitive environmental areas 11 km of the pipeline were buried using special techniques including insulation and refrigeration around the outside of the insulated pipe to prevent permafrost thaw. Bridges were used to conduct the pipeline across the larger rivers and the pipeline was buried beneath others.

The design of the aboveground pipeline supports is shown in Figure 16.51. Figure 16.52 is a photo of the pipeline. The pipe is supported by two vertical support members (VSMs) and a horizontal support beam. The VSMs are steel piles that are driven into the ground to depths ranging from 8 to 20 m. A critical component of the VSMs are sealed, 2-in.-diameter heat pipes containing anhydrous ammonia refrigerant. In the winter, the refrigerant evaporates from the lower part of the heat pipe and condenses at the top because the ground is warmer than the air. The heat is then released to the air by use of heat exchangers at the top of the VSMs. The evaporation of the refrigerant below the soil surface cools the permafrost, similar to the evaporation of sweat cooling the skin, to the point where the soil remains frozen

◀ FIGURE 16.52
Photo of aboveground section of pipeline near Paxson, Alaska. *Source:* Photo courtesy of Glenn Oliver.

during the warmer summer conditions. Thermal expansion and contraction of the pipe is accommodated at bends in the line at 250- to 600-m intervals. Connection of the pipe to the support beam allows up to 4 m of lateral movement of the line; vertical movement along the VSMs is also possible. This design protects against both ground thaw and disruption as well as earthquake ground motion. The equipment road running along the pipeline was constructed on a thick gravel mat to insulate the ground from temperature changes associated with the removal of vegetation.

To date, the pipeline has functioned as designed and disruption of the environment has been minimal for a project of this scope. Wildlife migration concerns were adequately addressed by the refrigerated buried sections and other measures. Oil production from Prudhoe Bay is now in decline. New reserves in Alaska have not been fully delineated, but the development of reserves in the Alaska National Wildlife Refuge (ANWR) is highly controversial. Major oil and gas reserves are also present in arctic Canada. Since the completion of the trans-Alaska pipeline, one other long Canadian pipeline in permafrost was constructed. Others in both the United States and Canada have been proposed, but not built. The success of the trans-Alaska pipeline was based on a knowledge of the geology of frozen ground. Future projects will continue to rely on geology to solve the unique problems of construction in cold climates.

Arid Regions and Deserts

Arid regions constitute such a large portion of the land area of the earth, between 25% and 40% depending upon the classification used, that it is important to understand a little about the unique processes and landforms in these environments. Aridity is based strictly upon the amount of precipitation received by a region. Although mean annual temperature is an important climatic variable because it influences the amount of evaporation that will occur, warm temperatures are not required for aridity. In fact, dry polar regions can be as arid as the deserts of the Middle East. In cold regions, the lack of vegetation is partly due to the low temperatures, in addition to the low precipitation. One classification of arid lands utilizes the ratio of precipitation to annual potential evapotranspiration, P/ETP (UNESCO, 1977). The four zones defined by this index are shown in Table 16.3.

Although mean annual temperature is not implicitly included in the classification, it has an indirect relationship. Because the amount of potential evapotranspiration increases with increasing temperatures, a certain amount of precipitation could yield a semi-arid climate in one area and an arid climate in a warmer region.

Arid Landscapes

The lack of precipitation in arid areas creates their most common characteristic—the lack or near lack of vegetation. To many, arid regions and deserts are synonymous—vast plains

Table 16.3	Classification of Arid Regions
P/ETP	**Classification**
<0.03	Hyperarid
0.03–0.20	Arid
0.20–0.50	Semi-arid
0.50–0.70	Subhumid

▲ FIGURE 16.53
The El-Qaa alluvial plain, south Sinai, Egypt, is a reg. Coarse gravel and cobble lag is concentrated at the surface by erosion of finer particles by wind. *Source:* Photo courtesy of the author.

covered with sand dunes. However, the topography and surface features of arid regions are surprisingly variable. It is true that plains and plateaus make up a large part of desert areas, but the surfaces of these regions can consist of rock outcrop, known as a *hammada*, by a layer of stones or gravel, known as a *reg* (Figure 16.53), or by vast seas of sand, known as *ergs* (Figure 5.4). Regs may consist of a concentrated surficial layer of closely spaced stones covering finer soils below. This kind of a surface is known as a *desert pavement*. The stones at the surface protect the finer soils beneath from wind erosion. The formation of desert pavements probably involves deflation, removal of the finer particles by wind, but also other processes such as freeze and thaw that preferentially move larger clasts to the surface.

Fault-block mountains in arid areas, such as those in the southwestern United States, are associated with characteristic landforms that develop in the intermontane basins (Figure 16.54). Along the mountain front, alluvial fans build up where there is an abundant source of sediment eroded from large drainage basins. In the absence of alluvial fans, the mountain front is commonly bordered by a pediment (Figure 16.54). Pediments are gently sloping surfaces cut on the same type of bedrock that makes up the mountain range. They may be overlain by a thin veneer of coarse alluvium. Their origin is still not completely understood and a wide variety of erosional processes have been

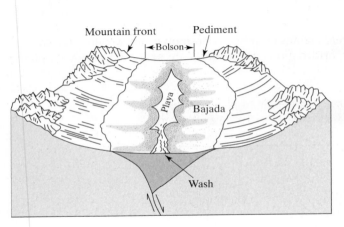

◄ FIGURE 16.54
Landforms of fault block mountains and basins in arid regions.

suggested for their formation. These include stream erosion, during the infrequent desert rainstorms; deep weathering during former, more humid climates; and erosional retreat of the mountain front.

The thin alluvium on pediments thickens toward the center of the basin, reaching thousands of meters in some cases. The alluvial deposits are basically a coalescing band of alluvial fans, running parallel to the mountain front. This depositional zone is called a *bajada*, and the entire depositional zone in the basin is known as a *bolson*. When it does rain in the Basin and Range province of the southwestern United States, runoff ponds to form temporary lakes in the center of the basin, because there are no through-flowing streams connecting the basins. The finest sediment carried out into the basin settles out in these playa lakes. The landforms of sandy deserts will be described after we consider the eolian (wind) processes acting on the desert surface.

Wind Erosion and Deposition

Wind Erosion and Transport

In the absence of significant vegetation, which protects and anchors the soil, wind can erode and transport huge quantities of sediment. The entrainment and transport of particles by wind is fundamentally similar to that of running water, but there is one major difference. Water, because it is so much denser than air, exerts a much larger drag force on particles lying on the surface and can transport much larger particles than air. The amount of drag, or shear stress, exerted by wind on particles on the land surface is related to the velocity of the wind by the following equation:

$$V_* = \sqrt{\tau/\rho} \qquad (16.3)$$

where V_* is known as the drag, or shear velocity, τ is the shear stress, and ρ is the density of air. The drag velocity, V_*, is actually the velocity gradient near the surface, rather than the actual wind velocity. It is related to the velocity at any point above the surface by equation (16.4).

$$V_z = 5.75V_* \log (z/k) \qquad (16.4)$$

V_z is the true wind velocity at a vertical distance z above the ground and k is a parameter related to the roughness of the surface. Depending upon the roughness, which involves the particle size and other factors such as vegetation, the wind velocity will rise from zero at a distance of k cm above the surface to some value dependent on the strength of the wind. When the wind velocity is plotted with distance above the surface, it yields a logarithmic relationship (Figure 16.55). These graphs are usually plotted on semi-log paper to make the relationship a straight line (Figure 16.55b). k can be determined as the distance from the origin to the point on the vertical axis where the two lines converge. The shear velocity, V_*, is the slope of the line on the semi-log plot. The line labeled $V_* = 61$ is a larger shear velocity and would therefore represent a higher wind velocity, according to equation (16.4). For a given wind blowing across the surface, the higher the k value, which means a rougher surface, the higher the shear velocity will be, for example, 61 compared to 26. The shear velocity, therefore, is really the gradient of the velocity above the ground. This is important because the size of particles that can be transported is a function of the shear velocity.

Figure 16.56 shows the relationship between the shear velocity, expressed here as V_{*_T}, or the critical threshold velocity necessary to move a particle, and the grain diameter. This plot requires some contemplation. First, there are several processes at work as wind blows over a surface. Because wind is a turbulent phenomenon, there are components of velocity in all directions, including directly upward, especially in turbulent eddies. When

▲ FIGURE 16.55

Relationship between height above the ground surface and wind velocity plotted on a linear scale (a) and a logarithmic vertical axis (b). The two curves represent winds of two different strengths. The slope of the lines on (b) is the shear velocity, V_*. The distance between the origin and the point on the vertical axis where the two lines converge is the parameter k in equation (16-4).

◄ FIGURE 16.56

Plot of critical shear velocity versus grain diameter. Fine particles are entrained by wind turbulence and transported in suspension. Coarser grains are primarily transported by saltation, which begins at the impact threshold. The impact threshold is reached at a lower velocity than the fluid threshold, which represents direct entrainment by the wind.

the upward velocity caused by the wind is greater than the fall velocity caused by the weight of the particle, particles are lifted off the surface and move upward into the air. They are then transported by suspension, which is the same process as occurs in running water. But because of the low density of air, suspension is only possible for very small particles, those that plot in the extreme left-hand side of Figure 16.56. The threshold of entrainment by suspension, a dashed line that rises to the left, actually increases as the particles get smaller. The reasons for this are similar to the behavior of fine particles in water.

▲ FIGURE 16.57
Trajectory of saltating grains launched by the impacts of descending grains.

Smaller particles tend to behave as aggregates rather than as individuals due to the attractive forces between them. In addition, particles this small are likely to lie entirely within the thickness of the layer in which velocity is zero ($<k$ cm). Once they are entrained at a certain shear velocity, however, they can remain in transport at considerably lower velocities. The dustbowl days of the 1930s are a good example of the suspension transport of fine particles for long distances.

As the grain size of surface particles increases, another process, saltation , comes into play to a much greater degree than it does in fluvial transport. At grain sizes greater than 0.1 mm, Figure 16.56 shows that there are two thresholds of movement. The higher one, the fluid threshold, indicates the shear velocity necessary to move a particle of a given size by shear alone. As these particles are launched into the air, they do not travel very far because of their greater weight. Their trajectory is similar to that shown in Figure 16.57. After a steep vertical rise, they impact the surface again at a low angle, not much more than a meter or so in the downwind direction. The momentum which they impart to particles at their point of collision makes it easier for the wind to move the impacted particles, so the impact threshold is lower than the fluid threshold. Thus, during a strong windstorm, there is layer of saltating grains, accounting for as much as 75% of the total sediment transport, within a height of several centimeters of the surface. Coarser sediment can also be moved by impact but remains in contact with the surface like bed load in a stream. This type of transport is known as *creep*. What is left at the surface after the amount of sediment moved by saltation and creep is a gravelly lag layer. Taken together, the three processes of suspension, saltation, and creep effectively sort the sediment on the desert surface. Upon erosion and deposition of windblown sediment, the surface tends to evolve into bare hammada, from which all sediment above bedrock is removed; gravelly reg, from which the sand and silt is removed; and ergs, in which the sand accumulates in dunes. The finer sizes transported by suspension can be carried downwind for great distances.

Erosion by wind includes two basic processes. *Deflation* is simply the removal of fine sediment by wind, leaving a *deflation hollow*. Deflation hollows can range in size to many kilometers in diameters, although it is sometimes difficult to eliminate the effects of other factors in producing the depression. *Abrasion* is the other process, the equivalent of sandblasting, in which sand grains carried by the wind abrade rock surfaces or clasts with which they come in contact. Cobbles or pebbles on the ground surface that show the effects of wind abrasion are known as *ventifacts*. Ventifacts can be recognized by planar facets that face into the wind and slope downward toward the effective wind direction. Many ventifacts have two or more facets, which intersect along sharp, linear ridges (Figure 16.58a). Other facets on ventifacts can be pitted, fluted, polished, or grooved (Figure 16.58b). Multiple facets can indicate the effects of prevailing winds from different directions, or alternatively, that the clast has been moved or rotated, exposing a new face to the wind.

(a)

(b)

◀ FIGURE 16.58
(a) Ventifact on an outwash plain, Iceland. (b) Polished and pitted facet on cobble, south Sinai, Egypt. *Source:* Photos courtesy of the author.

Larger landforms produced by wind erosion are called *yardangs*. These elongated, streamlined ridges range in relief from several meters to tens of meters. Their streamlined shape, with a length-to-width ratio of 4 or more reflects evolution toward a shape that provides the minimum resistance to the wind. Yardangs are best developed in weakly consolidated sedimentary rocks and they are thought to form by a combination of deflation and abrasion.

Wind Deposition and Dunes

Transport and deposition of sand produces a wide variety of landforms, with sand supply and wind direction and magnitude the major variables. Deposition of sand is enhanced by the presence of sand accumulations in dunes or other forms on the desert floor, because wind energy is absorbed by the transport of grains during saltation. Thus, any patch of sand with a downwind length of 4 to 6 m or more will have a net accumulation as sand arrives from upwind sources. Deposition is strongly influenced by obstacles to flow on the ground because wind accelerates around and over the obstacle. By contrast, a low-velocity zone is created in the lee of obstacles and streamlined mounds or ridges of sand are deposited in

▲ FIGURE 16.59
Sand shadow in the lee of a mound of sand stabilized by vegetation, south Sinai, Egypt. *Source:* Photo courtesy of the author.

these downwind locations. The size and shape of the resulting *sand shadows* (Figure 16.59) is dependent upon the size and dimensions of the obstacle.

The movement of gentle winds over a sand-covered surface will produce *wind ripples*, analogous to fluvial ripples (Figures 5.20, 5.21, and 14.15). Wind ripples are strongly associated with saltation because it is by this process that grains are set into motion by impact and moved downwind. Wind ripples therefore form low, wavelike forms transverse to the wind direction, with wavelengths of a meter or less. With an increase in wind velocity, more pronounced transport of sand takes place and *dunes* are formed. These landforms are also similar to fluvial dunes, with an asymmetric shape and a downwind direction of movement (Figure 16.60). Sand grains are transported up the gentle windward (stoss) side of the dune and are temporarily deposited near the top of the steeper lee face. Deposition occurs here because wind velocity drops as the wind passes over the crest of the dune. Deposition causes steepening of the lee face until a critical angle is reached. Grains then avalanche down the face, maintaining the lee face at the angle of repose. By erosion on the stoss side and deposition on the lee side, dunes migrate downwind, although the dune form remains constant (Figure 16.61). The steeply dipping lee-side cross beds (Figure 16.60) are preserved as lithified dune sequences and provide evidence for arid climates in the geologic record.

The major types of sand dunes are shown in Figure 16.62. *Barchans* have perhaps the most distinctive dune shape. These isolated forms develop on hard substrates where the

▲ FIGURE 16.60
Cross section of a sand dune showing asymmetry between the stoss and lee sides and internal configuration of slip-face cross beds.

▲ FIGURE 16.61
View from the top of a dune looking down the slip face at trees that are being buried as the dune advances; Indiana Dunes National Lakeshore. *Source:* Photo courtesy of the author.

sand supply is limited and the winds are unidirectional. The projections around the ends of the dunes are called *horns*, and the slip face has a concave shape in the downwind direction. As the sand supply increases, barchans begin to coalesce to form a more continuous ridge transverse to the wind direction, but one that still has projections like horns extending in the downwind direction. These dunes are called *barchanoid ridges*. There is probably a complete gradation between barchans, barchoid ridges, and *transverse dunes*, which are similar to barchanoid ridges without the horn segments.

When the wind direction is slightly variable, but still from the same general direction, *linear*, or *longitudinal, dunes* can develop. The trend of the dune may approximate the resultant of the various wind directions. These dunes are significant landforms that can be several meters high and several hundred meters in continuous length. In a transverse profile, the dunes are asymmetrical, with steeper slip faces along the crest of the dune. Sometimes slip-face segments alternate from side to side along the crest.

Dunes can be stabilized by vegetation with even a small increase in annual precipitation. Many arid and semi-arid environments are close to the boundary between active dunes and stabilized inactive dunes. In these areas, parts of stabilized dunes can be reactivated if the vegetation can be stripped away on part of the dune to form large scoop-shaped hollows, or *blowouts*. Blowouts have a variable orientation in their early stages of modification of older dunes but become more oriented with respect to the wind directions in more evolved forms. The blowouts become more pronounced and concentrated, leaving hornlike projections between them. The resultant forms look like barchans in plan view, but the horns are on the upwind side rather than the downwind side. These dunes are known as *parabolic dunes* (Figure 16.62). They can be differentiated on maps or air photos because of the steepness of their slip faces on the downwind side, although the slip face is now convex downwind instead of concave downwind like those of the barchans.

Star dunes are very large landforms, with ridges radiating in multiple directions from pointed crests or peaks. These dunes are the result of two or more divergent prevailing wind directions. In some areas, the wind directions are oriented exactly opposite from each other, with one being dominant and the other weaker. When this wind pattern is present, *reversing dunes* are observed. When the dominant wind prevails, the slip face is formed in the usual position. However, when the wind reverses during another season,

Barchan dunes

Blowout dunes

Linear dunes

Barchanoid ridge

Dome dunes

Star dunes

Transverse dunes

Parabolic dunes

Reversing dunes

▲ **FIGURE 16.62**
Basic types of eolian dunes, with arrows showing the most significant wind directions. *Source:* From
E. D. McKee, ed., 1979, *A Study of Global Sand Seas*. U.S. Geological Survey Professional Paper 1052.

the slip face is modified and partially converted to the opposite direction. One final dune
type is the *dome dune*, which consists of low mounds of sand without a slip face.

Large accumulations of wind-deposited sand are known as ergs (sand seas) or *sand
sheets*. Sand seas must be larger than 30,000 km² in area and can contain many types of
dunes in individual dune fields. Sand sheets are vast, featureless plains underlain by eolian sand that lacks distinct dune forms. Generally, the sand is thinner in sand sheets than
in ergs, sometimes only a few meters in thickness.

Loess

As mentioned earlier, wind is an effective sorting medium. In the process of erosion and
transportation by wind, silt and clay are separated from sand, which is deposited closer to
the source. The finest fraction is carried by winds high into the atmosphere and can be
transported around the earth many times. The ultimate depositional site for much of this
material is the oceans. The silt fraction of the eroded material forms significant deposits
known as *loess*. It blankets much of the landscape downstream from sources areas. In North
America, extensive loess deposits were formed during the Pleistocene, when vegetation

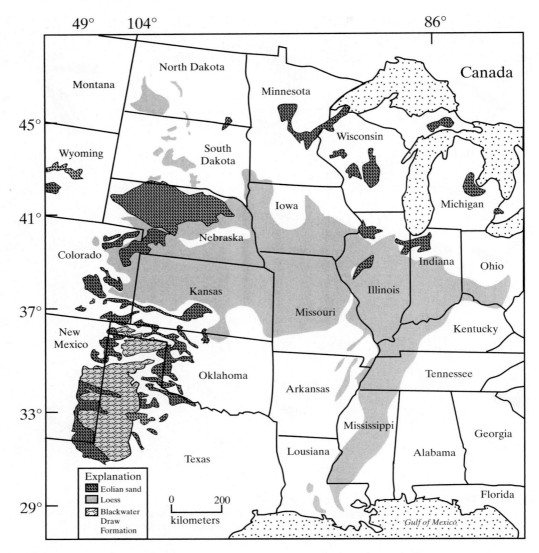

▲ **FIGURE 16.63**
Eolian deposits in the central United States. The loess deposits are mostly Pleistocene in age, although the eolian sand deposits may be Pleistocene and Holocene in age. The Blackwater Draw Formation is a sand sheet. *Source:* Reprinted from *The Quarternary Period in the United States*, A. Busacca et al., Eolian sediments, © 2004 with permission of Elsevier.

cover was decreased by colder climates. Very large areas of the central United States are covered by loess in the vicinity of the Missouri and Mississippi rivers (Figure 16.63). The source of this sediment was the abundant glacial outwash sediment carried by these rivers from the glacier fronts to the north. Loess is draped over the existing landscape, with the resulting topography being a more subdued version of the pre-loess topography. Stratigraphically, the loess deposits of the central United States are composed of multiple sheets corresponding to glacial periods, each one overlying an older one. Because of weak bonds of clay or water between the silt particles (Chapter 10), loess has the unusual engineering property of maintaining a vertical angle in slope cuts, as long as the material remains dry. If the surface is wetted, the clay bonds are weakened and the material may fail.

Case in Point 16.4
The Nebraska Sand Hills

Most of the active eolian deposits in the world today lie in modern deserts. During the Pleistocene, however, the southerly movement of continental glaciers in North America created cold, dry conditions at the margins of the ice fronts, along with higher winds. Wind erosion and deposition in the central United States was extensive during these periods (Figure 16.63). In addition to the areas of loess deposition associated with sources from outwash sediment from the major meltwater rivers, vast ergs and sand sheets were formed. The largest of these, covering an area of 50,000 km² and making it the largest sand sea in the Western Hemisphere, is the Nebraska Sand Hills (Figures 16.63 and 16.64).

The major dune types in the Sand Hills are shown in Figure 16.64. The most common are barchans and barchanoid ridges. The size of these dunes (up to 130 m in height) is similar to active dunes in the great deserts of today in Africa and the Middle East. This puts them into a category known as *megabarchans*. Loess plains on the southeastern (downwind) end of the Sand Hills are consistent with a transport direction from the northwest.

▲ FIGURE 16.64

Major dune types in the Nebraska Sand Hills. *Source:* From D. May, J. B. Swinehart, D. Loope, V. Souders, 1995, Late Quaternary fluvial and eolian sediments: Loup River Basin and the Nebraska Sand Hills, in *Geologic Field Trips in Nebraska and Adjacent Parts of Kansas and South Dakota: Parts of the 29th Annual Meetings of the North-Central and South-Central Sections, Geological Society of America*, C. A. Floweday, ed., Lincoln: Conservation and Survey Division, University of Nebraska–Lincoln, pp. 13–32.

▲ FIGURE 16.65
View of the Nebraska Sand Hills, which are stabilized with grassland vegetation under today's climatic conditions. Large ridge in background is megabarchanoid ridge. *Source:* Photo courtesy of the author.

Because today most of the Sand Hills and other dune areas are stabilized with vegetation (Figure 16.65), it was long assumed that these large dunes had not been active since the Pleistocene. However, recent work has shown that the dunes have been intermittently active in the Holocene, as recently as 3000 years before present. The very young age of some of these dunes illustrates the vulnerability of the landscape to climate change. The Holocene is considered to be a time of warm and relatively constant climatic temperatures. A mid-Holocene period of warmer and dryer conditions may prove to be the most important period in formation of the Sand Hills. The significance of these findings is that it may take only a minor decrease in precipitation to reactivate the Sand Hills. As we gather more and more information about past climates, it is important to keep in mind that relatively subtle climate changes can have significant impacts on landscapes that are close to thresholds of geomorphic change.

Summary and Conclusions

An understanding of glacial processes and landforms is important because of the large area covered by glaciers during the Pleistocene Epoch. The expansion and contraction of a glacier is determined by its mass balance—the relationship between accumulation and ablation. The position of the terminus is controlled by this relationship even though the ice may be flowing downslope continuously. Flow occurs by visco-plastic deformation and basal sliding. Both types of movement occur in wet-based glaciers, whereas only visco-plastic flow is characteristic of cold-based glaciers.

Glacial erosion includes abrasion and plucking by sliding glaciers. Glaciers frozen to their beds can incorporate huge slabs of bed material into the ice by thrusting. Eroded rock and sediment are transported and deposited at a later time. Deposition may take place beneath the ice to form lodgement till or above the ice to form ablation till. Other types of glacial drift are deposited by glacial meltwater streams as outwash or in proglacial lakes as lacustrine sediment.

Valley glaciers produce a characteristic set of erosional landforms in mountainous terrain. Chief among these is the U-shaped valley with oversteepened valley walls. Eroded sediment is deposited in terminal, lateral, and medial moraines. Modern glaciers cannot compare with the huge continental ice sheets that buried the landscape repeatedly during Pleistocene time. The landforms and deposits of these glaciers are the substrate upon which all human activities take place in northern latitudes.

Each type of glacial deposit has unique engineering characteristics. Glacial outwash deposits may contain valuable resources of aggregate material and groundwater. Glacial lacustrine sediment and tills present varied construction conditions depending upon their method of deposition, thickness, stratigraphy, and many other factors. Lodgement and ablation tills differ with respect to density, strength, and compressibility.

The factors that led to repeated glaciations extending into the midlatitudes during Pleistocene time are not fully known; however, it is thought likely that periodic perturbations in the earth's motion can explain the climatic cycles of cooling and warming. Long-term climatic changes were probably caused by lateral movement and elevation of the continents by plate interactions.

Permafrost causes severe engineering problems in cold regions, which cover more than 20% of the earth's surface. Most of these problems are related to instability of the active layer of the soil. Differential settlement and mass movement are common during thaw periods, and frost heave takes place during freezing of the active layer. These conditions require special design and construction procedures for almost all engineering projects.

Arid regions contain a wide variety of landscapes and landforms intimately associated with the climate under which they form. The surfaces of these areas include hammadas, regs, and ergs, depending on whether wind has removed all possible particles, deflated the finer clasts leaving only a coarse lag, or deposited thick accumulations of sand. The erosion and transport of sand is a function of the condition and roughness of the surface and the strength of the wind. When this sand is deposited, a variety of landforms can develop, including sand sheets and quite a few different types of dunes. Sand supply and the direction and variability of winds determine which specific type of dune is formed. Silt-size particles are transported farther than sand and come to rest in blanketlike sheets of loess that are draped over the existing topography. Dunes can be stabilized by vegetation and just as easily destabilized if the precipitation in an area decreases.

Problems

1. Summarize the mechanics of glacier movement.

2. What effect does the condition of the base of the glacier have upon movement?

3. Contrast the different types of glacial erosion.

4. How is till deposited?

5. What types of sediment other than till are associated with glaciers?

6. What topographic features indicate whether or not a mountainous area was glaciated during Pleistocene time?

7. What construction problems can be expected in glaciated terrain?

8. What is the best explanation for Pleistocene glaciations?

9. What landforms suggest the presence of permafrost?

10. Discuss the problems of construction in areas of frozen ground.

11. How are eolian sands identified in the rock record, and what types of evidence are used to infer that dunes were present in the environment of deposition?

12. What is the shear velocity that it would take to move a particle 0.6 mm in diameter? Explain the process by which this particle would be moved.

13. How is shear velocity related to the actual wind velocity?

14. Describe how differences in wind direction and sand supply result in the formation of different types of dunes.

References and Suggestions for Further Reading

ANTEVS, E. 1929. *Maps of the Pleistocene Glaciations. Geological Society of America Bulletin*, 40:635.

BAGNOLD, R. A. 1941. *The Physics of Blown Sand and Desert Dunes*. London: Methuen & Co., Ltd.

BENN, D. I., and D. J. A. EVANS. 1998. *Glaciers and Glaciation*. London: Arnold.

BLOOM, A. L. 1998. *Geomorphology, A Systematic Analysis of Late Cenozoic Landforms*. Upper Saddle River, N.J.: Prentice Hall, Inc.

BLUEMLE, J. P. 1991. *The Face of North Dakota*. North Dakota Geological Survey Educational Series 21.

BUSACCA, A., J. E. BEGÉT, H. W. MARKEWICH, D. R. MUHS, N. LANCASTER, and M. R. SWEENEY. 2004. Eolian sediments, in *The Quaternary Period in the United States: Developments in Quaternary Science* 1, A. R. Gillespie, S. C. Porter, and B. F. Atwater, eds. Amsterdam: Elsevier, pp. 275–309.

DAVIS, W. M. 1911. The Colorado Front Range. *Annals of the Association of American Geographers*, v. 1.

DROOFF, E. R., D. J. DOBBELS, and J. R. WHEELER. 1999. Role and performance of chemical grout in the underpinning of an active subway station, in *Proceedings of the Rapid Excavation and Tunnelling Conference*, D. E. Hilton, and K. Samuleson, eds. Society for Mining Metallurgy and Exportation, Chapter 24, pp. 382–402.

EASTERBROOK, D. J. 1993. *Surface Processes and Landforms*. New York: Macmillan.

EYLES, N. 1983. *Glacial Geology: An Introduction of Scientists and Engineers*. Elmsford, N.Y.: Pergamon Press.

FERRIANS, O. J., R. KACHADOORIAN, and G. W. GREENE. 1969. *Permafrost and Related Engineering Problems in Alaska*. U.S. Geological Survey Special Paper 678.

FLINT, R. F. 1945. *Glacial Map of North America*. Geological Society of America Special Paper 60.

GOYAL, B. B., P. K. DAS, C. K. SHAH, C. W. WOOD, and B. R. BRENNER. 2001. Underpinning of Red Line South Station, in *Engineering Geology of the Big Dig Project*, D. Bobrow and C. Daugherty, leaders, in *Guidebook for Geological Field Trips in New England*, D. P. West and R. H. Bailey, eds. 2001 Annual Meeting of the Geological Society of America.

JUDSON, S., M. E. KAUFFMAN, and L. D. LEET. 1987. *Physical Geology*, 7th ed. Upper Saddle River, N.J.: Prentice Hall, Inc.

KAYE, C. A. 1976. Beacon Hill end moraine, Boston: New explanation of an important urban feature, in *Urban Geomorphology*, D. A. Coates, ed. Geological Society of America Special Paper 174, pp. 1–20.

LEVERETT, F., and F. B. TAYLOR. 1915. *The Pleistocene of Indiana and Michigan and the History of the Great Lakes*. U.S. Geological Survey Monograph 53.

MAY, D., J. B. SWINEHART, D. LOOPE, and V. SOUDERS. 1995. Late Quaternary fluvial and eolian sediments: Loup River Basin and the Nebraska Sand Hills, in *Geologic Field Trips in Nebraska and Adjacent Parts of Kansas and South Dakota: Parts of the 29th Annual Meetings of the North-Central and South-Central Sections, Geological Society of America*, C. A. Floweday, ed. Lincoln: Conservation and Survey Division, University of Nebraska–Lincoln, pp. 13–32.

McKEE, E. D., ed. 1979. *A Study of Global Sand Seas*. U.S. Geological Survey Professional Paper 1052.

NEFF, T. L., K. K. SEE-THO, S. S. C. LIAO, and L. BEDINGDIELD. 1989. Integration of design and construction for Boston's Central Artery/Third Harbor Tunnel Project, in *Proceedings of the International Congress on Progress and Innovation in Tunnelling*, vol. 2, K. Y. Lo, ed., pp. 719–726. Toronto Canada, 1989.

PYE, K., and H. TSOAR, 1990. *Aeolian Sand and Sand Dunes*. London: Unwin Hyman, Ltd.

SUGDEN, D. E., and B. S. JOHN. 1976. *Glaciers and Landscape*. New York: John Wiley.

WILLIAMS, P. J. 1986. *Pipelines and Permafrost: Science in a Cold Climate*. Ottawa: Carleton University Press.

Units and Conversion Factors

Quantity	SI Unit	U.S. Customary Unit		Conversion Factors
Length, L	m	ft, in., yd, mi	1 in.	= 2.54 cm
			1 ft	= 0.3048 m
			1 yd	= 0.9144 m
			1 mi	= 1.6093 km
			1 m	= 3.2808 ft
			1 km	= 0.6214 mi
Area, L^2	m^2, ha	ft^2, $in.^2$, yd^2, mi^2, acre	1 $in.^2$	= 6.452 cm^2
			1 ft^2	= 0.0929 m^2
			1 acre	= 43,560 ft^2
				= 4046.85 m^2
				= 0.404685 ha
			1 mi^2	= 640 acre
				= 2.604 km^2
				= 259 ha
			1 m^2	= 10.764 ft^2
			1 ha	= 2.471 acre
				= 10,000 m^2
Volume, L^3	m^3, L	ft^3, yd^3, gal	1 ft^3	= 7.4805 U.S. gal
				= 0.02832 m^3
				= 28.32 L
			1 U.S. gal	= 0.134 ft^3
				= 0.003785 m^3
				= 3.785 L
			1 m^3	= 35.3147 ft^3
				= 264.172 gal
				= 1000 L
				= 10^6 cm^3
Mass, M	kg, g tonne (metric)	slug, lbm ton (short)	1 slug	= 14.59 kg
			1 kg	= 0.685 slug
			1 lbm	= 0.453592 kg
			1 kg	= 2.205 lbm
			1 ton (short)	= 2000 lbm
				= 907.2 kg
				= 0.9072 ton (metric)
			1 ton (metric)	= 1000 kg
				= 2204 lbm
				= 1.1023 ton (short)
Force, weight $\left(\dfrac{ML}{T^2}\right)$	N, dyne	lb, ton	1 N	= 0.2248 lb
			1 kN	= 224.8 lb
			1 lb	= 4.448 N
			1 dyne	= 1×10^{-5} N
				= 2.25×10^{-6} lb
				= 1 g-cm/s^2

Quantity	SI Unit	U.S. Customary Unit		Conversion Factors
Stress, pressure (F/L^2)	N/m^2, Pa, bar	lb/ft^2, atm $tons/ft^2$ psi, tsf	$1 \ N/m^2$ $1 \ Pa$ $1 \ bar$ $1 \ kN/m^2$ $1 \ lb/ft^2$ $1 \ psi$ $1 \ atm$	$= 0.0209 \ lb/ft^2$ $= 1 \ N/m^2$ $= 10^5 \ Pa$ $= 20.90 \ lb/ft^2$ $= 0.145 \ psi$ $= 47.88 \ N/m^2$ $= 6.895 \ kN/m^2$ $= 14.7 \ psi$ $= 1 \ bar$
Density (ρ) (M/L^3)	kg/m^3, g/cm^3	lbm/ft^3, $slugs/ft^3$	$1 \ Mg/m^3$ $1 \ lbm/ft^3$ $1 \ g/cm^3$ $1 \ kg/m^3$	$= 62.4 \ lbm/ft^3$ $= 0.0157 \ Mg/m^3$ $= 62.4 \ lbm/ft^3$ $= 0.062 \ lbm/ft^3$
Unit weight (γ) $(F/L^3, M/L^2T^2)$	kN/m^3 $dynes/cm^3$	lb/ft^3	$1 \ kN/m^3$ $1 \ lb/ft^3$	$= 6.36 \ lb/ft^3$ $= 0.0037 \ lb/in.^3$ $= 157 \ N/m^3$
Velocity (L/T)	m/s, cm/s	ft/s, ft/min mi/hr (mph)	$1 \ m/s$ $1 \ cm/s$ $1 \ ft/s$ $1 \ mi/hr$	$= 3.28 \ ft/s$ $= 1.97 \ ft/min$ $= 0.305 \ m/s$ $= 1.61 \ km/hr$
Acceleration (L/T^2)	m/s^2	ft/s^2	$1 \ m/s^2$ $1 \ ft/s^2$	$= 3.28 \ ft/s^2$ $= 0.3048 \ m/s^2$
Flow rate (Discharge) (L^3/T)	m^3/s	ft^3/s, ft^3/min U.S. gal/min	$1 \ ft^3/s$ $1 \ ft^3/min$ $1 \ gal/min$ $1 \ m^3/s$	$= 0.0283 \ m^3/s$ $= 448.8 \ gal/min$ $= 4.72 \times 10^{-4} \ m^3/s$ $= 7.4805 \ gal/min$ $= 6.31 \times 10^{-5} \ m^3/s$ $= 35.315 \ ft^3/s$ $= 2118.9 \ ft^3/min$
Energy (F-L)	Joule (N-m)	ft-lb, cal Btu, Kw-hr	$1 \ Btu$ $1 \ Joule$ $1 \ Kcal$	$= 778 \ ft\text{-}lb$ $= 252 \ cal$ $= 1 \ N\text{-}m$ $= 0.73756 \ ft\text{-}lb$ $= 4.185 \ kJ$

USEFUL CONSTANTS

$g = 9.8 \ m/s^2$
$\quad = 980 \ cm/s^2$
$\rho(water) = 1.0 \ g/cm^3$
$\quad\quad\quad = 1000 \ kg/m^3$
$\gamma(water) = 9.8 \times 10^3 \ N/m^3$
$\quad\quad\quad = 62.4 \ lb/ft^3$

INDEX

Bad Binding

UNIFIED SOIL CLASSIFICATION SYSTEM

Major Divisions			Group Symbols[a]	Typical Names	Field Identification Procedures (excluding particles larger than 75 mm and basing fractions on estimated weights)		
1	2		3	4	5		
Coarse-grained soils — More than half of material is *larger* than No. 200[b] (75 μm) sieve size.	Gravels — More than half of gravel fraction is larger than No. 4 sieve size (4.75 mm)	Clean gravels (little or no fines)	GW	Well-graded gravels, gravel sand mixtures, little or no fines.	Wide range in grain sizes and substantial amounts of all intermediate particle sizes.		
			GP	Poorly graded gravels, gravel-sand mixtures, little or no fines.	Predominantly one size or a range of sizes with some intermediate sizes missing.		
		Gravels with fines (appreciable amount of fines)	GM	Silty gravels, gravel-sand-silt mixtures.	Nonplastic fines or fines with low plasticity (for identification procedures see ML below).		
			GC	Clayey gravels, gravel-sand-clay mixtures.	Plastic fines (for identification procedures see CL below).		
	Sands — More than half of coarse fraction is smaller than No. 4 sieve size (4.75 mm)	Clean sands (little or no fines)	SW	Well-graded sands, gravelly sands, little or no fines.	Wide range in grain sizes and substantial amounts of all intermediate particle sizes.		
			SP	Poorly graded sands, gravelly sands, little or no fines.	Predominantly one size or a range of sizes with some intermediate sizes missing.		
		Sands with fines (appreciable amount of fines)	SM	Silty sands, sand-silt mixtures.	Nonplastic fines or fines with low plasticity (for identification procedures see ML below).		
			SC	Clayey sands, sand-clay mixtures.	Plastic fines (for identification procedures see CL below).		

(for visual classification, 5 mm may be used as equivalent to the No. 4 sieve size)

The No. 200 sieve size is about the smallest particle visible to the naked eye.

Major Divisions		Group Symbols[a]	Typical Names	Identification Procedures on Fraction Smaller than No. 40 Sieve Size		
				Dry strength (crushing characteristics)	Dilatancy (reaction to shaking)	Toughness (consistency near PL)
Fine-grained soils — More than half of material is *smaller* than No. 200 (75 μm) sieve size.	Silts and clays — Liquid limit less than 50	ML	Inorganic silts and very fine sands, rock flour, silty or clayey fine sands or clayey silts with slight plasticity.	None to slight	Quick to slow	None
		CL	Inorganic clays of low to medium plasticity, gravelly clays, sandy clays, silty clays, lean clays.	Medium to high	None to very slow	Medium
		OL	Organic silts and organic silty clays of low plasticity.	Slight to medium	Slow	Slight
	Silts and clays — Liquid limit greater than 50	MH	Inorganic silts, micaceous or diatomaceous fine sandy or silty soils, elastic silts.	Slight to medium	Slow to none	Slight to medium
		CH	Inorganic clays of high plasticity, fat clays.	High to very high	None	High
		OH	Organic clays of medium to high plasticity, organic silts.	Medium to high	None to very slow	Slight to medium
Highly organic soils		Pt	Peat and other highly organic soils	Readily identified by color, odor, spongy feel, and frequently by fibrous texture.		

After U.S. Army Engineer Waterways Experiment Station (1960), "The Unified Soil Classification System," *Technical Memorandum* No. 3–357, Appendix A, Characteristics of Soil Groups Pertaining to Embankments and Foundations, 1953, and Appendix B, Characteristics of Soil Groups Pertaining to Roads and Airfields, 1957; and A. K. Howard (1977), "Laboratory Classification of Soils—Unified Soil Classification System," *Earth Sciences Training Manual* No. 4, U.S. Bureau of Reclamation, Denver, 56 pp.

[a]Boundary classifications: soils possessing characteristics of two groups are designated by combinations of group symbols. For example: GW-GC well-graded gravel sand mixture with clay binder.

[b]All sieve sizes on this chart are U.S. Standard.